CAMBRIDGE MONOGRAPHS ON PHYSICS

Magnetic oscillations in metals

T0296958

Some pioneers

L. V. Shubnikov   W. J. de Haas   P. M. van Alphen
(1901–1945)   (1878–1960)   (1906–1967)
R. E. Peierls (b. 1907)   L. D. Landau (1908–1968)
L. Onsager (1903–1976)   I. M. Lifshitz (1917–1982)

Magnetic oscillations in metals

D. SHOENBERG, F.R.S.
Emeritus Professor of Physics, Cambridge University

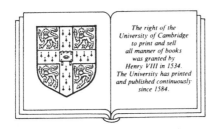

CAMBRIDGE UNIVERSITY PRESS

CAMBRIDGE

LONDON   NEW YORK   NEW ROCHELLE

MELBOURNE   SYDNEY

CAMBRIDGE UNIVERSITY PRESS
Cambridge, New York, Melbourne, Madrid, Cape Town, Singapore, São Paulo, Delhi

Cambridge University Press
The Edinburgh Building, Cambridge CB2 8RU, UK

Published in the United States of America by Cambridge University Press, New York

www.cambridge.org
Information on this title: www.cambridge.org/9780521118781

First published 1984
This digitally printed version 2009

A catalogue record for this publication is available from the British Library

Library of Congress Catalogue Card Number: 82-19762

ISBN 978-0-521-22480-2 hardback
ISBN 978-0-521-11878-1 paperback

Contents

Contents

Contents

Contents

Contents

Contents

Preface

It is just over 50 years ago that an oscillatory magnetic field dependence was first observed in the electrical resistance of bismuth by Shubnikov and de Haas and in the magnetization by de Haas and van Alphen. It was not long before Peierls showed how these effects could be understood in principle and, indeed, Landau had implicitly predicted oscillatory behaviour even before the experimental discovery, but the effects remained somewhat of a scientific curiosity for upward of 20 years. It was only in the 1950s with the observation of magnetic oscillations in many metals other than bismuth and the advent of improved theoretical understanding that it began to be realized that the effect was not only an aesthetically pleasing curiosity but potentially a powerful tool for understanding the electronic structure of metals.

During the following 20 years, exploitation of this possibility became somewhat of a 'band wagon' and with ever improving experimental and theoretical techniques, an immense amount of detailed information about the 'Fermiology' of individual metals has emerged. More recently the pace has slackened, though there are still many loose ends and unsolved problems, and during a sabbatical half year at the University of Waterloo in 1977, I felt the time was ripe for a new comprehensive review and somewhat light-heartedly embarked on it little thinking it would be five years before it was ready for the Press. These five years have been an educational experience for me in as far as I have had to tackle many problems *ab initio* in order to achieve the insight needed to give what I hope is a clear presentation at not too sophisticated a level.

As the title is intended to imply, the emphasis of the book is on the oscillations themselves and the pretty physics involved in understanding the many peculiarities of their behaviour. It is definitely *not* a book on band structure, though some examples are given in fair detail to illustrate how the oscillations have been exploited. For this purpose the examples chosen are relatively simple, both to avoid confusing the reader by too many excessively complicated facts and because it is mainly the simple examples which I know best. I have almost entirely avoided any exposition of the basic theory of band structures (i.e. how they are calculated from first principles), both because there is a vast body of literature on the subject to which the interested reader is referred and because it is a subject I only dimly understand.

In presenting the theory of the magnetic oscillations I start by taking for granted the concept of independent quasiparticles, and use the 'semi-classical' approach without any formal justification. Only at the end do I offer a brief and somewhat 'handwaving' explanation of why many-body effects do not in practice greatly affect the results. However in developing the consequences of the semi-classical approach I have tried to spell out in considerable detail just how the results follow from the starting assumptions, with particular emphasis on points I have myself found far from obvious in the original literature. In general I have tried either to preface or to conclude mathematical derivations by a descriptive account of what it is that the mathematics is aiming to do. I have made a good deal of use of appendices, both for dealing with mathematical detail whose intricacy would interrupt the main argument too seriously and for pursuing somewhat peripheral issues which contain interesting physics. It is debatable how much should be relegated to such appendices and how much remain in the main text and indeed it might be argued that the whole book should be merely an appendix to some more general treatise on metals, but I hope I have not erred too much in the direction of leaving matters of detail in the main text.

I have tried to give some historical perspective to the subject, first in the historical introduction and later throughout the book by reference to the earliest significant publication on each topic as well as to the most recent studies. I hope the authors of valuable work that falls between these two stools will forgive me for not quoting them explicitly, though usually their work is covered implicitly by reference to appropriate detailed reviews. I hope too that I shall be forgiven for placing too much emphasis on topics with which I have been particularly concerned, as compared with other topics of at least equal importance. Thus my chapter on magnetic interaction is much longer than that on magnetic breakdown and I have put more emphasis on the Fermiology of bismuth than on that of the transition metals or of metal compounds. My only excuse is that I am usually able to present matters more clearly when I have been personally involved, but to redress any imbalance I have taken care to give detailed references to topics that I have felt unable to treat adequately myself.

A great deal of the book is concerned with experiments designed to test the theory of the oscillations and their peculiarities. Often, however, agreement between theory and experiment has proved to be imperfect in one respect or another and I have tried to draw particular attention to such discrepancies and also to aspects of the theory which have not yet been adequately tested by experiment. I hope that my emphasis on such 'loose ends' will prove useful in opening up new lines of investigation in the future.

Some other aspects of the presentation need mention. I have deliberately introduced a considerable number of diagrams, partly to help understand what the formulae mean in visual form, partly to convey the aesthetic appeal of various kinds of oscillations as they appear in practice, and partly to give a flavour of the weird and wonderful variety of Fermi surfaces. The captions to the diagrams may sometimes seem excessively lengthy, but I have tried to include any information that may be relevant in grasping the meaning of the diagram, sometimes repeating information in the text to avoid the need for too much cross reference and sometimes adding details that could help the reader make his own deductions from the diagram. Similarly, detailed notes are appended to the tables where appropriate, and some of the tables collect together in a unified form information which is otherwise accessible only in a number of separate papers and often with a different form of presentation in each.

Notation is always a problem in covering a wide variety of topics, since there are insufficient letters in the English and Greek alphabets even with their upper case, lower case and script variants. The difficulty can be partially overcome by the use of lots of suffices and indices, but this easily leads to a considerable loss of transparency in the appearance of formulae and I have tried to avoid suffices and indices except where they seem positively helpful. Instead I have accepted the inevitable and to misquote Humpty Dumpty 'when I use a symbol it means just whatever I choose it to mean'. Thus k may be Boltzmann's constant, a wavenumber, a magnetic 'frequency' (as in $\sin kh$) or a serial number. At worst, the same symbol (e.g. c) may mean as many as eight different things but each new meaning is carefully defined when it occurs. In practice it is only rarely that such various meanings occur in the same context, and usually if they do it is easy to avoid confusion by a strategically placed warning or by using an appropriate suffix (e.g. k_B for Boltzmann's constant) only when there is such a clash. The various meanings of symbols together with often used abbreviations of names and technical terms are collected for reference on p. xvii.

Once again to avoid loss of transparency, I use a rather informal approach in dealing with vectors, usually emphasizing their vector nature by bold type only when it is essential to avoid confusion, and similarly the twiddle above a symbol to denote 'oscillatory part of' is often omitted when it is obviously implicit. Thus $\tilde{M}$ is often rendered as plain M when it is obvious that it is the oscillatory part that is relevant and the vector character of M is not particularly important. Often too, ordinary derivatives are used when it is not essential to stress the partial character (e.g. dy/dx may really mean $(\partial y/\partial x)_z$). Key equations are indicated by equation numbers in bold type, both when it is an equation which will be often quoted and in order that the reader can more easily recognise when he has reached the 'punch line' of any particular argument.

I hope my use of the Gaussian system of units will cause no serious difficulty to those brought up on the SI system (for an amusing commentary on units see Casimir 1968). Although my choice undoubtedly has something to do with the way I myself was brought up, it can be defended objectively on the grounds that in magnetic theory many relations assume a simpler and more meaningful form in the Gaussian system and moreover there was until recently some uncertainty about the definition of magnetization in the SI system. Anyone unfamiliar with the Gaussian system need only remember that it has the same unit, the Gauss (G), for magnetic field, magnetic induction and for magnetization, that the Gauss has the same magnitude as the Oersted (a unit not used in this book) and that $10^4 \, \text{G} = 10 \, \text{kG} = 1 \, \text{T}$ (Tesla). Other peculiarities, such as the use of centimetres instead of metres and grams instead of kilograms, should not offer too serious a stumbling block. Where any practical use is to be made of a theoretical formula, theoretical factors involving fundamental constants are usually given numerically in such a way that the answer will be in familiar units such as volts.

Finally it is my pleasant duty to acknowledge the great help I have had from many colleagues and friends in the preparation of the book. Above all I must thank Professor Sir Brian Pippard and Dr G. G. Lonzarich of the Cavendish Laboratory who have helped me over difficult places at nearly every stage of my task. The former's remarkable physical insight has provided almost immediate clarification on many occasions when I have been sorely puzzled, while the latter's willingness to discuss many points of detail and to clarify subtle points has also been a source of support and encouragement.

The following have provided valuable information and criticism on various particular topics and I list them roughly in the order in which these topics appear in the book. Acknowledgement of specific contributions is also made in the text where appropriate. Professors Sir Rudolf Peierls (Oxford), H. B. G. Casimir (Eindhoven), W. J. Huiskamp and G. J. van den Berg (Leiden) provided helpful recollections of the 'early days' (Chapter 1). The presentation of the theory of Chapter 2 was mainly worked out at the University of Waterloo, where I had the leisure to get to the bottom of a number of difficulties and also had useful discussions with Professor J. Vanderkooy. Since then I have benefited by visits to Bristol University, where Professor R. G. Chambers made helpful comments on an early draft of Chapter 2 and to Sussex University where Dr M. Springford has kept me up to date on his latest experiments and he, Dr M. Elliott and Mr P. Stamp have helped me to understand the rudiments of many-body theory and its relevance when the oscillations are observed in extreme conditions. Professors E. M. Lifshitz and M. I. Kaganov (Moscow) made helpful comments on my treatment of steady susceptibility (Appendix 4) and Mr. S. M. Hayden (Cavendish Laboratory) has made helpful critical comments on Chapters 2, 3 and 4. Professor P. J. Stiles (Brown University), Dr M. Pepper, Dr J. Wakabayashi (Cavendish Laboratory) and Professor D. J. Thouless (University of Washington) have educated me on various aspects of two-dimensional systems, enabling me to relate the ideal theory which I had worked out for myself, to the reality of actual experimental conditions. This topic really deserves a chapter to itself but is in fact distributed between Chapters 2, 3 and 4. Professor E. Maxwell (M.I.T.) drew my attention to early work on the temperature modulation method of detecting magnetic oscillations (Chapter 3). I have had valuable information and advice on questions of strain dependence of the Fermi surface from Professors E. Fawcett, M. J. G. Lee and J. M. Perz (Toronto), Dr I. M. Templeton (N.R.C. Ottawa) and Professor W. R. Datars and Dr M. Elliott (McMaster University). I first became acquainted with the problems of giant quantum oscillations during a short visit to Kyushu University in 1974, where discussion with Professor S. Mase stimulated me to work out the treatment given in Chapter 4.

Dr P. T. Coleridge and Dr I. M. Templeton (N.R.C. Ottawa) have given advice and criticism on a number of questions: the field modulation technique (Chapter 3), the detailed specifications of the Fermi surfaces of monovalent metals (Chapter 5) and the theory of dilute alloys (Chapters 5 and 8). Both have generously taken the time to carry out computations which have enabled me to present the data and their interpretation in a unified form. Mr C. M. M. Nex (Cavendish Laboratory) kindly computed the convolutions involved in the line shapes of fig. 4.7 and the Fermi surface sections of fig. 5.7 (e)–(h). I have had helpful correspondence from the following on a variety of points relevant to Chapter 5: Professor S. Berko (Brandeis University), Dr V. S. Edelman (Moscow), Dr M. R. Halse (Kent), Dr P. M. Holtham (Victoria, B.C.), the late Dr J.-P. Jan (N.R.C. Ottawa), Professor A. W. Overhauser (Purdue University), Dr M. G. Priestley (Bristol), Professor L. W. Roeland (Amsterdam) and Professor J. Trivisonno (John Carroll University, Cleveland, Ohio).

Professors J. J. Vuillemin and R. W. Stark (University of Arizona) have looked critically through Chapters 6 and 7 respectively and Professor A. V. Gold (University of British Columbia) kindly informed me in advance of publication of experiments on magnetic interaction relevant to Chapter 6 and discussed various points by correspondence. In writing about the effect of dislocations (Chapter 8) I have been greatly helped by discussions with Dr B. R. Watts (University of E. Anglia), especially as regards the interpretation of his own theoretical work. I owe much also to Dr A. Howie, Dr L. M. Brown (Cavendish Laboratory), Dr Z. S. Basinski (N.R.C. Ottawa and Cavendish Laboratory) and Professor F. N. Nabarro (Witwatersrand University and Cavendish Laboratory) for advice and information on the basics of dislocations, and to the last named for critical comments on my treatment in Chapter 8.

Last but not least, Mrs M. J. Apsley typed preliminary drafts of some of the early chapters, Mrs G. L. Lonzarich typed most of the later drafts of the whole book and also prepared the line drawings, and Mrs P. McCullagh helped out by typing the rest when time became short. I am most grateful to them for their skill in interpreting my scribbled copy and their patience in coping with my many changes of mind about details. I also have to thank Mr. K. Papworth for making photographic reproductions of many diagrams from original publications and Mr Jeremy Smith for his helpful subediting for the Press.

D. Shoenberg   March 1983

Some notes added in proof about recent developments appear on p. 566.

Symbols and abbreviations

a	area of cross-section of an arbitrary surface of constant energy at arbitrary k_H, MI parameter, i.e. $4\pi|\mathrm{d}\tilde{M}/\mathrm{d}B|$, sample dimension (e.g. radius).
A	extremal area of cross-section of FS, amplitude $|\tilde{M}|$.
A_s	diametral plane area of cross-section of spherical FS.
A_0, A_1 etc.	spin-symmetric coefficients in Fermi liquid theory
$\mathscr{A}$	area of cross-section of FS at arbitrary k_H.
b	small departure from reference value B_0 of B (i.e. $B - B_0$), magnitude of Burgers vector.
B	magnetic induction.
B	belly (usually as subscript).
c	velocity of light, torque per unit angle of twist, coupling constant of pick-up coil, calibration constant, specific heat, elastic modulus, coefficients in various expansions, atomic concentration.
C	heat capacity, curvature factor $(A'')^{-1/2}$, resultant amplitude of $\tilde{M}$ from two oscillations of nearly equal frequencies.
CDW	charge density wave.
d	average separation distance of dislocations.
dHvA	de Haas–van Alphen.
D	number of states on a Landau tube between κ and $\kappa + \delta\kappa$, distribution function, domain wall thickness, dislocation density, determinant of ellipsoid coefficients.
D	dog's bone.
$\mathscr{D}$	density of states.
e	electronic charge (occasionally e_0).
e_1, e_2, e_3	lengths of principal axes of FS electron ellipsoid of Bi.
E	energy of a system, electric field.
EE	electron–electron.
EP	electron–phonon, ellipsoidal parabolic.
ENP	ellipsoidal non-parabolic.

f frequency, Fourier transform, Fermi function, variable dHvA frequency.

F dHvA frequency, free energy ($E - TS$).

FE   free electron.

FS   Fermi surface.

$\mathscr{F}$ Friedel sum.

g spin-splitting factor.

G spin reduction factor (i.e. $R_s = \cos \pi S$).

h modulating magnetic field (amplitude h_0), small difference from a reference field H_0 (i.e. $H - H_0$).

$\hbar$ 'Dirac h', i.e. $h/2\pi$ if h is Planck's constant.

h_1, h_3 lengths of principal axes of FS hole ellipsoid of revolution in Bi.

H magnitude of magnetic field vector H.

H_e magnitude of applied or external magnetic field vector H_e.

H_0 reference magnetic field, magnetic breakdown parameter, scale field in Dingle factor $\exp(-H_0/H)$.

i current.

I current, various integrals.

$\mathscr{I}$ imaginary part of.

J current density, exchange integral.

J_k Bessel function of order k.

k Boltzmann constant, magnitude of electron wave vector k, i.e. wavenumber, harmonic number in context of time (e.g. $\cos k\omega t$), magnetic 'frequency' (i.e. $2\pi F/H_0^2$).

k_H component of k along H (usually denoted by κ).

k' component of k in plane normal to H.

k_F radius of spherical FS.

K repeat distance along an open orbit in k-space.

K_4, K_6 etc.   cubic harmonics.

KKR   Korringa, Kohn and Rostoker (method of band structure calculation).

$\mathscr{K}$ compressibility.

l sample length.

L effective path vector.

LK   Lifshitz–Kosevich.

m cyclotron mass.

m_a amplitude modulation parameter for oscillations of dM/dH.

m_b bare electron mass, cyclotron mass at bottom of non-parabolic band.

m_f frequency modulation parameter.

m_0 free electron mass.

M magnitude of magnetization vector M (i.e. magnetic moment per unit volume), but often also magnetic moment of an arbitrary volume V.

MI magnetic interaction.

MB magnetic breakdown.

$\mathcal{M}$ magnetic moment per unit area of a 2-D sample.

n number of electrons or quasiparticles per unit volume (but in 2-D context per unit area), limiting value of a serial integer r, demagnetizing coefficient divided by 4π.

N total number of electrons or quasiparticles, number of turns in a coil.

N neck.

NFE nearly free electron.

$\odot$ the basic circular orbit in MB theory.

OPW orthogonal plane wave (method of band structure calculation).

p magnitude of momentum vector p, pressure, harmonic number in dHvA oscillations, torque interaction parameter, partial correlation parameter in MI (also q), amplitude probability of MB.

P pitch of helical motion, electrical power, particle flux probability of MB.

P_l Legendre polynomial of lth order.

q magnitude of position vector q (conjugate to p), magnitude of phonon wave vector q, partial correlation parameter in MI (also p), amplitude probability of Bragg reflection.

Q quadratic function, particle flux probability of Bragg reflection (i.e. $(1 - P)$).

r arbitrary integer, orbit radius in real space (also R).

r_s average interelectronic separation in Bohr radius units.

Symbols and abbreviations

r_s^* 'corrected' value of r_s.

R reduction factor to multiply basic oscillatory dHvA amplitude with suffix to indicate cause (e.g. R_D for Dingle factor, R_S for spin factor, etc.), magnitude of position vector R in real space.

R   rosette (usually 4-R or 6-R).

R' projection of R on plane normal to H.

R_H component of R along H.

RBM   rigid band model.

$\mathscr{R}$ real part of.

s elastic compliance, sound velocity (when no confusion with elastic compliance, otherwise v), strain variable.

S area of 2-D sample, dimensionless spin parameter such that $R_S = \cos \pi S$, entropy, switching probability in MB, spin quantum number.

SdH   Shubnikov–de Haas.

SDW   spin density wave.

t time.

T absolute temperature, torque, switching probability in MB.

T_α, T_γ amplitudes of α and γ components of dHvA oscillations in Pb.

TI   torque interaction.

u ratio of sample dimension to skin depth (a/δ), relative amplitude of temperature oscillation (θ/T), Dingle reduction factor from phase smearing (also v), phase parameter which varies through a beat cycle, amplitude of oscillation of some quantity U.

U arbitrary oscillatory quantity.

v magnitude of electron velocity vector v, volume in k-space, voltage (but sometimes V), sound velocity (but sometimes s), Dingle reduction factor from phase smearing (also u).

V volume in real space (e.g. sample volume), voltage (usually v).

$V_{l,m,n}$ Fourier component of pseudopotential.

W double integral of $\mathscr{D}(x)$, thermal resistance, parameter determining cyclotron mass in EP model $((m_0/m)^2)$.

x Dingle temperature, continuous variable replacing quantized $(r + \tfrac{1}{2})$, reduced magnetic field variable (i.e. kh with $k = 2\pi F/H_0^2$).

X value of x at the FS (i.e. effectively F/H), domain repeat period, parameter in many-body analysis (see (9.30)).

y reduced form of $4\pi M$ (i.e. $4\pi kM$ with $k = 2\pi F/H_0^2$).

Y Young's modulus, slope of magnetization curve, i.e. $4\pi \mathrm{d}M/\mathrm{d}H_\mathrm{e}$.

Y_b value of Y at the flat bottom of an oscillation in strong MI conditions.

Y_b' value of Y_b in arbitrary units ($Y_\mathrm{b} = cY_\mathrm{b}'$ with a calibration constant c).

z argument of distribution function $D(z)$, value of $2\pi^2 pkT/\beta H$, half peak-to-peak amplitude of oscillation in $4\pi \mathrm{d}M/\mathrm{d}H_\mathrm{e}$.

z, z' values of Dingle reduction factors of opposite spins.

Z valency, reduced thermodynamic potential, i.e. $4\pi k^2\Omega$ with $k = 2\pi F/H_0^2$.

α area of R' orbit, value of $2\pi^2 k/\beta$, value of $2\pi^2 c\hbar/eH$ (i.e. $2\pi^2\eta$), Poisson's ratio, modified MI parameter (i.e. $a(1-n)$ (also denoted by a')).

α_0 modified MI parameter with no phase smearing (i.e. $a_0(1-n)$).

β $e\hbar/mc$.

β_0 value of β for $m = m_0$ (i.e. double Bohr magneton).

γ phase constant in quantization condition, Dingle reduction factor R_D for phase smearing (in MI context), amplitude of torque oscillation, parameter in calculation of mosaic structure reduction factor.

Γ imaginary part of self energy, sound attenuation coefficient.

δ Dirac delta function, value of $(X - (n + \tfrac{1}{2}))$ over one oscillation cycle, skin depth, parameter specifying distortion of nearly spherical FS.

Δ real part of self energy.

ε energy of an electron or quasiparticle, strain, dielectric constant of metal.

ε_g energy gap.

ζ chemical potential of a system, Fermi energy.

η scaling factor between R' and k orbits (i.e. $c\hbar/eH$), coupling constant between flux of a small sample in a coil and its magnetic moment.

η_l phase shift parameters in band structure calculation.

θ angular variable, amplitude of oscillatory temperature, reduced magnetic induction variable (i.e. kb with $k = 2\pi F/H_0^2$).

Θ Debye temperature.

κ component of k in H direction (i.e. k_H).

λ scaling parameter of distribution function, mass enhancement parameter $(m = (1 + \lambda)m_b)$, dimensionless field modulation parameter (i.e. $2\pi F h_0/H^2$), spin-orbit coupling parameter, correlation factor between phase departures ϕ and ϕ' in oscillations of two frequencies (i.e. $\phi' = \lambda\phi$).

Λ spin-orbit splitting in energy.

μ magnetic moment of electron orbit on FS, chemical potential (variable form of ζ), crystal anisotropy of F (i.e. $(1/F)(dF/d\theta)$), parameter in MI theory denoting $na_0\pi/(1 + a_0)$, parameter in theory of Dingle factor due to dislocations.

μ_a amplitude modulation parameter for oscillations of M.

ν phonon energy (i.e. $\hbar\omega$ if ω is angular frequency), amplitude of modulation of n (number of electrons per unit area in 2-D sample).

ρ density, parameter defining asymmetry of a GQO attenuation peak, ratio of anisotropy parameters (μ) of two dHvA oscillations of nearly equal frequency, electrical resistivity.

σ electrical conductivity, stress, characteristic parameter of strain distribution.

Σ complex self energy (i.e. $\Delta - i\Gamma$).

τ relaxation time (usually of electron, but occasionally in other contexts).

τ_ρ electron relaxation time inferred from resistivity ρ.

ϕ phase constant.

ϕ_l Friedel phase shifts.

Φ magnetic flux.

χ spin susceptibility.

χ_0 spin susceptibility of free electron gas.

ψ phase variable, sometimes denoting $2\pi p(F/H - \frac{1}{2}) \pm \pi/4$, sometimes $2\pi F/H$.

ω angular frequency.

ω_c cyclotron angular frequency (ω_0 in a special context).

Ω thermodynamic potential $F - N\zeta$ (often loosely called free energy).

Notes Some special meanings of symbols which are used only in isolated passages have been omitted. The suffices attached to symbols have been included only when it is necessary to explain their meanings. Entries corresponding to any particular symbol are listed roughly in order of their first appearance in the text. Throughout the book a wavy line over a symbol, e.g. as in $\tilde{M}$, indicates 'oscillatory part of'; this wavy line is, however, often omitted when it is obvious from the context that 'oscillatory' is implied.

1

Historical introduction

1.1. Early history

Soon after J. J. Thomson's discovery of the electron in 1897, Drude (1900) showed that most of the characteristic features of a metal could be understood, at least qualitatively, by supposing that some of the electrons were able to move freely through the metal, and a few years later Lorentz worked out the theory more rigorously on the basis of classical statistical mechanics. The outstanding quantitative success of this Drude–Lorentz theory was the explanation it gave of the Wiedemann–Franz law, the proportionality to absolute temperature T, of the ratio of thermal to electrical conductivity. Moreover the predicted constant of proportionality came out close to the experimental value (though less close in Lorentz's more rigorous calculation). However the theory was quite unable to explain why the free electrons did not make a large contribution to the specific heat and later, when electron spin had been discovered, it was not clear why the free electrons did not contribute a large paramagnetic susceptibility varying as $1/T$.

It is just over 50 years ago that Pauli (1927) made a major breakthrough by showing that if the recently discovered Fermi–Dirac statistics were used rather than classical statistics in working out the theory, the difficulty about spin susceptibility essentially disappeared. The calculated paramagnetic susceptibility then became independent of temperature and much feebler, roughly comparable to the experimental value. Not long afterwards Sommerfeld (1928) showed that application of the Fermi–Dirac statistics also removed the specific heat difficulty and other unsatisfactory features inherent in the classical theory* while still producing an explanation of the Wiedemann–Franz law (with an even better value of the constant of proportionality). From then on there was a flood of papers, particularly in *Zeitschrift für Physik*, exploiting the idea of applying quantum mechanics systematically to calculate all sorts of metallic properties.

Among these was a paper by L. Landau (1930) on the diamagnetism of

---

* For a review of the successes and shortcomings of the Drude–Lorentz theory and for detailed references see Wilson (1953a).

free electrons. He was then only 23 years old, but already so highly regarded in the Soviet Union that he was sent to study abroad at some of the leading centres of theoretical physics: Berlin, Copenhagen and Cambridge. It was in Cambridge that he produced the unexpected result that if quantum mechanics is properly applied to the orbital motion of metallic free electrons in a magnetic field, a feeble steady diamagnetic susceptibility is predicted which should be exactly one third of the spin paramagnetic susceptibility already calculated by Pauli. The paper also briefly points out that if the periodic field of the lattice is taken into account on the lines that Bloch (1928) had opened up, the calculations of orbital diamagnetism and spin paramagnetism should still be valid in principle, but the 1:3 ratio would be upset and Landau suggested that this might have something to do with the anomalously high diamagnetism of bismuth–as indeed later proved to be the case.

But Landau's paper contained yet another important idea. In essence this was a prediction that the magnetization at low temperatures should oscillate as the field varies, i.e. a prediction of the de Haas–van Alphen (dHvA) effect which is the main subject of this book. This prediction came in somewhat of a 'throwaway' remark, in which the possible effect is immediately dismissed as essentially unobservable. He points out that his use of the Euler–Maclaurin formula is justified only if H/T is small enough, and so should become invalid at sufficiently high fields and sufficiently low temperatures.

'In that case we should have a complicated, no longer linear dependence of magnetic moment on H, which would have a very strong periodicity in field. Because of this periodicity it would hardly be possible to observe this effect experimentally, since on account of the inhomogeneity of available fields there would always be an averaging.'*

In fact in dismissing the possibility of an experimentally observable effect Landau had overlooked two things. First that the kind of modifications of the free electron model which he himself had suggested might explain the anomalously high steady diamagnetism of bismuth, might also drastically enlarge the period of the predicted oscillations, thus making the demands on field homogeneity much less stringent. And secondly that even the stringent demands on homogeneity required to observe the short period characteristic of a monovalent metal (for which the free electron model is a fair approximation) are by no means impossible to meet. Many years later I asked Landau why he had made such a categorical remark about practical limitations. He replied that since he knew nothing about

---

* Translation from German.

experimental matters he had consulted Kapitza whom he was visiting at the time, and Kapitza had told him that the required homogeneity was impracticable. Fortunately what may have been true in 1930 was no longer a serious limitation for the technology of 30 years later and it is now quite a routine matter to observe the very short period oscillations that Landau had envisaged but dismissed.

By a rather remarkable coincidence the first experimental observation of an oscillatory magnetic behaviour was made by de Haas and van Alphen (1930*b*) within a month or two of Landau's paper, and it was in single crystal bismuth that they observed the effect (fig. 1.1). In the first communication (20 December 1930) which reported a clearly oscillatory pattern* they quote Landau's paper (among a number of others) as a source of information on the theory of magnetic susceptibility but it is evident that they either overlooked Landau's remark about oscillations or did not appreciate that it had any relevance to their observations.

The remarkable coincidence is that theoretical prediction and experimental observation of the oscillatory effect should have occurred almost simultaneously with neither side being aware of the other side's contribution. In fact the motive behind the Leiden experiments had nothing

Fig. 1.1. First observation of oscillatory field dependence of susceptibility in bismuth (de Haas and van Alphen 1930*b*).

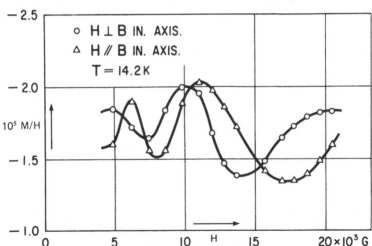

* In an earlier note (1930*a*) they reported a marked non-linearity in the relation between the magnetization *M* and the field *H*, but they realized that since the field varied considerably over the sample, the true behaviour might be more complicated, as indeed it proved to be when they used a much smaller sample with only a few % field variation over it.

to do with Landau's remark, but was based on a long standing hunch of de Haas that there should be a close correlation between diamagnetic susceptibility and the change of electrical resistance in a magnetic field. To quote from the introduction to their paper (1930*b*):

'The most detailed investigation was that of bismuth crystals, 1st because we had at our disposal a single crystal of extremely pure bismuth; 2nd because of the desirability of the examination of bismuth single crystals in connexion with the anomalous results of the resistance measurements by Schubnikow* and de Haas (1930) with these crystals. Because of the evident correlation (see de Haas (1914)) of the diamagnetic susceptibility with the change of resistance we were inclined to expect a dependence of the susceptibility on the field analogous to that found for the resistance. Further on we shall see that our expectation was fulfilled.'

A further paper by de Haas and van Alphen appeared in 1932 in which the experiments on bismuth were extended to higher fields (up to about 35 kG, but still only down to the relatively high temperature of 14.2 K), and a few measurements at liquid helium temperatures (down to 1.86 K) were reported in van Alphen's doctoral thesis (1933), but surprisingly no more work from Leiden has ever been reported on the effect since, in spite of its evident interest.

In the meantime Peierls (1933*b*) had developed a quantitative theory of the oscillations, essentially a theory for an isotropic gas of electrons whose effective mass m and Fermi energy ζ were treated as adjustable parameters. The theory is based on Landau's formulation of the energy levels in a magnetic field, which leads directly to the simple qualitative interpretation (see §2.1) that the oscillations occur as these 'Landau levels' pass through the Fermi level and become depopulated.[†] To quote Peierls:

'The quantitative calculation of the magnetic moment offers nothing new in principle; it is contained in Landau's expression for the thermodynamic potential at temperature T. For the limiting case $T = 0$ the evaluation of this formula is completely elementary but for finite T it leads to fairly time consuming integrations which were carried through for only a few values of the parameters and with low accuracy (errors of a few per cent).'[‡]

---

* This is the German transliteration of the Russian name. We shall use the English transliteration in referring to oscillatory variation of resistance as the Shubnikov–de Haas (SdH) effect.
† Although Peierls makes it clear that his calculations are based on Landau's earlier work, he does not make specific references to Landau's remark about the possibility of oscillations in principle, and it is probable that though he and Landau had discussed the problem of steady susceptibility together in detail in 1930 the question of the oscillations had not come up then and subsequently Peierls had not noticed the remark in the published paper. I should like to thank Professor Peierls for helpful correspondence on this point.
‡ Translation from German.

1.1. Early history

Peierls' calculations gave results (see fig. 1.2) which do indeed qualitatively resemble the experimental data if ζ (the Fermi energy) and m (the effective mass) are appropriately chosen, but the magnetizations come out nearly 70 times too small. This discrepancy later proved to be a consequence of neglect of the very strong anisotropy of the parameters, and Peierls' estimates of m (about 0.02 of a free electron mass) and of ζ (about 2×10^{-14} erg), which imply only 10^{-6} conduction electrons per atom, turn out to be roughly of the right order of magnitude. As has already been hinted, it is because of the very anomalous band structure of bismuth (which leads to such extreme values of the parameters) that the dHvA effect is so prominent and so easy to observe in bismuth.

The next step in development came from the experimental side. As a new

Fig. 1.2. Theoretical magnetization curves for an isotropic metal at various temperatures (a) $T = 0$, (b) $kT/\zeta = 0.1$, (c) $kT/\zeta = 0.15$, (d) $kT/\zeta = 0.2$ (Peierls 1933b); (c) and (d) are compared with experimental curves (de Haas and van Alphen 1930b) for Bi at 14.2 K and 20.4 K for H perpendicular to binary axis, shown by broken lines with displaced origin of M and scaled to match theoretical curves; the M and H axes for the experimental curves are shown at right and above. In the theoretical curves σ is a dimensionless parameter proportional to $-M$, β is eh/mc and ζ is the Fermi energy. If the 14.2 K curve is identified with that for $kT/\zeta = 0.15$ we find $\zeta \sim 1.3 \times 10^{-4}$ erg and $m/m_0 \sim 0.03$.

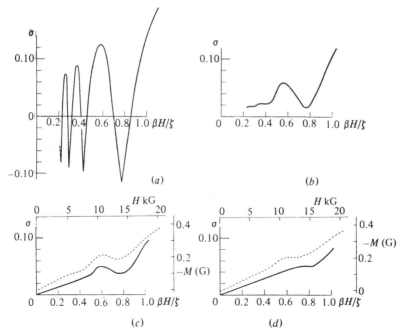

graduate student in the Royal Society Mond Laboratory in Cambridge I had been introduced to the mysteries of bismuth by my supervisor, P. Kapitza, who set me the problem of measuring its magnetostriction tensor (at room temperature) in 1932. I decided to follow this up with a study of the magnetic properties by the Faraday method and built a Sucksmith balance to measure the force on small single crystal samples in an inhomogeneous field. In due course the dHvA oscillations were observed and in collaboration with M. Z. Uddin (1936)* an investigation was made of how the oscillations were modified by alloying small quantities of other metals with the pure bismuth.

The modifications were striking in that very small additions of heterovalent elements (less than 0.08% Pb, 0.02% Sn or 0.01% Te) made quite big changes (of order 50%) in the periodicity of the oscillations. I was lucky that Peierls was on the spot as 'house theoretician' to the Mond Laboratory and was immediately able to interpret these changes in terms of his theory as being due to rather large relative changes in the small effective number of electrons of pure bismuth. He pointed out that the alloying was a bit like introducing or withdrawing a small volume of liquid into or from a bottle in which the level was in a very narrow neck. There could be quite a big change of level (i.e. of effective electron number) even though the volume change was only a very small fraction of the total (i.e. of all five valence electrons per atom).

Soon after this work, K. S. Krishnan paid a visit to Cambridge and described several ingenious methods he had developed to study magnetic anisotropy through the mechanical couple exerted on a crystal suspended in a uniform magnetic field (see for instance Krishnan and Banerjee 1935). It occurred to me that the couple might show up the dHvA oscillations much better than the Faraday force method that had been used before. The Faraday method requires the field to vary over the sample and this inevitably smoothes out the oscillations ('phase smearing'), particularly at low fields when the spread of field over the sample (proportional to H) becomes comparable to the field interval of the period of an oscillation (proportional to H^2). The torque method on the other hand requires a uniform field and so does not suffer from this phase smearing. To get a torque the metal must of course have magnetic anisotropy (later we shall see that the precise requirement is that the Fermi surface should be anisotropic) and bismuth certainly qualifies in that respect.

A convenient opportunity for carrying out this idea was provided by an invitation from Kapitza to work for a year in the Institute for Physical

* It was in this paper that the oscillations were first described as the 'de Haas–van Alphen effect'.

1.1. Early history

Problems in Moscow, which had just been built for him in 1936, after the Soviet authorities had refused him permission in 1934 to continue working in Cambridge. The torque experiment was just about simple enough to carry out in less than a year and, with the excellent facilities provided by the Institute, I was able to see striking oscillations within seven weeks of getting to Moscow (fig. 1.3). As expected, the oscillations could be followed to much lower fields than had been possible with the Faraday method* and a detailed study could be made of how the effect varied with crystal orientation, field and temperature. Once again I was lucky in having the guidance of a house theoretician. This time it was Landau and by what seems to have been a rather extraordinary coincidence he had just then developed a theory of the detailed form of the dHvA oscillations. I cannot now clearly recall whether he had been stimulated to do this by knowing of my experiments or whether it was pure coincidence. But I do remember clearly that one day he said quite casually that I might care to see a formula which I could test against every aspect of my experimental results.

This formula marked an important advance in two respects. First it was a completely explicit expression for the couple as a function of all the

Fig. 1.3. Oscillatory field dependence of torque for Bi at 4.2 K. The ordinate is C/H^2 where C is the torque per unit volume at field H. The field is at 9.7° from the binary axis in the binary-trigonal plane and the torque is measured about an axis normal to this plane (after Shoenberg 1939).

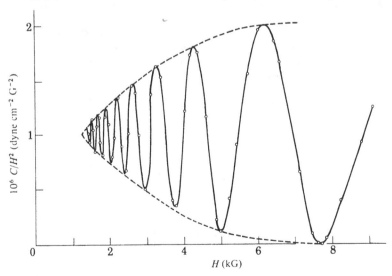

* The improvement of fig. 1.3 over fig. 1.1 reflects not only the absence of field inhomogeneity but also the lower temperature.

experimental variables: crystal orientation, field and temperature; and of the theoretical parameters, which were effectively the number of electrons and the mass tensor. Secondly it took proper account of the crystal symmetry by specifying four independent mass parameters. Indeed, just previously Blackman (1938) had already developed Peierls' theory further by taking into account the anisotropy and had considerably improved the agreement between theory and experiment, but he used a somewhat simplified model (with only three mass parameters) which did not fully allow for the possibilities offered by the actual crystal symmetry.

The existence of an explicit analytic formula was crucial both in providing a guide to what measurements were desirable and in their interpretation to deduce the basic parameters. With the Peierls and Blackman theories, even if they had allowed fully for the crystal symmetry, it would have been a much more difficult task to derive from the measurements the best values of the parameters, since complicated numerical computation was needed to derive each point on a magnetization curve for any one set of parameters. With modern computing methods optimization of the parameters could no doubt have been achieved, but in those days it would have been a daunting task. With Landau's formula, however, it was a relatively straightforward task. In fact the formula proved to be in excellent agreement with all the data except in two respects. First, to get a good fit to the amplitude dependence on field and temperature it seemed necessary to put $(T + x)$ rather than T into the formula, where x

Fig. 1.4. The Fermi surface of Bi consists of three electron ellipsoids and one hole ellipsoid. Sections of one of the electron ellipsoids by the $k_y k_z$ and $k_x k_z$ planes are shown approximately to scale; the other two are identical, but rotated about the k_z axis through $\pm 120°$. The hole ellipsoid is a spheroid of revolution about the k_z axis. The positions of the various ellipsoids in the Brillouin zone are shown in fig. 5.27.

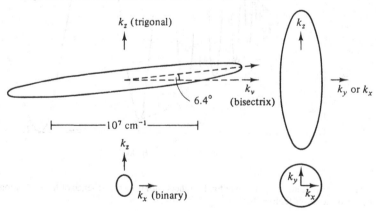

8

was a constant of about 1 K, and secondly the phase of the oscillations seemed to come out wrong by roughly 180°. In spite of these discrepancies – both of which got cleared up later – it was possible to make fairly consistent estimates of the mass tensor components and of the number of electrons. It should be emphasized that Landau had assumed (as had everyone earlier) that the surfaces of constant energy in k-space (including the Fermi surface) were ellipsoids and that the energy depended quadratically on k. With these assumptions the parameters that had been determined specified the Fermi surface and it is fair to claim that this was in fact the first experimental determination of a Fermi surface (see fig. 1.4).

Early in 1938 Landau was arrested during the Stalinist purges and was released only about a year later. In the meantime my experimental results were published both in Russian (1938) and in English (1939). In the Russian version all reference to Landau was deleted at the insistence of the laboratory administration, since it was at that time the inevitable practice to make a 'non-person' of an 'enemy of the people' (as anyone arrested was called) and it was impossible to mention such a person in print. However, the appropriate reference did appear in the English version together with an appendix in which Peierls reconstructed the derivation of Landau's formula from some rough notes which I was able to supply. Ironically, the English version appeared in print shortly after Landau was released.

1.2. 1947–60: The rise of Fermiology

The next development came only in 1947 after the war, with J. A. Marcus' discovery at Yale University of the dHvA effect in zinc. Nakhimovich (1941) had observed anomalies in the magnetoresistance of zinc and Marcus, arguing on the same lines as de Haas and van Alphen, that anomalous magnetoresistance might be associated with anomalous magnetic behaviour, decided to examine zinc. The hunch paid off and oscillations qualitatively similar to those found in bismuth were strikingly evident in the susceptibility as measured by a Curie balance. This was soon followed up by more detailed studies using the torque method (Sydoriak and Robinson 1949, Mackinnon 1949) and the results could be fitted roughly to Landau's ellipsoidal model though there were some discrepancies in detail.

In spite of an explicit comment by Peierls (1933b) that the dHvA effect should occur in all metals and especially in metals with a small number of free electrons, most experimenters (including the author) had tacitly assumed prior to 1947 that it was only in bismuth that the effect could be observed in practice. This complacent attitude was clearly no longer tenable after Marcus' discovery of the effect in zinc and in 1949 Verkin,

9

Lazarev and Rudenko in Kharkoff and I in Cambridge started looking for the effect in various other metals using the torque method.* It soon become evident that oscillations could be observed in nearly every polyvalent metal of which good single crystals were available. It was not too surprising that no effect could be seen in any monovalent metal, since rough calculations suggested that considerably higher fields would be needed to make the amplitude large enough to observe.

The various polyvalent metals showed a bewildering variety of behaviour, both in the range and multiplicity of the periodicities displayed and in the nature of the orientation dependence of the periodicities. It was evident that Landau's ellipsoidal model which had served so well for bismuth was usually inadequate to explain either the multiplicity of periods or their orientation dependence. However his formula did still accurately describe the uniform periodicity of the oscillations in reciprocal field and the form of the temperature dependence of the amplitude.

A significant theoretical advance was made at this time by Dingle (1952b), who pointed out that electron scattering should broaden the Landau levels and that this should lead to an extra amplitude reduction factor, which would have an affect approximately equivalent to that of a rise of temperature. This Dingle temperature, as the effective rise x of T came to be called, thus provided an explanation for the already mentioned discrepancy between the experimental results on bismuth and the original Landau formula. All the new results also required Dingle temperatures to be included in the formula in order to describe the field and temperature dependence of amplitude consistently. As will be discussed later (chapter 8), the Dingle reduction factor in pure metal samples is in fact less due to the Landau level broadening caused by electron scattering than to phase smearing caused by sample inhomogeneity. It is only in relatively impure samples that the effect of Landau level broadening is dominant.

It is convenient to mention here also another significant development of the theory since Landau's formula, which was to take account of the splitting of the Landau levels by electron spin. This was first considered by Akhieser (1939) for free electrons and sometime later, in a rather more general way, by Sondheimer and Wilson (1951) and Dingle (1952a).[†] The effect of the spin-splitting is effectively to introduce a phase difference between the oscillations from the spin-up and the spin-down electrons and the resultant amplitude is therefore in general multiplied by a reduction

---

* For full references up to 1952 see Shoenberg (1952a).
† Sondeimer and Wilson point out that Dingle's analysis came before theirs, though his paper appeared only later.

factor $\cos(\pi\Delta\varepsilon/\beta H)$, if $\Delta\varepsilon$ is the energy difference between the spin-up and spin-down levels and βH the energy difference between successive Landau levels (both at the Fermi surface). It is this spin factor which explains the 180° phase difference of the experimentally observed oscillations from the prediction of Landau's formula. This became apparent when Cohen and Blount (1960) showed that strong spin-orbit coupling should indeed make $\Delta\varepsilon$ nearly equal to βH for bismuth, thus producing a spin factor close to -1 just as for free electrons in Akhieser's theory. It was only a good deal later that other evidence for the validity of the spin factor became available.

By 1952 a considerable body of experimental data had been accumulated on the 13 polyvalent metals in which the effect had been observed (Shoenberg 1952a and references therein); only one, lead, had at that time given negative results. However, because the ellipsoidal model was usually inappropriate, only order of magnitude estimates of the numbers of effective electrons (i.e. the volumes of the Fermi surfaces) and the Fermi energies could be made, together with rather more accurate estimates of the effective (cyclotron) masses and their orientation dependence. Fortunately a major theoretical breakthrough occurred at this time, and as so often happens in the history of science, this development was made twice over, independently and almost simultaneously. In 1951 Onsager came for a sabbatical year to Cambridge from Yale and brought with him a very elegant and simple interpretation of the dHvA periodicity, in terms of the geometry of the Fermi surface (FS). Eventually he wrote up his idea in a three page paper (1952) which has become somewhat of a classic. Quite independently, I. M. Lifshitz, then in Kharkoff, had put forward essentially the same idea in an unpublished lecture to the Ukrainian Academy of Sciences in Kiev in 1950, and a little later he and Kosevich (1954, 1955) developed the idea into a detailed theory giving not only the frequency of the oscillations but also their amplitude and phase.

The Onsager–Lifshitz idea was based on a simple semi-classical treatment of how electrons move in a magnetic field, using the Bohr–Sommerfeld condition to quantize the motion. It then appears that the dHvA frequency F (i.e. the reciprocal of the period in $1/H$) is directly proportional to the extremal cross-sectional area A of the Fermi surface, and moreover that the constant of proportionality is a universal constant. The relation is in fact

$$F = (ch/2\pi e)A \qquad (1.1)$$

and its importance is, of course, that it provides a tool for accurate measurement of Fermi surfaces, though it took over five years before this tool began to be systematically used.

The difficulty in using the Onsager relation* to determine Fermi surfaces was partly experimental and partly theoretical. On the experimental side, the measurements were made in relatively low and not very homogeneous magnetic fields and the samples were often of poor quality, so that only rather low frequency oscillations showed up appreciably, corresponding to only small pieces of the FS. The oscillations associated with the major parts of the FS would have been of too feeble an amplitude and too high a frequency to be detected in the experimental conditions then in use. On the theoretical side, very little was then known about what the smaller parts of the FS of a polyvalent metal should look like. Band structure calculations were still too crude to make any reliable predictions about such fine details as the existence or non-existence of a small pocket of FS in a high-order zone. Thus there was little theoretical guidance for any sensible interpretation of what kind of FS could be constructed out of the observed small extremal areas.

Some guidance was indeed available from a very elegant geometrical theorem due to Lifshitz and Pogorelov (1954), which gave a precise and unique recipe for constructing a surface from a specification of its extremal areas A for all directions of the normal to the cross-section. This theorem was valid however only under severely limiting restrictions which, unfortunately, are rarely satisfied by any real FS. In spite of its elegance, the Lifshitz–Pogorelov theorem has not proved useful in practice. The only attempt to apply it systematically to experimental area data was by Gunnersen for aluminium (1957). He constructed a FS which consisted of three cushion-shaped surfaces and whose extremal areas were in fact reasonably consistent with the orientation dependence of the oscillation frequencies, but which later, in the light of more complete data and of guidance from new theoretical ideas about band structure, proved to be entirely wrong.

It was clear at this stage that it was highly desirable to extend the measurements to higher fields if further progress was to be made. Superconducting magnets had not yet been developed, while water-cooled magnets going to 50 kG or more were not easily available and moreover were unlikely to be easy to use because of the difficulty of stabilizing the field to better than a dHvA period. It occurred to me that the ideal solution might be to use a pulsed magnetic field, and to detect the magnetic oscillations by the oscillatory e.m.f. they induced in a suitable pick-up coil system. This technique (see chapter 3) first worked successfully in 1952

---

* This has come to be the accepted name for (1.1) and the quantization condition on which it is based; Lifshitz's contribution is remembered by referring to the general formula for the oscillation amplitude as the Lifshitz-Kosevich or LK formula (see (2.151)).

(Shoenberg 1952b) after about two years of development and the higher fields (typically 100 kG) made it possible to study much higher frequency oscillations and consequently much larger sheets of Fermi surface.

After a good deal of exploratory work (see Shoenberg 1953, 1957), the impulsive high field method was first systematically applied by Gold (1958), who studied in detail the orientation dependence of the rather complicated frequency spectrum of lead (see fig. 5.19). His results led him to an interpretation which proved to be an important landmark in the understanding of polyvalent metals. This interpretation was in terms of the Fermi surface of a nearly free electron (NFE) model – a model which at first sight seemed extremely implausible for a metal of high atomic number such as lead. In the reduced zone scheme the various pieces into which the free electron sphere of the model is cut up by the Brillouin zone planes are reconstituted according to the zone they come from. The resulting sheets of FS then provide a rough guide to what the FS might really be like if the periodic potential could be regarded as a relatively weak perturbation. The various sheets of the FS indicated by this NFE model are illustrated in fig. 5.15 and it can be seen that they offer a wealth of extremal areas, some of which Gold was able to associate plausibly with particular branches of the observed frequency spectrum. Some years later Anderson and Gold (1965) carried out a still more detailed study of lead (see fig. 5.19) with much improved techniques and confirmed the main features of Gold's earlier interpretation by finding many further branches of the spectrum predicted by the NFE model, but which had not been seen in the pioneer study.

The success of the NFE model was an important piece of evidence in the development of pseudopotential theory in the late 1950s. Heine (1957) had already found that a 'first principles' band structure calculation for aluminium gave a nearly free electron like FS* even though the periodic potential was much more than a feeble perturbation of the ideal free electron model. Thus the success of the NFE model for lead, whose periodic potential should be much stronger than in aluminium, suggested strongly that some new principle was involved. This principle was soon uncovered by the work of Phillips and Kleinman, Morrel Cohen and Heine and Harrison (for a detailed review see Heine, Cohen and Weaire 1970) which led to the concept of a *pseudopotential*. It turns out that by some mathematical juggling, the Schrödinger equation for the electrons moving

---

* However some of the details of this FS were very sensitive to the assumptions of the calculation. Some small cushion-shaped pieces of FS which seemed to fit plausibly with Gunnersen's dHvA data (see above) proved to be non-existent in the FS (still nearly free electron like) finally arrived at by a combination of theory and experiment (see §5.3.3.1).

13

in the real and strong periodic potential can be recast into what looks like a Schrödinger equation for electrons moving in a much feebler pseudo-potential. The recast Schrödinger equation is in terms of a modified wavefunction (a 'pseudo-wavefunction') but has, over the range of interest, energy eigenvalues nearly identical with those of the original problem. In this way the very complicated real problem reduces to the much simpler one of solving the Schrödinger equation for a NFE model with a feeble perturbing periodic potential.

Although the impulsive field technique was successful in showing up the oscillations associated with major pieces of the Fermi surfaces of polyvalent metals and thus paving the way for a much better understanding of their band structures, several years elapsed before the dHvA effect was discovered in any monovalent metal. With hindsight it is now evident that the reason for the negative results obtained in many preliminary attempts was partly because the effort was concentrated on the least promising metals and partly because poor quality samples were used. The very first attempts were made with sodium in the belief that it was the 'simplest' metal, though in fact because of its martensitic transformation at about 40 K it is difficult to make a crystal sample which is not badly damaged by cooling to helium temperatures. Then there were many unsuccessful attempts with copper crystals, which probably failed mostly because of poor quality and unsuitable crystal orientation. It was not then appreciated that for the noble metals, free electron like oscillations occur only over a fraction of all possible orientations, nor was it appreciated that of the three noble metals copper had the weakest amplitude of dHvA oscillations. It was a rather frustrating situation, rather like looking for a black cat in a dark room, without even having complete assurance that there was a cat.

Eventually the breakthrough came by a combination of luck and slightly faulty reasoning. It occurred to me that whiskers might provide much more perfect crystals than ones grown from the melt, and during a summer visit to the G. E. Research Laboratories at Schenectady in 1958 I was able to get hold of some copper whiskers which had been prepared by Mrs E. Fontanella. At last on 19 December 1958 one of these whiskers produced oscillations (fig. 1.5) which were unmistakeably above noise level and of just the expected dHvA frequency. Once a positive signal had been obtained it did not take long to improve the technique sufficiently to determine the Fermi surfaces of copper and also of silver and gold in fair detail (see §5.3.2). The slightly faulty reasoning lay in the argument that whiskers should be more perfect than crystals from the melt. In fact it is only extremely thin whiskers – too thin to be of much use – which are so perfect, while the whiskers I used were relatively thick (diameters of a few tenths mm) and

14

were probably little more perfect than crystals grown from the melt by the best techniques. There was little difficulty in making good enough silver and gold crystals from the melt and eventually better quality copper crystals from the melt also became available.

In parallel with the start of systematic exploitation of the dHvA effect to determine Fermi surfaces, there was a rapidly growing development of other methods.* Pippard (1954) had shown that the anomalous skin effect (the skin effect in conditions such that the mean free path was greater than the skin depth) for any given direction of the radio frequency (r.f.) field could be related to the geometrical properties of the FS.[†] Although the relation was a rather complicated one, involving the curvature of the surface, he suggested it might be used for determining the FS from systematic measurements of the anisotropy of the r.f. surface impedance in anomalous skin effect conditions. During a sabbatical visit to the University of Chicago he succeeded in determining the FS of copper in this way (Pippard 1957a) with sufficient precision to suggest that the surface probably bulged out from the free electron sphere enough to make contacts with the zone faces normal to ⟨111⟩. This was a very useful pioneer effort, particularly in providing guidance for what should be looked for in the later dHvA

Fig. 1.5. First observation of the dHvA effect in Cu (Shoenberg 1959). The sample was a ⟨111⟩ whisker at 1.1 K. During the sweep time of 0.6 ms, the field drops by about 250 G from 72 kG at the left; there are 28 oscillations across the picture, which implies $F = 5.8 \times 10^8$ G, almost exactly the accepted value today. The crowding of the oscillations as the field falls is because the field falls more and more rapidly away from its maximum.

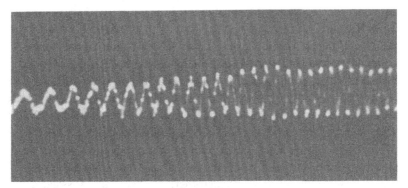

* For a brief review of the various methods, including some developed in the 1960s, see Shoenberg (1969b).
† Pippard's derivation of a general result was stimulated by an earlier calculation of Sondheimer (1954) of the anomalous skin effect for an ellipsoidal FS.

experiments. However the relation between the geometry of the FS and the anomalous skin effect is so complicated that it is usually hardly possible to go uniquely from the experimental data to the FS. Rather it is the other way round, that once the FS has been determined, it is possible to calculate what the anomalous skin effect should be, thus providing a useful check on the determination.

Pippard (1957*b*) was also instrumental in the development of the magnetoacoustic method of determining Fermi surfaces. This came about through his ingenious interpretation of the oscillations Bömmel (1955) had observed in the attenuation of ultrasonic waves passing through a tin crystal when the magnetic field was varied. He suggested that there should be increased attenuation whenever the classical electron orbit size could be fitted into an odd number of half wavelengths of the waves and that the periodicity could therefore be related to a 'caliper' dimension of the FS.

Other important contributions came from the Kharkoff theoretical group of I. M. Lifshitz. Lifshitz, Azbel and Kaganov (1956) showed how the orientation dependence of the magnetoresistance of a metal crystal could reveal the topology of the FS, i.e. could indicate whether or not sections of the FS by planes normal to the field direction contained any open as opposed to closed orbits. This idea soon began to be exploited by Alekseevski's experimental group in Moscow (see for instance Alekseevski and Gaidukov 1959) and often provided valuable guidance in the unravelling of complicated Fermi surfaces. Again, Azbel and Kaner (1956) invented the idea of a new kind of cyclotron resonance in which the magnetic field is parallel to the metal surface and there should be increased r.f. surface resistance whenever the r.f. was a multiple of the cyclotron frequency (i.e. the frequency of rotation of the electrons). Measurement of the periodicity of the corresponding oscillations in r.f. surface resistance should give the cyclotron frequency and the cyclotron mass and hence (as is explained in chapter 2) information on the differential properties of the FS. The effect was soon discovered by Fawcett (1956) and later studied in detail in a number of metals, particularly by.Kip and his group at Berkeley (see for instance Kip 1960). The cyclotron mass can, of course, also be determined from the temperature dependence of the dHvA amplitude.

Finally we should mention that many other physical properties besides the magnetization show oscillations as the field is varied, e.g. the temperature of the sample (magnetothermal effect), the physical dimensions (magnetostriction), the velocity of sound, the electrical properties (e.g. Shubnikov–de Haas effect) and the attenuation of sound. Some of these provide alternative techniques for determining the dHvA frequency and hence the FS, while others also provide additional information such as the stress dependence of the FS. These various oscillatory effects and what can be learnt from them are reviewed in chapter 4.

By 1960 a considerable activity had grown up in the investigation of the electronic structure of metals both theoretically through band structure calculations and experimentally by study of the various phenomena which could give information about it. This field of study gradually came to be called 'Fermiology', a convenient name for a subject which it is not very easy to define, and it is fair to say that it had by 1960 become an established branch of solid state physics. This was marked by an important conference on Fermi surfaces at Cooperstown, N.Y. at which both experimental and theoretical progress was reviewed in a way which stimulated mutual interaction (Harrison and Webb 1960).

1.3. After 1960: Fermiology comes of age

Progress after Cooperstown was surprisingly rapid on a number of fronts as Fermiology became increasingly popular and considerable improvements were made in both theoretical and experimental techniques. One of the hopes expressed at the conference was that it should soon be possible to apply many different techniques to a single metal and thus check that the FS was a concept which was independent of the particular phenomenon used to measure it. This hope was soon fulfilled for a number of metals, and indeed several new techniques such as Gantmakher's r.f. size effect (1962) and positron annihilation (see for instance Berko 1978) were developed which gave further confirmation. However it gradually became clear that usually the most versatile and accurate method of determination of the Fermi surfaces was in fact the dHvA effect (for a discussion see Shoenberg 1969b).

Added confidence in the de Haas–van Alphen method was provided by a rather general argument presented by Luttinger (1960a) at the conference, which showed that the constant of proportionality between dHvA frequency F and FS area A is indeed as (1.1) implies, a universal constant which should be unaffected by many-body interactions. This was contrary to suggestions by Falicov (1960) and Stern (1960) who had argued that the electronic charge e in (1.1) should be modified to an effective e^* which might be a few % different. There was indeed some experimental support for such a difference from the early measurements on the noble metals (Shoenberg 1960a), but this was soon shown (Shoenberg 1962) to be a spurious result, due to a small error in the calibration of the magnetic field.

A puzzling mystery mentioned at the Cooperstown conference was Priestley's observation that one of the frequencies he observed in magnesium was too high to make any obvious sense (see Shoenberg 1960b, p 80). It corresponded to a FS extremal area much larger than any which could be expected on the basis of plausible models. This mystery was soon solved in principle by Cohen and Falicov (1961) and worked out more fully by Blount (1962). It turned out that the electron was experiencing *magnetic breakdown*, a kind of quantum tunnelling between states in different zones of equal energy but separated by a small potential barrier. Thus effectively the electron could switch tracks from one orbit to another and in this way describe quite new and sometimes much larger orbits. This opened up a whole new field of investigation which was not only of considerable interest in its own right but provided valuable supplementary information about subtle details of the band structure (for a review see Stark and Falicov 1967*a* and also chapter 7).

Another peculiar aspect of the de Haas–van Alphen effect, which was observed at high fields in the noble metals was what appeared to be an anomalously high harmonic content in the oscillations (Shoenberg 1962). This also proved to be an effect of some interest, a kind of *magnetic interaction* due to the fact that the field seen by the electrons was the magnetic induction, B rather than the field, H. Ordinarily this can be ignored, and it was indeed ignored in the LK formula. However at high enough fields and low enough temperatures the amplitude of $4\pi\,dM/dB$ may become large enough for the difference between B and H to be important and this can result in various curious effects such as frequency mixing as well as modification of harmonic amplitudes (as compared with the prediction of the theory ignoring the magnetic interaction). The better understanding of this effect which developed during the 1960s helped in the clarification of various puzzles in the detailed interpretation of dHvA spectra and in some of the other applications mentioned below. A detailed review is given in chapter 6.

The main technical advance of the 1960s came from the introduction of superconducting magnets. Even though the available fields were rather less than those which could be reached impulsively, the fact that the field was steady could be exploited to improve greatly the signal to noise ratio and the precision of measurement. Instead of using the time variation of an impulsive field to induce an e.m.f., part of which was proportional to dM/dH, detection was achieved by superimposing a small alternating field on the steady field and observing how the pick-up signal amplitude varied as the steady field was slowly changed. One of my former students, J. K. Hulm, then at Westinghouse was a pioneer in the development of superconducting magnets and in a sentimental moment over a drink in a Pittsburgh bar he offered to present his old laboratory with a small 50 kG magnet. This, together with a visit to Cambridge of P. J. Stiles, a young

18

American post-doctoral researcher who had thought out just how the technique could be made to work, enabled us to be the first to use a superconducting magnet to study the dHvA effect (Shoenberg and Stiles 1964).

This kind of method was soon widely taken up and with the introduction of various ingenious new features, made possible by advances in electronic and computing techniques, it gradually became a precision tool for measuring dHvA spectra. Since Cooperstown the Fermi surfaces of nearly all the metals have been determined, some with great precision (as good as 1 in 10^3 in radius for copper). This has gone on in parallel with considerable advances in the techniques of band structure calculations and often the calculations and the interpretation of the dHvA data have gone hand-in-hand. For instance the theory has produced various ways of describing Fermi surfaces specified by just a few parameters (see chapter 5) and the experiments have made it possible both to check the reliability of the scheme and to determine the values of the parameters. With improvements in metallurgical methods it has also become possible to prepare single crystals of many metallic compounds of perfection and stoichiometric purity comparable to that achieved for elements. This has made possible the extension of the dHvA techniques to the study of the FS of many interesting new systems (for a review, see Sellmyer 1978).

During the last 10 or 15 years improvements in measuring techniques have made possible systematic study of other aspects of the dHvA oscillations than just the frequency. Thus the decay of amplitude with decreasing field yields the Dingle temperature x and systematic study of how x varies with concentration of the added component in a dilute alloy, gives the probability of scattering of the electron per added atom. This scattering probability is an average over the cyclotron orbit of the electron, but its orientation dependence can be deconvoluted to give the scattering probability at a *point* on the FS. This kind of information, together with information (from precise frequency measurements) on the slight FS changes due to alloying, is providing a valuable contribution to the understanding of the band structure of alloys (see for instance Springford 1971). As already mentioned, the Dingle temperature in relatively pure samples is often less due to electron scattering by impurities than to the presence of dislocations. These cause strain variation through the sample and, consequently, amplitude reduction through phase smearing of the oscillations. This kind of effect is as yet only partially understood and deserves further study. The whole question of Dingle temperatures and what can be learnt from them is discussed in chapter 8.

Measurements of the absolute amplitude and harmonic content of the oscillations can be analyzed with the help of the LK formula to give the spin reduction factor and hence the spin-splitting factor ('g-factor') of the conduction electrons, which in turn provides information about spin-orbit coupling and many-body interactions in the electron system. In the derivation of the g-factor from the experimental data, it turns out that the absolute phase of the oscillations is relevant in helping to resolve certain ambiguities of interpretation. For this reason phase and spin are dealt with together in chapter 9.

The study of magnetic oscillations is still being actively pursued and looks like continuing for a long time yet; certainly much remains to be done. With the increased sophistication of measuring techniques and improvements in sample preparation it is becoming possible to observe and make detailed studies of the oscillations in more complicated systems such as ferromagnetics and layer compounds and so to contribute to the understanding of their electronic structures. The more ready availability of higher fields and lower temperatures is making it possible to test the reliability of the theory in more extreme conditions and to reveal, for instance, departures from the standard independent particle theory specifically due to many-body effects. There are moreover quite a few topics which deserve further study because of inconsistencies between various experiments or apparent discrepancies with theory. Particular attention will be drawn to such topics at the relevant places in the book.

1.4. Plan of the book

The theory of the magnetic oscillations will be systematically developed in chapter 2 culminating in the LK formula for the oscillations of the free energy and the corresponding formulae for the oscillations of magnetization, Fermi energy and density of states. The presentation will be almost entirely in terms of independent particles, but the consequences of many-body interactions will be briefly outlined.

Chapter 3 will describe the experimental techniques for studying the dHvA effect, i.e. specifically the oscillations of M and its field derivatives. Other kinds of magnetic oscillations are reviewed in chapter 4: first oscillations of thermodynamic quantities (other than magnetization) which can be derived from the free energy (thermal and mechanical properties), then oscillations of other properties, in particular the Shubnikov–de Haas oscillations of electrical resistivity and the giant quantum oscillations of ultrasonic attenuation. For each effect we shall outline the relevant theory, the experimental techniques and occasionally some of the experimental results. How the size and shape of Fermi surfaces, their strain dependences and their differential properties (i.e. how the surfaces of constant energy

20

vary with energy at the Fermi energy) may be determined from measurements on dHvA and other magnetic oscillations is discussed in chapter 5. The results in this field are now so extensive that only a few relatively simple examples will be presented in detail, partly in order to illustrate the general principles and partly because the results will be needed for discussion of later topics in the book.

Chapters 6 and 7 deal with magnetic interaction and magnetic breakdown respectively–two effects not envisaged in the theory as presented in chapter 2. These effects are not only of interest in their own right, but in certain circumstances it is important that they are properly understood in the interpretation of oscillatory data, particularly amplitude data. Chapter 8 deals with Dingle temperatures and shows what can be learnt about electron relaxation times in dilute alloy samples where impurity scattering is dominant and about strain fields in pure samples, where phase smearing is the dominant cause of amplitude reduction. The book concludes with chapter 9 which reviews methods of measuring the absolute phase and of deriving the spin-splitting factor from suitable measurements of absolute amplitude and harmonic content.

Finally this is a convenient place to list a number of earlier reviews which may be consulted to give an overview of how the subject has developed during the last 30 years, for the elaboration of particular topics and for references not quoted explicitly in this book. These are Shoenberg (1957, 1965), Mercouroff (1967), Gold (1968), Cracknell and Wong (1973), Lengeler (1978) and also various articles in Cochran and Haering (1968), Ziman (1969) and Springford (1980).

2
Theory

2.1. Preliminary qualitative treatment

Before embarking on the detailed theory of the de Haas–van Alphen oscillations it is helpful first to give a qualitative (and in some respects semi-quantitative) explanation due to Chambers (1956) of how the oscillations come about. For this purpose we shall anticipate some of the detailed analysis given later and start from Onsager's result that in a magnetic field H the only permitted states lie on tubes in k-space–*Landau tubes* as we shall call them–specified by the condition of area quantization

$$a(\varepsilon, k_H) = (r + \tfrac{1}{2})2\pi e H/\hbar c \tag{2.1}$$

Here a is the area of cross-section of a Landau tube cut by a plane normal to H (i.e. for given k_H) and r is an integer; the actual cross-section of the tube is the intersection of a surface of constant energy ε with the plane. For a free electron gas whose surfaces of constant energy are spherical, the Landau tubes are simply circular cylinders with a common axis in the H direction. As will be shown later this follows directly from Landau's solution of the Schrödinger equation in a magnetic field, which gives the energy levels explicitly as

$$\varepsilon = (r + \tfrac{1}{2}) \frac{e\hbar H}{m_0 c} + \frac{\hbar^2 k_H^2}{2m_0}$$

where m_0 is the free electron mass, and there is no need to appeal to the more general argument.

Two examples of a set of Landau tubes are schematically illustrated in fig. 2.1 and at $T = 0$ only such states as lie within the FS will be occupied. It is now simple to understand why the energy and hence the magnetization should oscillate as the field is varied. If we consider the tube of largest area which will fall partially inside the FS, its occupied length will shrink as H increases and vanishes infinitely rapidly when the tube just parts company with the FS, i.e. whenever

$$a = A \tag{2.2}$$

where A is the *extremal* area of cross-section of the FS by planes normal

22

to H. Such vanishings of occupation happen periodically as tubes of successively smaller quantum numbers r pass through the FS and clearly such passings occur at equal intervals of $1/H$ given by

$$\Delta(1/H) = 2\pi e/(\hbar c A) \tag{2.3}$$

or with a frequency F, defined as $1/\Delta(1/H)$, given by

$$F = (c\hbar/2\pi e)A \tag{2.4}$$

At each vanishing we should expect the total energy E of the occupied states to undergo some sort of an anomaly (it turns out to be a cusp in the variation of dE/dH, i.e. of M, with H) and so we should expect oscillations of the energy and magnetization as H varies, with a periodicity in $1/H$ given by (2.3) or a frequency F given by the Onsager relation (2.4).

The detailed calculation of the amplitude of the oscillations will be the main problem of the bulk of this chapter, but we may note that some of the results we shall obtain later are qualitatively evident. The effect of a finite temperature is to blur out slightly the boundary between the occupied and unoccupied states which for $T = 0$ is at the FS and completely sharp. This blurring, which is over an energy range of order kT, reduces the abruptness with which the occupation vanishes and hence the amplitude of the oscillations. Clearly the reduction depends on how large the blurring is

Fig. 2.1. Schematic sketches of Landau tubes for (*a*) spherical surfaces of constant energy, (*b*) ellipsoidal surfaces of constant energy (direction of long axis shown by arrow). The FS is indicated by the broken curve and only the parts of the Landau tubes inside the FS are occupied at $T = 0$ (after Chambers 1956 and Gold 1968).

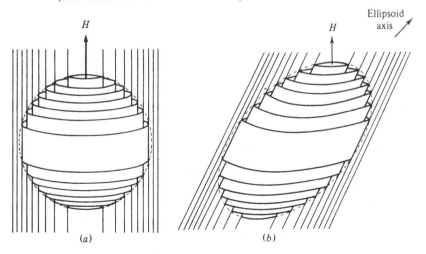

compared to the energy separation of neighbouring tubes, i.e. on the ratio $kT/\beta H$, where β is $e\hbar/mc$ and m is the cyclotron mass (to be defined later: see (2.21)). We shall see later that the amplitude of the pth harmonic of the oscillations is actually reduced by the factor $(2\pi^2 pkT/\beta H)/\sinh(2\pi^2 pkT/\beta H)$. According to the Uncertainty Principle the effect of a finite relaxation time τ is to blur the sharpness of the quantized states, i.e. to blur the Landau tubes, by something like $\hbar/\tau$ in energy and this too will reduce the amplitude by a factor depending on $\hbar/(\tau\beta H)$ which is the same as $1/\omega_c\tau$, where ω_c is the cyclotron frequency. The actual reduction factor of the pth harmonic amplitude (the Dingle factor as it is called) proves to be $\exp(-\pi p/\omega_c\tau)$.

The effect of electron spin is most simply thought of by supposing that effectively the spin-up and spin-down electrons have separate Landau tubes. This leads to a phase difference between the oscillations coming from the spin-up and spin-down electrons and hence a degree of interference. The reduction factor caused by this interference is just the cosine of half the phase difference and this comes out to be $\cos(\frac{1}{2}p\pi gm/m_0)$ for the pth harmonic, where g is the spin-splitting factor. Finally we notice that the abruptness with which the Landau tubes part company with the FS depends on how strongly the FS area varies with k_H around the extremum. Evidently the amplitude will be greater the more cylindrical is the FS at its extremum and the detailed calculation shows that the amplitude is proportional to $(d^2\mathscr{A}/dk_H^2)^{-1/2}$ where $\mathscr{A}$ is the area of the FS at some k_H.

We shall now set about a systematic exposition of the theory leading to the Lifshitz–Kosevich (LK) formula (1955) for the oscillatory part of the free energy of a system of independent quasiparticles, whose behaviour is specified by an arbitrary $\varepsilon(\mathbf{k})$ dependence of energy ε on wave vector $\mathbf{k}$. The presentation will be made in the following steps:

(1) Discussion of the electron dynamics in a magnetic field, based on the semi-classical approach used by Onsager. This relates the classical particle trajectories ('orbits') to the geometry of the constant energy surfaces in k-space. Finally, quantization of the motion gives the energy levels ε in Onsager's implicit form (2.1) and also determines the degeneracy of the levels.
(2) Calculation of the total energy E at $T=0$ (where it is of course also the free energy) for electrons whose k_H (which we shall denote by κ) lies in a narrow range $d\kappa$, i.e. essentially for a 2-dimensional (2-D) k-space. Although the resulting oscillatory effects are only a step in the calculation they have in fact some possible relevance to real 2-D systems and these will be pointed out.
(3) Integration over κ, to give the oscillations of the actual 3-D system, still at $T=0$.
(4) Use of the phase smearing concept to take account of finite temperature, electron scattering, inhomogeneities and electron spin. This finally leads to the standard LK formula for the oscillatory thermodynamic potential $\tilde{\Omega}$.

Although the theory is directed towards evaluation of the oscillatory thermodynamic potential $\tilde{\Omega}$, we also calculate at various stages the oscillatory magnetization $\tilde{M}$ and the oscillations in the density of states and the Fermi energy. At each point, the results for an arbitrary $\varepsilon(\mathbf{k})$ are illustrated by specializing to a free electron gas model whose properties can usually be derived in a more elementary way. Discussion of oscillations of the thermal and mechanical properties, which of course, like $\tilde{M}$, derive from $\tilde{\Omega}$ and of other kinds of oscillations, is deferred until chapter 4. The chapter finishes with a brief outline of many-body effects, which turn out not to modify the form of the LK formula appreciably except in extreme conditions, though the parameters of the formula may be appreciably modified.

2.2. Calculation of energy levels

2.2.1. Electron dynamics in a magnetic field*

The basic equation of the semi-classical approach is obtained by equating the rate of change of momentum, $\hbar \mathbf{k}$ to the Lorentz force, i.e.

$$\hbar \dot{\mathbf{k}} = \frac{-e}{c}(\mathbf{v} \times \mathbf{H}) \tag{2.5}$$

where $\mathbf{k}$ is the wave vector of the electron, $\mathbf{v}$ is the electron velocity and $-e$ is the electronic charge in e.s.u. (thus e is a positive quantity). The velocity $\mathbf{v}$ is related to the energy ε of the electron by

$$\mathbf{v} = \frac{1}{\hbar}\mathrm{grad}_k \, \varepsilon \tag{2.6}$$

Since the Lorentz force is perpendicular to $\mathbf{v}$ it can do no work on the electron and therefore ε remains constant as $\mathbf{k}$ changes with time. This is also evident from (2.6), which implies that $\mathbf{v}$ is normal to a surface of constant ε and consequently that $\dot{\mathbf{k}}$, which is perpendicular to $\mathbf{v}$, must be tangential to the constant ε surface. Since $\dot{\mathbf{k}}$ is also perpendicular to $\mathbf{H}$, it follows that the end of the $\mathbf{k}$ vector traces out an orbit which is the intersection of a constant ε surface and a plane normal to $\mathbf{H}$.

Integration of (2.5) w.r.t. time gives

$$\hbar(\mathbf{k} - \mathbf{k}_0) = \frac{-e}{c}(\mathbf{R} - \mathbf{R}_0) \times \mathbf{H} \tag{2.7}$$

---

* The treatment is based on Pippard's (1968) summary of the semi-classical method. Throughout this chapter we shall assume that the magnetization M is so small that the difference between B and H is negligible, as are also any shape effects due to the demagnetizing field. The necessary modifications to the theory if this assumption fails–as it does in certain conditions–are discussed in chapter 6.

Theory

where R is the position vector of the electron in its classical trajectory and the suffix zero indicates a value at some particular time. This means that the projection of the classical trajectory on to a plane normal to H is a scaled version of the trajectory of k, i.e. of a section of a constant ε surface by a plane normal to H (see fig. 2.2). Thus if R' denotes the projection of R on a plane normal to H,

$$|R' - R'_0| = \eta|k - k_0| \tag{2.8}$$

where η is a scaling factor given by

$$\eta = c\hbar/eH \tag{2.9}$$

Moreover, as can be seen from (2.7), the R' orbit is rotated by 90° in its plane w.r.t. the k orbit.

It should be emphasized that in general v has a component parallel as well as perpendicular to H so that in general the R trajectory will be helical, i.e. more complicated than the plane orbit trajectory of R'. The rate of advance along the helix, i.e. parallel to H, will in general vary as R' describes its plane trajectory, but will of course repeat periodically each time R' returns to the same value. The angular frequency with which the R' orbit is traced, the so-called *cyclotron frequency* ω_c, is easily calculated in terms of the geometry of the constant ε surfaces and the calculation also gives the pitch P of the classical helical motion.

From (2.5) and (2.6) we have that the time dt to traverse an element dk of the k orbit is

$$dt = \frac{c\hbar^2}{e}dk/(\text{grad}_k\varepsilon \times H) = \frac{c\hbar^2}{eH}dk/(\Delta\varepsilon/\Delta k'_n)$$

where $\Delta k'_n$ is the change of k' normal to the k orbit, corresponding to a change $\Delta\varepsilon$ of ε, as shown in fig. 2.2a (here k' is the component of k in the plane normal to H and quantities in plain type represent the magnitudes of the corresponding vectors). Thus

$$dt = \frac{c\hbar^2}{eH}\frac{dk\,\Delta k'_n}{\Delta\varepsilon} = \frac{c\hbar^2}{eH}\frac{\Delta(da)}{\Delta\varepsilon} = \frac{c\hbar^2}{eH}\left[\frac{\partial(da)}{\partial\varepsilon}\right]_\kappa \tag{2.10}$$

where da is the element of area of the k' orbit between two neighbouring radii from some arbitrary centre O and $\Delta(da)$ (which is the same as $d(\Delta a)$) is the change of this element for a change $\Delta\varepsilon$ of the energy, while keeping the component κ of k along H constant.

Integrating (2.10) over a complete orbit, we obtain the periodic time

26

2.2. Calculation of energy levels

Fig. 2.2. (a) (left) Sections at constant κ (i.e. in a plane normal to H) of surfaces of constant energy ε and $\varepsilon + \Delta\varepsilon$ and (right) the corresponding motion in real space projected on a plane normal to H (R' orbits); these R' orbits are scaled versions of the k' orbits but turned through 90°. (b) Sections at κ and $\kappa + \Delta\kappa$ (projected on a common plane) of a surface of constant energy ε such that these sections do not intersect. (c) As (b), but for a surface such that the sections intersect twice at A and B. The left-hand part of the area decreases with increase of κ more rapidly than the right-hand part increases. More complicated situations with four, six or any even number of intersections can also be envisaged. (d) Schematic graphs showing how the distance R_H parallel to H varies with time, corresponding to (b) and (c) and (lowest graph) for $\partial a / \partial \kappa = 0$ (cf. the simple example of fig. 2.3).

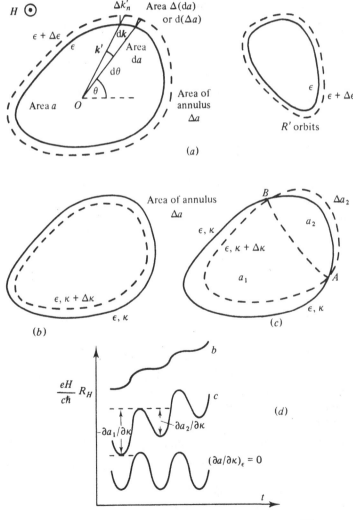

27

Theory

$2\pi/\omega_c$ and find

$$\omega_c = \frac{2\pi e H}{c\hbar^2} \bigg/ \left(\frac{\partial a}{\partial \varepsilon}\right)_\kappa \tag{2.11}$$

The nature of the helical motion in real space also follows from (2.10), since the advance dR_H of the electron in the H direction in time dt is

$$dR_H = v_H\, dt = \frac{1}{\hbar}\left(\frac{\partial \varepsilon}{\partial \kappa}\right)_{k'} dt = \frac{1}{\hbar}\left(\frac{\partial \varepsilon}{\partial \kappa}\right)_{da} dt \tag{2.12}$$

where the suffix k' means that in the differentiation the component of k perpendicular to H is to be kept constant. This is equivalent for a given $d\theta$ (see fig. 2.2) to keeping da constant (regarding da as a function of ε and κ). Substituting (2.10) for dt, (2.12) becomes

$$dR_H = \frac{c\hbar}{eH}\left(\frac{\partial \varepsilon}{\partial \kappa}\right)_{da} \left[\frac{\partial(da)}{\partial \varepsilon}\right]_\kappa = -\frac{c\hbar}{eH}\left[\frac{\partial(da)}{\partial \kappa}\right]_\varepsilon$$

and this can now be integrated over a complete cycle to give the pitch P of the helix as

$$P = -\frac{c\hbar}{eH}\left(\frac{\partial a}{\partial \kappa}\right)_\varepsilon \tag{2.13}$$

The integration can also be performed over any part of the cycle to get the displacement ΔR_H over part of the cycle.* If the sections at κ and $\kappa + d\kappa$ (for given ε) projected on a common plane are entirely non-intersecting, as in fig. 2.2b, dR_H has the same sign all round the orbit and the displacement along H increases monotonically in the helical motion. However in general the projected sections at κ and $\kappa + d\kappa$ may intersect an even number of times. For two intersections at A and B (as in fig. 2.2c) we have

$$\Delta R_H(B \to A) = -\frac{c\hbar}{eH}\left(\frac{\partial a_1}{\partial \kappa}\right)_\varepsilon, \qquad \Delta R_H(A \to B) = -\frac{c\hbar}{eH}\left(\frac{\partial a_2}{\partial \kappa}\right)_\varepsilon$$

The net displacement along H round a complete orbit is of course just the pitch P (since $(\partial a/\partial \kappa)_\varepsilon = (\partial a_1/\partial \kappa)_\varepsilon + (\partial a_2/\partial \kappa)_\varepsilon$) but the progress is oscillatory as shown in fig. 2.2d. Even in the non-intersecting case (b), for which the progress is monotonic, the *velocity* may be oscillatory as also illustrated in fig. 2.2d.

In the dHvA effect the 'extremal' orbits in k-space for which $(\partial a/\partial \kappa)_\varepsilon$ vanishes, are of particular interest. For such orbits P is zero, but R_H is

---

* I am indebted to Professor Pippard for the following argument which gives the amplitude of any oscillatory displacement in the field direction.

28

2.2. Calculation of energy levels

oscillatory as shown in (d) with a peak to peak swing of $-(c\hbar/eH)$ times $|\partial a_1/\partial\kappa|_\varepsilon$ or $|\partial a_2/\partial\kappa|_\varepsilon$ (these are of course equal when $(\partial a/\partial\kappa)_\varepsilon$ vanishes). Thus although the projection of the real space orbit on a plane normal to H has the same shape as the k-space orbit, the real space motion for the extremal case is not confined to a plane normal to H. It may for special cases lie in a plane, but this plane may be oblique to H; a simple example is shown in fig. 2.3.

Before going further it will be useful to illustrate the meaning of some of the above analysis in terms of the simplest concrete example, that of a free electron gas, for which the $\varepsilon(\boldsymbol{k})$ relation is of course explicit. This example will also serve to indicate the orders of magnitude of the relevant quantities.

Fig. 2.3. Motion for a circular cylindrical surface of constant energy with axis at angle θ to H and radius K. In k-space the motion is round the section AB normal to H which is an ellipse of semi-axes K, $K\sec\theta$; the arrows indicate the directions of the velocity vectors corresponding to A and B. In R-(real) space the motion is around a section normal to the axis of a circular cylinder of radius $\eta K \sec\theta$. The projection of this section on a plane normal to H is the ellipse shown, similar to the k-space section but scaled by the factor η and turned through $90°$. There is no net motion in the H (i.e. z) direction, but only a periodic 'yo-yo' motion of amplitude $\eta K \tan\theta$. The slightly more complicated case of an oblique ellipsoid of revolution, for which there is net motion superposed on the periodic motion if $k_z \neq 0$, is discussed in Appendix 1.

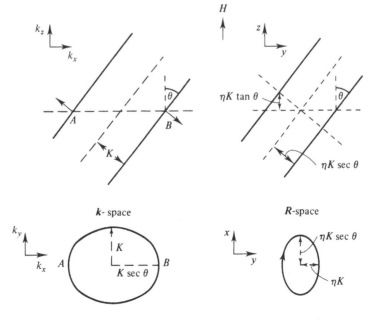

29

Theory

Thus for a gas of free electrons of mass m_0

$$\varepsilon = \hbar^2 k^2/2m_0 \qquad (2.14)$$

and the constant ε surfaces are simply spheres of radius k. The more general case of ellipsoidal constant energy surfaces is discussed in Appendix 1. If we write

$$k^2 = k'^2 + \kappa^2 \qquad (2.15)$$

we see that the orbits in k-space, being sections of the constant ε surfaces by planes of constant κ, are simply circles of radius k' and of area

$$a = \pi k'^2 \qquad (2.16)$$

In real space the orbits are also circles, of radius R' given by

$$R' = \eta k' = c\hbar k'/eH \qquad (2.17)$$

For this particular example, the 90° twist of the R' orbits with respect to the k orbits is of course irrelevant, because of the circular symmetry and the real space orbit for $\kappa = 0$ is in plane normal to **H**. For a free electron $\hbar k'/m_0$ is just v', the component of velocity perpendicular to H, so (2.17) reduces to

$$R' = m_0 v' c/eH \qquad (2.18)$$

which can of course be obtained much more simply by equating the Lorentz and centrifugal forces.

Combining (2.14), (2.15) and (2.16) we can express ε as an explicit function of a and κ:

$$\varepsilon = \frac{\hbar^2}{2m_0}\left(\frac{a}{\pi} + \kappa^2\right) \qquad (2.19)$$

We can now apply (2.11) to find ω_c. Thus from (2.19)

$$(\partial a/\partial \varepsilon)_\kappa = 2\pi m_0/\hbar^2$$

and

$$\omega_c = eH/m_0 c \qquad (2.20)$$

again an obvious result, since for a free electron orbit evidently $\omega_c = v'/R'$. For real metals, ω_c is given by the more general relation (2.11) and it is convenient to introduce a *cyclotron mass* m defined by

$$m = \frac{\hbar^2}{2\pi}\left(\frac{\partial a}{\partial \varepsilon}\right)_\kappa \qquad (2.21)$$

2.2 *Calculation of energy levels*

so that ω_c is given by substituting m for m_0 in (2.20), i.e.

$$\omega_c = eH/mc \tag{2.22}$$

For the free electron model the trajectory in real space is a regular helix, and the pitch as given by (2.13) and (2.18) is

$$P = 2\pi c h \kappa / eH = 2\pi v_H / \omega_c$$

yet again an obvious result.

Because of the nature of the Fermi distribution, we shall mostly be concerned with constant ε surfaces close to the FS, specified by the Fermi energy ζ, which for a free electron gas is

$$\zeta = \hbar^2 k_F^2 / 2m_0 \tag{2.23}$$

where k_F is the radius of the spherical FS. Since k_F is a reciprocal length, we should expect its order of magnitude to be that of a reciprocal atomic length, i.e. of order 10^8 cm^{-1}. In fact it is given by

$$k_F = (3\pi^2 n)^{1/3} \tag{2.24}$$

where n is the number of electrons per unit volume and if the electrons in monovalent metals (the alkalis and the noble metals) are treated as if they were free electrons, we find the values of k_F given in table 2.1, which do

Table 2.1. *Fermi radii and Fermi energies of the alkalis and noble metals*

Metal	Li	Na	K	Rb	Cs	Cu	Ag	Au
$10^{-8}k_F$(cm^{-1})	1.11	0.91	0.73	0.68	0.63	1.36	1.20	1.20
ζ(eV)	4.7	3.2	2.0	1.8	1.5	7.0	5.5	5.5

indeed confirm the intuitive order of magnitude estimate of 10^8 cm^{-1}; the values of ζ are typically a few electronvolts.

The scaling factor η which converts k dimensions into dimensions of real space is inversely proportional to H (see (2.9)). For the range $10^4 - 10^5$ G, which is usually appropriate for dHvA experiments, η varies from 6.6×10^{-12} to 6.6×10^{-13} cm^2. Thus the radii of free electron-like orbits typically range from about 10^{-4} cm at 10^5 G to 10^{-3} cm at 10^4 G. The cyclotron frequency ω_c for the free electron gas is 1.76×10^{11} s^{-1} at 10^4 G and 1.76×10^{12} s^{-1} at 10^5 G (corresponding to electromagnetic radiation of wavelengths 1.07 cm and 1.07 mm respectively).

Theory

The above estimates based on the free electron model do indeed provide quite a reasonable guide to what is actually observed for the monovalent metals. However, the polyvalent metals have much more complicated Fermi surfaces, sometimes consisting of quite a number of separate sheets; some of these sheets may be considerably smaller in their k dimensions than 10^8 cm^{-1} – even 100 times smaller – and moreover they are usually far from spherical. The orbits in real space are of course also correspondingly smaller. The cyclotron masses may also differ considerably from m_0. Thus for bismuth m/m_0 is as small as 1/100 for certain field orientations (and the cyclotron frequency correspondingly higher than for free electrons), while in some orbits for ferromagnetic metals m/m_0 may be as large as 10. Again, the Fermi energy for some of the FS sheets of a polyvalent metal (measured from the lowest state of the energy band corresponding to the sheet) may be very much less than the few electronvolts of the free electron model.

2.2.2. Quantization of the electron motion

We must now take account of the quantization of the electron motion, which restricts the permissible states and is, of course, the basic cause for the occurrence of the dHvA oscillations.

The Bohr–Sommerfeld quantization rule for a periodic motion is

$$\oint p \cdot \mathrm{d}q = (r + \gamma)2\pi\hbar \tag{2.25}$$

where p and q are the canonically conjugate momentum and position variables, r is an integer and the integration is to be taken round a complete cycle; the phase γ will be discussed below. For an electron in a magnetic field the appropriate meanings of p and q are

$$p = \hbar k - eA/c \qquad q = R' \tag{2.26}$$

where A is the vector potential of H and as before $-e$ is the electron charge.

Thus (2.25) becomes

$$\oint (\hbar k - eA/c)\,\mathrm{d}R' = (r + \gamma)2\pi\hbar \tag{2.27}$$

Using (2.7) to transform the first term in the integral (the constants k_0 and R'_0 contribute nothing in a complete cycle) and Stokes' theorem to transform the second, we find

$$H \cdot \oint (R \times \mathrm{d}R') - \int_S H \cdot \mathrm{d}S = (r + \gamma)2\pi\hbar c/e \tag{2.28}$$

where S denotes integration over the area of the orbit in real space and dS is a vector element of area of any surface having the R' orbit as rim. If we write $R = R' + R_H$, where R_H is the vector component of R along H, it is evident that R_H makes no contribution, since its contribution to the vector product is perpendicular to H. Thus, the first term reduces to $H|\oint(R' \times dR')|$ which is $2H\alpha$, if α is the area of the R' orbit in real space, and the second term is $-H\alpha$. In this way, (2.28) reduces to

$$\Phi \equiv H\alpha = (r + \gamma)2\pi\hbar c/e \qquad (2.29)$$

In other words the motion is quantized in such a way as to quantize the flux Φ of H through the R' orbit, in terms of the universal quantum of flux $2\pi\hbar c/e$ (which has the value $4.13 \times 10^{-7}\,\mathrm{G\,cm^2}$).

This pretty result can be rather more usefully expressed in terms of the area a of the orbit in k-space, i.e. the area of a section of a constant ε surface. Thus from (2.8) and (2.9) it follows that

$$\alpha = (c\hbar/eH)^2 a$$

and (2.29) becomes the result already quoted in (2.1):

$$a(\varepsilon, \kappa) = (r + \gamma)2\pi e H/c\hbar \qquad \textbf{(2.30)}$$

where we have emphasized that the area a depends on the energy ε of the particular constant energy surface concerned and on the value of κ at which the section is taken. This is the famous Onsager (1952) relation (derived independently by I.M. Lifshitz at about the same time but published only rather later (Lifshitz and Kosevich 1954)). It can be regarded as specifying implicitly the permitted energy levels ε in terms of H, κ, r and the band structure $\varepsilon(k)$ (which determines the surfaces of constant energy and hence the functional form of $a(\varepsilon, \kappa)$). The phase constant γ is exactly $\frac{1}{2}$ for a parabolic band (and in particular for free electrons) but in general γ departs slightly from $\frac{1}{2}$ by an amount depending on energy and H (Roth 1966). These slight departures prove to be unimportant for the dHvA oscillations but very relevant for the steady diamagnetic susceptibility. To avoid complicating the presentation we shall henceforth set $\gamma = \frac{1}{2}$ except in Appendix 4, where the departures and their relevance to the calculation of the steady susceptibility will be discussed.

The Onsager relation (2.30) evidently severely restricts the permissible values of k in a magnetic field and this restriction can be visualized by the helpful geometrical construction already discussed in outline in §2.1 (Chambers 1956). Thus for given r and H a definite value of a is specified by (2.30) and for given κ this specifies a definite energy ε, and a curve which is the plane section at κ of the surface of this particular constant energy. If

Theory

now κ is varied for some given H the plane curves will generate a family of tubes, each of constant area $a(r)$ specified by the Onsager relation (2.30) and the quantum number r. These Landau tubes are just an extension of the more familiar idea of Landau levels (as the energy levels for fixed κ are usually called). The meaning of the Onsager relation may then be expressed by saying that it restricts the permitted states in k space to lie on the Landau tubes.

For the free electron gas the Onsager relation applied to (2.19) immediately gives

$$\varepsilon = (r + \tfrac{1}{2})\beta_0 H + \hbar^2\kappa^2/2m_0 \tag{2.31}$$

where β_0 is a double Bohr magneton defined as

$$\beta_0 = e\hbar/m_0 c \tag{2.32}$$

Substituting the $\varepsilon(k)$ relation (2.14), we obtain from (2.31) the equation of the Landau tubes as

$$\frac{\hbar^2}{2m_0}(k_x^2 + k_y^2) = (r + \tfrac{1}{2})\beta_0 H \tag{2.33}$$

(k_x, k_y are introduced here as the coordinates in the plane normal to H). Thus for this simple case the Landau tubes are co-axial circular cylinders parallel to H, with cross-sectional areas given by (2.30) with $\gamma = \tfrac{1}{2}$. They are illustrated in fig. 2.1, which also shows the form of the Landau tubes if the surfaces of constant energy are ellipsoids (see Appendix 1). In general there is no requirement that the tubes should run parallel to H or that they should be cylindrical.

From either (2.29) or (2.30) it is evident that typically the quantum numbers r are large. Thus if we take $k_F \sim 10^8\,\mathrm{cm}^{-1}$, a is of order $\pi \times 10^{16}\,\mathrm{cm}^{-2}$ and we find r of order 3×10^3 at $10^5\,\mathrm{G}$ and 3×10^4 at $10^4\,\mathrm{G}$. The fact that r is large plays an important role in all the calculations; in particular it makes it possible to calculate differences associated with changes of r to a good degree of approximation by appropriate differentiation. An important example is the calculation of the energy separation between neighbouring Landau levels, i.e. between the energies at given κ for successive quantum numbers r. Thus we have, using the Onsager relation

$$(\Delta\varepsilon)_{\Delta r=1} = \left(\frac{\partial\varepsilon}{\partial a}\right)_\kappa (\Delta a)_{\Delta r=1} = \frac{2\pi e H}{c\hbar(\partial a/\partial\varepsilon)_\kappa} = \hbar\omega_c = \beta H \tag{2.34}$$

where ω_c is defined by (2.11) and β by (2.32), but with the cyclotron mass

2.2. Calculation of energy levels

substituted for the free electron mass, i.e.

$$\beta = eh/mc \tag{2.35}$$

The quantum mechanical significance of (2.34) is that the cyclotron frequency ω_c is associated with a transition between the two energy levels ε_{r+1} and ε_r and the relation $(\Delta\varepsilon)_{\Delta r = 1} = \hbar\omega_c$ is in principle always exact. However the relation between $(\Delta\varepsilon)_{\Delta r = 1}$ and $(\partial a/\partial\varepsilon)_\kappa$ relies on the correspondence principle, which relates quantum behaviour at high quantum numbers with classical motion. At low r the correspondence principle begins to break down and if $(\partial a/\partial\varepsilon)$ varies with ε, and ε_{r+1} and ε_r are appreciably different, (2.34) is no longer precise, since it does not specify the ε at which $(\partial a/\partial\varepsilon)_\kappa$ should be evaluated. For the special case of the free electron model (and as shown in Appendix 1, for any parabolic band) $(\partial a/\partial\varepsilon)_\kappa$ is independent of ε and (2.34) defines the level separation exactly for all r.

One more point needs to be considered to complete this discussion of the energy levels in a magnetic field. This is the *degeneracy* of the levels. For $H = 0$ the number of states per unit volume of k-space* (bearing in mind that there are two spin states for each k state) is $V/4\pi^3$ where V is the real space volume of the metal. In the presence of a magnetic field, the permissible states, as we have seen, are restricted to be on the Landau tubes, and the annular cross-sectional area between neighbouring tubes is given by (2.30) as

$$\Delta a = 2\pi eH/c\hbar \tag{2.36}$$

Thus on average the number D of states on a length of tube† lying between κ and $\kappa + d\kappa$ must be

$$D = \Delta a \, d\kappa \, V/4\pi^3 = eHV \, d\kappa/2\pi^2 c\hbar \tag{2.37}$$

Although this is rather an informal derivation of D, the result is in fact rigorously true. It is certainly true for a free electron gas, as emerges from the exact solution of the Schrödinger equation, and since as Lifshitz and Kosevich (1955) point out, D is independent of the form of $\varepsilon(k)$, it must also be true for an arbitrary $\varepsilon(k)$.

To recapitulate then, the energy levels ε in a magnetic field H are given implicitly in terms of a quantum number r and the continuous variable κ,

---

* Occasionally it is appropriate to discuss the spin-up and spin-down states separately, as for instance in discussing the constant energy surfaces of ferromagnetic metals, and it is important to remember that the number of states if only a single spin is considered for given k is *half the* $V/4\pi^3$ quoted for two spins.
† The length of the tube need not of course be parallel to H (see for instance fig. 2.1b) but the volume of the annular space will always be $\Delta a d\kappa$.

35

through the quantization relation (2.30) for a, the area of cross-section of the constant ε surface. The form of the function $a(\varepsilon, \kappa)$ is determined by the geometry of the surfaces of constant energy for $H = 0$. For given r (i.e. for a given Landau tube), the energy levels are degenerate in the sense that there are D states between κ and $\kappa + d\kappa$, where D is given by (2.37).

2.3. Calculation of the free energy

In most of the derivations of thermodynamic quantities it will be the chemical potential ζ (equal to the Fermi energy) rather than the number of electrons N, which appears explicitly in the calculations, and it therefore proves more convenient to calculate the thermodynamic potential defined by

$$\Omega = F - N\zeta \tag{2.38}$$

rather than the free energy F defined as

$$F = E - TS \tag{2.39}$$

where E is the internal energy and S the entropy of the system.* As elaborated in Appendix 2, this is because in the differentiation w.r.t. H to obtain the magnetic moment M, N must be held constant if F is differentiated, which is awkward since N enters into the calculation only implicitly. If, however, Ω is differentiated it is ζ which must be held constant and this greatly simplifies the calculation.

More explicitly, the vector magnetic moment is given by

$$M = -(\text{grad}_H\Omega)_\zeta \tag{2.40}$$

or in more practical terms, the components of M parallel and perpendicular to H are

$$M_\| = -(\partial\Omega/\partial H)_\zeta \tag{2.41}$$

$$M_\perp = -\frac{1}{H}(\partial\Omega/\partial\theta)_{\zeta,H} \tag{2.42}$$

In (2.42) we keep the magnitude of H constant and consider the variation of Ω with the direction of H; here θ is an angle specifying the direction of H in the plane in which Ω varies most rapidly with the direction of H. It must be emphasized that although M is derived from Ω by differentiation at

---

* Strictly speaking all quantities are for an arbitrary volume V; thus M denotes magnetic moment rather than magnetization, but we shall often ignore this distinction when it is obvious what is meant.

2.3. Calculation of the free energy

constant ζ, this does *not* imply that ζ is in fact constant as H varies, and indeed we shall see later that associated with the dHvA oscillations of M there is in general an oscillatory variation of ζ as H varies. Other important quantities such as entropy, strain and electron number can also be derived by appropriate differentiation of Ω, and these will be discussed later.

As is shown in standard texts (e.g. Landau and Lifshitz 1980), for a system obeying Fermi–Dirac statistics and having states of energy ε, the thermo-dynamic potential is given by

$$\Omega = -kT \sum \ln(1 + e^{(\zeta - \varepsilon)/kT}) \tag{2.43}$$

where the summation is over all possible states (i.e. allowing properly for any degeneracies by appropriate repetitions, and replacing summation by integration where ε varies continuously). For the system of energy levels specified by (2.30) with the degeneracy specified by (2.37), this becomes

$$\Omega = -kT \int_{-\infty}^{\infty} d\kappa \left(\frac{eHV}{2\pi^2 ch}\right) \sum_r \ln(1 + e^{(\zeta - \varepsilon_r)/kT}) \tag{2.44}$$

where we now specify ε_r as the solution of the implicit equation (2.30) for quantum number r (the energy ε_r is of course also a function of H and κ).

In effect (2.44) completes the formulation of the problem of how to calculate the magnetization and other thermodynamic properties of the metal and it remains only to do the mathematics. In order to help bring out the physical meaning of the mathematics we shall break down the calculation into a number of simpler steps. Thus we shall first carry through the calculation only at $T = 0$ and only for $\delta\Omega$, the contribution to Ω from the Landau tubes between κ and $\kappa + \delta\kappa$. We shall then perform the integration over κ and finally, by appropriate convolutions, take into account not only the effect of finite temperature, but also of electron spin, electron scattering and sample inhomogeneity.

2.3.1. Calculation at $T = 0$ for a two-dimensional slab of k-space

For $T = 0$, the thermodynamic potential Ω becomes simply $E - N\zeta$ (see (2.38) and (2.39)) and the contribution $\delta\Omega$ to (2.44) reduces to

$$\delta\Omega = \delta\kappa \left(\frac{eHV}{2\pi^2 ch}\right) \sum_{r=0}^{n} (\varepsilon_r - \zeta) \equiv D \sum_{r=0}^{n} (\varepsilon_r - \zeta) \tag{2.45}$$

where D is the degeneracy factor of (2.37). The summation over r is now to

Theory

be taken only over such values of r for which $\varepsilon_r < \zeta$; we shall always use n to denote the highest value of r and enumerate the states in increasing order of energy (so that $r = 0$ is the state of lowest energy). This summation can be worked out either by the Poisson summation formula, first used for this purpose by Landau (1939), or by the Euler–Maclaurin formula which was used for the calculation of the steady diamagnetism (also by Landau (1930)). We shall follow the latter method since it brings out rather more clearly where the various terms in the answer come from; the more usual derivation by the Poisson method is given in Appendix 3.

To the degree of approximation we shall need, the Euler–Maclaurin formula gives the sum from $r = 0$ to $r = n$ of a function $f(r)$ of an integer r as

$$\sum_0^n f(r) = \int_0^n f(r)\,dr + \tfrac{1}{2}[f(n) + f(0)] + \tfrac{1}{12}[f'(n) - f'(0)] \quad (2.46)$$

where on the right-hand side r is to be treated as a continuous variable. The next higher approximation involves terms in d^3f/dr^3 and in our context these are of order n^2 times smaller than the smallest terms we shall retain. It should be noticed that if f is a polynomial of order less than 3, (2.46) is exact.

Applying (2.46) to (2.45) we have

$$\frac{\delta\Omega}{D} = \int_0^n (\varepsilon_r - \zeta)\,dr + \tfrac{1}{2}(\varepsilon_n - \zeta) + \tfrac{1}{2}(\varepsilon_0 - \zeta)$$

$$+ \tfrac{1}{12}\left[\left(\frac{\partial\varepsilon}{\partial r}\right)_{r=n} - \left(\frac{\partial\varepsilon}{\partial r}\right)_{r=0}\right] \quad (2.47)$$

where n is the highest value of r for which $\varepsilon_r < \zeta$. We now introduce a continuous variable x in place of the discrete variable $r + \tfrac{1}{2}$, i.e. x is defined as

$$a(\varepsilon, \kappa) = x(2\pi eH/c\hbar) \quad (2.48)$$

and also a parameter X which specifies the value x assumes at the FS, i.e.

$$\mathscr{A}(\kappa) \equiv a(\zeta, \kappa) = X(2\pi eH/c\hbar) \quad (2.49)$$

and

$$\varepsilon(X, \kappa) = \zeta \quad (2.50)$$

Thus the condition $\varepsilon_r < \zeta$ can be expressed also as $(r + \tfrac{1}{2}) < X$.

We shall also need the relation

$$\left(\frac{\partial\varepsilon}{\partial x}\right)_\kappa = \frac{(\partial a/\partial x)_\kappa}{(\partial a/\partial\varepsilon)_\kappa} = \frac{2\pi eH/c\hbar}{2\pi m/\hbar^2} = \beta H \quad (2.51)$$

38

2.3. Calculation of the free energy

where m is the cyclotron mass as defined in (2.21) and β is as defined in (2.35); m and β are of course in general functions of x.

If we now change the variable in (2.47) from r to x, it becomes

$$\frac{\delta\Omega}{D} = \int_{\frac{1}{2}}^{n+\frac{1}{2}} [\varepsilon(x) - \zeta] \, dx + \tfrac{1}{2}[\varepsilon(n - \tfrac{1}{2}) - \zeta] + \tfrac{1}{2}[\varepsilon(\tfrac{1}{2}) - \zeta]$$

$$+ \tfrac{1}{12}[\beta(n + \tfrac{1}{2})H - \beta(\tfrac{1}{2})H] \tag{2.52}$$

where the notation $\varepsilon(y)$ and $\beta(y)$ means the values for $x = y$, or

$$\frac{\delta\Omega}{D} = \int_{0}^{X} [\varepsilon(x) - \zeta] \, dx + \delta(\tfrac{1}{2}\beta H \delta) - \tfrac{1}{2}[\varepsilon(0) + \tfrac{1}{4}\beta(0)H - \zeta]$$

$$- \tfrac{1}{2}\beta H \delta + \tfrac{1}{2}[\varepsilon(0) - \zeta] + \tfrac{1}{4}\beta(0)H + \tfrac{1}{12}[\beta - \beta(0)]H \tag{2.53}$$

where the two terms following the integral come from the slight changes of the limits of integration and (2.50) has been used; small energy differences have been evaluated by use of (2.51). We have also introduced the notation

$$\delta = X - (n + \tfrac{1}{2}) \quad \text{and} \quad \beta = \beta(X) \tag{2.54}$$

and we have ignored terms of order $(\partial\beta/\partial x)H$ compared with terms of order βH; this is again equivalent to ignoring $1/n$ compared with unity.

Collecting terms we find

$$\delta\Omega = \delta\kappa \left(\frac{eHV}{2\pi^2 c\hbar}\right) \left\{ \int_{0}^{X} [\varepsilon(x) - \zeta] \, dx + \tfrac{1}{24}\beta(0)H \right.$$

$$\left. + \tfrac{1}{2}\beta H(\delta^2 - \delta + \tfrac{1}{6}) \right\} \tag{2.55}$$

Let us now consider the significance of the various terms of (2.55) in turn. Since ε is in fact a function of xH, as can be seen from (2.48), the integral on the r.h.s. must be proportional to $1/H$ and so its contribution to $\delta\Omega$ is independent of H. In fact it is not difficult to see that its contribution to $\delta\Omega$ is simply the value of $\delta\Omega$ in the absence of H (see Appendix 4). This is however no longer true if we take into account the spin magnetic moment of the electron, which we have so far ignored. As shown in Appendix 4, if the spin degeneracy is lifted, this term is modified in such a way as to contribute a negative term proportional to H^2, and this gives a susceptibility which reduces to the Pauli spin susceptibility if the spin-splitting is symmetrical about each level.

The second term gives a positive contribution to $\delta\Omega$ proportional to H^2 and leads to a steady diamagnetic susceptibility. Here, however, it turns out

39

that our assumption of $\gamma = \frac{1}{2}$ is not an adequate approximation, except for a parabolic band; a more exact calculation based on Roth's (1966) better approximation for γ is given in Appendix 4 and leads to the correct formula for the steady diamagnetic susceptibility.

Thus as far as the oscillatory magnetic behaviour is concerned, the first two terms of (2.55) are irrelevant and we now consider only the contribution of the last term, which we shall denote by $\delta\tilde{\Omega}$. Substituting for δ, this oscillatory term becomes

$$\delta\tilde{\Omega} = \delta\kappa \frac{e\beta H^2 V}{4\pi^2 ch} \{[X - (n + \tfrac{1}{2})]^2 - [(X - (n + \tfrac{1}{2})] + \tfrac{1}{6}\} \qquad (2.56)$$

which is valid for $(n + \frac{1}{2}) \leqslant X \leqslant (n + \frac{3}{2})$; as soon as X crosses the boundaries of this range, the n in the formula must be increased or decreased by 1. The variation of $\delta\tilde{\Omega}$ as a function of X (i.e. of $1/H$) is shown in fig. 2.4. It is clearly periodic in X with period 1 or, equivalently, periodic in $1/H$ with period $2\pi e/ch\mathscr{A}$. It is important to note that the oscillations are completely characterized by the electronic structure at the FS, i.e. for given κ, by the area $\mathscr{A}$ of the cross section of the FS and by β which is a measure of the derivative $(\partial a/\partial \varepsilon)$ at the FS.

The oscillations can be Fourier analyzed into a harmonic series and it is easily shown that (2.56) is equivalent to

$$\delta\tilde{\Omega} = \frac{\delta\kappa\, e\beta H^2 V}{4\pi^2 ch} \sum_{p=1}^{\infty} \frac{1}{\pi^2 p^2} \cos 2\pi p(X - \tfrac{1}{2}) \qquad (2.57)$$

The factor in front of the summation on the r.h.s. of (2.57) can also be

Fig. 2.4. Periodic variation with $X(= \mathscr{A}ch/2\pi eH)$ at constant ζ of the thermodynamic potential $\delta\tilde{\Omega}$ of a slab of k-space $\delta\kappa$ thick containing δN_0 electrons in the absence of a field (see (2.56)). The scaling factor $2X/\beta H\delta N_0$ of $\delta\tilde{\Omega}$ varies as X^2 as H varies, but if $X \gg 1$, this variation is not significant over a few oscillations. The broken curve is the fundamental component in the Fourier analysis of the periodic variation (see (2.57)).

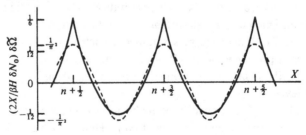

2.3. Calculation of the free energy

expressed as $\frac{1}{2}D\beta H$ or $\frac{1}{2}(\delta N_0)\,\beta H/X$, where

$$\delta N_0 = XD = \delta\kappa\,\mathscr{A}V/4\pi^3 \tag{2.58}$$

Here δN_0 is simply the number of electrons in the two-dimensional slab of area $\mathscr{A}$ and thickness $\delta\kappa$ at $H = 0$; as we shall see below, the actual number δN in the slab oscillates as H varies. The fundamental component of the oscillation is also shown in fig. 2.4. This way of expressing the result will prove particularly useful in the further development of the theory, when we have to integrate over κ bearing in mind that X is a function of κ, and later still when we have to consider various forms of phase smearing.

2.3.2. Magnetization and number of electrons for a two-dimensional slab of *k*-space

The magnetic moment $\delta\tilde{M}$ parallel to H is given by

$$\delta\tilde{M} = -(\partial(\delta\tilde{\Omega})/\partial H)_\zeta \tag{2.59}$$

and this can be expressed either explicitly as a 'saw-tooth' variation by differentiating (2.56) or as a Fourier series by differentiating (2.57). In either case, the factor proportional to H^2, which determines the amplitude of the periodic behaviour of $\delta\tilde{\Omega}$, can be ignored in the differentiation, since it would lead to a term of order $1/X$ times the term coming from the bracketed parts of (2.56) or (2.57).

Starting from (2.56) and using (2.49) to evaluate $(\partial X/\partial H)_\zeta$, we find for $n + \frac{1}{2} \leqslant X \leqslant n + \frac{3}{2}$

$$\delta\tilde{M} = \frac{\beta\mathscr{A}V\,\delta\kappa}{4\pi^3}[X - (n+1)] = \beta\,\delta N_0[X - (n+1)] \tag{2.60}$$

where δN_0 is the number of particles in the two-dimensional slab for $H = 0$, as given by (2.58) for any particular value of $\mathscr{A}$ (i.e. of ζ); however the particular value of $\mathscr{A}$ (or ζ) may, if we wish, vary with H. The result (2.60) corresponds to a 'saw-tooth' variation, as illustrated in fig. 2.5. It is instructive to note that the saw-tooth variation can be thought of as a 'staircase' made up of the discontinuities as each successive level passes through the Fermi level superposed on a compensating steady variation due to the field dependence of all the underlying levels.

Similarly starting from (2.57) (or equivalently from Fourier analysis of (2.60)), we have

$$\delta\tilde{M} = -\beta\,\delta N_0 \sum_{p=1}^{\infty} \frac{\sin 2\pi p(X - \frac{1}{2})}{\pi p} \tag{2.61}$$

41

Theory

The contribution to $\delta\tilde{M}$ of the fundamental frequency ($p = 1$) is also shown in fig. 2.5; its amplitude is $2/\pi$ times the 'amplitude' of the saw-tooth.

It is convenient at this point to discuss the question of the number of electrons in the two-dimensional slab. The number δN can be written down immediately as

$$\delta N = (n + 1)D = (n + 1)\delta N_0/X \tag{2.62}$$

(valid for $n + \frac{1}{2} \leqslant X \leqslant n + \frac{3}{2}$) since we know that the $(n + 1)$ levels ($r = 0$ to n) are each occupied D times over.

If ζ remains constant as H varies, this represents a saw-tooth variation of δN (fig. 2.6) about the mean value δN_0 with discontinuities of approximately $\delta N_0/n$ each time a new Landau level has to be filled as X increases (H decreases). It is instructive to derive this result also from the thermodynamic relation

$$\delta N = -[\partial(\delta\Omega)/\partial\zeta]_H \tag{2.63}$$

with $\delta\Omega$ as given by (2.55). Using $(\partial X/\partial\zeta)_H = 1/\beta H$, we find for $n + \frac{1}{2} \leqslant X \leqslant n + \frac{3}{2}$

$$\delta N = D(X - (X - (n + 1))) \tag{2.64}$$

which is of course identical with (2.62), but provides a useful breakdown of δN into two parts. The first is XD which is just δN_0 and comes from the first term in (2.55), the second provides the saw-tooth part $\delta\tilde{N}$ of δN shown in fig. 2.6, i.e.

$$\delta\tilde{N} \equiv \delta N - \delta N_0 = -D[X - (n + 1)]$$

$$= -\frac{\delta N_0}{X}[X - (n + 1)] \equiv -\frac{\delta N_0}{X}(\delta - \tfrac{1}{2}) \tag{2.65}$$

where δ is as defined in (2.54).

Fig. 2.5. Periodic variation of $\delta\tilde{M}$ with X at constant ζ (see (2.60) and (2.61)) for same slab as in fig. 2.4.

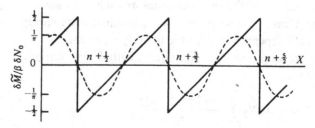

42

2.3. Calculation of the free energy

As important feature of the saw-tooth variation $\delta\tilde{N}$ is that it is directly proportional to $\delta\tilde{M}$ as can be seen from (2.60); thus we have

$$\delta\tilde{N} = -\frac{\delta\tilde{M}}{\beta H} = -\frac{2\pi m H}{\hbar^2\mathscr{A}}\delta\tilde{M} \tag{2.66}$$

Since, as we shall see in §2.3.6, the main contribution to $\tilde{M}$ for a three-dimensional FS comes from only a narrow range of κ for which $\mathscr{A}$ is close to its extremal value A, (2.66) may be integrated over $d\kappa$ to give the same relation between $\tilde{N}$ and $\tilde{M}$, but with $\mathscr{A}$ replaced by A. A simple physical interpretation of (2.66) was pointed out by Pippard (1968). The magnetic moment μ of an electron at the FS going round its classical orbit is given by

$$\mu = -e\alpha\omega_c/2\pi c \tag{2.67}$$

where α is the area of the orbit (in real space). If we now use (2.9) and (2.22) this becomes

$$\mu = -\frac{e}{c}\left(\frac{ch}{eH}\right)^2 \mathscr{A}\left(\frac{eH}{2\pi mc}\right) = \frac{-\hbar^2\mathscr{A}}{2\pi m H} = -X\beta \tag{2.68}$$

and we see that

$$\delta\tilde{M} = \mu\,\delta\tilde{N} \tag{2.69}$$

Thus the saw-tooth oscillations of $\delta\tilde{M}$ at constant ζ can be thought of as due to the oscillations $\delta\tilde{N}$ in the *number* of electrons, each electron carrying the magnetic moment μ, (since this variable number is coming from electrons close to the Fermi energy).

The main application of this analysis will be to three-dimensional metals and we shall then have to add together the contributions from all the different slices $\delta\kappa$ of the FS. The saw-tooth variations of δN within any one slice are supplied from other slices (which all have the same ζ) and there is

Fig. 2.6. Periodic variation of number of electrons δN with X at constant ζ for same slab as in fig. 2.4 (see (2.65) and (2.66)). The scaling factor varies as X, but for $X \gg 1$ this is unimportant over a few oscillations.

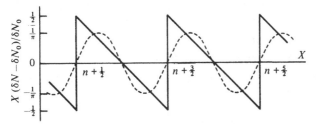

no difficulty in arranging to keep the *total* number of electrons within a sheet or sheets of the FS constant (or equal to the number of holes in other sheets of the FS) by suitably varying the Fermi energy ζ. As we shall see in §2.4, the oscillatory variation of ζ required to achieve this is quite small, so that arguments based on assuming a constant ζ are in fact only slightly in error and moreover there is no difficulty in applying an appropriate correction if required.

2.3.3. Calculations for constant number of electrons (a digression)

There is some interest in considering to what extent the above two-dimensional analysis could be applied to a real physical two-dimensional system, i.e. one in which the energy is independent of κ. For such a system we might wish to impose the restriction that the number of particles in the system is to be held exactly constant, and here we run into the difficulty that the saw-tooth variation of δN given by (2.65) appears to be quite inevitable and cannot be suppressed by any compensating variation of the Fermi energy ζ. In fact this difficulty is built into our formulation, which assumes that the highest occupied level is always *below* the Fermi energy and therefore fully occupied at $T = 0$. This assumption is in fact invalid if δN is to be kept constant and the correct treatment for constant δN requires more careful consideration of how the limit $T = 0$ should be taken.

The analysis leads to a quite simple and plausible way out of the difficulty. This is that at $T = 0$ the Fermi energy is exactly *equal* to the energy of the highest occupied level and this level is then only *partially* occupied (in contrast to the implication of (2.45) that all levels are fully occupied) with an occupancy which varies from 0 to D as the field is decreased from a value at which the given number of electrons fits exactly into n levels to the value at which it fits exactly into $n + 1$ levels. Thus for constant δN, as successive Landau levels fill up (with falling field), ζ 'sticks' to the highest occupied level and shows a saw-tooth field dependence. The detailed analysis shows that at low enough T the partial occupation of the top level is achieved by ζ departing from it slightly, being at first slightly below this level, so that the Fermi distribution demands small occupancy and then crossing it to come slightly above, thus eventually producing almost complete occupancy.

Since we are concerned with constant δN, the calculation at $T = 0$ is most simply carried out by calculating δE rather than $\delta \Omega = (\delta E - \zeta \delta N)$. The magnetization δM and the Fermi energy ζ are then given by

$$\delta M = -[\partial(\delta E)/\partial H]_{\delta N} \quad \text{and} \quad \zeta = -[\partial(\delta E)/\partial(\delta N)]_H \qquad (2.70)$$

2.3. Calculation of the free energy

Assuming that $\zeta = \varepsilon_n$ so that the nth level can be partially occupied, we have

$$\delta E = D \sum_{r=0}^{n-1} \varepsilon_r + (\delta N - nD)\varepsilon_n \tag{2.71}$$

with the lowest n levels ($r = 0$ to $n - 1$) each having occupancy D and the $(n + 1)$th level ($r = n$) containing the balance $(\delta N - nD)$ of the total number of electrons. This relation is valid from the value of field which makes $nD = \delta N$, down to that which makes $(n + 1)D = \delta N$. It is convenient here to introduce a continuous variable X_0 such that

$$X_0 - \mathscr{A}_0\, ch/2\pi eH = \delta N/D \tag{2.72}$$

and

$$\varepsilon(X_0) = \zeta_0 \tag{2.73}$$

where $\mathscr{A}_0$ is the area of the FS and ζ_0 the Fermi energy, for $H = 0$. Our previous variable X, proportional to $\mathscr{A}$ is no longer appropriate as a specification of the field, since $\mathscr{A}$ itself now varies in saw-tooth manner as H and ζ vary; the relation between X and X_0 is shown in fig. 2.7. In terms of X_0, the range of validity of (2.71) is simply

$$n < X_0 < n + 1 \tag{2.74}$$

By use of the Euler–Maclaurin formula (see (2.46)), (2.7) can be expressed as

$$\delta E = D \left\{ \int_0^{n-1} \varepsilon_r\, dr + \tfrac{1}{2}(\varepsilon_{n-1} + \varepsilon_0) \right.$$

$$\left. + \tfrac{1}{12}[\beta H - \beta(0)H] + (\delta N - nD)\varepsilon_n \right\} \tag{2.75}$$

or in terms of the continuous variable x corresponding to $r + \tfrac{1}{2}$ (see 2.48),

$$\delta E = D \left\{ \int_{1/2}^{n-1/2} \varepsilon(x)\, dx + \tfrac{1}{2}[\varepsilon(n - \tfrac{1}{2}) + \varepsilon(\tfrac{1}{2})] + \tfrac{1}{12}[\beta H - \beta(0)H] \right.$$

$$\left. + (\delta N - nD)\varepsilon(n + \tfrac{1}{2}) \right\} \tag{2.76}$$

If we now slightly adjust the limits of integration at both ends and adjust the values of x by appropriate corrections (in the same way as was done to arrive at (2.53) from (2.52)) and also make use of (2.72) and (2.73) we find

$$\delta E = D \left\{ \int_0^{X_0} \varepsilon(x)\, dx - \zeta_0(X_0 - n + \tfrac{1}{2}) + \tfrac{1}{2}\beta H(X_0 - n + \tfrac{1}{2})^2 \right.$$

$$- \tfrac{1}{2}\varepsilon(0) - \tfrac{1}{8}\beta(0)H + \tfrac{1}{2}\zeta_0 - \tfrac{1}{2}\beta H(X_0 - n + \tfrac{1}{2}) + \tfrac{1}{2}\varepsilon(0) + \tfrac{1}{4}\beta(0)H$$

$$\left. + \tfrac{1}{12}[\beta H - \beta(0)H] + (X_0 - n)[\zeta_0 - \beta H(X_0 - n - \tfrac{1}{2})] \right\} \tag{2.77}$$

45

Theory

After collecting terms this becomes

$$\delta E = D\left\{\int_0^{X_0} \varepsilon(x)\,dx + \tfrac{1}{24}[\beta H + \beta(0)H]\right.$$

$$\left. - \tfrac{1}{2}\beta H[(X_0 - n)^2 - (X_0 - n) + \tfrac{1}{6}]\right\} \qquad (2.78)$$

The various terms of (2.78) can be interpreted in a similar way to those of (2.55). The last term,

$$\delta\tilde{E} = -\frac{1}{2}\frac{\beta H\,\delta N}{X_0}[(X_0 - n)^2 - (X_0 - n) + \tfrac{1}{6}] \qquad (2.79)$$

which has been adjusted (by adding the $\frac{1}{6}$ term and subtracting it from the earlier terms) to average to zero over a complete cycle of variation of X_0 (from n to $n + 1$), is similar to that of (2.56) for the case of constant ζ (see fig. 2.4) but differs in two important respects. First it has the opposite sign and secondly it has its cusps at integral values of X_0, rather than (as in fig. 2.4) for $X = r + \frac{1}{2}$.

Correspondingly the oscillatory magnetization δM is given by

$$\delta\tilde{M} = -\beta\,\delta N[X_0 - (n + \tfrac{1}{2})] = \frac{\beta\,\delta N}{\pi}\sum_{p=1}^{\infty}\sin\frac{(2\pi p X_0)}{p} \qquad \textbf{(2.80)}$$

Fig. 2.7. Illustrating the relation between $X(= \mathscr{A}ch/2\pi eH)$ and X_0 $(= \mathscr{A}_0 ch/2\pi eH)$ as H varies for a constant number δN of electrons in the slab of fig. 2.4. As X_0 varies from one integer to the next the Fermi energy ζ sticks to the highest occupied level while it is being filled in such a way as to keep X constant at a half odd integral value. Once the next integer is reached, ζ jumps to the next higher level and sticks there while it is being filled, and correspondingly X jumps to the next half odd integar.

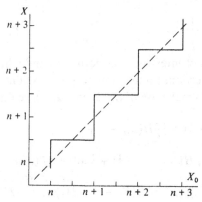

2.3. Calculation of the free energy

which, as shown in fig. 2.9, resembles the saw-tooth of fig. 2.5 for constant ζ, but has the opposite sign and has its discontinuities at integral values of X_0. It should be noticed that the simple physical picture in the constant ζ case, of $\delta\tilde{M}$ arising from the oscillatory part $\delta\tilde{N}$ of the number of electrons, no longer applies in the constant δN case. Rather the $\delta\tilde{M}$ oscillations appear to correspond to the number of electrons in the partially occupied level (each having a moment μ as defined by (2.68)).

Fig. 2.8. Analogue of fig. 2.4 for constant number δN of electrons in the slab instead of constant ζ (see (2.79)). The variable is now the energy $\delta\tilde{E}$ (scaled by $2X_0/\beta H d N$, which varies insignificantly over a few cycles if $X_0 \gg 1$) and it is plotted as a function of $X_0 (= \mathscr{A}_0 ch/2\pi eH)$. The curve is identical with that of fig. 2.4 but turned upside down and displaced by half an integer.

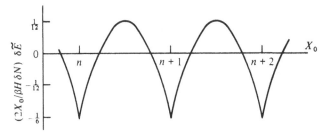

Fig. 2.9. Analogue of fig. 2.5, i.e. plot of $\delta\tilde{M}/\beta\delta N$ against X_0 for constant δN (see (2.80)).

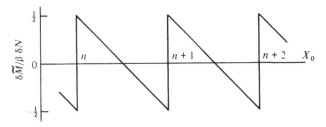

Fig. 2.10. Variation of $(\zeta - \zeta_0)/\beta H$ with X_0 for constant δN (see 2.82)). This plot is identical with that of fig. 2.9, and apart from a shift of half an integer also with that of fig. 2.6.

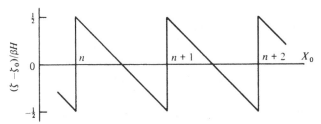

Theory

The steadily varying magnetization (other than due to spin) is given by differentiating the second term of (2.78) w.r.t. H and corresponds to a diamagnetic susceptibility, which, as before, is in general incorrect because the result is modified if the slight non-constancy of the phase γ is properly taken into account (see Appendix 4). For the special case of free electrons the result should, however, be correct and it can be seen that the steady diamagnetic susceptibility for constant δN is just twice as large as for constant ζ (see also §2.3.5).

As already intuitively deduced, the Fermi energy ζ for constant δN has to 'stick' to the partially occupied highest level ε_n and so has a saw-tooth character. The form of the saw-tooth (see fig. 2.10) is given by

$$\zeta = \varepsilon_n = \varepsilon(n + \tfrac{1}{2}) = \zeta_0 - \beta H[X_0 - (n + \tfrac{1}{2})] \tag{2.81}$$

or

$$\zeta - \zeta_0 = -\beta H[X_0 - (n + \tfrac{1}{2})] \tag{2.82}$$

This can also be obtained from the relation $\zeta = [\partial(\delta E)/\partial(\delta N)]_H$, bearing in mind that δN enters into the expression (2.78) for δE only through X_0 and ζ_0.

2.3.4. Application to real 2-D systems (digression continued)

As already mentioned, our calculations for a thin slice of k-space have some application to real physical systems in which the electrons are constrained to move in two dimensions. Among such 2-D systems are (a) electrons held by image forces to the surface of liquid helium (Grimes 1978) and (b) various semiconductor device arrangements, such as the inversion layer in a silicon MOSFET and heterostructures such as $GaAs-Al_xGa_{1-x}As$ (for reviews see Pepper 1977; Ando, Fowler and Stern 1982). Although in many ways (a) provides the simplest 2-D system and has an electron density n which is independent of magnetic field, it is of little use for study of oscillations because the electron density cannot be made much higher than $n \sim 10^9 \, \text{cm}^{-2}$. For such a low density the degeneracy temperature (given by $\pi \hbar^2 n/mk$) is only of order $3 \times 10^{-2} \, \text{K}$, so extremely low temperatures would be required to achieve a quantum rather than a classical system and moreover even if experiments were made at the necessary low temperatures the oscillation amplitudes would be too small to measure with present techniques.

In the semiconductor device systems the electron density, which is either controlled by a gate voltage (in the MOSFET and very recently also in the heterostructure (Narita *et al.* 1981)) or determined by the composition of

48

2.3. Calculation of the free energy

the heterostructure, can be made much greater: typically $n \sim 10^{12} - 10^{13}\,\mathrm{cm}^{-2}$. However, for these systems comparison with our idealized calculations is not quite straightforward because of various complications, such as electron scattering, the occurrence in certain circumstances of localized states and the structure of the depletion region between the inversion layer and the underlying semiconductor. Thus it is not always clear whether the system more closely resembles one in which ζ is constant or one in which n is constant, as H is varied. A detailed discussion of real 2-D systems is beyond the scope of this book, so we shall do little more than indicate the orders of magnitude of the oscillatory frequencies and amplitudes to be expected, which are much the same whichever constraint is assumed.

For a 2-D system of area S in real space the number of electrons is given by

$$N = S\mathscr{A}/2\pi^2 \quad \text{or} \quad n = \mathscr{A}/2\pi^2 \tag{2.83}$$

where $\mathscr{A}$ is the area of the 2-D FS. If N is independent of H a suffix zero should be added to $\mathscr{A}$, while if ζ (and $\mathscr{A}$) is independent of H a suffix zero should be added to N, where the zero suffix denotes value at $H = 0$. The result (2.83) can also be derived by thinking of the FS as 3-D and having the form of a very long cylinder in the H direction, with the length κ given by

$$\kappa = 2\pi/t$$

where t is the real space thickness of the film. If we then integrate (2.58) over κ, we immediately obtain (2.83). It is convenient to note here that for the 2-D system the degeneracy factor D of (2.37) becomes

$$D = eHS/\pi ch \tag{2.84}$$

All the formulae of the previous two sections can now be interpreted as properties of the idealized 2-D system at $T = 0$ if we simply remove the δs from the various formulae referring originally to a slab $\delta\kappa$. Thus we replace δM by M (the net magnetic moment), δN by N (total number), etc. The period $\Delta(1/H)$ of the saw-tooth oscillations of M and of ζ (if N is constant) or of N (if ζ is constant) is related to n, the number of electrons per unit area (i.e. N/S) by

$$\Delta(1/H) = 2\pi e/ch\mathscr{A} = e/\pi chn \tag{2.85}$$

or expressed as a frequency F,

$$F = 1/\Delta(1/H) = \pi chn/e \tag{2.86}$$

Theory

Since n can be as high as $10^{13}\,\mathrm{cm}^{-2}$, F can be as high as $2 \times 10^6\,\mathrm{G}$, which is quite comparable to typical values for small 3-D Fermi surfaces.

The amplitude of the M oscillations as given by either (2.61) or (2.80) is ideally $\beta n/\pi$ for the fundamental, but in practice is much reduced by various effects. As discussed in Appendix 7 the amplitude proves to be very small as compared with that of typical 3-D metal samples, but by using the special techniques outlined in §3.4.2.1 and §3.4.2.5 it should be possible to detect the oscillations and their study may help to improve understanding of the various 2-D systems. P. J. Stiles at Brown University is at present attempting such an experiment.

If the number of electrons is constant (or nearly so) as H is varied, the Fermi energy ζ should oscillate with an amplitude or order $\frac{1}{2}\beta H$ (see (2.82)), which is of order $10^{-3}\,\mathrm{eV}$ for typical conditions ($m/m_0 = 0.2$, $H = 2 \times 10^4\,\mathrm{G}$) and might be detected by the method outlined in §4.4. So far the only oscillatory properties that have been observed are in the galvomagnetic behaviour. A strong Shubnikov–de Haas effect was first observed in a silicon inversion layer (Fowler, Fang, Howard and Stiles 1966) and has been much studied since. The theory of this effect and the accompanying peculiarities of the Hall effect will be outlined in §4.5 and some of the spectacular recent results will be briefly presented there.

2.3.5. 2-D results for parabolic band

Before going on to 3-D systems it is useful to write down the rather simpler forms assumed by our 2-D results for the special case of a parabolic band (e.g. for free electrons) for which

$$\beta = \beta(0) \qquad \varepsilon(x) = x\beta H \tag{2.87}$$

(here $\varepsilon(x)$ and ζ are measured from the bottom of the band, i.e. from $\varepsilon(0)$). *For constant ζ*

$$\zeta = X\beta H \quad \text{and} \quad XD = \delta N_0 \tag{2.88}$$

Thus for $n + \frac{1}{2} < X < n + \frac{3}{2}$, (2.55) reduces to

$$\delta\Omega = -\tfrac{1}{2}\zeta\,\delta N_0 + \frac{\beta^2 H^2\,\delta N_0}{24\zeta} + \frac{1}{2}\frac{\beta^2 H^2\,\delta N_0}{\zeta}$$
$$\times \{[X - (n + \tfrac{1}{2})]^2 - [X - (n + \tfrac{1}{2})] + \tfrac{1}{6}\} \tag{2.89}$$

which reduces to

$$\delta\Omega = \beta H\,\delta N_0 \left\{\frac{(n + 1)^2}{2X} - (n + 1)\right\} \tag{2.90}$$

50

2.3. Calculation of the free energy

The result (2.90) may also be obtained directly from simple summation of the explicit energy levels, i.e. from

$$\delta\Omega = D\left\{\sum_0^n [(r + \tfrac{1}{2})\beta H - \zeta]\right\} \qquad (2.91)$$

The form (2.89) is however more meaningful than (2.90) since it separates out the zero field value of $\delta\Omega$ (first term), the part associated with a linear magnetization (second term) and the purely oscillatory part $\delta\tilde{\Omega}$ (third term).

The corresponding magnetization is

$$\delta M = -\frac{\beta^2 H \, \delta N_0}{12\zeta} + \beta \, \delta N_0 [X - (n + 1)] - \frac{\beta^2 H \, \delta N_0}{\zeta}$$

$$\times \{[X - (n + \tfrac{1}{2})]^2 - [X - (n + \tfrac{1}{2})] + \tfrac{1}{6}\} \qquad (2.92)$$

The first term corresponds to the two-dimensional Landau steady diamagnetic susceptibility and the second is the main oscillatory part $\delta\tilde{M}$, which has exactly the same form as the result (2.60) for an arbitrary $\varepsilon(k)$. The third term, which comes from differentiating the H^2 factor outside the last bracket on the r.h.s. of (2.89), is also oscillatory but is of order $1/n$ times the main oscillatory term and was ignored in the general formula (2.60). The calculation leading to (2.60) involved several approximations in which terms of this order were ignored, so there would have been little point in retaining this particular term in the final answer. For the free electron case, however, the calculation is exact and the retention of the small extra term is justified; it may be noticed that it is exactly comparable in magnitude with the steady diamagnetism term. The relative magnitudes of the three terms are illustrated in fig. 2.11.

The number of electrons δN as given by (2.64) is

$$\delta N = \delta N_0 - \frac{\delta N_0}{X}(X - (n + 1)) = \frac{(n + 1)}{X}\delta N_0 \qquad (2.93)$$

For a constant number of electrons δN

We have (see (2.87)), with all energies measured from $\varepsilon(0)$

$$\zeta_0 = X_0 \beta H \qquad X_0 D = \delta N \qquad (2.94)$$

so (2.78) becomes

$$\delta E = \tfrac{1}{2}\zeta_0 \, \delta N + \frac{\beta^2 H^2 \, \delta N}{12\zeta_0} - \frac{\beta^2 H^2 \, \delta N}{2\zeta_0} [(X_0 - n)^2 - (X_0 - n) + \tfrac{1}{6}] \qquad (2.95)$$

51

Theory

which reduces to

$$\delta E = \beta H \, \delta N \left\{ \left(n + \frac{1}{2} \right) - \frac{n(n+1)}{2X_0} \right\} \qquad (2.96)$$

The form (2.96) may also be obtained directly from simple summation of the explicit energy levels, i.e. from

$$\delta E = D \sum_{0}^{n-1} (r + \tfrac{1}{2})\beta H + (\delta N - nD)(n + \tfrac{1}{2})\beta H \qquad (2.97)$$

As with (2.89), the form (2.95) has the advantage that it separates out the constant value of δE (first term), the part associated with a linear magnetization (second term) and the purely oscillatory part (third term).

Fig. 2.11. Showing the various contributions in (2.92) to δM, the magnetization of a slab of k-space $\delta \kappa$ thick for free electrons at constant ζ and $T = 0$. For this model $X = \zeta / \beta H$, where ζ is measured from the level corresponding to $X = 0$ and the number δN of electrons adjusts itself to keep ζ constant as H varies; $\delta N = \delta N_0$ for $H = 0$. The saw-tooth curve (full line) is identical with that of fig. 2.4 (the second term of (2.92)); the oscillatory curve (broken) whose amplitude decays as $1/X$ comes from the third term and was ignored in fig. 2.4 since it was assumed that $X \gg 1$. The smoothly decaying curve (dotted) is the steady diamagnetism coming from the first term and is also negligible compared with the main term when $X \gg 1$. For $X < \frac{1}{2}$ the lowest quantum state is above ζ so there can be no electrons ($\delta N = 0$) according to this model (which is of course, somewhat unrealistic in this limit) and the total δM must vanish.

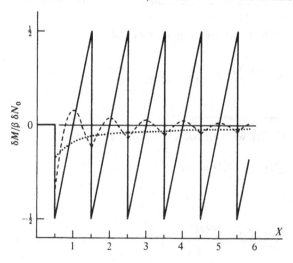

2.3. Calculation of the free energy

The magnetization is given by

$$\delta M = -\frac{\beta^2 H \delta N}{6\zeta_0} - \beta \delta N[X_0 - (n + \tfrac{1}{2})] + \frac{\beta^2 H \delta N}{\zeta_0}$$

$$\times [(X_0 - n)^2 - (X_0 - n) + \tfrac{1}{6}] \tag{2.98}$$

The steady diamagnetic susceptibility is seen to be twice as large for constant δN as for constant ζ, and, as in (2.92), we have included an extra oscillatory term, of order $1/n$ times the main oscillatory term, but comparable in amplitude to the steady diamagnetic susceptibility term.

The oscillation in ζ, (2.83), becomes

$$\zeta - \zeta_0 = -\beta H[X_0 - (n + \tfrac{1}{2})] = -\zeta_0\left(1 - \frac{(n + \tfrac{1}{2})}{X_0}\right) \tag{2.99}$$

2.3.6. Integration over κ

We now take the next step of going from two to three dimensions in k-space. Since the chemical potential ζ must be the same for all electrons within the FS, we need to integrate over κ treating ζ as a constant. Thus the appropriate expression to integrate is (2.57) and we find

$$\tilde{\Omega} = \frac{eH^2 V}{4\pi^2 ch} \int \beta \, d\kappa \sum_{p=1}^{\infty} \frac{1}{\pi^2 p^2} \cos\{2\pi p[X(\kappa) - \tfrac{1}{2}]\} \tag{2.100}$$

It may perhaps be helpful to recall here that X is defined by (2.49) as $X = \mathscr{A}ch/2\pi eH$. The limits of integration are the values of κ for which the plane $\kappa =$ constant is tangential to the FS, or if the FS is a continuous tube repeating periodically between zone planes, the limits are the values of κ at the zone planes. As we shall see directly, however, it is usually a quite adequate approximation to set the limits as $\pm \infty$ if κ is measured from the value which makes X (i.e. the area) have its maximum or minimum value X_0. In fact if $X_0 \gg 1$, only a small part around $\kappa = 0$ of the full range of integration contributes appreciably to the integral and outside this range the detailed form of $X(\kappa)$ and the limits of integration are immaterial. For the same reason β, which in general is also a function of κ, can be treated as a constant having the value of β at $\kappa = 0$, and can therefore be taken outside the integral. To proceed further we make a Taylor expansion of X about $\kappa = 0$, i.e.

$$X = X_0 \pm \tfrac{1}{2}X''\kappa^2 \pm \tfrac{1}{24}X^{IV}\kappa^4 \pm \cdots \tag{2.101}$$

where

$$X'' = |\partial^2 X/\partial\kappa^2|_{\kappa=0}, \qquad X^{IV} = |\partial^4 X/\partial\kappa^4|_{\kappa=0} \tag{2.102}$$

and the $\pm$ signs are chosen according to the signs of the derivatives (i.e. for the κ^2 term the $+$ sign implies that X_0 is a minimum and the $-$ sign that it is a maximum).* Usually it is not necessary to go beyond the κ^2 term and we shall assume this is adequate here; the special situation for which X'' is small or zero is discussed in Appendix 5. The contribution of the pth term of the integral in (2.100) is on this assumption (omitting the factors β and $1/\pi^2 p^2$)

$$I_p = \int_{-\infty}^{\infty} d\kappa \cos[2\pi p(X(\kappa) - \tfrac{1}{8})]$$

$$= 2\int_0^{\infty} d\kappa \cos[2\pi p(X_0 \pm \tfrac{1}{2}X''\kappa^2 - \tfrac{1}{8})] \qquad (2.103)$$

The justification for replacing the actual upper limit, say κ_0, by ∞ is that usually

$$\tfrac{1}{2}X''\kappa_0^2 \gg 1 \qquad (2.104)$$

This is certainly true for an ellipsoidal FS, since then it is exactly true (see Appendix 1) that

$$\tfrac{1}{2}X''\kappa_0^2 = X_0 \qquad (2.105)$$

and we are all along supposing that $X_0 \gg 1$. Thus there is little contribution to the integral except from values of κ such that $\kappa/\kappa_0 \gtrsim X_0^{-1/2}$; the integrand over the rest of the range of κ, whether terminated at κ_0 or at ∞, changes sign rapidly as κ grows and so contributes little to the integral. This argument is elaborated in Appendix 5 and it is shown that even when X_0 is as small as 2, there is little error made by integrating to infinity if the FS is ellipsoidal.

To evaluate I_p in (2.103) it is convenient to introduce a variable u defined by

$$u^2 = 2pX''\kappa^2 \qquad (2.106)$$

and (2.103) becomes

$$I_p = \left(\frac{2}{pX''}\right)^{1/2} \left\{ \cos[2\pi p(X_0 - \tfrac{1}{8})] \int_0^{\infty} \cos\tfrac{1}{2}\pi u^2 \, du \right.$$

$$\left. \mp \sin[2\pi p(X_0 - \tfrac{1}{8})] \int_0^{\infty} \sin\tfrac{1}{2}\pi u^2 \, du \right\} \qquad (2.107)$$

But

$$\int_0^{\infty} \cos\tfrac{1}{2}\pi u^2 \, du = \int_0^{\infty} \sin\tfrac{1}{2}\pi u^2 \, du = \tfrac{1}{2} \qquad (2.108)$$

---

* We have assumed that X is an even function of κ; if it is not, terms in κ^3 would need to be considered as well.

2.3. Calculation of the free energy

so (2.107) becomes

$$I_p = (pX'')^{-1/2} \cos[2\pi p(X_0 - \tfrac{1}{2}) \pm \tfrac{1}{4}\pi] \tag{2.109}$$

and we see that the net result of the integration is to change the phase of each harmonic by $\pm\tfrac{1}{4}\pi$ ($+$ if X_0 is a minimum and $-$ if it is a maximum) and to multiply the amplitude by the factor $(pX'')^{-1/2}$. The significance of this factor can be appreciated by the example of an ellipsoidal FS for which X'' is given by (2.105), so that the factor is simply $\kappa_0/(2pX_0)^{1/2}$ as compared with the factor $2\kappa_0$ which would be appropriate if all the slices of the FS had the same value of X (as for a cylindrical FS). Thus it is as if only a fraction $1/(8pX_0)^{1/2}$ of the range $2\kappa_0$ contributed coherently to the integration and the rest of the range was completely ineffective owing to dephasing.

The integration leading to (2.109) can also be carried out graphically by an amplitude–phase curve in which the amplitude and phase of the integral up to any desired upper limit are given by the length and inclination of a chord of the curve, which in this case is a Cornu spiral. This method is described in Appendix 5 and it used to discuss, for an ellipsoidal FS, the possible error involved in setting the upper limit of the integral infinite; the error turns out to be negligible almost up to the quantum limit (i.e. for X_0 as small as 1).

If we now sum over all the harmonics (i.e. over p), we find

$$\tilde{\Omega} = \left(\frac{e}{2\pi c\hbar}\right)^{3/2} \frac{\beta H^{5/2}}{\pi^2(A'')^{1/2}} \sum_{p=1}^{\infty} \frac{1}{p^{5/2}} \cos\left[2\pi p\left(\frac{F}{H} - \frac{1}{2}\right) \pm \frac{\pi}{4}\right] \tag{2.110}$$

where we have introduced the de Haas–van Alphen frequency F defined (as in (2.4)) by

$$F = (c\hbar/2\pi e)A = X_0 H \tag{2.111}$$

and A is the extremal area of the Fermi surface, i.e. the maximum or minimum of $\mathscr{A}$ with respect to κ; we have also introduced the notation

$$A'' = |\partial^2\mathscr{A}/\partial\kappa^2|_{\kappa=0} = (2\pi eH/c\hbar)X'' \tag{2.112}$$

The $1/H$ dependence of $\tilde{\Omega}H^{-5/2}$ is illustrated in fig. 2.12 and it can be seen that the dephasing brought about by integrating over κ has smoothed out the cusps of the two-dimensional variation of fig. 2.4 and also introduced a left–right asymmetry into the oscillations because the dephasing is only in one sense. In algebraic terms this smoothing can be thought of as a consequence of the extra $p^{1/2}$ factor in the denominator of (2.110) as compared with (2.57), i.e. of the relative weakening of the higher har-

55

monics. As already pointed out, the absolute amplitude of the oscillations per electron is considerably smaller for the whole FS than for the two-dimensional slice of it; for free electrons (see below) the reduction is by a factor of order $(H/F)^{1/2}$. The expressions for $\tilde{M}$ follow as before by differentiation w.r.t H for $M_{\parallel}$ or w.r.t an angular variable θ (see (2.40) to (2.42)), such as to give the greatest variation $\partial F/\partial\theta$ for $M_{\perp}$. We obtain

$$\tilde{M}_{\parallel} = -\left(\frac{e}{c\hbar}\right)^{3/2} \frac{\beta F H^{1/2} V}{2^{1/2}\pi^{5/2}(A'')^{1/2}} \sum_{p=1}^{\infty} \frac{1}{p^{3/2}} \sin\left[2\pi p\left(\frac{F}{H} - \frac{1}{2}\right) \pm \frac{\pi}{4}\right]$$

$$\tag{2.113}$$

$$\tilde{M}_{\perp} = -\frac{1}{F}\frac{\partial F}{\partial\theta}\tilde{M}_{\parallel} \tag{2.114}$$

where, as before, in the differentiations we ignore the small contributions coming from the $H^{5/2}$ term in (2.110) or from the θ dependence of $\beta/(A'')^{1/2}$. The oscillatory variation of $\tilde{M}H^{-1/2}$ is shown in fig. 2.13 and as with the $\tilde{\Omega}$ oscillations, the integration over κ has somewhat smoothed the asperities

Fig. 2.12. $X_0^{5/2}\tilde{\Omega}/N_0\zeta$ as a function of $X_0(=F/H)$ at $T=0$. The scale of ordinates is given for a free electron metal; for an arbitrary electronic structure the scale would be modified, but the form of the oscillations would be the same. As can be seen from (2.115), the plotted quantity is $3/(2^{7/2}\pi^2)$ times the Fourier sum of (2.110) or (2.115).

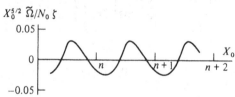

Fig. 2.13. $X_0^{1/2}\tilde{M}/N_0\beta$ as a function of $X_0(=F/H)$ at $T=0$. Same remarks apply as for fig. 2.12 regarding the scale of ordinates. The plotted quantity is $3/(2^{5/2}\pi)$ times the Fourier sum in (2.113) or the first Fourier sum in (2.116).

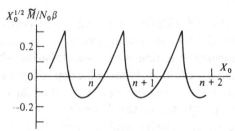

2.3. Calculation of the free energy

of the two-dimensional result, the discontinuities of fig. 2.5 being modified to cusps.

Finally we may note that for free electrons or, more generally, for a parabolic band, the factor in front of the summation in (2.110) simplifies and since, for a sphere, the area extremum is a maximum we obtain

$$\tilde{\Omega} = \left(\frac{eH}{c}\right)^{5/2} \frac{V}{4\pi^4 m h^{1/2}} \sum_{p=1}^{\infty} \frac{1}{p^{5/2}} \cos\left[2\pi p\left(\frac{F}{H} - \frac{1}{2}\right) - \frac{\pi}{4}\right] \quad (2.115)$$

bearing in mind that $F = \zeta/\beta$ (see (A1.19)) and $N_0 = (V/3\pi^2)(2m\zeta/\hbar^2)^{3/2}$, the factor in front of the summation can be expressed more meaningfully as $(3/2^{7/2}\pi^2)(N_0\zeta)(H/F)^{5/2}$ and we have also

$$\tilde{M}_{\parallel} = -\frac{3}{2^{5/2}\pi} N_0 \beta \left(\frac{H}{F}\right)^{1/2} \left\{ \sum_{p=1}^{\infty} \frac{1}{p^{3/2}} \sin\left[2\pi p\left(\frac{F}{H} - \frac{1}{2}\right) - \frac{\pi}{4}\right] \right.$$

$$\left. + \frac{5}{4\pi}\left(\frac{H}{F}\right) \sum_{p=1}^{\infty} \frac{1}{p^{5/2}} \cos\left[2\pi p\left(\frac{F}{H} - \frac{1}{2}\right) - \frac{\pi}{4}\right] \right\} \quad (2.116)$$

The second summation in (2.116) may be neglected for large (F/H) but becomes significant as we approach the quantum limit.

These results, first obtained by Landau (1939) are a remarkably good approximation to the exact results which can be explicitly given for this special case (see Appendix 2). The comparison is shown graphically for $\tilde{M}_{\parallel}$ at low values of F/H in fig. 2.14 (see p. 69), which also shows how $\tilde{M}_{\parallel}$ is modified if the number of electrons rather than the Fermi energy is kept constant as H varies–it is only for low F/H that the difference is appreciable.

2.3.7. Phase smearing (Shoenberg 1969a)

Evidently the oscillatory variations (2.110) and (2.113) represent a highly idealized situation and we have yet to introduce not only such essential features as finite temperature T, finite electron relaxation time τ and electron spin, but also more practical complications such as sample and field inhomogeneity. It turns out that all of these features can be introduced as particular examples of 'phase smearing' in the sense that they are equivalent in their effect to the superposition of oscillations like (2.110) in which the frequency F, or equivalently the phase, is varied over a small range around the value corresponding to the idealized situation. The nature of the variation is chosen appropriately to the particular feature under consideration. We shall now show that such a phase smearing reduces the amplitude of each periodic term in (2.110) (or any derivative of it, such as

Theory

(2.113)) by a factor which is very simply related to the appropriate distribution of phase over which the smearing is made.*

Let us simplify the notation by the substitution

$$\psi = 2\pi p \left(\frac{F}{H} - \frac{1}{2} \right) \pm \frac{\pi}{4} \tag{2.117}$$

so that a typical periodic term in (2.110) is simply $\cos \psi$. The effect of superimposing a distribution of such terms with a range of values of F/H is then to replace $\cos \psi$ by

$$I = \int_{-\infty}^{\infty} \cos(\psi + \phi) \, D(\phi/\lambda) \, d\phi \Big/ \int_{-\infty}^{\infty} D(\phi/\lambda) \, d\phi \tag{2.118}$$

where ϕ is the phase departure from the standard value ψ and $D(\phi/\lambda)$ is the distribution function of the phase smearing, defined so that the probability of ϕ lying between ϕ and $\phi + d\phi$ is proportional to $D(\phi/\lambda) \, d\phi$. The scaling parameter λ will be characteristic of the particular kind of phase smearing concerned; its precise definition need not concern us here, but in general it will be defined in such a way as to make $D(1)$ appreciably less than 1; in other words λ is something like the width of the phase smearing distribution function. The denominator in (2.118) is effectively a normalizing factor and takes care of the constant of proportionality between $D(\phi/\lambda)$ and the true probability.

It is evident that if $\lambda \ll 1$ only very small values of ϕ will contribute appreciably to the numerator in (2.118) so that to a good approximation $\cos(\psi + \phi)$ can be replaced by $\cos \psi$ and taken outside the integral; for this case then I reduces to $\cos \psi$ and the phase smearing is unimportant. If however $\lambda \gtrsim 1$, there is considerable dephasing and the resultant I may be smaller than $\cos \psi$. In general (2.118) can be evaluated by writing it as

$$I = \mathscr{R} e^{i\psi} \int_{-\infty}^{\infty} e^{i\phi} \, D(\phi/\lambda) \, d\phi \Big/ \int_{-\infty}^{\infty} D(\phi/\lambda) \, d\phi \tag{2.119}$$

or, putting

$$z = \phi/\lambda$$

we have

$$I = \mathscr{R} e^{i\psi} \int_{-\infty}^{\infty} e^{i\lambda z} D(z) \, dz \Big/ \int_{-\infty}^{\infty} D(z) \, dz \tag{2.120}$$

or

$$I = \mathscr{R} \{ [f(\lambda)/f(0)] e^{i\psi} \} \tag{2.121}$$

---

* We are of course throughout assuming that the gradual field dependence of the *amplitude* may be ignored. This is an approximation which is usually quite adequate except at low values of F/H.

58

2.3. Calculation of the free energy

where

$$f(\lambda) = \int_{-\infty}^{\infty} e^{i\lambda z} D(z)\, dz \qquad (2.122)$$

which is simply the Fourier transform of $D(z)$ w.r.t. λ.

Thus the effect of the phase smearing is to multiply the amplitude by the reduction factor

$$R = |f(\lambda)|/f(0) \qquad \textbf{(2.123)}$$

and in general also to modify the phase. If, however, $D(z)$ is a symmetrical function of z, as it will be in nearly all the slightly idealized situations we shall consider, then $f(\lambda)$ is real and there is exactly *no* phase shift. However, it should not be forgotten that there may be practical situations in which $D(z)$ is not exactly symmetrical and then there will be a phase shift whose tangent is given by the ratio of the imaginary to the real parts of $f(\lambda)$.

For convenience we collect together in table 2.2 the values of $f(\lambda)/f(0)$ for the various functions $D(z)$ we shall need to consider. The calculation of the Fourier transforms quoted in the table is elementary, except for that of $1/(1 + \cosh z)$, which requires contour integration.

An important feature of this approach to phase smearing is that it leads simply to a multiplication factor $R(p)$ for the pth harmonic term and so each kind of phase smearing can be treated independently (to the approximation in which the smearing of the modulation factor multiplying each periodic term can be ignored). We shall now consider the various problems in turn.

2.3.7.1. Effect of finite temperature

At finite temperature T, the probability of occupation of a state of energy ε is the Fermi function

$$f(\varepsilon) = 1/(1 + e^{(\varepsilon - \zeta)/kT}) \qquad (2.124)$$

Table 2.2. *Fourier transforms of some useful functions*

Description	$D(z)$	$f(\lambda)/f(0)$
Negative derivative of Fermi function	$1/(1 + \cosh z)$	$\pi\lambda/\sinh(\pi\lambda)$
Lorentzian	$1/(1 + z^2)$	$e^{-\lambda}$
Gaussian	e^{-z^2}	$e^{-\frac{1}{4}\lambda^2}$
Top hat	$D(z) = \frac{1}{2}$ for $-1 < z < 1$ $D(z) = 0$ elsewhere	$\sin \lambda/\lambda$

59

and intuitively it would seem that the actual metal, with Fermi energy ζ at temperature T, could be thought of as equivalent to a *distribution* of hypothetical metals, all at $T = 0$, with a range of Fermi energies μ spread around ζ and distributed in such a way that a fraction $f(\varepsilon)$ of the hypothetical metals have $\mu > \varepsilon$ (and will therefore have this state occupied). It is not difficult to see that the required distribution is such that $-[df(\mu)/d\mu]\,d\mu$ have Fermi energies between μ and $\mu + d\mu$. With such a distribution, the fraction having $\mu > \varepsilon$ is

$$-\int_{\varepsilon}^{\infty} [df(\mu)/d\mu]\,d\mu$$

which is indeed $f(\varepsilon)$. A more formal justification of this intuitive approach is given in §2.5, where it is shown that for any Fermi–Dirac system, a convolution of the thermodynamical potential $\Omega(0)$ at $T = 0$ with the distribution function $-df(\mu)/d\mu$, does indeed correctly give $\Omega(T)$.

The distribution function $-df(\mu)/d\mu$ can be expressed as

$$-\frac{df(\mu)}{d\mu} = \frac{1}{2kT[1 + \cosh(\mu - \zeta)/kT]} \tag{2.125}$$

and its Fourier transform is given in table 2.2. Since the dHvA frequency F is a function of ζ, a spread of μ is equivalent to a smearing of the phase ψ. For a change $(\mu - \zeta)$ of Fermi energy the corresponding change ϕ of the phase is

$$\phi = \frac{2\pi p}{H}(\mu - \zeta)\frac{dF}{d\zeta} = \frac{2\pi p}{H}\frac{c\hbar}{2\pi e}\frac{dA}{d\zeta}(\mu - \zeta)$$

$$= \frac{2\pi p}{\beta H}(\mu - \zeta) \tag{2.126}$$

Thus

$$\frac{\mu - \zeta}{kT} = \frac{\beta H \phi}{2\pi p k T} \equiv \frac{\phi}{\lambda} \tag{2.127}$$

with λ defined as $2\pi p k T/\beta H$.

The effect of a finite temperature is therefore equivalent to a phase smearing with the distribution function $1/[1 + \cosh(\phi/\lambda)]$ and from (2.123) and table 2.2 we see that this has the effect of multiplying the amplitude by a reduction factor R_T given by

$$R_T = \frac{\pi\lambda}{\sinh \pi\lambda} = \frac{2\pi^2 p k T/\beta H}{\sinh(2\pi^2 p k T/\beta H)} \tag{2.128}$$

To our usual approximation of differentiating only the relatively rapidly

2.3. Calculation of the free energy

varying periodic factors, the same reduction factor R_T is valid also for derivatives such as the magnetization. This factor approaches 1 for $2\pi^2 pkT/\beta H \ll 1$, as of course it should (negligible phase smearing), while for $2\pi^2 pkT \gtrsim 1$ it approaches

$$R_T = \frac{4\pi^2 pkT}{\beta H} \exp(-2\pi^2 pkT/\beta H) \qquad (2.129)$$

We may note that for practical conditions of T and H and for a cyclotron mass m not very different from the free electron mass m_0 (e.g. as in a noble or an alkali metal), (2.129) is usually a very good approximation. The reduction factor (2.129) may be expressed numerically as

$$R_T = 294p(m/m_0)(T/H)\exp[-147p(m/m_0)T/H] \qquad (2.130)$$

if H is given in kG (if H is in Tesla the numbers become 29.4 and 14.7). For copper, $m/m_0 \sim 1.5$, so for $T = 1\,\text{K}$, $H = 50\,\text{kG}$ and $p = 1$, we have $R_T \sim 0.1$, while for $p = 2$, $R_T = 0.002$. Thus, for these typical conditions, it is also a good approximation to treat the oscillations as simple harmonic. As we shall see below, the reduction factors due to other kinds of phase smearing still further reduce the harmonic content of typical dHvA oscillations.

2.3.7.2 Effect of finite relaxation time

If the electrons have a finite relaxation time, τ, due to scattering, the otherwise sharp quantum levels ε_r (for given κ) become broadened in accordance with the uncertainty principle and, as was first shown by Dingle (1952b), this leads to a reduction of the oscillation amplitude. The broadening can be described by a Lorentzian distribution function, such that the probability of the energy lying between ε and $\varepsilon + \text{d}\varepsilon$ is proportional to

$$\frac{\text{d}\varepsilon}{(\varepsilon - \varepsilon_r)^2 + (\hbar/2\tau)^2} \qquad (2.131)^*$$

If τ can be regarded as independent of ε, the effect of this broadening is equivalent to that of a spread of Fermi energy μ around its true value ζ, with a probability that it should lie between μ and $\mu + \text{d}\mu$ proportional to

$$\frac{\text{d}\mu}{(\mu - \zeta)^2 + (\hbar/2\tau)^2} \qquad (2.132)$$

* As was first pointed out by Brailsford (1966), Dingle used an unconventional definition of τ, equivalent to twice that implied in (2.131). This has caused some confusion in the literature on relaxation times deduced from observed Dingle temperatures.

Theory

Just as before, a spread of Fermi energy causes a spread of F through the dependence of FS area on Fermi energy and so the level broadening is equivalent to a phase smearing. As before, the phase departure ϕ is related to $(\mu - \zeta)$ by (2.126) and so the distribution function of phase smearing can be expressed as

$$D(\phi/\lambda) = 1/[1 + (\phi/\lambda)^2] \qquad (2.133)$$

with

$$\lambda = \pi p\hbar/\beta H\tau \qquad (2.134)$$

From table 2.2 it follows that the 'Dingle' reduction factor R_D due to the level broadening is

$$R_D = e^{-\pi p\hbar/\beta H\tau} = e^{-\pi p/\omega_c\tau} \qquad (2.135)$$

It is instructive to note that R_D is just the reduction of the wave amplitude of an electron which has completed p circuits of a cyclotron orbit, i.e. the square root of the fraction of the electrons which have survived (i.e. have not been scattered). This plausible result finds a deeper interpretation in the analysis of Falicov and Stachowiak (1966).

As can be seen from (2.129) the effect of this reduction factor is similar (but not quite identical, because of the extra factor T in (2.129) and because (2.129) is in any case an approximation) to that of a rise of temperature x given by

$$x = \hbar/2\pi k\tau \qquad (2.136)$$

Thus the Dingle factor can be expressed as

$$R_D = \exp(-2\pi^2 pkx/\beta H) \qquad \mathbf{(2.137)}$$

and x is known as the Dingle temperature. Typically, Dingle temperatures due to impurity scattering are of order $10–100\,\text{K}$ per atomic % of added impurity (see table 8.1); in practice it is difficult to observe values much above $5\,\text{K}$ because the oscillation amplitude becomes too small to detect.

It should be noticed that a rather drastic simplifying assumption was made at the start of the calculation, namely τ is independent of ε. Without this assumption the equivalence of broadened energy levels and sharp Fermi energy to sharp levels and broadened Fermi energy is no longer quite valid. However this lack of validity is in practice not very important. This can be seen as a consequence of the argument outlined in §2.3.2, which shows that for the two-dimensional slice at $T = 0$ it is, in a sense, only the single Landau level that is passing through the Fermi energy which contributes to a particular discontinuity in M. When the integration over κ

2.3. Calculation of the free energy

is made this is no longer quite true, but it is still effectively only a few Landau tubes close to the one actually passing through the Fermi surface which contribute appreciably to any particular oscillation.

At finite T and for finite relaxation time, rather more tubes are involved than would be the case at $T = 0$, but in fact the number of tubes can never be large. This follows in a rather general way because the relevant phase smearing distribution functions $D(\phi/\lambda)$ fall off rapidly beyond $\phi \sim \lambda$. Since the reduction factors R_T or R_D become very small when λ is more than say π, it follows that the 'effective' range (i.e. $\pm \lambda$) of ϕ for an observable oscillation amplitude is only of order 2π, or one oscillation. In other words the broadening due to finite temperature or finite relaxation time effectively only brings in the immediately neighbouring one or two tubes to the one which is crossing the Fermi surface. Thus it is only the broadening of these few tubes, close to the one leaving the FS, which matters in the calculation and if we suppose that τ does not vary too rapidly from one tube to the next, the assumption of a τ independent of ε is probably a reasonable approximation.*

2.3.7.3. Other damping effects

In practice the oscillation amplitude is often damped more by phase smearing due to sample inhomogeneity than by the finite relaxation time due to electron scattering.[†] Such inhomogeneity may be in virtue of mosaic structure, which implies a small spread of crystal orientation and therefore of F, or in virtue of varying strains in the sample (basically due to defects such as dislocations), which again cause a small spread of F. The phase shift ϕ associated with a change ΔF of F (due to whatever cause) will be given by

$$\phi = 2\pi p \Delta F / H \qquad (2.138)$$

Thus if, for instance, the distribution function of ΔF is Lorentzian, proportional to

$$1/\{1 + [\Delta F/(\Delta F)_0]^2\} \qquad (2.139)$$

the distribution function of ϕ will be of the form (2.133) with λ given by

$$\lambda = 2\pi p (\Delta F)_0 / H \qquad (2.140)$$

---

* The discussion given here is rather oversimplified; a proper discussion involves detailed consideration of the electron scattering mechanism in a magnetic field (see for instance reviews by Středa (1974) and Coleridge (1980)).

† However sample inhomogeneity is inevitably accompanied by electron scattering (usually low angle scattering) and it is not obvious whether or not the scattering needs to be considered as an effect additional to the phase smearing; here we consider only the phase smearing.

and the reduction factor $e^{-\lambda}$ would have the same form as the Dingle factor R_D of (2.137), but with

$$x = \beta(\Delta F)_0/\pi k. \qquad (2.141)$$

As will be discussed in more detail in chapter 8, there is no *a priori* reason to expect the distribution of ΔF to be Lorentzian, but in practice the exponential form of the reduction factor (2.137) often seems to be valid over a wide range of fields. This perhaps means no more than that the assumption of a Lorentzian distribution is an adequate approximation in the experimental conditions concerned. Typical Dingle temperatures due to inhomogeneity effects are found to be of order 1 K or less.

It should also be noticed that implicit in all the discussion of this section is the assumption that the size of the classical orbit is small compared to the scale over which F varies appreciably. At sufficiently low fields or for inhomogeneity on a sufficiently fine scale, it will be shown in chapter 8 that the arguments used so far break down and that in the opposite limit of orbits *large* compared with the scale of inhomogeneity, the reduction factor assumes the exponential form (2.137) whatever the distribution function of the inhomogeneity.

To conclude this section one other source of damping can conveniently be mentioned here. This is phase smearing due to field inhomogeneity. If for instance there is small linear variation of H along one of the dimensions of a sample,* the corresponding phase distribution would be the 'top hat' of table 2.2 with

$$\lambda = \pi p F \Delta H / H^2 \qquad (2.142)$$

where ΔH is the total change of field along the sample. The reduction factor is then

$$R_{\Delta H} = \sin \lambda / \lambda \qquad (2.143)$$

However this is a topic relevant to the technique of observing the de Haas-van Alphen effect rather than to the basic factors within the sample which determine the amplitude of the oscillations. We shall return to the effect of field inhomogeneity in chapter 3, which deals with experimental techniques. Finally it should be mentioned that the top hat distribution can also be relevant to the case of a crystal bent slightly into the arc of a circle; unless it is close to a symmetry orientation it can easily be seen that the reduction factor is again $\sin \lambda/\lambda$, but with $\Delta F/H$ replacing $F\Delta H/H^2$ in (2.142), where ΔF is the spread of F over the sample.

---

* We assume the idealized situation of a constant area of cross section of the sample normal to the direction of field variation and a sharp cut off at each end.

2.3. Calculation of the free energy

2.3.7.4. Effect of electron spin

In a magnetic field H the spin degeneracy of the energy levels is lifted and each level is split in two becoming

$$\varepsilon \pm \tfrac{1}{2}\Delta\varepsilon \qquad (2.144)$$

where

$$\Delta\varepsilon = \tfrac{1}{2}g\beta_0 H \qquad (2.145)$$

Here $\beta_0 = e\hbar/m_0 c = 2$ Bohr magnetons and g is called the spin-splitting factor, defined by the relation (2.145); for free electrons $g = g_0 = 2.0023$, which for most purposes can be taken as 2. The effect of this splitting is exactly equivalent to a phase difference ϕ given by

$$\phi = 2\pi\Delta\varepsilon/\beta H \equiv 2\pi S \qquad (2.146)$$

between the oscillations coming from the spin-up and the spin-down electrons (or p times as much for the pth harmonic). This is because one full oscillation (or a phase shift of 2π) corresponds to the passage of two successive levels, βH apart, through the Fermi level, and $\Delta\varepsilon/\beta H$ is the fraction of the 2π phase shift corresponding to the spin splitting $\Delta\varepsilon$. The superposition of the spin-up and spin-down oscillations will therefore produce a net amplitude which is the original amplitude multiplied by a reduction factor R_s, which for the pth harmonic is given by

$$R_s = \cos(\tfrac{1}{2}p\phi) = \cos(p\pi\Delta\varepsilon/\beta H) \qquad (2.147)$$

or

$$R_s = \cos(\tfrac{1}{2}p\pi g m/m_0) \equiv \cos(p\pi S) \qquad \mathbf{(2.148)}$$

This result was first obtained by Sondheimer and Wilson (1951) and Dingle (1952a).

This simple argument can be formalized by noticing that the situation of spin-split energy levels $\varepsilon \pm \tfrac{1}{2}\Delta\varepsilon$ passing through the Fermi energy ζ is exactly equivalent to that of unsplit levels ε passing through two different Fermi energies $\zeta \pm \tfrac{1}{2}\Delta\varepsilon$ (according as the spin is down or up).* Thus any periodic term $\cos \psi$ becomes

$$\frac{1}{2}\left\{\cos\left(\psi + \frac{1}{2}\Delta\varepsilon\frac{\mathrm{d}\psi}{\mathrm{d}\zeta}\right) + \cos\left(\psi - \tfrac{1}{2}\Delta\varepsilon\frac{\mathrm{d}\psi}{\mathrm{d}\zeta}\right)\right\} \qquad (2.149)$$

---

* Indeed both pictures can be regarded as equally valid. If potential energy is included, the Fermi energy is the same for spin-up and spin-down electrons (which have different energies) but if these are thought of as sub-bands the bottoms of the two sub-bands are no longer the same. If potential energy is *not* included the bottoms of the two sub-bands coincide (and so do the spin-up and spin-down energy levels) but the Fermi energies differ.

Theory

But

$$\frac{d\psi}{d\zeta} = \frac{pch}{eH}\frac{dA}{d\zeta} = \frac{2\pi p}{\beta H}$$

So (2.149) reduces to

$$\cos[\tfrac{1}{2}p\pi g(m/m_0)]\cos\psi, \tag{2.150}$$

corresponding to just the reduction factor R_s of (2.148).

The form of R_s can also be deduced by the general approach of Fourier transforming a distribution function

$$D(z) = \delta(z - \tfrac{1}{2}) + \delta(z + \tfrac{1}{2})$$

with respect to a λ given by

$$\lambda = 2\pi\Delta\varepsilon/\beta H$$

However, it is hardly necessary to use such a sledgehammer method for deriving a result which is so intuitively obvious.

We may note that for free electrons ($g = 2, m = m_0$) the reduction factor becomes simply $(-1)^p$, corresponding to a spin-splitting exactly equal to the Landau level spacing (Akhieser 1939). As we shall see later, this result may still be approximately true for bands in which $m/m_0 \ll 1$, but with strong spin-orbit coupling, which makes $\tfrac{1}{2}gm/m_0 \sim 1$. A factor $(-1)^p$ is exactly equivalent to a phase shift of $p\pi$ for the pth harmonic.

2.3.8. Discussion of the Lifshitz–Kosevich formula for $\tilde{\Omega}$ and $\tilde{M}$

The final formula for $\tilde{\Omega}$ after inserting the reduction factors R_T, R_D and R_s into the 'ideal' formula (2.110) is

$$\tilde{\Omega} = \left(\frac{e}{2\pi ch}\right)^{3/2}\frac{2kTH^{3/2}V}{(A'')^{1/2}}\sum_{p=1}^{\infty}\frac{\exp(-2\pi^2 pkx/\beta H)\cos[\tfrac{1}{2}p\pi g(m/m_0)]}{p^{3/2}\sinh(2\pi^2 pkT/\beta H)}$$

$$\times \cos\left[2\pi p\left(\frac{F}{H} - \frac{1}{2}\right) \pm \frac{\pi}{4}\right] \tag{2.151}$$

and correspondingly

$$\tilde{M}_{\parallel} = -\left(\frac{e}{ch}\right)^{3/2}\frac{2FkTV}{(2\pi HA'')^{1/2}}\sum_{p=1}^{\infty}\frac{\exp(-2\pi^2 pkx/\beta H)\cos[\tfrac{1}{2}p\pi g(m/m_0)]}{p^{1/2}\sinh(2\pi^2 pkT/\beta H)}$$

$$\times \sin\left[2\pi p\left(\frac{F}{H} - \frac{1}{2}\right) \pm \frac{\pi}{4}\right] \tag{2.152}$$

while $\tilde{M}_{\perp}$ is, as before, given by (2.114).

66

These results (apart from the Dingle factor) were first given by Lifshitz and Kosevich (1954, 1955) and we shall often refer to them and other results derived from them as the 'LK formula'. In general, Fermi surfaces are complicated and usually there is more than one cross-section of extremal area and correspondingly more than one F, β (or m), g and x for any given field direction. Thus the total oscillatory quantity is the sum of a number of contributions, each of which has the LK form but with different sets of parameters. Although the formula has been calculated assuming ζ constant, we shall show in §2.4 that it is very little affected by taking account of the weak oscillations of ζ.

As we shall discuss in §2.6, the form of the LK results is unchanged if the theory is generalized to take account of many-body interactions, except in extreme conditions. The LK formula does however ignore two complications which can sometimes be of practical importance: these are magnetic interaction and magnetic breakdown, both briefly mentioned in §1.3. We defer detailed treatment of these effects to chapters 6 and 7 respectively, so that we can consider the applications of the LK formula without worrying about these complications until we have to. About half the rest of the book (particularly chapters 5, 8 and 9) is, in fact, concerned with the experimental verification of the LK formula, the determination of the various parameters, and their exploitation to learn about basic properties of metals. A brief summary of the programme has already been given in §1.4.

2.4. Oscillations of Fermi energy

As we saw in §2.3.3 (see (2.82)) the oscillations of Fermi energy required to keep the number of electrons constant in a two-dimensional slice of FS at $T = 0$ are of quite appreciable amplitude. We shall now show that for the whole three-dimensional Fermi surface the oscillations of ζ are much feebler, even at $T = 0$, and usually of no consequence.

The total number of electrons is given by

$$N = -(\partial\Omega/\partial\zeta)_H \tag{2.153}$$

and this can be written as*

$$N = N_0(\zeta) + \tilde{N}(\zeta) \tag{2.154}$$

where

$$\tilde{N}(\zeta) = -(\partial\tilde{\Omega}/\mathrm{d}\zeta)_H \tag{2.155}$$

---

* We ignore the small field-dependent, but non-oscillatory contribution which comes from the second term of (2.55) after integration over κ.

67

Theory

and $N_0(\zeta)$ is the number of electrons for given ζ at $H = 0$; (this is derived from Ω_0, which comes from the integral over κ of the 1st term in (2.55), together, where appropriate, with contributions from other bands). It is now a simple matter to keep N constant as $\tilde{N}$ oscillates, by choosing an appropriate small oscillation $\tilde{\zeta}$ of Fermi energy which will vary $N_0(\zeta)$ by the small amount necessary to cancel the $\tilde{N}$ oscillation. Ignoring the effect of $\tilde{\zeta}$ itself on $\tilde{N}$, as it is justified to a good approximation, the appropriate $\tilde{\zeta}$ is given by

$$\tilde{\zeta}\partial N_0(\zeta)/\partial\zeta + \tilde{N}(\zeta) = 0 \qquad (2.156)$$

Now $\partial N_0(\zeta)/\partial\zeta$ is simply $\mathcal{D}(\zeta)$, the density of states at the Fermi level, and (2.66) integrated over κ gives

$$\tilde{N} = -(2\pi mH/\hbar^2 A)\tilde{M} = -H\tilde{M}/F\beta$$

Thus

$$\tilde{\zeta} = -\tilde{N}/\mathcal{D}(\zeta) = -H\tilde{M}/F\beta\mathcal{D}(\zeta) \qquad (2.157)$$

This formula was first derived by Kaganov, Lifshitz and Sinelnikov (1957). We can estimate the order of magnitude of $\tilde{\zeta}$ by noting that

$$\mathcal{D}(\zeta) \sim N_0/\zeta \qquad (2.158)$$

(for a parabolic band $\mathcal{D}(\zeta)$ is exactly $\frac{3}{2}N_0(\zeta)/\zeta$), so that

$$\frac{\tilde{\zeta}}{\zeta} \sim \frac{\tilde{N}}{N_0} \qquad (2.159)$$

For the two-dimensional case at $T = 0$, $\tilde{N}/N_0$ is of order $1/X$, i.e. $(F/H)^{-1}$ (see (2.65)), but the integration over κ introduces a reduction factor of order $(F/H)^{-1/2}$ (see discussion following (2.109)), so we should expect that at $T = 0$

$$\tilde{\zeta}/\zeta \sim (F/H)^{-3/2} \qquad (2.160)$$

For $F/H \sim 10^4$ (typical for a large FS), this is only 10^{-6} and of course if the spin reduction factors R_T, R_D and R_s are taken into account it could well be a good deal smaller. Since ζ is typically of the order of a few electron volts the oscillations should not exceed a microvolt or so. In exceptional situations, however, such as in bismuth, where the total number of electrons and holes is very small and much smaller values of F/H may be relevant, the oscillations may reach an amplitude of nearly a millivolt. The question of observing the oscillations of ζ through the corresponding oscillations in contact potential will be discussed later (§4.4).

68

2.5. *Oscillations of density of states*

In principle the oscillations of ζ imply that all the parameters in the formula for the thermodynamic potential, and in particular, the dHvA frequency F, are themselves oscillatory functions of H, which in turn implies that the true harmonic content of the oscillations will be modified from what it appears to be for constant parameters. In practice, however, because of the smallness of $\tilde{\zeta}/\zeta$, the relative amplitudes of oscillation of the parameters are of the same small order of magnitude (e.g. $\tilde{F}/F \sim 10^{-6}$ or less) and the effect on the harmonic content is usually entirely negligible. Figure 2.14 illustrates for the example of free electrons how little difference it makes whether ζ or N is kept constant. In fact no appreciable error is usually involved in treating all the parameters F, β (or m), g and x as constants. However, as will be briefly discussed in §4.4, Bi provides an exception.

2.5. Oscillations of density of states

It is convenient to discuss the density of states and in particular its oscillatory field dependence here for two reasons. First, because the effect of many-body interactions, to be discussed in §2.6, can be formulated most directly in terms of the density of states and secondly because the oscillations of the density of states are closely related to the oscillations

Fig. 2.14. $M/N_0\beta$ as a function of $X_0(=\zeta/\beta H = F/H)$ for a spinless free electron gas at $T = 0$ for low X_0 computed from (*a*) the exact expression (A2.18) for ζ constant (full curve), (*b*) the exact implicit expression for N constant (as explained in Appendix 2); note that X_0 is now defined as $\zeta_0/\beta H$ (broken curve) and (*c*) from the Fourier sums (2.116) with an extra term $-1/8X_0$ to allow for the steady diamagnetism (dotted curve). On the scale of the diagram the differences between (*c*) and (*a*) are discernible only below about $X_0 = 0.7$.

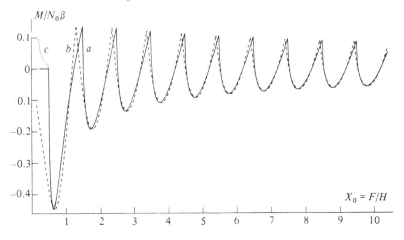

of resistivity (Shubnikov–de Haas effect) which we shall discuss in chapter 4.

We shall start by calculating the density of states $\mathcal{D}(\varepsilon)$ from first principles in the same sort of way as was done in §2.3 to obtain the thermodynamic potential Ω. The contribution $\delta\mathcal{D}(\varepsilon)$ from a 2-D slice of k-space, of thickness $d\kappa$ is given by

$$\delta\mathcal{D}(\varepsilon) = \left(\frac{eHVd\kappa}{2\pi^2 c\hbar}\right) \sum_{r=0}^{\infty} \left(\frac{da}{d\varepsilon}\right)_{\kappa} \delta\left[a(\varepsilon,\kappa) - (r+\tfrac{1}{2})\frac{2\pi eH}{c\hbar}\right] \qquad (2.161)$$

since permitted states occur only when the argument of the δ-function vanishes. The factor $(da/d\varepsilon)_{\kappa}$ is needed to normalize the δ-function, i.e. to ensure that if (2.161) is integrated up to some value ε of the energy, the result will be just the degeneracy factor (the first bracket of (2.161)) times the number of Landau tubes whose energy for the given κ is less than ε.

If we use (2.21) to express $(da/d\varepsilon)_{\kappa}$ in terms of the cyclotron mass $m(\varepsilon)$ and introduce, as before, the variable x defined by (2.48), we find

$$\delta\mathcal{D}(\varepsilon) = \frac{m(\varepsilon)Vd\kappa}{2\pi^2\hbar^2} \sum_{r=0}^{\infty} \delta[x - (r+\tfrac{1}{2})] \qquad (2.162)$$

This can now be Fourier analyzed to give

$$\delta\mathcal{D}(\varepsilon) = \frac{m(\varepsilon)Vd\kappa}{2\pi^2\hbar^2} \left\{1 + 2\sum_{p=1}^{\infty} \cos[2\pi p(x - \tfrac{1}{2})]\right\} \qquad (2.163)$$

and in particular for $\varepsilon = \zeta$, x becomes $X(=F/H)$ and $m(\varepsilon)$ becomes m, the cyclotron mass at the Fermi surface, so

$$\delta\mathcal{D}(\zeta) = \frac{mVd\kappa}{2\pi^2\hbar^2} \left\{1 + 2\sum_{p=1}^{\infty} \cos[2\pi p(X - \tfrac{1}{2})]\right\} \qquad (2.164)$$

It can be seen that $\delta\mathcal{D}(\varepsilon)$ consists of a steady part $\delta\mathcal{D}_0(\varepsilon)$ and an oscillatory part $\delta\tilde{\mathcal{D}}(\varepsilon)$. The steady part can be expressed as

$$\delta\mathcal{D}_0(\varepsilon) = \left(\frac{d^2v(\kappa)}{d\varepsilon\, d\kappa}\right)\frac{Vd\kappa}{4\pi^3}$$

where $v(\kappa)$ is the volume of the surface of constant energy ε up to the plane $k_z = \kappa$. Thus, integrating over κ, we find as we should expect,

$$\mathcal{D}_0(\varepsilon) = (dv/d\varepsilon)V/4\pi^3 \qquad (2.165)$$

where v is now the total volume of the surface of constant energy ε.

It is helpful to note that for spherical constant energy surfaces with

2.5. Oscillations of density of states

cyclotron mass m, the density of states at the Fermi energy reduces to

$$\mathcal{D}_0(\zeta) = (mV/\pi^2\hbar^2)(3\pi^2 N/V)^{1/3} \tag{2.165a}$$

if there are N electrons in the volume V.

As regards the oscillatory part $\tilde{\mathcal{D}}(\varepsilon)$, the integration over κ is carried out by exactly the same method as was used in going from $\delta\tilde{\Omega}$ to $\tilde{\Omega}$. If we compare (2.164) for $\delta\tilde{\mathcal{D}}(\zeta)$ with the expression (2.57) for $\delta\tilde{\Omega}$, it is evident that $\delta\tilde{\mathcal{D}}(\zeta)$ is proportional to $d^2(\delta\tilde{\Omega})/dx^2$, which in turn is proportional to $d\tilde{M}/dH$. This proportionality is exactly preserved in the integration over κ and it is easily shown that

$$\tilde{\mathcal{D}}(\zeta) = \frac{1}{\beta^2 X^2} d\tilde{M}/dH \tag{2.166}$$

Thus $\tilde{\mathcal{D}}(\zeta)$ is in phase with $d\tilde{M}/dH$, but in quadrature with $\tilde{M}$.

Level broadening and spin produce exactly the same reduction factors R_D and R_s as before, but of course the factor R_T must *not* be included, since this is concerned with the probability of *occupation* of the states rather than with the states themselves. We thus find

$$\tilde{\mathcal{D}}(\zeta) = \left(\frac{2eH}{ch}\right)^{1/2} \frac{mVR_D R_s}{\pi^{3/2}\hbar^2(A'')^{1/2}} \sum_{p=1}^{\infty} \frac{1}{p^{1/2}} \cos\left[2\pi p\left(\frac{F}{H} - \frac{1}{2}\right) \pm \frac{\pi}{4}\right] \tag{2.167}$$

The density of states $\mathcal{D}(\varepsilon)$ for an arbitrary energy ε is given by essentially the same formula except that F is determined by the extremal area of the surface of energy ε, rather than by A, that of the FS, and similarly m refers to $m(\varepsilon)$ and $(A'')^{1/2}$ to the ε surface. The order of magnitude of $\tilde{\mathcal{D}}(\zeta)$ will be discussed in §4.5 and we shall see that it is only a small fraction of $\mathcal{D}_0(\zeta)$–at best of order $(H/2F)^{1/2}$ times $\mathcal{D}_0(\zeta)$.

In fact the relations (2.166) and (2.167) can be quite simply deduced from a more general formulation, without the need for going through our 'first principles' argument. If we go back to the basic definition of Ω in (2.43), the sum can be replaced by an integral if $\mathcal{D}(\varepsilon)$ is introduced, giving

$$\Omega(\zeta, T) = -kT \int_{-\infty}^{\infty} \mathcal{D}(\varepsilon)\ln[1 + \exp(\zeta - \varepsilon)/kT]\, d\varepsilon \tag{2.168}$$

For $T = 0$, this becomes

$$\Omega(\zeta, 0) = \int_{-\infty}^{\zeta} (\varepsilon - \zeta)\mathcal{D}(\varepsilon)\, d\varepsilon \tag{2.169}$$

71

from which it immediately follows, by differentiating twice, that

$$\mathscr{D}(\zeta) = -\partial^2\Omega(\zeta,0)/\partial\zeta^2 \tag{2.170}$$

This can also be derived from the total number of particles N, which is given both by $-(\partial\Omega/\partial\zeta)$ (see (2.153)) and by

$$N_{T=0} = \int_{-\infty}^{\zeta} \mathscr{D}(\varepsilon)\,d\varepsilon$$

If, as usual, we differentiate only the oscillatory part of Ω in (2.170) and remember that $\partial^2/\partial\zeta^2$ applied to $\cos[2\pi p(X - \frac{1}{2}) \pm \pi/4]$ is just $[(dX/d\zeta)/X]^2$ times $\partial^2/\partial H^2$ and that $dX/d\zeta = 1/\beta$, we immediately recover the result (2.166).

In the next section we shall make use of some of the general relations between Ω and $\mathscr{D}(\varepsilon)$ to discuss the consequences of many-body interactions, but before concluding this section it is convenient to show how (2.168) can be used to give a formal justification for the phase smearing method of introducing the effect of temperature (§2.3.7.1).

If (2.168) is integrated by parts we have, since the first term of the integration vanishes at both limits:

$$\Omega(\zeta, T) = -\int_{-\infty}^{\infty} d\varepsilon\, f(\varepsilon - \zeta) \int_{-\infty}^{\varepsilon} \mathscr{D}(y)\,dy \tag{2.171}$$

where $f(x)$ is the Fermi function $(e^{x/kT} + 1)^{-1}$. Integrating a second time, again by parts, the first term again vanishes at both limits and we have

$$\Omega(\zeta, T) = \int_{-\infty}^{\infty} \frac{\partial f(\varepsilon - \zeta)}{\partial \varepsilon} W(\varepsilon)\,d\varepsilon \tag{2.172}$$

where $W(\varepsilon)$ is the double integral of $\mathscr{D}(\varepsilon)$, i.e.

$$W(\varepsilon) = \int_{-\infty}^{\varepsilon} dx \int_{-\infty}^{x} \mathscr{D}(y)\,dy \tag{2.173}$$

But from (2.170) it follows that

$$W(\mu) = -\Omega(\mu, 0) \tag{2.174}$$

(it is convenient to use μ instead of ε or ζ here, in order to conform to the notation of §2.3.7.1). Thus (2.172) can be expressed as

$$\Omega(\zeta, T) = -\int_{-\infty}^{\infty} \frac{\partial f(\mu - \zeta)}{\partial \mu}\Omega(\mu, 0)\,d\mu \tag{2.175}$$

This is precisely the convolution which was assumed intuitively in §2.3.7.1

and, of course, as applied to an oscillatory Ω whose frequency varies with ζ or μ, is equivalent to a phase smearing of the oscillations.

2.6. Many-body interactions

The basic theory of the effect of many-body interactions of the electrons (with each other or with phonons or magnons) is beyond the scope of this book (for detailed discussions from various points of view see for instance Pines and Nozières 1966, Grimvall 1981 and Mahan 1981). However, some idea of how the independent particle treatment presented earlier in this chapter is modified by the interactions can be given by merely quoting basic theoretical results without derivation, and showing how they affect the dHvA oscillations. As has already been mentioned, it turns out that the LK formula remains valid except in extreme conditions, but the parameters that enter into the formula have to be modified quite appreciably.

Our simplified presentation inevitably slides over various subtle points in the theory and omits many qualifications but one point should be emphasized at the start. This is that much of the theory really applies rigorously only to an isotropic metal. Thus, although our treatment appears to deal with arbitrarily shaped constant energy surfaces, this generality may be in fact somewhat spurious and the application to a strongly anisotropic metal such as Hg may be only qualitatively valid.

2.6.1. The self energy concept

The basic result of many-body interactions that concern us here is that the energies $\varepsilon(\mathbf{k})$ of the independent particle model are modified in two ways. First the independent particle energy $\varepsilon(\mathbf{k})$ is shifted to a value ε (the so-called dynamical quasiparticle energy) which is $\varepsilon(\mathbf{k}) + \Delta(\varepsilon - \zeta)$, and secondly it is broadened in Lorentzian fashion with a characteristic parameter $\Gamma(\varepsilon - \zeta)$. The relation between ε and $\varepsilon(\mathbf{k})$ for the electron–phonon interaction is shown schematically in fig. 2.15, and graphs of Δ and Γ for the concrete example of Hg are shown in fig. 2.16.

The oscillatory density of states $\tilde{\mathscr{D}}(\varepsilon)$ for the non-interacting particles is modified by the shift and the broadening to $\tilde{\mathscr{D}}'(\varepsilon)$ for the interacting system, and $\tilde{\mathscr{D}}'(\varepsilon)$ is given by

$$\tilde{\mathscr{D}}'(\varepsilon) = \frac{\Gamma(y)}{\pi} \int_{-\infty}^{\infty} \frac{\tilde{\mathscr{D}}(\varepsilon + x)\,dx}{[\Delta(y) + x]^2 + [\Gamma(y)]^2} \qquad (2.176)$$

where we have put

$$y = \varepsilon - \zeta$$

Theory

and the factor $\Gamma(y)/\pi$ is, of course, the normalizing factor of the Lorentzian distribution regarded as a function of x. The oscillatory part of $\mathcal{D}(\varepsilon)$, as given by (2.167) with ζ replaced by ε, can be expressed as

$$\tilde{\mathcal{D}}(\varepsilon) = \mathcal{R} \sum_{p=1}^{\infty} c_p e^{i(p\psi(\varepsilon) + \varphi)} \tag{2.177}$$

where

$$\psi(\varepsilon) = \psi(\zeta) + (\varepsilon - \zeta)\,\mathrm{d}\psi/\mathrm{d}\varepsilon = \psi(\zeta) + 2\pi y/\beta H \tag{2.178}$$

and c_p is shorthand for the factors in (2.167). (We shall in practice be concerned only with values of ε so close to ζ that the approximation (2.178) is adequate.) Thus (with ψ denoting $\psi(\zeta)$ in what follows)

$$\tilde{\mathcal{D}}'(\varepsilon) = \mathcal{R} \sum_{p=1}^{\infty} c_p e^{i(p\psi + \phi)} \frac{\Gamma(y)}{\pi} \int_{-\infty}^{\infty} \frac{e^{2\pi i p(y + x)/\beta H}\,\mathrm{d}x}{(\Delta(y) + x)^2 + (\Gamma(y))^2} \tag{2.179}$$

If we substitute

$$u = (x + \Delta(y))/\Gamma(y) \tag{2.180}$$

the r.h.s. of (2.179) is seen to be proportional to the Fourier transform of a

Fig. 2.15. Schematic graph of ε (including electron—phonon interaction) versus $\varepsilon(k)$ (without interaction). The broadening of the curve as ε and $\varepsilon(k)$ depart from the Fermi energy ζ is intended to indicate schematically the imaginary part of the self energy which is essentially equivalent to a smearing. The energy differences $\varepsilon - \zeta$ and $\varepsilon(k) - \zeta$ shown in the diagram are of the order of typical phonon energies (after Wilkins 1980).

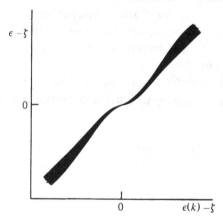

2.6. Many-body interactions

Lorentzian and using the formula of table 2.2, (2.179) eventually reduces to

$$\tilde{\mathscr{D}}'(\varepsilon) = \mathscr{R} \sum_{p=1}^{\infty} c_p e^{i(p\psi + \phi)} \exp\{2\pi i p[y - \Delta(y) + i\Gamma(y)]/\beta H\} \qquad \textbf{(2.181)}$$

Thus the 'renormalized' density of states, as $\mathscr{D}'$ is called, is obtained from $\mathscr{D}$

Fig. 2.16. Graphs of $-\Delta$ (full) and Γ (broken) versus $(\varepsilon - \zeta)$ at 0, 5 and 10 K for Hg (after Elliott 1979). Note that 1 meV is equivalent to kT for $T = 11.6$ K.

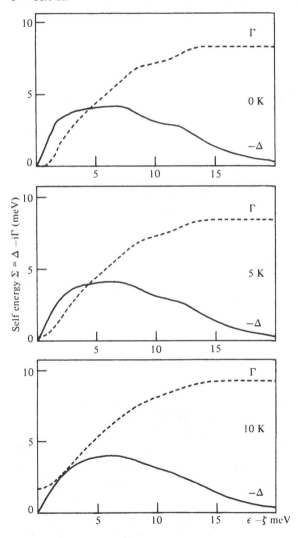

simply by subtracting the complex quantity $\Delta(y) - i\Gamma(y)$ from the energy y measured from the Fermi energy. This complex quantity is called the *self energy* and is denoted by $\Sigma(y)$. It should be noticed that $\tilde{\mathscr{D}}'(\varepsilon)$ depends on ζ as well as ε (since $y = \varepsilon - \zeta$) and we shall sometimes emphasise this by the notation $\tilde{\mathscr{D}}'(\varepsilon, \zeta)$.

We shall now discuss how this additional self energy modifies the oscillatory thermodynamic potential derived for non-interacting particles. At first sight the most direct procedure would seem to be to replace $\mathscr{D}(\varepsilon)$ in (2.168) by the renormalized $\tilde{\mathscr{D}}'(\varepsilon)$ of (2.181), but this is not quite correct* because (2.168) is no longer valid if (as for $\tilde{\mathscr{D}}'(\varepsilon, \zeta)$) the density of states depends on ζ as well as ε. The formula for $\tilde{N}$ is, however, still valid, i.e.

$$\tilde{N} = \int_{-\infty}^{\infty} \tilde{\mathscr{D}}'(\varepsilon, \zeta) f(\varepsilon - \zeta) \, d\varepsilon \tag{2.182}$$

where f is the Fermi function, and so is the relation $\tilde{N} = -(\partial\tilde{\Omega}/\partial\zeta)$, so we can obtain $\tilde{\Omega}$ by integrating (2.182) to give

$$\tilde{\Omega} = -\int_{-\infty}^{\zeta} d\mu \int_{-\infty}^{\infty} \tilde{\mathscr{D}}'(\varepsilon, \mu) f(\varepsilon - \mu) \, d\varepsilon \tag{2.183}$$

Substituting (2.181) for $\tilde{\mathscr{D}}'(\varepsilon, \zeta)$ and recalling that $d\psi(\mu)/d\mu = 2\pi/\beta H$ we get on inserting the appropriate expression for c_p (i.e. (2.167) with R_D and R_s omitted)

$$\tilde{\Omega} = \mathscr{R} \sum_{p=1}^{\infty} \left(\frac{eH}{2\pi ch}\right)^{3/2} \frac{V}{(A'')^{1/2}} \frac{2i}{\pi p^{3/2}} e^{i(p\psi + \phi)}$$

$$\times \int_{-\infty}^{\infty} \frac{\exp\left\{\dfrac{2\pi i p}{\beta H}[y - \Delta(y) + i\Gamma(y)]\right\} dy}{e^{y/kT} + 1} \tag{2.184}$$

This is the basic result from which the effects of many-body interactions can be deduced.

It is important to note that if the self energy terms in (2.184) are omitted, the integration can be done by parts and leads to the standard convolution discussed in §2.3.7.1 for the effect of temperature in the independent particle model; (2.184) then exactly reproduces the standard LK formula (2.151), apart from the Dingle damping factor, as of course it should. Indeed the LK formula for $\tilde{\Omega}$ is often (e.g. Dingle 1952*a*) derived by starting from the

---

* Elliott, Ellis and Springford (1980) appear to ignore this and nevertheless arrive at the same answer (2.184) as given here. A close examination of their analysis shows, however, that this is a result of a compensating error in dealing with one of the integrations involved.

density of states and using (2.168), which for non-interacting particles, is equivalent to (2.183).

To discuss the significance of (2.184) we need to consider the concrete form of the complex energy Σ. Ideally this should include all the types of interaction together, but in practice each interaction is treated separately and their effects combined. Because different kinds of interaction do rather different things, this is probably a reasonable approximation. Before going into any detail about the various interactions it is useful first to note a few simple properties of the result (2.184). We have already mentioned that if $\Delta = \Gamma = 0$, i.e. if there is no interaction, (2.184) leads to the standard independent particle result. Another simple special case of practical interest is if both Δ and Γ are approximately constant (i.e. independent of y) over the sensitive range of integration*. We then again recover the standard LK formula, but modified by the extra factors $e^{-2\pi p \Gamma/\beta H}$ and $e^{2\pi i p \Delta/\beta H}$. The first of these is just the Dingle factor R_D which can be derived in this way with $\Gamma = \hbar/2\tau$, if it has not already been included in the density of states (as in (2.167)) and the second is equivalent to a small shift of the Fermi energy (from ζ to $\zeta - \Delta$) and a corresponding small change of the dHvA frequency F. Such a shift of F was considered by Brailsford (1966) in his derivation of the Dingle factor due to impurity scattering. It is important to note that the β in the Dingle factor is determined by the mass m prior to introducing any interactions. This is often called the 'bare' mass and denoted by m_b, though in this section and in Appendix 6 we shall continue to denote it by m.

The first real novelty occurs if $\Delta(y)$ varies linearly with y, i.e.

$$\Delta(y) = -\lambda y \tag{2.185}$$

and $\Gamma(y)$ can be ignored. This has the effect of changing the β inside the integral of (2.184) to $\beta/(1 + \lambda)$ but otherwise the development of the analysis is exactly as for independent particles. The integral of (2.184) reduces to $\pi k T/i \sinh(2\pi^2 p k T(1 + \lambda)/\beta H)$ and the final result is that the temperature reduction factor R_T for the pth harmonic is

$$R_T = (2\pi^2 p k T/\beta^* H)/\sinh(2\pi^2 p k T/\beta^* H) \tag{2.186}$$

where

$$\beta^* = \beta/(1 + \lambda) = e\hbar/m^*c \quad \text{with} \quad m^* = (1 + \lambda)m \tag{2.187}$$

The mass m^* is usually described as the 'renormalized' or 'enhanced' mass and we see that the only consequence of having a linearly varying $\Delta(y)$ is

---

* A non vanishing $\Delta(0)$ is really only possible for a non-spherical FS and it should be emphasized that methods of calculating Δ and Γ for anisotropic metals have not yet been developed.

that the renormalized mass must be substituted for the 'bare' mass. Elsewhere in the book (apart from this section and Appendix 6) we shall drop the stars and use m to mean the enhanced mass and β to mean the reduced moment.

We shall now consider separately the effects of electron–electron (EE) and electron–phonon (EP) interactions.

2.6.1.1. Electron–electron interaction (Luttinger, 1960*b*)

In terms of the self energy concept the main effects are

(a) $\Delta(0)$ does not in general vanish and may vary with the direction of k. This implies that the FS extremal areas are changed from what they would have been without interaction. Another way of seeing this is to think of the non-interacting FS as being defined by the k vectors such that $\varepsilon(k) = \zeta$; the effect of the interaction is to change this to

$$\varepsilon(k) - \Delta(0) = \zeta \tag{2.188}$$

which means that the FS is modified if $\Delta(0)$ varies with the direction of k. Luttinger (1960*b*) shows, however, that $\Delta(0)$ must be a function of k of such a form that the volume of the FS is unchanged. Evidently it is only for non-spherical symmetry that it is possible to modify the surface without modifying the volume.

(b) The variation of $\Delta(y) - \Delta(0)$ is sufficiently linear and $\Gamma(y)$ is sufficiently small, for the argument leading to (2.186) to apply, so that in general the mass is modified. The appropriate electron–electron value of λ in (2.185) for simple metals and, consequently, the mass enhancement, is usually small. Typically λ is of order a few times 10^{-2}.* As we shall see below, λ for EP interaction can be much larger – typically of order 1.

(c) $\Delta(y)$ can be strongly spin-dependent and this has the effect of modifying the spin-splitting, i.e. the g-factor of the non-interacting states. The modification can be quite considerable as we shall see in chapter 9 – for instance from $g = 2$ to $g = 2.8$ for potassium. Another way of looking at this g-factor enhancement is to think of the spin susceptibility being enhanced by an interaction field, as in the molecular field theory of ferromagnetism. If this field, essentially arising from an electron exchange interaction, can be described as αM, then the spin susceptibility is modified from a non-interacting value χ to $\chi/(1 - \alpha\chi)$.

In practice the effect of electron–electron interactions is estimated without explicit calculation of the self energy but rather by way of Fermi

---

* However, for metals such as Pd and Pt, where spin fluctuations are important, λ for EE interaction can be a good deal larger and sometimes even larger than for EP interaction.

liquid theory combined with basic band structure theory. We shall refer to some of these estimates in discussing the experimental results on spin-splitting in chapter 9.

2.6.1.2. Electron–phonon interaction (Fowler and Prange (1965), Engelsberg and Simpson (1970))

The self energy associated with EP interaction is independent of spin and moreover $\Delta(0) = 0$, so in contrast to the EE interaction it does not affect the spin reduction factor R_s or the Fermi surface itself,* but only the amplitude of the dHvA oscillations. The reduction in amplitude, which comes essentially from the integral in (2.184) is determined by the forms of $\Delta(y)$ and $\Gamma(y)$ for EP interaction, which can be expressed in terms of the phonon density of states $F(v)$, the electron–phonon coupling interaction $\alpha(v)$ and the Fermi and Bose distributions.

These forms are set out in Appendix 6 and here we shall give only a qualitative discussion partly based on their limiting properties and partly on the results of the more general calculation in the Appendix. Graphs of $\Delta(y)$ and $\Gamma(y)$ as calculated for the specific form of $\alpha^2(v) F(v)$ for Hg are shown in Fig. 2.16. It should be noted that $\Delta(-y) = -\Delta(y)$ but $\Gamma(-y) = \Gamma(y)$.

The form of (2.184) shows that only values of y over a range of a few times βH around the origin will contribute appreciably to the integral. At sufficiently low T, $-\Delta(y)$ is linear in y with a slope λ, independent of T, given by (A6.11) i.e.

$$\lambda = 2 \int_0^\infty \frac{\alpha^2(v)F(v)\,\mathrm{d}v}{v} \tag{2.189}$$

while $\Gamma(y)$ is negligible over the relevant range of y. Thus for sufficiently low H (for the conditions of Hg, say below $10\,\mathrm{kG}$, for which $\beta H \sim 2\,\mathrm{meV}$ at the relevant orientation) we have exactly the situation envisaged earlier i.e. a simple mass enhancement by a factor $(1 + \lambda)$, with a temperature reduction factor given by (2.186).

Another simple limiting situation (J. J. Hopfield, private communication) is at extremely high T $(kT > \text{phonon frequencies})$, though this is a somewhat unrealistic limit since for such high temperatures the dHvA amplitude would be quite unobservable. As shown in Appendix 6 (see

---

* This statement however ignores higher order effects in the electron–phonon coupling. Thus the Fermi surface may be slightly modified even at $T = 0$ and may change slightly with temperature (see §5.5). Also there may be slight effects on the spin-splitting through modification of spin-orbit coupling.

Theory

(A6.13)), $\Gamma(y)$ is then independent of y and given by

$$\Gamma(y) = \pi k T \lambda \qquad (2.190)$$

while $\Delta(y)$ is negligible. It is as if the mass were no longer enhanced but there was a considerable scattering by phonons. However it is easily seen that the form of the amplitude is governed by exactly the same formula as at low H and low T. The $\Gamma(y)$ factor in (2.184) can be taken outside the integral giving $\exp(-2\pi^2 pkT\lambda/\beta H)$ while for $\Delta = 0$ the remaining term in the integral is exactly as for independent particles of bare mass. Thus the integral reduces to

$$\frac{\pi k T}{i} \exp\left(\frac{-2\pi^2 pkT\lambda}{\beta H}\right) \bigg/ \sinh(2\pi^2 pkT/\beta H)$$

$$\simeq \frac{2\pi k T}{i} \exp\left[\frac{-2\pi^2 pkT(1 + \lambda)}{\beta H}\right]$$

since $kT \gg \beta H$. At $T = 0$ the integral in (2.184) reduces to $\beta H/2\pi i p(1 + \lambda)$ (limiting form of $\pi k T/i \sinh(2\pi^2 pkT(1 + \lambda)/\beta H)$), so the reduction factor R_T (given by the ratio of high T to $T = 0$ values) is

$$R_T = [4\pi^2 pkT(1 + \lambda)/\beta H] \exp[-2\pi^2 pkT(1 + \lambda)/\beta H] \qquad (2.191)$$

which is exactly the form (2.186) assumes for $2\pi^2 pkT/\beta^* H \gg 1$.

Although the low T, low H and the high T reduction factors have identical forms, the behaviour for intermediate conditions cannot be so simply derived from the look of the formula. At first sight it would seem that as T is raised, the scattering of electrons by phonons would cause an appreciable reduction of amplitude in addition to the reduction factor R_T of (2.186). Thus without the many-body theory one might have expected a T dependent Dingle reduction factor of the form $\exp(-2\pi p\Gamma(0)/\beta H)$, with $\Gamma(0)$ varying roughly as T^3. For Hg this should be quite a large effect but, as discussed in chapter 8, experiments by Palin (1972) proved entirely negative (Palin was unaware of the many-body theory when he started his experiments). If one examines the form of (2.184) it is not at all obvious why this should be so. Indeed since the initial slope of the graph of $-\Delta(y)$ against y actually increases* by something like 10% between $T = 0$ and 5 K, before decreasing at higher T, and $\Gamma(y)$ increases with y, it would seem that the reduction might be even stronger than with a simple Dingle factor based

---

* Direct evidence for an increase in the mass enhancement at intermediate temperatures was found in cyclotron resonance experiments in Pb by Goy and Castaing (1973), though they were unable to show the predicted subsequent decrease because the damping became too severe at higher T.

on $\Gamma(0)$. It proves, however, that the integration in (2.184) is more subtle that it looks and our intuitive reasoning, which does give the right answers for the two limits of low T, low H and high T, proves quite wrong for intermediate conditions.* The correct analysis was first made by Fowler and Prange (1965) and worked out in more detail by Engelsberg and Simpson (1970), who knew of Palin's preliminary results and were able to deal quantitatively with the conditions of his experiments. The essentials of the analysis are given in Appendix 6.

Although unfortunately it does not seem possible to see intuitively how it comes about, it proves that, except in extreme conditions of high H and low T, the reduction factor retains almost exactly the same form as in (2.186), i.e. as if the usual LK formula applied, but with the mass enhanced by a constant factor $(1 + \lambda)$. The detailed calculation also shows that for sufficiently high fields (the necessary H increases with T) slight but appreciable deviations should occur and, as discussed in chapter 8, Elliott *et al.* (1980) have indeed found some evidence for deviations of this kind in experiments on Hg.

2.6.2. Summary of §2.6

From this brief survey it will be seen that most of the effects of many body interactions are unspectacular in as far as they do not alter the analytic form of the LK formula but merely change the parameters that enter into the formula. Specifically, EE interaction may change the FS itself, the cyclotron mass (usually only slightly) and the g-factor, while the EP interaction changes only the mass. The difficulty about checking such predictions is that there is usually no direct way of measuring the 'bare' parameters (except perhaps averaged values of mass from optical data), so that usually the only check is a comparison between the measured 'renormalized' parameters and the predictions of a combination of band structure and many-body theory. Since these theories nearly always involve many simplifying assumptions, such checks are at best only semi-quantitative.

The most interesting qualitative consequence of many-body effects is for the EP interaction – the result that even for strong scattering by phonons there should be no increase in Dingle temperature. This rather exact 'compensation' of the extra scattering by a reduction of the renormalized mass towards the bare mass is not intuitively obvious, and Palin's negative

---

* The fallacy of the intuitive reasoning is probably that at higher T the range of y which contributes appreciably to the integral is controlled more by kT than by βH and so brings in values of y large enough for $-\Delta(y)$ to fall below λy.

result in looking for an increase of Dingle temperature as T rises provides a striking confirmation of the theory. Experiments such as those of Elliott *et al.*, to detect the predicted slight deviations from the LK formula at high fields are also of particular interest in providing evidence as to how reliable are the simplifications implicit in the present form of the many-body theory.

3

Observation of the de Haas–van Alphen effect

3.1. Introduction

In principle every physical property of a metal should show oscillations as the magnetic field is varied and many such effects have in fact been experimentally detected. In this chapter we shall discuss in some detail the various experimental methods for studying the de Haas–van Alphen effect proper, i.e. oscillations of the magnetization and its field derivatives. An account of the oscillations of other physical properties and what can be learned from them will be given in chapter 4.

It is important to remember that any assessment of experimental methods depends very much on the state of technology at the time; thus new developments in the techniques of achieving low temperatures and high magnetic fields and of electronics and data processing may well change the picture as drastically in the future as they have done in the past. In our account of experimental methods we shall concentrate on the present state of the art and pass fairly rapidly over methods which have now been superseded.

3.2. Orders of magnitude of de Haas–van Alphen amplitudes

In order to appreciate the experimental possibilities of observing the oscillations of the magnetic properties it is useful to have an idea of the orders of magnitude involved. Magnetic properties are commonly thought of in terms of susceptibility defined either as M/H or dM/dH, and the LK formula (2.152) for the fundamental ($p = 1$) gives for the oscillatory parts of these susceptibilities expressed in numerical form*

$$\frac{M}{H} = -2.602 \times 10^{-6} \left(\frac{2\pi}{A''}\right)^{1/2} \frac{GFT}{H^{3/2}} \frac{\exp(-\alpha x/H)}{\sinh(\alpha T/H)} \sin\left(\frac{2\pi F}{H} + \phi\right) \quad (3.1)$$

---

* Although we shall usually be referring only to the *parallel* component of the *oscillatory* component of M, we shall for simplicity drop the ∥ suffix and the wavy superscript except where it is necessary to distinguish between $M_\parallel$ and $M_\perp$ or between oscillatory and steady parts. For simplicity we omit the summation over different frequencies F, but this summation must be borne in mind. The formulae assume CGS units, and M is here defined as per unit volume.

83

$$\frac{\mathrm{d}M}{\mathrm{d}H} = 1.635 \times 10^{-5} \left(\frac{2\pi}{A''}\right)^{1/2} \frac{GF^2 T}{H^{5/2}} \frac{\exp(-\alpha x/H)}{\sinh(\alpha T/H)} \cos\left(\frac{2\pi F}{H} + \phi\right) \qquad (3.2)$$

where

$$G = \cos\left(\frac{\pi}{2} g \frac{m}{m_0}\right), \qquad \alpha = 1.47(m/m_0) \times 10^5 \, GK^{-1} \qquad (3.3)$$

and ϕ is the phase constant of (2.152) which will not concern us here. As usual, in the differentiation of M we have retained only the contribution from the periodic term, so that $|\mathrm{d}M/\mathrm{d}H|$ is simply $2\pi F/H$ times $|M/H|$ (i.e. $2\pi n$ times, if n is the quantum number of the highest occupied Landau level).

Some numerical results based on these formulae are presented and discussed in Appendix 7 on the basis of (a) a typical metal with a large and free electron-like Fermi surface and (b) bismuth, which provides an example of the opposite extreme of a small and highly anisotropic Fermi surface. A rough guide is also given to the values of the 'anisotropy factor' $(1/F)(\mathrm{d}F/\mathrm{d}\theta)$ of (2.114) which determines the ratio of $M_\perp$ to $M_\parallel$ and the appendix also includes a brief discussion of the orders of magnitude for a '2-D metal' (see §2.3.4). The following general conclusions of relevance to experimental methods emerge:

(1) The oscillations are essentially a *low temperature* phenomenon, in the sense that liquid helium temperatures or lower are usually essential. Even for the extreme example of bismuth, the amplitude falls off drastically above 20 K (liquid hydrogen).

(2) For large FS the oscillations are also a *high field* phenomenon, in the sense that at 1 K the amplitude falls off drastically below about 20 kG. Even if dilution refrigerator temperatures of say 0.1 K were used, the observation of the oscillations at low fields would be seriously hampered by resolution difficulties (the period of an oscillation at 10 kG is only of order 0.2 G). For the extreme case of bismuth, however, because of the much smaller Fermi surface (and much smaller cyclotron mass), the oscillations have appreciable amplitude to fields as low as 200 G at 1 K.

(3) Typical amplitudes of M/H in optimum conditions of H and T are of order 10^{-5} for metals with both large and small FS. This is comparable to, or if anything larger than, typical steady susceptibilities $(10^{-6}–10^{-5})$ of feebly magnetic solids. Thus traditional methods of measuring steady susceptibilities should be amply sensitive for the oscillatory susceptibilities. However, as will be discussed in more detail below, the best known traditional method, due originally to Faraday, suffers from the disadvantage that it inevitably damps down the oscillations because of the phase smearing due to the inhomogeneous field involved.

3.2. Orders of magnitude of de Haas–van Alphen amplitudes

(4) The amplitudes of dM/dH are much larger* than those of M/H (the ratio is 2π times the quantum number of the highest occupied Landau level) and inductive methods such as are appropriate for relatively strong magnetism (e.g. superconductors and ferromagnetics) prove to be feasible for the study of the oscillations. During the last decade one such method, in which the field is periodically modulated, has developed into the most sensitive, precise and versatile of all.

(5) Small sheets of the FS of polyvalent metals have complicated shapes, which leads to strong anisotropy, with $(1/F)(dF/d\theta)$ comparable to 1. This means that $M_\perp$ may be quite comparable to $M_\parallel$. One method of measuring $M_\perp$ which has been used a great deal in the study of Fermi surfaces, is to measure the *torque* on the crystal (which is $HM_\perp$ per unit volume), and in some circumstances this method may, even today, rival the field modulation method.

(6) The oscillations of magnetic moment per unit area $\mathscr{M}$ of 2-D silicon inversion layer samples should be very feeble and are unlikely to be observable using standard techniques, though it is quite possible that a technique especially suited to this problem (see §3.4.2.3) may improve the situation appreciably.

One final general remark needs to be made. This is that the dHvA oscillations may be studied at a variety of different levels and the experimental technique will vary accordingly. Thus at the crudest level it may be merely a question of detecting the oscillations and roughly estimating the period, in which case all that matters is to maximise the signal to noise ratio. If, however, precision frequency measurements are needed – as in determination of a FS – it is essential to be able to follow the oscillations over a large range of fields and to have a sufficiently accurate field measurement technique. Again, for optimum sensitivity in the field modulation technique, the modulation should be of amplitude comparable with a period of the oscillation (as will be discussed in detail later), but such high modulation amplitude introduces complications if the object is to study the line shape of the oscillations. Similarly, more care in the choice of modulation frequency and amplitude is required if the aim is accurate measurement of the absolute rather than the relative amplitude of the dHvA oscillations. Thus in comparing the merits of the various methods it will be important to bear in mind which particular feature of the oscillations it is desired to study.

Common to all the methods of measurement to be discussed below is the problem of sample preparation. We shall not describe the metallurgical techniques for producing high purity strain-free single crystals beyond mentioning that usually it pays to cut the sample to size and shape either by

---

* However as we shall see later, if $4\pi|dM/dH|$ becomes comparable to unity, magnetic interaction becomes important. The line shape of the oscillations is then modified and becomes dependent on sample shape, and the amplitude of the fundamental of dM/dH is limited to roughly $1/4\pi$.

acid-saw or spark erosion and that great care is essential in handling and mounting the sample. Although at first sight it would seem to be advantageous to make the sample as large as possible, there are in fact several limiting factors. First the sample must be confined to a sufficiently homogeneous region of the magnet (i.e. such that the field spread over the sample is much less than H^2/F). Secondly in the field modulation method (see §3.4.2.1) little is gained by increasing the linear dimensions beyond the skin depth corresponding to the modulation frequency. Finally there is a general tendency in most methods for the 'noise' level to go up and the sensitivity (i.e. per unit volume) to go down, as the sample size increases. Which of these considerations dominates depends on the particular technique, but usually it is not advantageous to use samples of more than a few mm in linear dimensions. Likewise it is not usually practicable to use samples of dimensions much smaller than a few tenths of a mm because of handling difficulties.

How to classify the various methods in a logical way is to some extent a matter of taste, but a convenient division into two broad categories is according as the method does or does not require the magnetization to vary with time in order to produce a signal. We shall refer to these as *dynamic* and *static* methods respectively

3.3. Static methods

3.3.1. Faraday–Curie method

The classical method of measuring a weak magnetic moment is that of Faraday and Curie, in which the mechanical force on a sample in an inhomogeneous field is measured. As already pointed out, the drawback of this method is that the inhomogeneity of the field causes phase smearing and kills the oscillations if the quantum number n of the highest occupied Landau level (i.e. F/H) is too high. On the basis of plausible assumptions, the highest value of n which can be observed is given by

$$n \sim [(|M|/H)L^2 F^2/f_0]^{1/3} \tag{3.4}$$

where the sample has linear dimensions of order L and volume of order L^3 and f_0 is the smallest force which can be detected. If we take $L \sim 0.2$ cm and $f_0 \sim 1$ dyne (these are to some extent linked, in as far as f_0 increases with L) and $|M|/H \sim 10^{-5}$ (though in reality of course $|M|/H$ falls off as n increases), we find that

$$n \sim 5000 \quad \text{for} \quad F \sim 5 \times 10^8 \, \text{G (large FS)}$$

$$n \sim 5 \quad \text{for} \quad F \sim 1.4 \times 10^4 \text{G (small FS)}$$

3.3. Static methods

This means that the oscillations for a large FS would become observable only at fields above about 10^5 G, while for a small FS only a few oscillations below the quantum limit would survive the phase smearing. It should be noted that in this method the experimental signal (i.e. the force) varies roughly as MH^2 and so falls off particularly rapidly with H, quite apart from phase smearing.

In fact, as already mentioned, the force method was the one used by de Haas and van Alphen (1930a, b, 1932) in their discovery of the effect in bismuth (see fig. 1.1) and subsequently by Shoenberg and Uddin (1936), again in the study of bismuth and its dilute alloys, and by Marcus (1947) in his discovery of the effect in the 'needle' of zinc ($F \sim 1.5 \times 10^4$G). In these various experiments, only a few oscillations were indeed observed, in reasonable agreement with the rough estimate (3.4). Because of the intrinsic inhomogeneity limitation, the complexity of the techniques required to measure small forces on a sample at low temperatures, and the availability of more convenient methods, it is unlikely that the force method will find much application in the future for studying the oscillations.

3.3.2. Torque method

As already mentioned, if the FS is anisotropic (in the sense that its extremal area normal to H varies with the direction of H), there will be a torque T on the crystal, given $M_\perp HV$, where $M_\perp$ is the component of M perpendicular to H. The component of this torque about any particular axis perpendicular to H is

$$T = \frac{-1}{F}\frac{dF}{d\theta} M_\parallel HV \tag{3.5}$$

where $M_\parallel$ is the parallel component of M (as given by (3.1)), θ is an angle specifying the direction of H in the plane normal to the chosen axis and V is the volume of the sample.

If we take H as 3×10^4 G, $(1/F)(dF/d\theta)$ as 10^{-1}, (see table A7.2), $|M_\parallel|/H$ as 10^{-5} (see table A7.1) and V as 10^{-2} cm^3, we find

$$|T| \sim 10 \text{ dyne cm}$$

as a typical value for the amplitude of the oscillatory torque, though of course it may be much smaller in less favourable conditions. It turns out that even with quite simple apparatus a torque of this magnitude is easily measurable with a precision of better than 1%, while with more sophisticated equipment the performance can be considerably improved.

As compared with the Faraday method, the torque method has the merit that no field inhomogeneity is required and that the equipment involved in measuring a torque is generally rather simpler and more robust than that for measuring a force. On the other hand the method has 'blind spots' whenever the field lies along crystal symmetry directions. For such orientations the torque vanishes, either because $dF/d\theta$ vanishes, or because there are exactly cancelling contributions from symmetry related sheets of the Fermi surface. In practice this is not really a very serious snag, since usually the torque oscillations can still be followed to within a very small angle of a symmetry direction and the blind spot can be bridged by only a slight interpolation. Another disadvantage of the torque method as compared with some of the others discussed below is that because of the factor H in (3.5), the torque amplitude falls off more rapidly as H decreases than the amplitude of M itself (though in this respect it is better than the force method, which, as we have seen, has effectively a factor H^2 in the observable amplitude).

The apparatus used in the first application of the torque method to dHvA studies (Shoenberg 1939) is illustrated in fig. 3.1 and it can be seen that it was indeed simple. A torque acting on the sample s twists the mirror m slightly against the torsional restraint of the short beryllium-bronze wire

Fig. 3.1. Apparatus to measure torque (Shoenberg 1939); the labels are explained in the text.

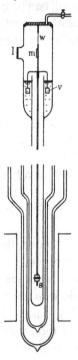

w and so displaces the reflected image of an illuminated slit formed by the lens l on a distant scale; damping of vibrations is provided by the vanes v dipping in an oil bath. The position of the image on the scale could be located with a precision of order 0.02 mm by means of a travelling microscope, and since a torque of 10 dyne cm produced a deflection of order 6 mm (corresponding to a twist of 3×10^{-3} radians) quite reasonable precision was obtained (see fig. 1.3 for some typical results).

One complicating feature of torque measurement is that under the action of a torque the sample turns slightly and thus the measurements are not made at exactly constant orientation. If, as is generally the case for the dHvA effect, the torque is itself a function of orientation, the 'line-shape' of the oscillations is modified from that of oscillations at constant orientation. The greater the twist per unit torque of the suspension, the more serious is the modification of the oscillations and for a weak enough suspension the system becomes unstable (Shoenberg 1952a, Vanderkooy and Datars 1968). As will be shown in chapter 6, the parameter p which determines the seriousness of this 'torque interaction effect', as it is sometimes called, is:

$$p = 2\pi \frac{\gamma}{c} \frac{F}{H} \left(\frac{1}{F} \frac{dF}{d\theta} \right) \qquad (3.6)$$

where c is the torque per unit angle of twist of the suspension and γ is the amplitude of the torque oscillation. If $p \ll 1$, torque interaction is negligible, but if $p \geqslant 1$, instability sets in. In the simple equipment used in the early experiments c was of order 3×10^3 dyne cm per radian, so that for $\gamma \sim 10$ dyne cm, $(1/F)(dF/d\theta) \sim 1$, and $F/H \sim 5$ (as for Bi) we find $p \sim 10^{-1}$ implying that torque interaction was appreciable, though not serious. However for other metals, for which F/H could be much higher, torque interaction sometimes becomes a serious nuisance. As can be seen from (3.6) it is always possible in principle to reduce torque interaction by increasing c, i.e. by stiffening the suspension. Provided there is no instability, the effect can then be completely eliminated by rotating the magnetic field (or the whole suspension) slightly to keep the sample in a constant orientation relative to the field. The loss of sensitivity caused by increasing c can to some extent be made up by increasing the magnification of the angular response, but with the simple system of the original apparatus, the possibilities (such as observing the reflected image with a microscope) are evidently limited.

As the torque method began to be more widely used, more sophisticated techniques of measurement were introduced and these have largely removed the difficulty of torque interaction, as well as greatly reducing the labour of taking data by making possible continuous recording of the output signal which measures the torque. The improvements have been of two kinds: first the use of different methods of sensing the angle of twist of the suspension and secondly the use of electronic feedback to increase the *effective* torque constant of the suspension.

Usually the novel methods of angle sensing have been combined with feedback, but one ingenious method of angle sensing developed by Griessen (1973) is of such sensitivity that c can be made large enough without the use of feedback and it is convenient to outline this method before discussing the principles of feedback systems. In Griessen's method the sample is rigidly attached to one plate of a capacitor which is held in place by stiff springs. A torque acting on the sample has the effect of very slightly displacing the plate and so changing the capacity of the capacitor. This change of capacity can be measured very sensitively using the same kind of bridge techniques as have been developed for measuring small capacity changes brought about by minute changes of length, as in magnetostriction and thermal expansion. Griessen achieved a torque constant of order 4×10^6 dyne cm per radian, which is more than 10^3 times that achieved by the simple equipment in which the magnification of the angular twist was provided by optical lever and microscope. This improvement is amply sufficient to make torque interaction negligible and the capacitor method is particularly convenient if it is desired to measure magnetostriction simultaneously with torque (see §4.3.3). However, the precision measurement of capacity is not the easiest of techniques and some of the feedback techniques which we shall now describe are rather simpler and more versatile in practice.

The essential features of any feedback scheme are illustrated in fig. 3.2. The sample s is rigidly attached to an angle sensing device A and to a device B which develops a torque to balance the torque acting on s. The output

Fig. 3.2. Schematic feedback system to measure torque with negligible sample twist; the labels are explained in the text.

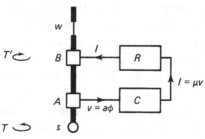

of A is a voltage v proportional to the small angle ϕ through which the sample twists, so that

$$v = a\phi \tag{3.7}$$

and this voltage is fed into a suitable electronic network C which converts it into a current I (d.c.) which is fed into B. We shall suppose that I is proportional to v, i.e.

$$I = \mu v \tag{3.8}$$

and that the balancing torque T' developed by B is proportional to I so that

$$T' = bI \tag{3.9}$$

The combination of s, A and B is held by a very weak suspension w with a very small torque constant c (which ideally could be zero). The balance condition when a torque T acts on s is thus

$$T = T' + c\phi = bI + c\phi = (\mu ab + c)\phi \tag{3.10}$$

and we see that if $c \ll \mu ab$, the *effective* torque constant of the suspension is μab, which can in principle be made as large as desired and, in particular, large enough to make torque interaction negligible.

From (3.10) we have

$$I = T/b(1 + c/\mu ab) \simeq T/b \tag{3.11}$$

and we see that a direct measure of the torque is provided by the current I, which can be continuously recorded on a chart recorder R to show how T varies as the field or sample orientation is varied.

In the first realization of this kind of scheme (Croft, Donahoe and Love 1955), the device A was a split photocell which responded to motion of a spot of light reflected from a mirror attached to the suspension, while B was simply a moving coil galvanometer, i.e. a coil suspended in the field of a small permanent magnet. In a later scheme successfully used by Condon and Marcus (1964), a single coil performed the function of both A and B. This coil acted as an angle sensing device through its coupling with a fixed coil carrying a.c. (the frequency was 200 kHz) and it simultaneously acted as a galvanometer coil in the field of a permanent magnet, so that a torque was exerted on it when the current I flowed through it. The angle sensing method in this scheme is simpler and less subject to complications than the earlier photoelectric method.

An ingenious further development by Vanderkooy (1969) again combined the angle sensing and torque balancing functions in a single coil, but used a novel method of producing the balancing torque. This was to use the main magnet field H itself, rather than the field of a galvanometer permanent magnet, to produce the balancing torque. A photograph of the device is shown in fig. 3.3 and one important merit of the device is immediately apparent. This is that since there is no longer any need for a long rigid connection reaching outside the high field and low temperature environment of the sample, the whole device can be made much more compact. Moreover because of this compactness the axis of the suspension is no longer restricted to be vertical but can be set in any direction, and in particular it can be horizontal (as in the illustration). This is a merit of Griessen's device also.

The central feature of the device is the coil A which is suspended by a taut silk fibre with almost negligible torque constant and in which the sample is rigidly fixed. The large coils CC on either side of this coil are fed with a.c. (the frequency was 10 kHz) and induce an alternating e.m.f. in A if it turns from an orientation normal to that of CC. Any such induced e.m.f. v is amplified by a lock-in amplifier to give a current I (d.c.) which is fed back into A; this in turn is acted on by the main magnet field H to balance the torque T which it is desired to measure.

An additional refinement of the device is that the coil A can, if desired, be set with its axis at *arbitrary* angle θ to the axis of CC, rather than only at 90°. This is achieved by feeding into A a suitable alternating e.m.f. x of the same

Fig. 3.3. Vanderkooy's (1969) feedback torque measuring device.

3.3. Static methods

frequency and phase as the induced voltage v so that the total voltage input to the amplifier is $(v + x)$ rather than v. Evidently A will then set itself at an angle θ such that $(v + x)$ vanishes (if the torsion of the suspension can be neglected.) With this refinement the simple theory of (3.7)–(3.11) needs some slight modifications in view of the additional e.m.f. x and the possibility of angles θ which are no longer small.

If for simplicity we neglect the torsional constant c of the silk fibre, it is easy to show that (3.10) and (3.11) are modified to

$$T = \phi\mu ab\cos^2\theta \quad \text{and} \quad I = T/b\cos\theta \tag{3.12}$$

where θ is the angle to which the coil is set by the voltage x, in the absence of the torque T, i.e.

$$\sin\theta = -x/a \tag{3.13}$$

(it is assumed throughout that the extra twist ϕ is small compared with θ). Yet another advantage of the Vanderkooy device is that since b is proportional to the main field H, the current I effectively measures $M_\perp$ itself (i.e. T/H) rather than $M_\perp H$. Thus the general drawback mentioned earlier that the torque falls off more rapidly with H than does M, is avoided.

Originally the torque method had seemed to be essentially a 'low-field' method since a vertical axis of suspension requires a horizontal magnetic field such as the field in the gap of an iron electromagnet, and this cannot usually exceed 3×10^4 G or so. With the development of superconducting magnets, rather higher fields, say up to 5×10^4 G, became possible with the suspension inserted in the gap between two superconducting coils, but this was still a good deal lower than the 10^5 G or more available at the centre of a continuous superconducting coil, or the even higher fields available in the pulsed field technique. Thus, since the Vanderkooy device with a horizontal suspension can be inserted into the centre of a vertical superconducting magnet, it extends the application of the torque method into the 'high-field' region. This is an important consideration, since a limitation to 'low fields' implies that only the dHvA oscillations associated with the relatively small extremal areas of a complicated FS can be observed, rather than the higher frequency oscillations associated with larger extremal areas of major parts of a FS, which require higher fields for their observation.* Some dHvA oscillations obtained with the Vanderkooy device are shown in fig. 3.4.

The actual performance of any of the methods we have outlined above depends very much on how it has been set up in practice (e.g. the care taken

---

* However the distinction is not hard and fast. Thus Joseph and Thorsen (1965) could with a sensitive torque meter study the high frequency belly oscillations of the noble metals in the top fields (~ 40 kG) of a large iron electromagnet.

to eliminate vibrational and electronic noise and drift) and probably the original published performance data could today be improved by exploiting advances in electronic technology. Thus it is hardly worth while attempting any detailed comparison of the various methods – and it would indeed be difficult since often information about design details and performance is only partially available. However two general orders of magnitude may be quoted. First, effective torque constants greater than 10^6 dyne cm/radian are easily achieved by the feedback methods as well as by Griessen's capacitor method, so that torque interaction can be made completely negligible. Secondly the 'noise level' of the output signal can easily be made low enough to correspond to a torque of 10^{-2} dyne cm for fairly large samples and with care as low as 10^{-4} dyne cm for smaller samples. If we suppose that the minimum oscillation amplitude which can be detected reliably is three times the noise, we might estimate that $|T|_{min} \sim 3 \times 10^{-3}$ dyne cm for a volume of say 0.3 cm^3. If the field is 3×10^4 G, this corresponds to

$$(|M|/H)_{min} \sim 10^{-11} \qquad (3.14)$$

if we assume $M \sim M_\perp$. It must be emphasized that this is an extremely rough estimate and it might well prove that the torque method is capable of even greater sensitivity, but it will be helpful to have this estimate in mind when considering the other types of method to be discussed below.

Fig. 3.4. Torque dHvA oscillations in Au at 1.2 K obtained with the Vanderkooy device (Shoenberg and Vanderkooy 1970). The Au sample had its $\langle 111 \rangle$ axis at 8.5° to H; the main oscillations are from the neck with the much faster belly oscillations just visible (they are damped down because the field is varied too fast in relation to the instrumental relaxation time). The field runs from 74 kG (at left) to 69 kG; the torque amplitude at left is approximately 16 dyne cm (the current $I \sim 2$ mA) corresponding to $|\tilde{M}_\perp| \sim 1.2 \times 10^{-2}$ (sample volume 0.018 cm^3).

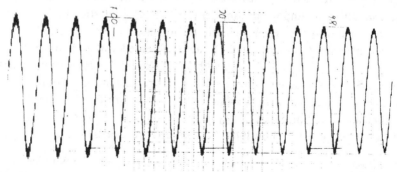

3.3. Static methods

3.3.3. Foner method (1959, 1975)

In this method the sample is vibrated over a small range in a homogeneous magnetic field in such a way that the flux linkage of its magnetic moment with a coil system varies periodically with time. The amplitude of the alternating e.m.f. induced in the coil system then provides a measure of M, and the method has the merit that the signal is directly proportional to M and so does not fall off as rapidly with H as in the force method or in some of the other methods to be described. The method is particularly suitable where the dHvA effect is nearly isotropic, so that the torque method (§3.3.2) fails, or of very low frequency, so that the modulation method (§3.4.2) may be too insensitive. It has been successfully used (Wampler and Springford 1972, Brignall 1974) to study the dHvA effect in some semi-metallic compounds (n-InSb and n-InAs) and also in Bi up to and beyond the quantum limit (Brignall and Shoenberg 1974).

Details of the method may be found in the papers cited above, but a few general points may be mentioned here. The signal to noise ratio of the device is very sensitive to the design of the detecting coil system. Close coupling of the sample to the system is obviously important, to increase the signal, but it is equally important to reduce noise by balancing the coils both against small field changes and against small movement. Clearly the coil system should be mounted as rigidly as possible to avoid any vibrating movement being transmitted from the mechanism which drives the sample, but inevitably some vibration does come through and its effect can be greatly reduced by suitable balancing of the coil system.

Since it is difficult to maintain an exactly constant amplitude of vibration, a convenient trick it is to null the output signal by passing a suitable small direct current through a small coil rigidly attached to the sample; i.e. effectively to produce a compensating magnetic moment. If the geometry of the coil is sufficiently well defined, this current provides an accurate absolute measure of the magnetic moment. Finally, it may be mentioned that in the de Haas–van Alphen experiments on n-InSb etc. carried out so far, the minimum detectable amplitude of M was typically of order 2×10^{-4} G, at fields of order 5×10^4 G, so that $(|M|/H)_{min}$ was of order 4×10^{-9}. Foner (1975) points out that with careful attention to the design of the coil system and the electronics of the detecting system, values of $|M|/H$ as small as 10^{-12} can be measured, but even with such a high sensitivity the method is not as simple as some of the others discussed below and probably its main application is to the special situations mentioned earlier.

3.3.4. Miscellaneous static methods

In the Foner method the modification of the magnetic field close to the sample is revealed by moving the sample relative to a fixed coil system, but there are quite a few other possibilities of detecting the modification of the field (and hence the magnetic moment of the sample). The field close to a magnetized sample departs from the applied field H by an amount of order $4\pi M$ (depending on the exact geometry) and in the dHvA effect this departure oscillates with an amplitude which is typically a few G or less for H of order 3×10^4 G. Thus any method capable of measuring small field changes of this order in the presence of the much larger main field could in principle be developed to show up the de Haas–van Alphen effect. The magnetoresistance of a suitably placed Bi probe has been used by Gold and Schmor (1976) and n.m.r. of a suitable material might provide another possible technique. Recently developed superconducting devices (e.g. 'squids') and older ones, such as the flux gate magnetometer based on the sensitivity of highly permeable ferromagnetics to small magnetic fields, also offer possibilities, though there are difficulties in using such techniques in the presence of a high ambient field. One such method was recently used by de Wilde and Meredith (1976), who overcame the difficulty of the high field environment by using a superconducting flux transformer to transmit the small field changes at the sample to a remote and magnetically shielded location where they were measured by a commercial flux gate detector. The method is, however, rather complicated in detail and not particularly sensitive, so it is unlikely to be competitive with the torque or modulation methods.

3.4. Dynamic methods

3.4.1. Pulsed field technique (Shoenberg 1952b, 1953, 1962)

During the early 1950s superconducting magnets were not yet available and static magnetic fields with spatial homogeneity and temporal steadiness adequate for observing high frequency de Haas–van Alphen oscillations were available only in iron electromagnets and limited to something like 2 or 3×10^4 G. It is true that static fields of up to 10^5 G were available in water-cooled coils in a few special installations, but generator ripple and poor homogeneity would have made their use difficult. It was to overcome these limitations and open up the exploration of higher frequency de Haas–van Alphen oscillations (i.e. of larger extremal areas of Fermi surfaces) that the pulsed field technique was developed. This technique kills two birds with one stone. Not only does it provide much higher magnetic fields, but through the rapid variation of the field with time it provides a novel method of detecting the de Haas–van Alphen oscillations. If the sample is placed in one of a balanced pair of pick-up coils, an e.m.f. v will appear across the coils proportional to dM/dt, i.e.

$$v = c\frac{dM}{dH}\frac{dH}{dt} \tag{3.15}$$

where c is an appropriate coupling constant, and this e.m.f. will oscillate with the dHvA oscillations of dM/dH as H varies with time. In the realization of this idea the pulsed field (typically rising to 1 or 2×10^5 G in a time of order 10 ms) was produced by discharging a large capacitor bank (typically 4000 μF charged to 2 kV) through a coil cooled in liquid air (to reduce its resistance). The e.m.f. generated by the pick-up coil was passed through an amplifier which incorporated a high pass filter (thus conveniently removing the relatively slowly changing e.m.f. due to residual lack of balance of the pick-up coils) and displayed on a CRO which was triggered synchronously with the field pulse. Some typical oscillations are shown in fig. 3.5.

The detailed design of the pulsed field coil involves compromises between a number of conflicting considerations. Thus the time of rise of the field to its peak must not be so short that induced eddy currents make the field appreciably inhomogeneous over the sample cross section (or raise the sample temperature). Lengthening the rise time, however, reduces the peak field which can be achieved. The requirement of adequate field homogeneity over the sample length also conflicts with the aim of getting the highest possible peak field. Evidently the shorter and thinner the sample can be made, the less drastic is the reduction of peak field which is necessitated by these considerations. But here again a compromise has to be struck because of the difficulties of avoiding handling damage to the sample if it is too short and thin. In practice it is difficult to work with samples of length much below a few mm and of diameter much below 0.2 mm, and this sets a limit of something like 2×10^5 G to the peak field and requires a rise time of at least a few ms. Another important consideration in the design of the coil is its mechanical strength against the bursting forces of the high magnetic field. Even to achieve the relatively modest peak field of 2×10^5 G, the coil must be very carefully wound and impregnated with a suitable cement to ensure rigidity. It is indeed mechanical strength which sets the ultimate limit (of order 6×10^5 G) to the peak field which can be achieved if the requirements of homogeneity and slow rise time are completely relaxed.

Although we shall not go into the practical details of the actual working of the method it is worth mentioning briefly some of the tricks which have been used to improve its accuracy and versatility. It was noticed in the early

Observation of the dHvA effect

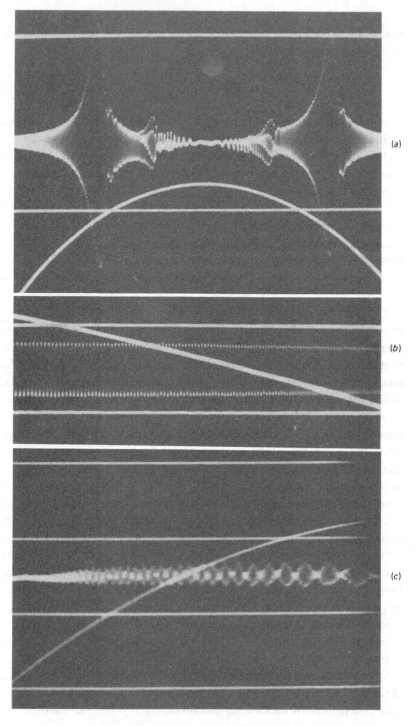

(a)

(b)

(c)

98

3.4. Dynamic methods

Fig. 3.5. Oscillograms by impulsive field method (Shoenberg 1962). (*a*) Cu ⟨111⟩ belly oscillations showing resonant blips of fundamental and harmonic. The curved trace shows the 'variation of *H* and the two horizontal calibration lines indicate 107 and 112 kG; about 3 ms across picture. (*b*) As (*a*) but without resonance and on expanded scale (about 1 ms across picture). The calibration lines are at 107.0 and 109.3 kG. (*c*) Illustrating beat method; the beats are between two Cu samples–one with *H* along ⟨111⟩ and the other close to ⟨100⟩; they differ by about 3% in frequency. Calibration lines at 93, 105, 116 and 128 kG; about 5 ms across picture.

experiments (e.g. as in fig. 3.5*a*) that the envelope of the oscillations usually showed rather sharp 'blips' on both sides of the peak field (at the maximum field the signal amplitude as given by (3.5), of course, vanishes, since $dH/dt = 0$). It was soon realized that this came about because the pick-up circuit had sufficient capacity (mostly in the cable to the amplifier) to have a resonant frequency f_0 (typically of order 10^5 Hz) and the peaks occurred when the frequency f of the oscillations in time just matched f_0 (i.e. once for rising field and once for falling field). Later this feature was deliberately exploited to act as a kind of spectrometer when several de Haas–van Alphen frequencies were present. The resonant frequency f_0 could be adjusted to any convenient value by deliberately adding capacitors to the pick-up coil, and also, when desired, the resonance could be suppressed by inserting a cathode follower between the pick-up coils and the cable.

It is easy to see that

$$f = \frac{F}{H^2} \left| \frac{dH}{dt} \right| \tag{3.16}$$

and if we make the approximation that the field varies parabolically with time in the neighbourhood of the field peak, i.e.

$$H = H_0 [1 - (t/\tau)^2] \tag{3.17}$$

where the origin of *t* is taken at the peak and τ is a scale time, approximately equal to the time from the start of the pulse to the peak, it follows that the time interval Δt between the resonant peaks for rising and falling field is given approximately by

$$\Delta t / \tau = H_0 f_0 \tau / F \tag{3.18}$$

provided $\Delta t / \tau$ is not too large. This means that to this approximation the resonant blips display a linear spectrum of $1/F$. If we put in some typical figures, say $f_0 = 5 \times 10^4$ Hz, $\tau = 5 \times 10^{-3}$ s, $H_0 = 10^5$ G, $F = 5 \times 10^8$ G,

99

we find $\Delta t/\tau = 0.05$ and we see that the spectrometer action would work reasonably well for a range of F down to say 5×10^7 G.

In practice this spectrometer action provides a useful technique for a preliminary sorting out of the dHvA frequencies in a complicated situation such as illustrated in fig. 3.6 (Phillips and Gold 1969). With care it can be used quantitatively for comparison of F values, though a number of corrections have to be taken into account if an accuracy of order 1% in relative values of F is to be achieved (Shoenberg 1962).

In order to make absolute measurements of F, the CRO is triggered to start sweeping at some predetermined time after the start of the field pulse and the sweep is made rapid enough to display as many of the oscillations as can conveniently be resolved (see for instance fig. 3.5b). A twin beam CRO is used and the second beam monitors the e.m.f. across a small resistor in the field coil circuit, i.e. in effect monitors the field. Calibration lines are superimposed on the photograph so that the field at any instant can be obtained by interpolation and the frequency F can then be deduced from the values of the initial and final fields for a given number of oscillations. With care, the absolute frequency can be determined to rather better than 1% in this way, though this would be difficult to achieve if several fairly close frequencies were simultaneously present.

For a metal with strong anisotropy, such that the variation of F with orientation is comparable with F itself, this kind of accuracy in the determination of F may be sufficient for the purpose of getting a fair idea of

Fig. 3.6. Spectroscopic action of impulsive field method (Phillips and Gold 1969). The resonant blips correspond to the various frequencies and harmonics of Pb with H along $\langle 111 \rangle$. Peak field 165 kG; Calibration lines at intervals of 26.7 kG (lowest at $H = 0$).

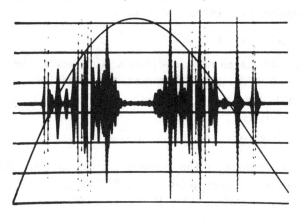

100

the size and shape of the FS. For a more nearly spherical FS, however, for which F varies by only a few % or less, such accuracy is no longer adequate to show up the *departures* from isotropy with any precision. To deal with this situation a special technique was devised in which the dHvA oscillations from a fixed reference sample were superimposed on those from the sample under study, giving rise to beats of the kind shown in fig. 3.5c. Evidently the beat frequency shows up any change of F as the sample orientation is varied more sensitively than direct observations of the oscillations from the sample alone and in fact it proved that changes of F as small as 0.2% could be detected. This precision was just about adequate for determining the significant features of the FS of the noble metals but the FS of the lighter alkali metals are still more nearly spherical and the small departures from isotropy had to wait for the development of the more versatile field modulation method.

Some other developments or variants of the pulsed field method will now be mentioned. The first is the use of high speed analogue to digital conversion of the information contained in the oscillatory e.m.f., with automatic recording in a memory (Panousis and Gold 1969). This makes possible the use of more powerful analytical techniques for determination of the amplitudes and frequencies of the oscillations and it may come into its own in special situations where it is essential to have the highest possible field, such as only the pulse method can provide. The second is one in which as usual the condenser discharge is used for producing the high field, but the detection of the dHvA effect does not make use of the time variation of the main field. Instead, the oscillations of magnetic susceptibility are detected by the effect they have on the high frequency self-inductance of a coil (or the mutual inductance of a pair of coils) round the sample. The high frequency has to be high enough for at least a few cycles to occur in the time of one dHvA oscillation and something like 10^7 Hz proves suitable; the oscillations in the relevant inductance are displayed on the CRO. This variant (which in principle is rather similar to the field modulation technique to be described below) has two advantages:

(1) The oscillations no longer disappear at the peak field.
(2) Because of the high frequency, it is only the surface properties of the specimen which are sampled; thus if the field varies appreciably with depth into the specimen (due to induced eddy currents) this no longer matters so much.

This idea has been successfully tried (Shoenberg 1955) but never systematically developed.

Two variants of the method of producing the time variable field should also be mentioned. In one (Gold and Van Schyndel 1981) a 'triangular ramp', going up to 1 kG is superimposed on the steady field of a superconducting magnet, giving $dH/dt \sim 10^3\,\mathrm{G\,s^{-1}}$ and repeated periodically once a second. The output of the pick-up coil is integrated to give M and the noise is reduced by superimposing many identical periods of the ramp. The other (Arko *et al.* 1978) is based on the high field installation of the University of Amsterdam and follows the signal of the pick-up coil as the field decays (in a time of order 1 s) from 400 kG.

Finally we may note that the sensitivity achieved in typical pulsed field experiments (see Shoenberg 1962) was such that the smallest detectable value of $|dM/dH|$ was of order 10^{-4}. For $F/H = 5 \times 10^3$, this corresponds to a minimum amplitude of $|M|/H$ of order 3×10^{-9}. No doubt this could be improved by more careful attention to elimination of noise but, as we shall see below, the rather simpler field modulation method can be made a good deal more sensitive.

3.4.2. Modulation methods

The development of superconducting magnets for the first time made available extremely steady and homogeneous high magnetic fields (initially only up to 5×10^4 G or so, but later up to and beyond 10^5 G) which could be varied slowly and smoothly at will. This opened up the possibility of developing inductive methods for studying the de Haas–van Alphen effect at high fields in conditions more favourable, and in particular more sensitive, than those of the pulsed field method. In such inductive methods the sample is put into one of a balanced pair of pick-up coils, but instead of inducing an e.m.f. through the continuous variation of the main field H_0, as in the pulsed field method, the magnetization M is made to vary periodically with time with small amplitude at some suitable frequency ω and the induced e.m.f. at the same frequency or at a higher harmonic $k\omega$ is examined. The great advantage of such modulation methods is that phase sensitive detection of the induced e.m.f. can be used, thus greatly improving the signal to noise ratio and so greatly increasing the effective sensitivity of detection.

3.4.2.1. Field modulation

In the field modulation method the periodic variation of M is achieved by superimposing a small periodic field $h_0 \cos \omega t$ on H_0, and since this is the method which has been most thoroughly developed we shall devote most attention to it. It is, however, not the only possibility and we shall see later that there are possible advantages in varying M through modulation of the sample temperature rather than the field, an idea originally demonstrated as feasible by Oder and Maxwell (1965) but not yet developed to its full

3.4. Dynamic methods

potential. Another possibility which has advantages for special purposes, as we shall see later, is to modulate the field direction relative to the sample, as can be done if the small periodic field has a component perpendicular to the main field H_0. This is effectively equivalent to modulating the sample orientation. Modulation of the pressure (or more generally the stress) acting on the sample is yet another possibility.

We shall now give a general account of the field modulation method. This was first introduced by Shoenberg and Stiles (1964) and very soon developed and improved in a number of places. For more comprehensive reviews, see Goldstein, Williamson and Foner (1965) and Stark and Windmiller (1968). In order not to obscure the basic principles, we shall start by outlining how the method works in its simplest form, with the modulating field h varying as $h_0 \cos \omega t$ and parallel to the steady field H, with h_0 small compared to the field interval of a dHvA oscillation and with the modulation frequency ω low enough for induced eddy currents in the sample to be ignored.

The voltage induced in the pick-up coil is again given by (3.15), as for the pulsed field, but now dH/dt is simply $-h_0 \omega \sin \omega t$ instead of the derivative associated with the rapid variation of H in the pulse; moreover the time dependence of dM/dH must be taken into account if h_0 is appreciable. Thus

$$v = -ch_0 \omega \sin \omega t \left\{ \frac{dM}{dH} + \frac{d^2M}{dH^2} h_0 \cos \omega t \cdots \right.$$
$$\left. + \frac{d^k M}{dH^k} \frac{h_0^{k-1}}{(k-1)!} (\cos \omega t)^{k-1} + \cdots \right\} \quad (3.19)$$

where all the derivatives are to be taken at $H = H_0$. Consequently if the variation of M with H is non-linear, v contains not only the fundamental frequency ω but also higher harmonics, i.e.

$$v = -c\omega \left\{ h_0 \frac{dM}{dH} \sin \omega t + \tfrac{1}{2} h_0^2 \frac{d^2M}{dH^2} \sin 2\omega t \cdots \right.$$
$$\left. + \frac{h_0^k}{2^{k-1}(k-1)!} \frac{d^k M}{dH^k} \sin k\omega t + \cdots \right\} \quad (3.20)$$

Here only the lowest power of h_0 has been retained in the amplitude of each harmonic. If we now make the specific assumption that the $M-H$ relation is

$$M = A \sin \left(\frac{2\pi F}{H} + \phi \right) \quad (3.21)$$

and we write the field interval of one dHvA oscillation as $\Delta H = H^2/F$,

103

Observation of the dHvA effect

(3.20) becomes

$$v = -c\omega A \left\{ \left(\frac{2\pi h_0}{\Delta H}\right) \sin\left(\frac{2\pi F}{H} + \phi - \frac{\pi}{2}\right) \sin \omega t \right.$$

$$+ \frac{1}{2}\left(\frac{2\pi h_0}{\Delta H}\right)^2 \sin\left(\frac{2\pi F}{H} + \phi - \pi\right) \sin 2\omega t$$

$$\left. + \cdots \frac{1}{2^{k-1}(k-1)!}\left(\frac{2\pi h_0}{\Delta H}\right)^k \sin\left(\frac{2\pi F}{H} + \phi - \frac{k\pi}{2}\right) \sin k\omega t + \cdots \right\}$$

(3.22)

and we see that if, as we have assumed, $h_0 \ll \Delta H$, the harmonic amplitudes are weak, each successive harmonic being weaker by a factor of order $(2\pi h_0/\Delta H)$.

The signal to noise ratio is greatly improved if a phase-sensitive detector (PSD) is used to detect the signal and, as will be discussed below, there are considerable advantages in tuning the PSD to a harmonic of ω (usually 2ω) rather than to ω itself. But if the amplitude of such a harmonic is to be appreciable, the modulation level must be increased to such an extent that the condition for (3.22) to be a good approximation, i.e. $(2\pi h_0/\Delta H) \ll 1$, is no longer valid.

We shall now show how (3.20) or (3.22) can be generalized for arbitrary $(2\pi h_0/\Delta H)$. However this generalization is only possible if a specific form of the relation between M and H is assumed and we shall assume the simple harmonic relation of (3.21). If then we put in the time dependence of H explicitly we have

$$M(t) = A \sin\left\{\frac{2\pi F}{H_0 + h_0 \cos \omega t} + \phi\right\}$$

(3.23)

Or, since $h_0 \ll H_0$, even though $2\pi h_0/\Delta H$ may not be small,

$$M(t) = A\left\{\sin\left(\frac{2\pi F}{H_0} + \phi\right)\cos(\lambda \cos \omega t)\right.$$

$$\left. - \cos\left(\frac{2\pi F}{H_0} + \phi\right)\sin(\lambda \cos \omega t)\right\}$$

(3.24)

where

$$\lambda = \frac{2\pi F h_0}{H_0^2} \equiv \frac{2\pi h_0}{\Delta H}$$

(3.25)

The functions $\cos(\lambda \cos \omega t)$ and $\sin(\lambda \cos \omega t)$ can be expressed as sums of

104

3.4. Dynamic methods

harmonics $\cos k\omega t$ with coefficients proportional to the Bessel functions $J_k(\lambda)$ and after a little rearrangement (3.24) reduces to

$$M(t) = A \left\{ J_0(\lambda) \sin\left(\frac{2\pi F}{H} + \phi\right) \right.$$

$$\left. + 2 \sum_{k=1}^{\infty} J_k(\lambda) \cos k\omega t \sin\left(\frac{2\pi F}{H} + \phi - \frac{k\pi}{2}\right) \right\} \quad (3.26)$$

Here and in what follows, to avoid cumbersome notation, we omit the suffix zero from H_0 except where confusion might arise between the time dependent field H and the steady field H_0. The induced e.m.f. corresponding to (3.26) is

$$v = c\frac{dM}{dt} = -2c\omega A \sum_{k=1}^{\infty} k J_k(\lambda) \sin\left(\frac{2\pi F}{H} + \phi - \frac{k\pi}{2}\right) \sin k\omega t$$

$$(3.27)$$

This, of course, reduces to the simpler expression (3.22) if $\lambda \ll 1$, (i.e. if $(2\pi h_0/\Delta H) \ll 1$), since then $J_k(\lambda) \simeq \lambda^k/2^k k!$, but for arbitrary λ, interesting new features appear.

Figure 3.7 shows the variation of $J_k(\lambda)$ with λ for a few values of k and we see that by suitable choice of modulation amplitude (i.e. of λ) the amplitude of any particular harmonic of v can be considerably enhanced. If we denote the amplitude of the kth harmonic, i.e. the coefficient of $\sin k\omega t$ in (3.27) by v_k, the amplitude $|v_k|$ of the dHvA oscillation of v_k as field or orientation is varied, has a maximum value given by

$$|v_k|_{\max} = \alpha_k c\omega A \quad (3.28)$$

Fig. 3.7. Graphs of $J_k(\lambda)$ for $k = 1, 2, 3$ and 4.

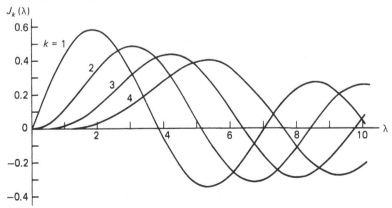

Observation of the dHvA effect

where α_k is a number given by

$$\alpha_k = 2k(J_k(\lambda))_{max} \qquad (3.29)$$

whose variation with k is shown in table 3.1, together with the variation of $\lambda_{max}(k)$, the value of λ required to produce the first maximum of amplitude. It can be seen that α_k is of order 1 for the fundamental, of order 2 for $k = 2$ and then rises gradually with k to a value of about 6 for $k = 10$, so that detection at a harmonic produces more signal, the higher the harmonic. However the modulation amplitude has to be somewhat increased to achieve this (from $h_0 = 0.29\,\Delta H$ for $k = 1$ to $h_0 = 1.88\,\Delta H$ for $k = 10$). Except for special purposes, such as dHvA frequency discrimination (see (4) on p. 108), little is gained by going beyond $k = 2$, since the higher modulation amplitude required to reach the Bessel maximum as k is increased can sometimes prove a limiting consideration, especially for low F. It should be noted that if optimum modulation is used (for detection of any harmonic) the measured quantity is really M, rather than d^kM/dH^k which is characteristic of small modulation.

Although the result (3.27) is based on the special assumption of simple harmonic dHvA oscillations (i.e. (3.21)), it is easy to generalize it by summing the contributions from the individual Fourier components of the actual wave form. Thus if the wave form is given by

$$M = \sum_{r=1}^{\infty} A_r \sin\left(\frac{2\pi rF}{H} + \phi_r\right) \qquad (3.30)$$

Table 3.1. Values of α_k and $\lambda_{max}(k)$ (see (3.29))

k	α_k	$\lambda_{max}(k)$
1	1.16	1.8
2	1.95	3.1
3	2.61	4.2
4	3.19	5.3
5	3.75	6.4
6	4.24	7.5
7	4.73	8.6
8	5.19	9.6
9	5.63	10.7
10	6.05	11.8

3.4. Dynamic methods

the amplitude v_k of $\sin k\omega t$ in the generalization of (3.27) is

$$v_k = -2c\omega k \sum_{r=1}^{\infty} A_r J_k(r\lambda) \sin\left(\frac{2\pi r F}{H} + \phi_r - \frac{k\pi}{2}\right) \qquad (3.31)$$

If there are several frequencies F, say $F_1, F_2, \ldots F_j$ etc., (3.31) is immediately generalized by adding an extra suffix j to A_r, λ, F and ϕ_r and then summing over j.

It can be seen that the wave form of v_k (as a function of $1/H$) will in general be quite different from that of M both because the coefficients $J_k(r\lambda)$ may differ greatly from each other and because for odd k the sine in (3.30) becomes a cosine of the same argument in (3.31). In general, to reconstruct the original wave form (3.30), the output v_k as a function of $1/H$ must be Fourier analyzed, each Fourier coefficient divided by $J_k(r\lambda)$ and the Fourier components then resynthesized with the appropriate phases ϕ_r as determined from the Fourier analysis. For the special case of sufficiently weak modulation, however, $J_k(r\lambda)$ is proportional to r^k and v_k has exactly the wave form of $d^k M/dH^k$; in fact it is easy to check that in this limit (3.31) reduces exactly to the kth term of (3.20). In practice the amplitudes of the higher harmonics are too feeble to be useful in conditions of weak modulation and so it is only dM/dH (and perhaps $d^2 M/dH^2$) whose wave forms can be directly recorded without distortion by detecting at ω (or 2ω) with weak modulation. It is easily shown that if $J_1(r\lambda)$ is to differ from $\frac{1}{2}r\lambda$ by less than 1%, $r\lambda$ should be <0.28; this then is a criterion for the modulation to be weak enough for v_1 to give faithful reproduction of dM/dH.

As has already been mentioned, detection at a harmonic rather than at the fundamental modulation frequency has a number of advantages, which we shall now briefly review:

(1) If the pick-up coils are not quite perfectly balanced, the e.m.f. fed into the PSD contains a considerable $\sin \omega t$ component not associated with the magnetic properties of the sample and this is conveniently suppressed by detecting at a harmonic of ω.

(2) Even at low modulation frequencies, eddy currents induced in the sample upset the balance of the pick-up coils to an extent which depends on the steady magnetic field because of magnetoresistance in the sample. Consequently the dHvA oscillations if detected at the fundamental frequency ω effectively appear on a sloping base line, which can be very inconvenient. This effect can be much reduced by proper choice of the phase setting of the PSD, but it is almost completely eliminated by detection at the 2ω or higher harmonic frequency.

(3) One of the main sources of noise in the output signal comes from e.m.f.s induced by vibrations of the pick-up coil in the steady magnetic field. One cause of such vibrations which is difficult to eliminate completely is the interaction of the modulating current with the steady field and this produces noise predominantly at the fundamental frequency ω. Thus the signal to noise ratio of the

107

system can be considerably improved by detection at a harmonic rather than at the fundamental.

(4) The Bessel function dependence of the output signal on modulating field can be exploited in various ways to help sort out a complicated dHvA frequency spectrum. This frequency discrimination is the more effective the higher the harmonic of ω used in detection. The parameter λ which determines the 'effective' modulation is proportional to F and so if there is a dominating *low frequency* F in the spectrum it can be suppressed by adjusting the modulation field to make its λ come on the initial very flat part of the Bessel function curve, while the λ of a higher dHvA frequency comes near the Bessel maximum. Alternatively, a dominant *high frequency* F can be suppressed by making its λ come at a zero of the Bessel function. In suitable circumstances several dHvA frequencies can be simultaneously suppressed, or at least partially suppressed, by using more than one Bessel zero and choosing a high harmonic for which the λs of the Bessel zeros are just in the ratio of the Fs to be eliminated. In all these tricks the required modulation amplitudes depend of course on the value of H and if it is desired to preserve a constant value of λ as H is varied, the modulation current must be varied as H^2.

So far we have assumed that both the modulation field and the pick-up coil axis are parallel to the steady field H_0, but there are interesting possibilities in setting either or both the modulation field and the pick-up coil axis directions at an angle to H_0. Stark and Windmiller (1968) have reviewed these possibilities and shown how they can be exploited to discriminate between frequencies in a complicated dHvA spectrum. Since, however, such discrimination can usually be achieved rather more simply by varying the level of modulation in the parallel configuration, we shall give only a brief account. If the modulation field h is at an angle θ to the steady field H_0 the *direction* of H, and hence the magnitude of F, is modulated as well as the magnitude of H. It is not difficult to show that (3.24) will still be valid if the definition of λ is modified from (3.25) to

$$\lambda = \frac{2\pi F h_0}{H_0^2}\left[\cos\theta - \sin\theta\left(\frac{1}{F}\frac{dF}{d\theta}\right)\right] \tag{3.32}$$

Thus by suitable choice of θ, λ, and hence the signal, can be made to vanish for some particular dHvA frequency. Moreover since $dF/d\theta$ depends on the plane in which the rotation θ is made, it is possible by suitable choice of both θ and the direction of this plane to make λ vanish for any two different dHvA frequencies simultaneously.

Variation of θ can be achieved without mechanically turning the coil, by using two coils at right angles and adjusting the ratio of the currents carried by them (Edelman, Volsky and Khaikin 1966). A variant of this technique is to use equal currents in the two coils with a phase difference of $\pi/2$, i.e. effectively circularly polarized modulation (Volsky and Teplinski 1970); the

108

effective direction of modulation can then be varied by adjusting the phase of the PSD. A drawback of such methods, however, is that a couple acts on the transverse modulation coil causing vibration problems at high modulation.

Similarly, by setting the direction of the pick-up coil at some angle ϕ to H, the coil picks up not simply $M_{\parallel}$ but a combination of $M_{\parallel}$ and $M_{\perp}$. This means effectively that all previous formulae for the output of the pick-up coil must be multiplied by a factor

$$\cos \phi - \frac{1}{F}\frac{\mathrm{d}F}{\mathrm{d}\phi}\sin \phi$$

and again, by suitable choice of ϕ and of the plane in which the tilt is made, it is possible to make the output vanish for any two dHvA frequencies. Thus by suitable choice of θ, ϕ and the respective planes in which modulation and pick-up are rotated it is possible to suppress four different dHvA frequencies simultaneously. As already mentioned, suppression of a dominant frequency (or several simultaneously) can be useful in showing up a much weaker one. A particular virtue of the technique of tilting the modulation and pick-up coils is that it can discriminate between two nearly equal frequencies, provided the angular dependences of F are sufficiently different.

Apart from an occasional brief mention, we have up to now ignored the complication of eddy currents induced in the sample by the modulating field. If the frequency is low enough, in the sense that the skin depth δ, given by

$$\delta = 1/(2\pi\omega\sigma)^{1/2} \qquad (3.33)$$

is large compared with some relevant sample dimension a (e.g. the radius of a cylinder or sphere or the thickness of a plate) i.e. if

$$u \equiv a/\delta \ll 1 \qquad (3.34)$$

then the neglect of eddy currents is indeed justified. We shall now consider in broad outline how the theory leading to (3.27), according to which the output of the pick-up coil is proportional to ω, is modified when ω is increased sufficiently to invalidate (3.34), so that eddy currents can no longer be ignored. The special case of a cylindrical sample has been discussed in detail by Knecht, Lonzarich, Perz and Shoenberg (1977) and although in what follows we shall quote results only for this particular geometry, the qualitative features of the results remain true for other geometries.

The first noticeable effect of raising the modulation frequency ω is that the *phase* of the fundamental in the pick-up coil signal begins to depart from being in quadrature to that of the modulating field, i.e. an appreciable $\cos \omega t$ term begins to appear in addition to the $\sin \omega t$ term of (3.22). For increase of ω beyond the value for which $a/\delta \sim \frac{1}{2}$, the *amplitude* of the signal begins to fall off appreciably from the strict proportionality to ω given by (3.20) or (3.22). Eventually, beyond $a/\delta \sim 3$, the phase shift settles at about 45° from quadrature and the amplitude approaches $(\delta/\sqrt{2}a)$ times that given by (3.20) or (3.22). This is, of course, the 'skin effect' limit, in which effectively only a fraction of order δ/a of the volume of the sample 'feels' the modulation. Since $\delta \propto \omega^{-1/2}$ (see (3.33)) this means that in this skin effect limit the amplitude increases only as $\omega^{1/2}$ rather than as ω. The form of the frequency dependence of the output e.m.f. amplitude and phase for 'weak' modulation ($\lambda \ll 1$) and small $\mathrm{d}M/\mathrm{d}H$ is shown in fig. 3.8.

For 'strong' modulation ($\lambda \gtrsim 1$) the effect of eddy currents is rather complicated in detail, particularly in the frequency range intermediate between $a/\delta \ll 1$ and $a/\delta \gg 1$. Nevertheless it remains qualitatively true that for $a/\delta > 1$ the output is as if a fraction of order δ/a of the sample

Fig. 3.8. Effect of eddy currents on amplitude per turn and phase of the output $v \cos(\omega t - \phi)$ of a pick-up coil closely wound on a long cylindrical sample (radius a, conductivity σ) for field modulation $h_0 \cos \omega t$; only the voltage caused by the sample magnetization is included in v. The scaling of v is chosen to make the ordinate a function of the dimensionless variable a/δ, where δ is the skin depth $(2\pi\omega\sigma)^{-1/2}$.

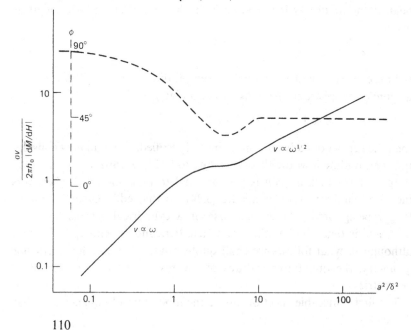

volume is effective, and in particular, apart from numerical factors of order 1, the maximum output amplitude is given by δ/a times that given by (3.28) in the absence of eddy currents. The phase of the output e.m.f. for strong modulation varies in a complicated way not only with ω but also with the modulation amplitude (i.e. with λ) and one consequence of this phase variation is that the value of λ for which the amplitude vanishes (what would be the 'Bessel zero' in the absence of eddy currents), depends in general on the phase setting of the PSD and on ω.

In the quantitative theory for arbitrary a/δ and λ, the amplitudes and phases of the harmonics of the output e.m.f. are given by integrals which have to be computed for each pair of values of a/δ and λ, but even this rather complicated formulation is of only limited validity for a number of reasons. First, the results are quantitatively valid only for the particular cylindrical geometry assumed. Secondly we have ignored the Hall effect, which gives rise to helicon resonances and so may considerably distort the predicted frequency dependence in the region of high ω, for both weak and strong modulation. Thirdly the theoretical predictions are modified appreciably in practice by sample inhomogeneity. For these reasons the detailed theoretical results, except in particular limiting situations, should be used with caution and are generally more of qualitative than quantitative value.

We have already mentioned the consequence of eddy currents that the balance condition of the pick-up coil system becomes sensitive to the sample resistivity. This introduces a background e.m.f. which, because of magnetoresistance, not only varies steadily with H (as mentioned earlier, this variation can be much reduced by detection at a harmonic of ω), but also may vary in oscillatory fashion with H in virtue of the Shubnikov–de Haas effect. The parameter $|\Delta\sigma|/\sigma$, describing the SdH oscillations, is usually much smaller than the parameter $|4\pi \, dM/dH|$ of the dHvA oscillations and so, even at high modulation frequencies, the observed oscillations of the output signal as H varies are predominantly due to the dHvA effect. However, in special circumstances, such as magnetic breakdown (see chapter 7) and in semimetals near the quantum limit, this may no longer be true and it is then possible, by detection at a sufficiently high modulation frequency, to observe oscillations to which both the SdH and dHvA effects contribute comparably, while at a modulation frequency low enough for eddy currents to be negligible, only the oscillations due to the dHvA effect appear. From such observations at different modulation frequencies the two effects can be separated (Lonzarich and Holtham 1975).

We may now usefully discuss some broad principles which determine the optimal design of the field modulation technique. If the aim is to study the wave form and absolute amplitude of the oscillations, some sacrifice of sensitivity is usually necessary. Thus it is simplest to avoid the complications of eddy currents as far as possible by using a sufficiently low modulation frequency. A convenient criterion in this respect (Knecht 1975) is that if detection is at the fundamental with weak modulation, the phase of the output should not depart by more than 8° from quadrature with the phase of the modulating field. For a cylindrical sample this ensures that the amplitude of that part of the e.m.f. which comes from the magnetic properties of the sample does not depart by more than 1% from strict proportionality to ω. This criterion implies that $a/\delta < 0.53$ if a is the sample radius. In practice this means that for $a \sim 0.5$ mm, the frequency $f(=\omega/2\pi)$ must be well below 100 Hz, and exceptionally even below 10 Hz, for a pure metal with little magnetoresistance (e.g. an alkali) but can be a good deal higher if there is considerable magnetoresistance (as for noble metals, except at particular orientations). For detection at the kth harmonic of ω a similar criterion can be formulated, and it turns out that ω must be roughly k times lower than for detection at the fundamental, if the amplitude is not to depart by more than 1% from proportionality to ω.

If the main aim of the experiment is to study the dHvA F spectrum rather than absolute amplitude and wave form, the emphasis of the design would be on optimum sensitivity, in the sense of signal to noise ratio, and the choice of modulation frequency and amplitude would be guided by somewhat different and often conflicting considerations. As regards ω, the signal at first increases linearly with ω, while the purely electronic noise goes down as $1/\omega$. However, typically beyond $f \sim 100$ Hz the noise no longer improves and in any case there may also be noise of mechanical origin with a complicated spectrum showing resonant peaks at quite low frequencies. Usually the signal to noise ratio may be somewhat improved by going above 100 Hz but not too much, because eventually the eddy currents enter the skin effect regime (signal proportional to $\omega^{1/2}$) and moreover the capacitative impedance of the coil begins to be important in reducing the fraction of the signal reaching the input of the PSD. Another consideration which may sometimes be relevant is that the inductive impedance of the pick-up coil goes up as ω and as the square of the number of turns; thus to maintain impedance matching, the number of turns may have to be reduced if ω is increased.

It is difficult to give any general recipe for the optimum frequency because so much depends on the nature and dimensions of the sample and on the coil design (see below), but usually the best signal to noise ratio is obtained for f somewhere between 50 and 5000 Hz. In the pioneer experiments of Shoenberg and Stiles (1964) the question of optimum

112

conditions was not really considered and the modulation frequency $(1.6 \times 10^6 \, \mathrm{Hz})$ was chosen simply as high as conveniently possible to get as high a voltage output as possible. However, in later developments of the field modulation technique it soon became evident that much better signal to noise ratios could be obtained with rather lower modulation frequencies.

As regards the modulation amplitude, h_0, clearly the largest signal will be for h_0 corresponding to the Bessel function maximum, but this may not be practicable if F is low. Thus for $F \sim 10^7 \, \mathrm{G}$ the optimum h_0 for a field of $50 \, \mathrm{kG}$ would be about $125 \, \mathrm{G}$ for 2ω detection and such a high amplitude would not be easy to achieve at a frequency much higher than $10^3 \, \mathrm{Hz}$. It may be noted that the use of a modulation coil wound with wire which remains superconducting in the high steady field can help in reaching high modulation amplitudes. Even if the Bessel maximum can be reached, it is not necessarily advantageous to do so since the 'noise' caused by vibrations associated with interaction of the modulation current and the main field may increase more rapidly with h_0 than the output signal.

The design of pick-up coil depends very much on whether sensitivity or reliable calibration is the main aim and also on the kind of sample that is used and on whether or not the sample orientation needs to be varied. Some typical arrangements are illustrated in fig. 3.9. If relatively large samples are available, such as a disc several mm in diameter and a mm or so thick (as in (a)) or a cylinder several mm long and a mm or so in diameter (as in (b)), a coil closely wound round the middle of the sample gives close coupling and the flux of the H field (which is balanced out by the second coil) is as small as it can be made. The magnitude of this H flux is one of the factors which determine the 'noise' if for instance the pick-up coil system vibrates in the direction of H or if the axes of the two coils slightly change their relative directions in vibration. Thus (a) and (b), in which the H flux is almost entirely restricted to the sample cross-section, are better from the noise point of view than (c) and (d), where a great deal of H flux is contained in the space around the sample. The advantages of the (c) and (d) types of system, especially for rather compact small samples (e.g. a sphere of a mm or so diameter) are that they can accommodate a variety of shapes and sizes of sample, that samples can more easily be changed and (as we shall see below) absolute calibration is more easily possible.

Type (d) has two advantages over (c) (though at the sacrifice of some sensitivity). First that the sample orientation can be more easily varied without having to rotate the pick-up coils also and secondly that its balance depends less critically on its position in the magnet so that any vibration produces less noise. This type of noise can, of course, be reduced by making the whole coil system as small as possible so that the field is as homogeneous as possible over the coil region. Type (*e*) or its variant (*f*) is an arrangement suggested by G. G. Lonzarich (private communication) and in a different variant (*g*), independently by P. J. Stiles (private communication). It is designed to achieve close coupling with the magnetization normal to a thin flat plate sample; the coil can be constructed by depositing an evaporated layer on the plate through a suitable mask; a compensating coil can be laid down on the sample over a thick insulating layer, so that it has little flux linkage with the magnetization. It should be particularly suitable for samples of layer compounds or for detecting the dHvA effect in a 2-D system (see §2.3.4 and below) and the rigid bond between sample and coil should make it particularly noise-free.

Fig. 3.9. Schematic sketches of various types of pick-up coil systems. For each type, c is the coil flux-linked to the sample s and c' is the compensating coil only weakly flux-linked to s. (*a*) c is closely wound round the central part of a thin disc sample, (*b*) as (*a*) but for a cylindrical sample, (*c*) system with small sample, (*d*) as (*c*) but with different arrangement of c' ((*d*) is sometimes varied by breaking each of the coils into a Helmholtz-like pair), (*e*) top and side view of a spiral coil laid down by evaporation on thin sample normal to *H*, (*f*) as (*e*) but with only a single annular turn, (*g*) if the sample itself (shaded area) can be divided into many strips laids down on a substrate (only four strips are shown, but many more can be achieved) the coil can be evaporated along the broken lines; if connection is made at *BB'* the coil is flux-linked to the *gaps* between the sample strips but if at *AA'* it is flux-linked to the strips themselves. Note that c' and the side views are not shown in (*f*) and (*g*).

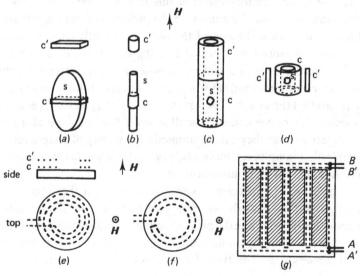

114

3.4. Dynamic methods

For the close coupling arrangement of (a) or (b) the flux Φ through a coil of N turns due to the sample magnetization M (or more precisely the component of M parallel to the coil axis) is

$$\Phi = 4\pi M N S (1 - n) \qquad (3.35)$$

where S is the appropriate cross-sectional area of the sample and $4\pi n$ is the demagnetizing coefficient of the sample. Thus the coupling constant c of (3.15) or (3.19) is given by

$$c = -4\pi N S (1 - n) \qquad (3.36)$$

In practice plate-like and cylinder-like samples cannot be made exactly in the form of perfect ellipsoids so that n is not very well defined* and S may not be well defined either. Thus in practice it would be difficult with arrangements (a) or (b) to achieve an accuracy of better than a few per cent in an *absolute* determination of M.

For (c) and (d), however, if the sample is small enough, the value of Φ does not depend on the sample shape, but only on its net dipole moment MV, where V is the sample volume, which can be more precisely determined (e.g. by weighing) than an area of cross-section. There is a simple general reciprocity relation which shows that in e.m.u. the flux in any coil per unit dipole at a point P is identical with the field at P per unit current in the coil. Thus if η_1 is the field per unit current in the pick-up coil at the position of the sample, the flux Φ can be neatly expressed as

$$\Phi = \eta_1 M V \qquad (3.37)$$

provided the sample is small enough. It is easy to see that (3.37) is roughly equivalent to (3.35) by recalling that, for a coil of length L large compared with its diameter, $\eta_1 \sim 4\pi N/L$, so if the sample is a cylinder of length l and cross-section area S, $\Phi \sim 4\pi M N S (l/L)$, which is equivalent to (3.35) if demagnetization is ignored and a fraction (l/L) of the pick-up coil turns is considered effective.

The constant η_1 and also η_2, the field per unit current of the modulating coil, may be determined with good accuracy by an ingenious method due to Knecht (1975), based on the vanishing of the dHvA signal amplitude at the 'Bessel zero' for a particular value of the modulating current. In order that the method should be reliable, the complications of eddy currents are best avoided by working at low modulation frequency and it is simplest to detect

---

* For a platenoid whose diameter is five times its thickness $n \simeq 0.16$, and the uncertainty in the appropriate value of $(1 - n)$ for a disc with rounded edges may well amount to a few %. For an ovoid of length five times its diameter $n \simeq 0.05$ and the uncertainty in the appropriate value of $(1 - n)$ for a cylinder with rounded ends is probably only of order 1%.

at the fundamental, i.e. $k = 1$. A sample is chosen which has an accurately known dHvA frequency F, and if λ_0 is the value of λ which makes the Bessel function $J_k(\lambda)$ vanish $\left(\text{for } J_1(\lambda), \lambda_0 = 3.83\right)$, and i_0 is the modulating current amplitude for this Bessel zero, we have

$$\eta_2 = \lambda_0 H^2 / 2\pi F i_0 \tag{3.38}$$

If now the roles of the pick-up and modulating coils are exchanged and a Bessel zero is found for the modulating current in what is normally the pick-up coil, the constant η_1 can be determined in exactly the same way. It will be noticed that this method of calibration requires the sample to be small enough, but as shown in Appendix 8, the coupling constants η_1 and η_2 can be allowed to vary by up to 10% over the sample before the error in calibration by this method approaches 1%. The error is even smaller if the modulation field constant η_2 is uniform over the sample, or if the measurements of magnetization are made at the Bessel maximum.

To discuss the flux linked with a pick-up coil of type (e), it is helpful first to consider the distribution of field just above a thin flat sample (thickness t, area S, typical linear dimension a) magnetized normal to its plane, and we consider specifically how the field normal to the sample varies with distance x away from the projection of the sample edge in a plane at small distance d above the sample, such that $t < d \ll a$. This field is exactly that of a current $i = Mt$ (or $\mathscr{M}$, the dipole moment per unit area of the sample) flowing round the rim; thus the normal component of the field rises from zero at $x = 0$ to a peak of order i/d at $x = d$ (the sign is opposite according as we move into or out of the region above the sample) and then falls off approximately as $2i/x$ as x increases well beyond d.

The important consequence of this is that the flux normal to the sample is mostly concentrated in a region close to the sample edge and so depends essentially on the perimeter P of the sample rather than its area S. In fact the flux Φ within an area bounded by the projection of the sample edge ($x = 0$) and a contour x ($> d$, but $\ll a$), is given roughly by:

$$\Phi = 2Pi \ln(x/d) = 2P\mathscr{M} \ln(x/d) \tag{3.39}$$

The sign of the flux is opposite according as the contour is towards the inside or outside of the sample. By considering the concrete example of a circular disc of radius a this flux can be compared with the *total* flux within the sample (using the table given in Pidduck 1925, pp. 186–92). From (3.39), the value of Φ for the annulus between $x = 0$ and $x = a/10$ is about 4.6 Pi if $d = 10^{-2} a$, while the total flux over the whole circle of radius a works out to be about 9.4 Pi; thus an annulus of one fifth the sample area carries about half the total flux through the sample.

116

3.4. Dynamic methods

If the noise of the system is proportional to the area of the pick-up coil there is some advantage in using a single annular loop of the kind shown in fig. 3.9(f) rather than, say, a ten turn spiral of the kind shown in fig. 3.9(e). For the example given above, the ratio of flux/area for the annular coil is about three times that for the ten turn spiral and about two times that for a single turn on the rim, so it can be seen that the advantage is only a modest one. In fact it is more likely that the system noise will *not* vary appreciably with the coil area, especially since it is difficult to arrange many turns, so that the area is inevitably much smaller than with an ordinary pick-up coil.

If we consider this opposite limit, in which the coil area does not appreciably determine the noise, it is obviously advantageous to lay down a spiral coil of as many turns as possible, bearing in mind that it is only the turns fairly close to the rim that contribute. As already mentioned, it is difficult to lay down many turns by evaporation, especially if all the turns are to be close to the edge, but the difficulty can to some extent be avoided by an ingenious method proposed by Stiles. This method is particularly applicable to the problem of detecting the dHvA oscillations in the 2-D system of a Si inversion layer, which can itself be prepared by deposition techniques. The idea is to prepare the sample in N separate strips, as in fig. 3.9(g), (N can be made as large as 20 or possibly larger) and to deposit a pick-up coil which is flux linked to all the gaps between strips. If there are N strips, the flux linkage is given by (3.39) as approximately

$$\Phi = 4Na\mathcal{M}\ln(x/d) \qquad (3.40)$$

i.e. effectively as if there were N 'annular' turns of width x on top of each other of the kind shown in fig. 3.9(f), though it should be noticed that the area of a coil of N turns similar to that of fig. 3.10(f) is four times that of fig. 3.9(g). An interesting feature of Stiles' system is that with the alternative connection indicated in fig. 3.9(g) the flux linkage is with the strips themselves instead of with the gaps between them (the flux linkage with the strips is given by the same formula but with x meaning the strip width instead of the gap width).

An important point in the design of a pick-up coil of types (a) to (d) is that it pays to wind it with as fine a wire as practicable; with sufficient care even 56 SWG copper wire can be used. The use of fine wire allows the number of turns to be greatly increased without greatly increasing the area of the outermost turns, so that any noise associated with 'stray' pick-up does not increase as rapidly with N as it would for thicker wire. Noise can also be reduced by careful design in the construction and mounting arrangements of both the pick-up and modulation coils. The modulation coil should be mounted as rigidly as possible to reduce the generation of mechanical vibration and it is best to avoid any physical contact between modulation and pick-up coils to avoid transmission of such residual vibration as is generated. It is also important to arrange the leads non-inductively and to anchor them to both coils as firmly as possible. However, if provision has to be made for rotating the coils, the extra flexibility required in the arrangement of the leads inevitably introduces some extra noise. As mentioned earlier, however, variation of the effective coil axis direction can be achieved without actually turning it by using two coils at right angles and combining their contributions in proportions which can be varied by a suitable external circuit.

Other important sources of 'noise' which can be greatly reduced by careful design come from the circuits used to drive the modulating current and to amplify the output signal. We shall not go into any detail here, but merely mention the importance of using a pure enough source of modulation, of adequate grounding of various parts of the circuits, and of interposing a step-up transformer between the pick-up coil and the PSD.

It is convenient to mention here some design features concerned with control of the main field and with the recording and analysis of data. To produce oscillations at a fixed orientation, H has to be swept and the slower the sweep, the longer the time constant τ which can be used in the PSD to reduce noise. Since one dHvA oscillation must not be swept in a time much less than 10τ if distortion is to be avoided, it is hardly practicable to use a time constant of much more than $10\,s$ if many oscillations are to be swept and this sets a limit to the noise reduction that can be achieved. Another approach is to use a shorter time constant but to sweep between exactly the same initial and final fields a large number of times and then to use signal averaging to reduce the effective noise. This last approach, which has the advantage of reducing 'drift' effects, is only possible if the fields at which signals are recorded are extremely reproducible and this is only possible in practice if the field is continuously monitored with the precision of n.m.r. For some purposes it is convenient to arrange that $1/H$ rather than H itself should sweep uniformly with time; the dHvA oscillations then appear as periodic *in time*. Extra selectivity (as against other F values) and also noise reduction can then be achieved by using an amplifier tuned to the appropriate time frequency.

If oscillations are to be observed at fixed H by varying the orientation, it is important that H should be sufficiently constant and this is conveniently achieved by using the persistent mode of a superconducting magnet. It is also important that rotation should be sufficiently smooth and slow to match the time constant of the PSD, and this is not easy to achieve. A smoother variation of orientation can be achieved if the whole magnet can be rotated rather than the sample, but this is only possible using a split coil superconducting magnet or an iron electromagnet (with considerable sacrifice of maximum field, particularly with the latter).

In the early applications of the field modulation technique the output of the PSD was displayed on a chart recorder (either an $X-Y$ recorder in which X was effectively either H or $1/H$, or a recorder which ran at uniform speed so that X was simply time) and the data were read off the graphs and analyzed subsequently. Greater accuracy can, however, now be achieved with less labour from the experimenter, by digitizing the output at frequent intervals and either feeding the digitized data directly into a computer for analysis or storing it to be analyzed later. In either case a simultaneous graphical record on a chart recorder still serves as a helpful visual monitor of what is happening. Some useful discussion of detailed schemes may be found in Alles and Higgins (1973, 1975), Templeton (1975) and one particular scheme is outlined in fig. 3.10 by way of illustration; an example of a recorded set of oscillations together with the computed frequency spectrum is shown in fig. 3.11.

The ultimate sensitivity of detection of M by the field modulation technique evidently depends critically on the detailed design, and on technological limitations which change rapidly as the years roll on, but nevertheless some numerical estimates based on practical experience with the scheme of fig. 3.10 may be helpful. If the close coupling scheme of fig. 3.9(a) is used and detection is at the Bessel maximum of the $p = 2$ harmonic of the modulation frequency f, the output voltage amplitude $\tilde{v}$ from the N turn pick-up coil is given by (3.28) and (3.35) as

$$\tilde{v} = 2 \times 2\pi f \times 4\pi \tilde{M} \times NS \times 10^{-8}\,\text{V} \tag{3.41}$$

(we have used $2(J_2(\lambda))_{\text{max}} \simeq 1$ and put $n = 0$; cA of (3.28) is effectively $|\Phi|$). As we have seen in the earlier discussions, the adjustable parameters f, N, S and the smallest detectable $v(\equiv |\tilde{v}|)$ are linked in various ways. Thus the highest possible f is limited (a) by eddy currents, which depend on S and (b) by the requirement of large modulation amplitude; the optimum value may be anywhere between 50 and 5000 Hz and we shall assume 500 Hz. The number of turns N is limited by the fact that the outer turns are progressively more poorly coupled, so that eventually increasing N increases the noise more than the signal; in practice even using 56 SWG wire, 2000 turns is about as much as can be usefully achieved. The area S might be $10^{-2}\,\text{cm}^2$ (say 4 mm $\times \frac{1}{4}$ mm) bearing in mind that f would have to be lower if the thickness of the plate-like sample were increased and that part of the noise increases with S. Finally the minimum v which can be detected with suitable signal processing might, on a conservative estimate,

Observation of the dHvA effect

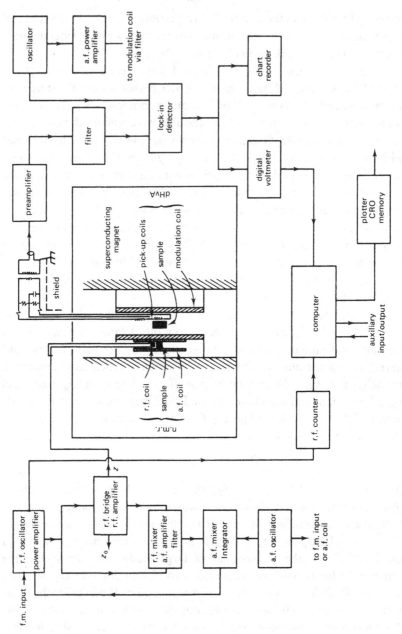

3.4. Dynamic methods

Fig. 3.10. Simplified block diagram of field modulation equipment used by Dr G. G. Lonzarich in the Cavendish Laboratory.

Notes (1) The dHvA modulation coil, wound with NbTi superconducting wire suitable for a.c. use, is rigidly attached to the core of the superconducting magnet; the coil volume is only a few % of the magnet core volume. (2) Accurate compensation of the pick-up coil system is achieved by the external circuit shown. The low noise transformer (kept at 4.2 K) amplifies the signal about 100 times before it enters the preamplifier which has an equivalent input noise of about $3 \times 10^{-10} \text{ V Hz}^{-1/2}$. (3) The filter passes only the required harmonic of ω (usually 2ω). (4) The oscillator and power amplifier are designed for low noise and low harmonic distortion. (5) The digital voltmeter typically measures to a precision of 1 part in 10^4 (noise level $\sim 1 \text{ mV}$). (6) The n.m.r. sample consists of a few mm^3 of powdered Cu. The r.f. coil is closely coupled to the sample and of comparable size; the r.f. varies from about 20 to 100 MHz; a.f. modulation (a few G at 4 kHz) is generated by a long thin coil designed to produce negligible field at the position of the dHvA sample. The n.m.r. sample is about 7 mm away from the dHvA sample; the magnet field at the two positions differs by only about 1 G at 10^5 G. (7) The n.m.r. detection system is similar to that of Hulbert (1976) and is designed to track the magnetic field over a wide range by varying the r.f. For simplicity the diagram shows that either frequency modulation of the r.f. or a.f. amplitude modulation of the magnet field can be used, but actually a somewhat more complicated circuit arrangement is used which combines both kinds of modulation and thus improves tracking stability and sensitivity. (8) The computer is a Hewlett-Packard 9825 calculator which records the radio frequency (from the r.f. counter), the digitized dHvA signal and any other variables of interest (temperature, modulation amplitude etc.). It controls the operation of the magnet sweep unit and also analyzes the data fed into it, conveying the results to a plotter, a CRO or to the memory as required. Frequently used types of analysis are correction of data for background and non-linearity, Fourier analysis of dHvA oscillations according to appropriate algorithms, and least squares fitting.

be somewhere between 3×10^{-12} and 3×10^{-11} V; we shall assume 10^{-11} V. If these numbers are put into (3.41) we find for the minimum observable dHvA amplitude

$$M_{\min} \sim 2 \times 10^{-9} \text{ G, and for } H = 2 \times 10^4 \text{ G,}$$

$$(M/H)_{\min} \sim 10^{-13} \tag{3.42}$$

Although this estimate is extremely rough it suggests that, with the most favourable geometry, the field modulation method should be able to detect oscillations 10^8 times feebler than the $M/H \sim 10^{-5}$ typical of a 'good' dHvA effect. The modulation method also compares favourably with the torque method, though the apparent factor of 100 between (3.42) and (3.14) may be hardly significant in view of the many rather arbitrary assumptions. However, quite apart from the somewhat poorer ultimate sensitivity of the torque method, the fact that it is a 'mechanical' method makes it rather more difficult to set up at its optimum than the purely electromagnetic field modulation method. The torque method is moreover only sensitive when the FS is strongly anisotropic and of course it cannot be used at all for ferromagnetic materials because of the predominance of torques due to the anisotropy of the ferromagnetic magnetization.

For a two-dimensional system with $p = 2$ harmonic modulation and a coil such as that of fig. 3.9(g), the output amplitude $\tilde{v}$ in volts is given by (3.28) and (3.40) as:

$$\tilde{v} = 2 \times 2\pi f \times 4aN \ln(x/d) \times \mathcal{M} \times 10^{-8} \qquad (3.43)$$

If we put $N = 20$, $a = 0.3\,\text{cm}$, $x/d = 10$ and again suppose the smallest detectable $v(\equiv |\tilde{v}|)$ to be $10^{-11}\,\text{V}$ and $f = 500\,\text{Hz}$, we find

$$\mathcal{M}_{\text{min}} \sim 3 \times 10^{-9}\,\text{G\,cm} \qquad (3.44)$$

Fig. 3.11. Example of dHVA oscillations (upper trace) obtained by the scheme of fig. 3.10 with a crystal of Ni_3Al and H along $\langle 100 \rangle$; the field varies uniformly from $43.60\,\text{kG}$ at the left to $72.95\,\text{kG}$ at the right. The lower trace shows a Fourier analysis of the oscillations with peaks at 3.08, 3.25, 5.57, 5.84, 6.01, 6.18 and $6.37 \times 10^6\,\text{G}$.

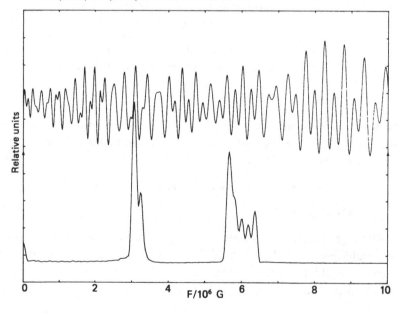

3.4. Dynamic methods

This suggests that it might just be possible to detect the estimated $|\tilde{\mathcal{M}}| \sim 10^{-8}$ G cm for the Si inversion layer system (see table A7.3). In fact, however, the position is rather more favourable than the 'just' suggests. First the possibility of modulating electron density rather than field (see §3.4.2.3) makes the achievement of the Bessel maximum more feasible (for field modulation, a modulation amplitude of order 500 G would be required with $H \sim 5 \times 10^4$ G and $F \sim 2 \times 10^6$ G). Secondly f could probably be made much higher than 500 Hz, since the poor conductivity of Si removes the eddy current difficulty and thirdly the compact design of sample and pick-up coil should reduce the noise level and make possible detection of a lower level than the 10^{-11} V we have assumed. Thus the detection of the 2-D dHvA oscillations would appear to be quite feasible and the outcome of Stiles' experiments is awaited with interest.

To conclude this survey of the field modulation method we should mention briefly that several variants have been proposed and successfully used, though sometimes it is doubtful if the potential advantages of these variants really outweigh the extra complications involved. Often the difficulties they were designed to deal with have since been overcome by more careful attention to the detailed design of the straightforward method. One variant (Lee 1966) is to modulate the field simultaneously at *two* frequencies, a relatively high frequency ω and a much lower frequency Ω. The signal is then passed through *two* PSDs, the first tuned to detect a small range of frequencies round 2ω (broad enough to include side bands at $2\omega \pm \Omega$) and the second tuned to detect around Ω. In this way drift and noise coming through the first PSD at 2ω is much reduced by the second PSD.

Another variant used successfully by Volsky (1962, 1964a), actually before the development of the field modulation method, was to put the sample in an n.m.r. spectrometer and observe the variations of the radio-frequency impedance due to dHvA oscillations of the magnetic permeability. In fact, the n.m.r. spectrometer used an r.f. of 5×10^6 Hz and a low frequency modulation at 30 Hz, so apart from the technique of detection and the geometry of the modulating fields it is not very different from Lee's variant. It should be noticed that the theory of the modulation method implicitly assumes that orbit sizes are always small compared to the skin depth. This may not be true for large F orbits with r.f. modulation and consequently the use of r.f. tends to emphasize oscillations of F low enough for the orbits to be contained within the skin depth.

Volsky and Petrashov (1970) have also developed an interesting technique of observing dHvA oscillations by their influence on the frequency of helicon resonance in a suitable sample. This technique has the

merit that under suitable conditions the relative changes of frequency give
dM/dB absolutely without involving any calibration constants, but rather
special conditions are needed to observe a suitable helicon resonance.

3.4.2.2. Temperature modulation

The idea of temperature modulation was first applied to measurement of
the dHvA effect by Oder and Maxwell (1965) who pointed out that it has
several advantages over the field modulation method. Thus there is no
longer the problem of noise caused by mechanical vibrations associated
with a modulating coil in the large steady field and there is no longer the
problem of balancing out e.m.f. in the pick-up coil system due to lack of
exactly zero mutual inductance with the modulating coil. Moreover the
method offers the possibility–at least in principle–of direct determination
of cyclotron mass from the amplitude of the signal induced in the pick-up
coil. As we shall now show, the method should have an ultimate sensitivity
for detecting dHvA oscillations comparable to that of field modulation and
it is to be hoped that this possibility will be exploited before long.

As was shown in §2.3.7.1 (see (2.128)) the temperature enters into the
dHvA effect only through an amplitude reducing factor for each periodic
term in the oscillatory magnetization. Thus if we denote any individual
periodic term (i.e. a particular harmonic of any dHvA frequency) by $\tilde{M}$ at
temperature T and by $\tilde{M}_0$ at $T = 0$, we have from (2.128)

$$\tilde{M} = \tilde{M}_0 f(z)$$

where   (3.45)

$$f(z) = z/\sinh z \quad \text{and} \quad z = 2\pi^2 pkT/\beta H$$

for the pth harmonic (usually we shall consider only the fundamental,
$p = 1$).

If now the temperature is modulated at angular frequency ω with
amplitude θ, we must replace T in (3.45) by $T + \theta\cos\omega t$ and z by
$z(1 + u\cos\omega t)$ where

$$u = \theta/T$$ (3.46)

Thus (3.45) becomes

$$\tilde{M} = \tilde{M}_0\left\{f(z) + zf'(z)u\cos\omega t + \tfrac{1}{2}z^2f''(z)u^2\cos^2\omega t\cdots\right.$$

$$\left. + \frac{1}{k!}z^k f^{(k)}(z)u^k\cos^k\omega t\cdots\right\}$$ (3.47)

124

and this can be written as a sum of harmonic terms $\cos k\omega t$, in which each coefficient is given by a power series in u. This is analogous to the procedure leading to (3.20) for field modulation. For field modulation, each power series in the modulation parameter λ can be summed to give a Bessel function $J_k(\lambda)$, but for temperature modulation the sum of each power series in the modulation parameter u is not expressible in terms of any standard functions. For simplicity we give the resulting expression only as far as $\cos 2\omega t$. This is

$$
\frac{\tilde{M}}{\tilde{M}_0} = f(z) + u\cos \omega t \left\{ zf'(z) + \frac{z^3 f^{(3)}(z)}{1!2!}\left(\frac{u}{2}\right)^2 \right.
$$

$$
\left. + \cdots + \frac{z^{2r+1} f^{(2r+1)}(z)}{r!(r+1)!}\left(\frac{u}{2}\right)^{2r} + \cdots \right\}
$$

$$
+ \frac{u^2}{4}\cos 2\omega t \left\{ z^2 f''(z) + \frac{z^4 f^{(4)}(z)}{1!3!}\left(\frac{u}{2}\right)^2 \right.
$$

$$
\left. + \cdots + \frac{z^{2r} f^{(2r+2)}(z)}{r!(r+2)!}\left(\frac{u}{2}\right)^{2r} + \cdots \right\} + \cdots \qquad (3.48)
$$

If, as in the field modulation method, a pick-up coil is coupled to the sample giving an e.m.f.

$$
v = c\mathrm{d}\tilde{M}/\mathrm{d}t
$$

this e.m.f. is (with $\tilde{M}$ now for $u = 0$)

$$
v = -c\tilde{M}\omega \left\{ u\sin \omega t \left[\frac{zf'(z)}{f(z)} + \frac{z^3 f^{(3)}(z)}{f(z)}\frac{u^2}{8} + \cdots \right] \right.
$$

$$
\left. + \frac{u^2}{2}\sin 2\omega t \left[\frac{z^2 f''(z)}{f(z)} + \frac{z^4 f^{(4)}(z)}{f(z)}\frac{u^2}{24} + \cdots \right] + \cdots \right\} \qquad (3.49)
$$

In practice it would probably be difficult to make u more than 0.2 or so (because the oscillating amplitude θ is inevitably accompanied by a steady rise of temperature, which is at least as large) so that for practical values of z ($\lesssim 5$) only the leading terms in the series expansions are likely to be significant. The function $-zf'(z)$ has a maximum of 0.65 for $z \doteqdot 2.5$, while $z^2 f''(z)$ has a maximum of 1.6 for $z \doteqdot 4$, so if a temperature modulation amplitude of $u = 0.2$ is possible, we find that the highest practicable amplitudes v_{1T} and v_{2T} of the rectified ω and 2ω signals are something like

$$
v_{1T} \sim 0.1\, c\omega M_0, \qquad v_{2T} \sim 3 \times 10^{-2}\, c\omega M_0 \qquad (3.50)
$$

where we have written M_0 for $|\tilde{M}_0|$. These may be compared with the

125

highest practicable amplitudes v_{1H} and v_{2H} for field modulation (assuming a typical situation with $z \sim 2$ so that $z/\sinh z \sim 0.5$). From (3.28) and (3.29) we find

$$v_{1H} \sim 0.5 \, c\omega M_0 \qquad v_{2H} \sim c\omega M_0 \qquad (3.51)$$

Although these figures show that thermal modulation produces at best rather weaker signals than can be achieved by field modulation, the comparison is in reality a little unfair because it takes no account of the reduction in noise level which can almost certainly be achieved by using temperature rather than field modulation. We have also implicitly assumed in this comparison that the optimum sample dimensions (which enter into the coupling constant c, since M is defined as per unit volume, so that c is proportional to volume V) and modulation frequency ω are much the same whether we modulate the temperature or the field. Although the design parameters are limited by somewhat different considerations in the two techniques, the assumption of comparable values of sample dimensions and of comparable ω for the two techniques does in fact prove to be reasonable.

At this point we may mention that in Oder and Maxwell's pioneer experiments the sensitivity achieved was a good deal smaller than that which might be achieved on the design considerations discussed above. Their modulation frequency was only 100 Hz as compared with perhaps 10^3 Hz which might be used and their amplitude of temperature modulation was only 10^{-2} K as compared with the $\frac{1}{2}$ K or so implied in (3.50), so it is not surprising that their limit of detection corresponded to $M_0 \sim 10^{-6}$ G rather than the 10^{-8} G implied if it is supposed that the smallest $|\tilde{M}|$ detectable by temperature modulation is about five times larger than that detectable by field modulation and (as in (3.42)) that $|\tilde{M}|_{\min} \sim 2 \times 10^{-9}$ G is the limit of the field modulation technique. It remains to be seen if the rough design considerations we have discussed above will prove realistic in practice.

As was mentioned earlier, in principle the temperature modulation technique offers the possibility of determining the cyclotron mass m from the amplitude of the signal measured at a single combination of temperature and field. If detection is at the fundamental of ω and the amplitude of temperature modulation is small, the absolute amplitude of v is given by (3.49) as

$$v_{1T} = c|\tilde{M}|\omega \frac{\theta}{T} \left\{ \frac{z}{\tanh z} - 1 \right\} \qquad (3.52)$$

Thus if all the other factors were known, $((z/\tanh z) - 1)$, (which for $z \gg 1$ is

3.4. Dynamic methods

approximately $(z - 1)$) could be determined and hence z and m. There are two possibilities of eliminating some of the awkward factors (in particular the factor $c|\tilde{M}|$):

(a) To measure the amplitude v_{2T} of the signal for detection at 2ω as well as v_{1T}. Since

$$v_{2T} = \tfrac{1}{2}c|\tilde{M}|\omega\left(\frac{\theta}{T}\right)^2\left\{\frac{2z^2}{\tanh^2 z} - \frac{2z}{\tanh z} - z^2\right\}$$

we see that the combination $c|\tilde{M}|\omega$ is eliminated by taking the ratio of v_{2T}/v_{1T}. We find

$$\frac{v_{2T}}{v_{1T}} = \frac{1}{2}\frac{\theta}{T}\left\{\frac{2z^2}{\tanh^2 z} - \frac{2z}{\tanh z} - z^2\right\}\Big/\left\{\frac{z}{\tanh z} - 1\right\} \tag{3.53}$$

(note that for $z \gg 1$ this is approximately $\tfrac{1}{2}(\theta/T)[z(z - 2)/(z - 1)]$).

(b) If, with the same pick-up coils and the same modulation frequency, the field is modulated instead of the temperature, the signal amplitude is given by (3.27) and for ω detection, with field modulation parameter λ,

$$v_{1H} = 2c|\tilde{M}|\omega J_1(\lambda).$$

Thus the ratio of the signals for temperature and field modulation is

$$v_{1T}/v_{1H} = \frac{1}{2}\frac{\theta}{T}\left(\frac{z}{\tanh z} - 1\right)\Big/J_1(\lambda) \tag{3.54}$$

However both (a) and (b) require the absolute amplitude θ of the temperature modulation to be accurately measured if z, and so m, is to be determined. This may prove difficult in practice.

A rather different possibility (again in principle) of determining z and so m, is to note that $f''(z)$ changes sign for $z = 1.6$. Thus if v_{2T} were observed (for low modulation amplitude) as H or T was varied to carry z through the critical value for which the signal vanishes, say for $H = H_0$, $T = T_0$, we should have

$$2\pi^2 kT_0/(e\hbar/mc)H_0 = 1.6$$

and so m could be determined. Unfortunately this method shares with method (a) above one awkward snag. This is that all our formulae are based on the assumption that the temperature modulation is strictly simple harmonic, and in reality this would probably be difficult to achieve because the thermal constants involved (the heat capacity of the sample and thermal resistance of the heat link) are strongly temperature dependent. If the temperature modulation contains any higher harmonics, then of course these will drastically modify the harmonics in the output signal. If, for instance, the modulation is described by $z(1 + u\cos\omega t + u'\cos(2\omega t + \psi))$, (3.49) will contain an extra term $2u'zf'(z)\sin(2\omega t + \psi)$ which may well be

127

appreciable compared with the sin $2\omega t$ term in (3.49). Possibly this difficulty might be circumvented by careful monitoring of the waveform of the temperature modulation so that some allowance could be made for the actual harmonic content, but it would be difficult to achieve any high accuracy from use of the second harmonic in the output signal. This snag and the difficulty of accurately measuring θ, the amplitude of temperature modulation, make it improbable that the method will ever compete with the more conventional methods of determining cyclotron mass (i.e. temperature variation of dHvA oscillation amplitudes or cyclotron resonance). This, however, does not mean that temperature modulation may not prove competitive as a sensitive method of detecting dHvA oscillations.

Finally it should be noted that much of the analysis of the temperature modulation technique can be easily adapted to *any* temperature dependent magnetization M even if it is not oscillatory in character. For instance the method can be very suitable (and has indeed been used by Legkostupov (1971)) to study the temperature dependence of ferromagnetic magnetization, and in particular the very slight dependence at low temperature. All that is needed to adapt the formulae is to redefine z, for instance as $z = T/T_c$, where T_c is the Curie temperature, and to treat $f(z)$ as an unknown function to be determined. With these modifications (3.49) is still applicable and by suitable measurements the form of the derivatives of $f(z)$ can be determined.

3.4.2.3. Modulation of gate voltage in a 2-D system

In the 2-D system provided by a silicon inversion layer the number of electrons per unit area which determines both the amplitude and frequency of the dHvA effect, can be varied by changing the gate voltage of the device. Thus modulation of the gate voltage is equivalent to modulating the number of electrons – a possibility which is not available for the more usual 3-D situation of a metal. We express the fundamental oscillation of the dipole moment $\mathscr{M}$ per unit area as

$$\mathscr{M} = \mathscr{M}_0 \sin \alpha n \qquad (3.55)$$

where n is the number of electrons per unit area and

$$\alpha = 2\pi^2 c\hbar/eH \qquad (3.56)$$

(see (2.61) or (2.80) and (2.86)) and now suppose n is modulated, i.e.

$$n = n_0 + v \cos \omega t \qquad (3.57)$$

In order not to complicate the calculation we ignore the fact that $\mathscr{M}_0$ is itself proportional to n and consider only the modulation of the dHvA frequency (it is easily shown that this is a good approximation if $\alpha n \gg 1$, and it is still a reasonable approximation for the practical situation in which αn may be only two or three times 2π). We then find analogously to (3.26) (omitting the term independent of t and noting that the minus of (3.24) becomes a plus in the present context)

$$\mathscr{M}(t) = 2\mathscr{M}_0 \sum_{n=1} J_k(\alpha v) \cos k\omega t \sin\left(\alpha n_0 + \frac{k\pi}{2}\right) \tag{3.58}$$

and the induced voltage v in the pick-up coil is

$$v = c'\,\mathrm{d}\mathscr{M}/\mathrm{d}t \equiv \sum v_k \sin k\omega t \tag{3.59}$$

so we find in analogy to (3.27)

$$|v_k| = 2c'\omega k\mathscr{M}_0 J_k(\alpha v) \tag{3.60}$$

It follows that the maximum signal in any pick-up coil for this kind of modulation is identical with that for field modulation. The important difference, however, is that it should be much easier to modulate the gate voltage up to the Bessel maximum than it would be to modulate the field. For instance with 2ω detection, the requirement to reach the maximum of $J_2(\alpha v)$ is $\alpha v = 3$, i.e, the modulation amplitude corresponds to a gate voltage which would shift a dHvA oscillation by about half a period. This corresponds to an amplitude of something like a volt or so which is much easier to achieve than an amplitude of 500 G or more in field modulation. Thus the estimate of (3.44) that $\mathscr{M}_{\min} \sim 3 \times 10^{-9}\,\mathrm{G\,cm}$ – a few times less than the expected magnitude of $\mathscr{M}$ – is more likely to be realistic with gate voltage modulation than with field modulation.

3.5. Dependence of frequency on stress

As we shall see in chapter 5, studies of the strain dependence of the Fermi surface provide valuable auxiliary information for the understanding of band structure. This strain dependence can either be inferred indirectly from the oscillatory magnetostriction and oscillatory sound velocity, as described in detail in chapter 4, or directly by actually applying a stress and observing the change of de Haas–van Alphen frequency. In this section we shall give a brief review of the experimental problems involved in using the direct method.

First we must note that the effects of stress are rather slight. As a very rough guide we may suppose the relative change of frequency to be comparable to the relative dimensional change of the sample, i.e. to the strain. Thus, even a pressure as high as 10 kbar ($10^{10}\,\mathrm{dyne\,cm}^{-2}$) which reduces the volume of a metal by something like 1% or so will typically produce a change of only 1% in F, while for other types of stress, plastic deformation limits the permissible strain and consequent change in F to less than 0.1%. The most sensitive and reliable way of measuring such small effects is to follow the changes in dHvA phase as the stress is applied. Since typically a phase change of one full oscillation corresponds to a frequency change of 1 in 10^3 or less, and much smaller phase changes can be reliably measured, it is possible to achieve good accuracy with quite modest stresses.

So far most of the direct experiments have been on hydrostatic pressure dependence. Here the phase shift method has the advantage that its high sensitivity makes possible the use of liquid helium as the pressure transmitting agent in spite of the limit of 25 bar set on the pressure by the solidification of helium at about 1 K (75 bar at 3 K, but the higher temperature usually outweighs the advantage of higher pressure). This technique was pioneered by Templeton (1966) who used the persistent mode of a superconducting magnet to provide the necessary field stability and was able to detect a phase shift as small as 10^{-3} of a full oscillation (see fig. 3.12). This meant that the actual phase shift of a 'belly' oscillation in copper (of order 0.2 of an oscillation for the highest pressure) could be measured to rather better than 1%. An ingenious feature of his experiment was the measurement of the relative phase shift of the much slower 'neck' oscillations which appear together with the 'belly' oscillations for H along $\langle 111 \rangle$ (see §5.3.2). In fact the change $\Delta F_N/F_N$ of the neck low frequency is a good deal greater than the change $\Delta F_B/F_B$ of the belly high frequency and the difference shows up if the same few slow oscillations are swept at different pressures, by an appearance of the fast oscillations sliding slightly relative to the slow ones. From the amount of 'sliding' the difference $\Delta F_N/F_N - \Delta F_B/F_B$ can be easily determined. This type of method has since been refined in various ways; for details see Templeton (1974, 1981).

The alternative method, of going to much higher pressures to produce much larger frequency changes (e.g. Schirber 1970, Melz 1966) suffers from two disadvantages. First the pressure has to be transmitted through solid helium and it is difficult to ensure that a purely hydrostatic pressure is applied to the sample, and secondly it is no longer possible to use the accurate phase shift method because it is not practicable to follow what is happening to the oscillations while the pressure is being applied. Instead the frequency must be measured absolutely at each pressure and if the change of F with pressure is only 1% or so, the absolute accuracy must be of order 1 in 10^4 to achieve 1% accuracy in ΔF. This 'brute force' method is, however, unavoidable if, as in the alkalis, the volume change at high pressures no longer varies linearly with pressure and it is desired to study how the FS changes in this regime.

130

3.5. Dependence of frequency on stress

The effects of stresses other than hydrostatic pressure are rather more difficult to study directly and only a few experiments have been made on the effect of uniaxial tension and compression (Verkin and Dmitrenko 1958, Shoenberg and Watts 1967, Gamble and Watts 1973, Perz and Hum 1971, Gerstein and Elbaum 1973, Mayers and Watts 1978). In most of these experiments the phase shift technique is used to determine the very small changes in frequency that can be tolerated before the crystal undergoes irreversible damage by plastic deformation. The practical difficulties are in ensuring that the applied stress is reliably transmitted to the sample (for instance without loss by friction through a seal), in producing and measuring a uniform stress (this implies having an accurately cylindrical

Fig. 3.12. Effect of hydrostatic pressure on belly oscillations in Au $\langle 111 \rangle$ at about 50 kG (Templeton 1966). The superconducting magnet is kept in the persistent mode (i.e. constant H) and the pressure is increased and reduced in seven steps each of about 3.5 bars; between successive steps an exact cycle is swept out by passing current through an auxiliary coil (1 period $\sim$ 5 G). The steps in phase correspond to about 0.01 of a cycle or $\Delta F/F \sim 10^{-6}$ (since $F/H \sim 10^4$) giving d $\ln F/dp \sim 3 \times 10^{-13}$ dyne^{-1} cm^2 (cf. 2.9 $\pm$ 0.1 $\times$ 10^{-13} from detailed analysis of the data).

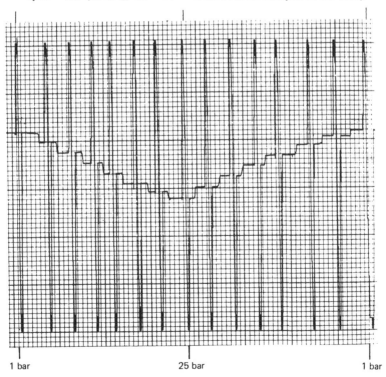

1 bar   25 bar   1 bar

131

sample) and in avoiding any rotation of the sample when stress is applied. The techniques have improved considerably since the experiment was first tried, but for determining strain dependence of the FS the direct method is still hardly as accurate and is also not as flexible in its ability to yield general strain derivatives, as the indirect methods to be described in chapter 4.

4

Other oscillatory effects

4.1. Introduction

In this chapter we shall review some of the other properties of a metal which oscillate as the magnetic field is varied. These fall into two categories. The first comprises essentially thermodynamic properties and their oscillatory variation with field can be derived directly from the oscillatory part $\tilde{\Omega}$ of the thermodynamic potential. In this category fall the magnetic properties, i.e. the de Haas–van Alphen effect, which we have already discussed, the thermal properties (the sample temperature and specific heat), the 'mechanical' properties (the sample dimensions, i.e. magnetostriction and elastic properties) and the chemical potential (i.e. oscillations in the Fermi energy).

The second comprises properties which involve non-equilibrium properties and so cannot be derived from the thermodynamic potential alone. Their oscillatory variation with H, however, is due to the same basic cause, the passage of the Landau tubes through the Fermi surface, and so shows essentially the same periodicity as do the oscillations of thermodynamic properties. Since the theory of these non-equilibrium properties inevitably involves more difficult considerations than those involved in thermodynamic properties, we shall not attempt more than rather simplified treatments, and only two effects: oscillations in electrical resistivity (Shubnikov–de Haas effect) and oscillations in attenuation of ultrasonic waves (including the so-called 'giant quantum oscillations'), will be discussed in any detail. Oscillations in other properties, such as optical properties and nuclear magnetic resonance will be only briefly mentioned.

Finally for the sake of completeness we shall mention a number of other oscillatory effects whose periodicity is determined by quite different considerations, but we shall do little more than indicate the nature of the relevant mechanisms.

4.2. Oscillatory thermal properties

Since the entropy S is given by the temperature derivative of the thermodynamical potential Ω, S and its derivatives will, like Ω, have components which oscillate as the field H is varied. The most direct manifestations of this oscillatory dependence are oscillations in the temperature of a thermally isolated sample (*magnetothermal oscillations*) and in the specific heat *c*. Both effects are small in the sense that the amplitude of the oscillations is usually less than 0.1% of the steady temperature or the steady specific heat, but because of the great sensitivity of low temperature thermometry (changes ΔT as small as 10^{-7} K or perhaps even 10^{-8} K can be detected), the oscillations can be fairly easily observed. The magnetothermal oscillations have indeed been used to determine Fermi surfaces through measurements of the orientation dependence of the frequency, though we shall see that probably they are not quite competitive as regards sensitivity with the oscillations of magnetic properties. It should be noticed that the theory which follows is in some ways closely analogous to that of the temperature modulation technique for studying the dHvA effect. Essentially this is because the temperature derivative of the oscillatory magnetisation is involved in both the phenomena concerned.

4.2.1. Magnetothermal oscillations

The magnitude of the temperature change for an adiabatic change of field is most simply obtained from the thermodynamic relation

$$\left(\frac{\partial T}{\partial H}\right)_S = -\left(\frac{\partial S}{\partial H}\right)_T \bigg/ \left(\frac{\partial S}{\partial T}\right)_H = -\frac{T}{c}\left(\frac{\partial M}{\partial T}\right)_H \tag{4.1}$$

where *c* is the specific heat per unit volume defined by

$$c = T(\partial S/\partial T)_H \tag{4.2}$$

Thus, considering only the oscillatory part $\widetilde{\Delta T}$ of the temperature change, and bearing in mind that the changes in *T* and *c* are always very small compared with their mean values,

$$\widetilde{\Delta T} = -\frac{T}{c}\int^H (\partial \tilde{M}/\partial T)\,\mathrm{d}H \tag{4.3}$$

As usual, it is convenient to confine our attention to a single oscillatory term, say the *p*th harmonic of a particular dHvA frequency *F*, for which

$$\tilde{M} = A\sin(2\pi pF/H + \phi) \tag{4.4}$$

and we shall again use the notation of (3.45)

$$\tilde{M} = \tilde{M}_0 f(z), \quad z = 2\pi^2 pkT/\beta H, \quad f(z) = z/\sinh z \tag{4.5}$$

134

4.2. Oscillatory thermal properties

together with

$$\tilde{\Omega} = \tilde{\Omega}_0 f(z) = -\frac{H^2 A}{2\pi pF} \cos(2\pi pF/H + \phi) \tag{4.6}$$

where $\tilde{\Omega}$ is the oscillatory part of the thermodynamic potential as given by (2.151) and $\tilde{\Omega}_0$ is its value for $T = 0$. We then easily find from (4.3), with the usual approximation, valid for $F/H \gg 1$, that only the oscillatory factor needs to be differentiated w.r.t. H.

$$\widetilde{\Delta T} = \frac{zf'(z)}{cf(z)}\tilde{\Omega} \quad \text{and} \quad |\widetilde{\Delta T}| = -\frac{H^2}{2\pi pFc}\frac{zf'(z)}{f(z)}|\tilde{M}| \tag{4.7}$$

The phase of $\widetilde{\Delta T}$ differs by π from that of $\tilde{\Omega}$ since $f'(z)$ is negative and is in quadrature to that of $\tilde{M}_0$ or $\tilde{M}$.

Before discussing the general result (4.7) we shall first consider two limiting situations for which $zf'(z)/f(z)$ assumes relatively simple analytic forms:

(a) $z \ll 1$

   (This is nearly valid for H along the binary axis in Bi at $T = 1$ K, $H \gtrsim 5$ kG, though (4.7) is then no longer a very good approximation because F/H is no longer very large.)

$$\frac{zf'(z)}{f(z)} \equiv \left(1 - \frac{z}{\tanh z}\right) \simeq -\frac{1}{3}z^2 = -\frac{1}{3}p^2z_1^2 \tag{4.8}$$

where z_1 is the value of z for $p = 1$. Thus

$$|\widetilde{\Delta T}| = \frac{p}{6\pi}\frac{H^2z_1^2}{Fc}|\tilde{M}| = \frac{2\pi^3k^2T^2}{3\beta^2Fc}p|\tilde{M}| \tag{4.9}$$

In this limit, then, it can be seen that the pth harmonic is p times stronger in $\widetilde{\Delta T}$ than in $\tilde{M}$. If we express $|\tilde{M}|$ as $(H^2/2\pi pF)|d\tilde{M}/dH|$, (4.9) becomes

$$|\widetilde{\Delta T}| = \frac{\pi^2k^2T^2H^2}{z\beta^2F^2c}\left|\frac{d\tilde{M}}{dH}\right| \tag{4.10}$$

which shows that the relative harmonic content of $\widetilde{\Delta T}$ in this limit is the same as that of $d\tilde{M}/dH$. The H^2 factor, however, means that for $z \ll 1$ the ΔT oscillations increase roughly as $H^{1/2}$ in contrast to those of $d\tilde{M}/dH$ which vary roughly as $H^{-3/2}$ (ignoring the Dingle factor). These features can be qualitatively seen in fig. 4.1. Since βF (which is proportional to F/m) does not vary greatly from one part of the FS to another, the relative amplitudes of oscillations of different F's for given H should be similar for $\widetilde{\Delta T}$ and for $d\tilde{M}/dH$.

(b) $z \gg 1$

   We now have

$$zf'(z)/f(z) \simeq (1 - z) \simeq -z = -pz_1$$

135

Other oscillatory effects

Fig. 4.1. (a) Magnetothermal oscillations in Bi for H along a binary axis (Kunzler *et al.* 1962) $T \sim 1.3$ K. (b) dHvA oscillations of dM/dH, $T \sim 0.6$ K (unpublished data of Barklie and Shoenberg 1974). The rate of sweep was not uniform, so distance along chart is not quite linear in H; the fields of the various oscillations are as marked in (a). Spin-splitting in the last oscillation (and the next to last in (b)) is clearly visible. Comparison of (a) and (b) illustrates the similarity of line shape in the two effects and the difference of the H dependence of amplitude. However, the comparison can only be qualitative since (a) and (b) are at different temperatures on different crystals and moreover $z \ll 1$ is true for only the last two or three oscillations before the quantum limit (at about 15 kG). Further illustrations of the oscillations of dM/dH in Bi are shown in figs. 8.8 and 8.9.

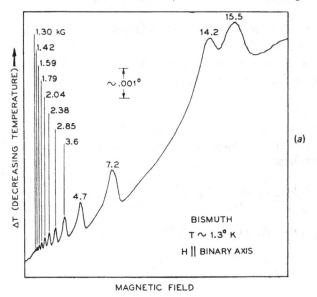

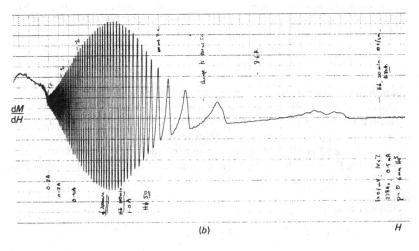

136

so that

$$|\widetilde{\Delta T}| \simeq \frac{H^2 z_1}{2\pi Fc}|\tilde{M}| = \frac{\pi k T H}{\beta Fc}|\tilde{M}| \qquad (4.11)$$

In this limit then (and this is the usual practical situation), the oscillations of $\widetilde{\Delta T}$ have a harmonic content which is the same as that of $\tilde{M}$. However, usually in the limit $z \gg 1$, the harmonic content is anyhow very small, so the p-dependence of $\widetilde{\Delta T}$ is only of academic interest. Since βF does not vary greatly over the FS, the relative amplitudes for different Fs are comparable in $\widetilde{\Delta T}$ and $\tilde{M}$.

Estimation of the order of magnitude of the amplitude of the temperature oscillations and evaluation of the conditions needed to produce the largest amplitude is not quite straightforward because of the variety of c values for different metals and at different temperatures. These questions are discussed briefly in Appendix 9 and it proves that the largest amplitude of $\widetilde{\Delta T}$ ranges from about 10^{-3} K for Bi (at $T \sim 0.1$ K) to a few times 10^{-4} K for the noble metals (at $T \sim 1$ K). These orders of magnitude indicate that the effect can be easily observed, since temperature changes of 10^{-7} K were detectable in the experiments of Halloran and Kunzler (1968) and the development of thermometric and electronic techniques since then would probably set the limit of detection even lower.

The experimental technique for observing magnetothermal oscillations is fairly simple and needs no special description, but one general point should be mentioned. This is that the sensitivity may be appreciably enhanced by field modulation and indeed a detection limit of 10^{-7} K or smaller can only be achieved in this way. The theory of field modulation for the magnetothermal effect is closely similar to that discussed earlier for the dHvA effect (§3.4.2.1) and if detection is at the kth harmonic of the modulation frequency it is easily shown that the amplitude of the $k\omega t$ term in $\widetilde{\Delta T}$ for a given field H is $2J_k(\lambda)$ times the value of ΔT at that field without modulation, where λ is the modulation parameter defined by (3.25). Thus, as H is varied, the oscillations of the amplitude of the $k\omega t$ term in the modulation technique have amplitudes $2J_k(\lambda)$ times those of $\widetilde{\Delta T}$ observed without modulation. The maximum values of $2J_k(\lambda)$ for $k = 1, 2, 3$ are 1.164, 0.972 and 0.868 respectively, so at first sight modulation might appear to produce little gain. However, modulation has the important advantage that it makes possible the use of phase sensitive detection and so greatly reduces the noise level. Moreover the use of modulation means that the thermal relaxation time τ of the link between the sample and the liquid helium bath may be greatly reduced. Indeed it need only satisfy the condition $\omega\tau \gg 1$, rather than the condition $\tau \gg t_0$, where t_0 is the time to sweep H over one dHvA oscillation. Since t_0 is usually at least a few seconds, while $1/\omega$ can be of the order of milliseconds, this means that the requirements on thermal isolation can be greatly reduced if field modulation is used.

The experimental techniques for studying magnetothermal oscillations are perhaps marginally simpler than those of observing the conventional dHvA effect, though if sensitivity is to be pushed to the limit the difference in simplicity is not great. However if the ultimate sensitivity of $\widetilde{\Delta T}$ is translated into one of $\tilde{M}$ it turns out that direct measurement of $\tilde{M}$ is usually appreciably more sensitive. Thus (as discussed in Appendix 9), if we suppose the limit of detection of $|\Delta T|$ is 10^{-8} K, this corresponds to a limit of a few times 10^{-6} G for detection of $|\tilde{M}|$ in a large FS metal, which is several orders of magnitude higher than can be achieved by magnetic techniques. The magnetothermal oscillations are only a little more competitive for Bi, for which the limiting equivalent $|\tilde{M}|$ which can be detected might be as small as a few times 10^{-7} G. One disadvantage of the magnetothermal method as compared with some of the dHvA methods is that it is not simple to make the method absolute because of the difficulty of determining the specific heat c sufficiently precisely. Part of the difficulty is that the c in the formulae is not just a property of the sample alone, but contains contributions, which may be difficult to assess, from the heat capacities of the sample mounting and the thermal link.

Finally it should be mentioned that, just as with the temperature modulation technique, the magnetothermal effect combined with field modulation might provide a useful method for studying dM/dT in non-oscillatory situations and in particular for ferromagnets. The main difficulty would be to eliminate spurious signals from the resistance thermometer, for instance due to induced e.m.f.s and magnetoresistance; eddy current heating effects would occur at 2ω and so could easily be eliminated. Such spurious signals are of course less important in the observation of oscillatory effects since, provided they are not overwhelming, they modify only the base line of the oscillations.

4.2.2. Oscillations in specific heat

The oscillatory part $\Delta \tilde{c}$ of the specific heat is given by

$$\Delta \tilde{c} = T(\partial \tilde{S}/\partial T)_H = -T\frac{\partial}{\partial T}\left(\frac{\partial \tilde{\Omega}}{\partial T}\right)_H$$

and if (4.6) is used, this reduces to

$$\Delta \tilde{c} = \frac{-z^2 f''(z)}{Tf(z)}\tilde{\Omega} \tag{4.12}$$

138

4.2. Oscillatory thermal properties

The order of magnitude of the effect is most readily appreciated by noting that

$$\frac{|\Delta\tilde{c}|/c}{|\widetilde{\Delta T}|/T} = \left|\frac{zf''(z)}{f'(z)}\right| \tag{4.13}$$

This ratio vanishes for $z = 1.6$ and then rises to 1.56 for $z = 3$; for higher values of z (for which, however, the oscillation amplitude of $\Delta\tilde{c}$ or $\widetilde{\Delta T}$ falls off as ze^{-z}) the ratio is approximately $z(z-2)/(z-1)$. Thus bearing in mind the earlier estimates of $|\widetilde{\Delta T}|/T$ we see that the oscillations of c amount at best to a few percent (for Bi) but are usually 10 or 100 times smaller.

The oscillations in c have been observed experimentally in Be by Sullivan and Seidel (1967) by a method somewhat similar to that involved in the temperature modulation technique for observing the dHvA effect. Thus if a sample of heat capacity C and separated from the liquid helium bath by a link of thermal resistance W is heated by an a.c. source of power $P(1 + \cos\omega t)$, the amplitude θ of the alternating temperature rise of the sample is given by

$$\theta = PW/(1 + \omega^2\tau^2)^{1/2} \tag{4.14}$$

where $\tau = CW$. Thus if $\omega\tau$ is made $\gg 1$, we find

$$\theta = P/C\omega \quad \text{and} \quad |\tilde{\theta}| = P|\widetilde{\Delta C}|/C^2\omega$$

so that an oscillatory variation of C shows up as an oscillatory variation of θ if the field H is slowly swept. In Sullivan and Seidel's experiment τ was 1.2 s and $\omega/2\pi$ was 10 Hz.

As can be seen from fig. 4.2, oscillations of c with an amplitude approaching 1% of c were convincingly demonstrated. The oscillations $\tilde{\theta}$ are particularly large for Be partly because it has a fairly low specific heat ($\sim 5 \times 10^2$ erg cm^{-3} K^{-1} at 1.5 K) and partly because the relevant part

Fig. 4.2. Specific heat oscillations in Be; H along hexagonal axis, $T \sim 1.5$ K (Sullivan and Seidel 1967).

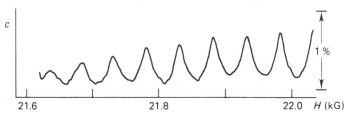

of the FS (the 'cigar') is nearly cylindrical, so that the curvature factor in the oscillatory amplitude is particularly large.

As with the magnetothermal effect, the sensitivity of detection of specific heat oscillations could be enhanced by field modulation, but if the a.c. heating method of Sullivan and Seidel were used, the field modulation would have to be carried out at a frequency appreciably lower than that of the a.c. heating.

4.3. Oscillatory mechanical effects

Corresponding to the oscillatory field dependence of the thermodynamic potential, there must also be an oscillatory field dependence of its derivative with respect to stress, i.e. of its strain under a fixed stress. This is the oscillatory *magnetostriction* and clearly its amplitude will be proportional to the stress derivative of the de Haas–van Alphen frequency F. This then provides a method of studying the distortion of the Fermi surface when stress is applied to the metal. Similarly the derivative of strain with respect to stress, i.e. the elastic compliance, must also show an oscillatory field dependence which can be observed as an oscillatory field dependence of the velocity of sound. This provides yet another handle on the stress dependence of the FS.

Oscillatory magnetostriction was first observed by Green and Chandrasekhar (1963) in Bi, and oscillations in the velocity of sound by Mavroides, Lax, Button and Shapira (1962), again in Bi. Both effects have since been observed in many other metals. As we shall see below, it turns out that observations on the magnetostriction oscillations, on the velocity of sound oscillations and on the direct dHvA oscillations combined together in various ways enable the stress or strain dependence of the FS to be studied in detail. Such techniques have proved to be rather more powerful and versatile than the direct approach of observing the change of F when a static stress is applied to the sample (see §3.5), though in some respects the two approaches are complementary. We shall now discuss the general principles of the oscillatory effects and show how the basic information on stress dependence of the FS can be extracted. The application of these principles to individual metals is sometimes a little complicated because of the complications of elastic anisotropy, so we shall not discuss the practical application of the principles in too much detail. Further details may be found in a recent review by Fawcett, Griessen, Joss, Lee and Perz (1980).

4.3.1. Oscillatory magnetostriction

If we generalize (2.151) to the case of a number of different dHvA frequencies and take account of the elastic properties explicitly, we can

140

express the thermodynamic potential Ω as a function of the stress tensor σ

$$\Omega = \tilde{\Omega} + \Omega_0 \equiv \sum_r \Omega_r \cos\left(\frac{2\pi F_r}{H} + \varphi_r\right) - \tfrac{1}{2}(s_{ikpq})_0 \sigma_{ik}\sigma_{pq} \qquad (4.15)$$

where $(s_{ikpq})_0$ is the elastic compliance tensor in the absence of a field and the F_r notation can be taken to include the harmonics of any basic frequency as well as the different basic frequencies.

The strain ε_{ik} for given stress σ_{pq} is then given by

$$\varepsilon_{ik} = -\partial\Omega/\partial\sigma_{ik} = \tilde{\varepsilon}_{ik} + (s_{ikpq})_0 \sigma_{pq} \qquad (4.16)$$

where

$$\tilde{\varepsilon}_{ik} = -\partial\tilde{\Omega}/\partial\sigma_{ik} = \sum_r \Omega_r \frac{2\pi}{H}\frac{\partial F_r}{\partial\sigma_{ik}}\sin\left(\frac{2\pi F_r + \phi_r}{H}\right) \qquad (4.17)$$

If the oscillatory magnetization $\tilde{M}$ is similarly expressed as

$$\tilde{M} = \sum_r M_r \sin\left(\frac{2\pi F_r}{H} + \phi_r\right) = \sum \tilde{M}_r \qquad (4.18)$$

we have that

$$M_r = -\frac{2\pi F_r}{H^2}\Omega_r \qquad (4.19)$$

so

$$\tilde{\varepsilon}_{ik} = -\sum_r (\partial \ln F_r/\partial\sigma_{ik})\tilde{M}_r H \qquad \mathbf{(4.20)}$$

The relation of the actual oscillatory change $\widetilde{\Delta l}$ of a sample dimension l to ε_{ik} is discussed later, but evidently $\widetilde{\Delta l}/l$ is of the same order of magnitude as the largest component of $\tilde{\varepsilon}_{ik}$ and we shall refer loosely to $\tilde{\varepsilon}_{ik}$ as the oscillatory magnetostriction, though strictly oscillatory magnetostriction means $\widetilde{\Delta l}/l$. It should be noticed that if there is only a single basic F (so that r refers only to the various harmonics) $\partial \ln F_r/\partial\sigma_{ik}$ is independent of r and (4.20) simplifies to

$$\tilde{\varepsilon}_{ik} = -(\partial \ln F/\partial\sigma_{ik})\tilde{M}H \qquad (4.21)$$

In general, however, $\partial \ln F_r/\partial\sigma_{ik}$ will be different (sometimes very different) for the different basic F_rs corresponding to different FS sections. Thus the spectrum of the oscillatory magnetostriction may show relative amplitudes which are quite different from those in the spectrum of the dHvA effect (i.e. of $\tilde{M}$). In particular, 'small' extremal areas of the FS, such as the necks in the monovalent metals or that occur more generally in polyvalent metals,

141

are much more sensitive to stress than 'large' free electron-like extrema, and the corresponding low frequencies show up relatively more strongly than the high frequencies in the magnetostriction oscillations as compared to the magnetization oscillations.

The order of magnitude of $|\tilde{\varepsilon}_{ik}|$ is easily estimated for a free electron-like spherical FS, the volume of which is simply inversely proportional to the volume V of the sample as the stress is varied. For this simple model, since F is proportional to the area A of the diametral section of the sphere, we have

$$\partial \ln F / \partial \sigma_{ik} = -\tfrac{2}{3} \partial \ln V / \partial \sigma_{ik} \qquad (4.22)$$

If for concreteness we consider a uniaxial tension σ per unit area, (4.22) becomes

$$\partial \ln F / \partial \sigma = -\tfrac{2}{3}(1 - \alpha)/Y \qquad (4.23)$$

where Y is Young's modulus and α is Poisson's ratio. For the noble metals this gives

$$\partial \ln F / \partial \sigma \sim 5 \times 10^{-13} \, \text{dyne}^{-1} \, \text{cm}^2$$

Direct experiments on the effect of uniaxial tension (Shoenberg and Watts 1967) show that this is indeed a reasonable estimate for the bellies of the noble metals, which can be described qualitatively by the free electron model. For the 'necks' however, $\partial \ln F / \partial \sigma$ is a good deal larger, typically $10^{-11} \, \text{dyne}^{-1} \, \text{cm}^2$, while for Bi, with H along the binary axis, it is larger still, roughly $3 \times 10^{-10} \, \text{dyne}^{-1} \, \text{cm}^2$ (Brandt and Ryabenko 1959). If we combine these estimates (collected in table 4.1) with those of $|\tilde{M}|$ (see table A7.1), assuming that for a noble metal at $H = 10^5 \, \text{G}$, $|\tilde{M}| = 0.5 \, \text{G}$ for a belly and $|\tilde{M}| = 0.1 \, \text{G}$ for a neck*, while for Bi at $H = 5 \times 10^3 \, \text{G}$, $|\tilde{M}| = 0.05 \, \text{G}$, we find the estimates of the oscillatory magnetostriction $|\tilde{\varepsilon}|$ shown in table 4.1.

Since the sample length l must be kept small–typically a few mm–to avoid phase smearing from field inhomogeneity, the amplitudes $\widetilde{\Delta l}$ are typically only a few $10^{-8} \, \text{cm}$. Such small length changes can be observed with a signal to noise ratio as high as 10^3 by making the change of length change the capacity of a condenser or the resonant condition of a microwave cavity. We shall not describe the details of the sophisticated electronic techniques required to achieve the necessary sensitivity (references to papers describing these techniques may be found in the reviews

---

* This estimate is not indicated in table A7.1, but is based on data for Au in Shoenberg and Vuillemin (1966). The amplitudes for Cu and Ag necks are rather smaller.

142

Table 4.1. Rough estimates of $\partial \ln F/\partial\sigma$, $|\tilde{\varepsilon}|$ and $\tilde{v}/v$

	H (G)	F (G)	$\partial \ln F/\partial\sigma$ (dyne^{-1} cm^2)	$|\tilde{\varepsilon}|$	Y (dyne cm^{-2})	$\tilde{v}/v$
Noble metal belly	10^5	5×10^8	5×10^{-13}	3×10^{-8}	10^{12}	2×10^{-4}
Noble metal neck	10^5	2×10^7	10^{-11}	10^{-7}	10^{12}	5×10^{-4}
Bismuth (H parallel to binary axis)	5×10^3	1.4×10^4	3×10^{-10}	10^{-7}	3×10^{11}	10^{-4}

of Chandrasekhar and Fawcett (1971) and Fawcett *et al.* (1980)). The signal to noise ratio could be improved by field modulation but it would be difficult to avoid skin effect complications because the samples used for magnetostriction are usually of rather larger diameter than dHvA samples.

As already indicated, the interest of studying oscillatory magnetostriction is that it provides the possibility of studying how the FS is distorted by stress or strain, but before discussing how (4.20) can be used to extract such information from oscillatory magnetostriction data, we shall first discuss the oscillatory field dependence of the elastic constants and the velocity of sound.

4.3.2. Oscillatory elastic properties and velocity of sound

If we differentiate (4.16) with respect to σ_{pq} we find

$$s_{ikpq} = (s_{ikpq})_0 + \tilde{s}_{ikpq}$$

where

$$\tilde{s}_{ikpq} \equiv \frac{\partial \tilde{\varepsilon}_{ik}}{\partial \sigma_{pq}} = \sum_r \left(\frac{\partial \ln F_r}{\partial \sigma_{ik}}\right)\left(\frac{\partial \ln F_r}{\partial \sigma_{pq}}\right)\frac{d\tilde{M}_r}{dH}H^2 \tag{4.24}$$

i.e. an oscillatory field dependence of the elastic compliance s_{ikpq}. Alternatively we can describe this oscillatory behaviour in terms of oscillations in c_{ikpq}, the elastic moduli. This is most simply done by starting with a potential Ω' defined in terms of strain as

$$\Omega' = \tilde{\Omega} + \tfrac{1}{2}(c_{ikpq})_0 \varepsilon_{ik}\varepsilon_{pq} \tag{4.25}$$

where $(c_{ikpq})_0$ is the elastic stiffness tensor. The stress for given strain is then $\partial\Omega'/\partial\varepsilon_{ik}$ and has an oscillatory component $\tilde{\sigma}_{ik}$ as well as the steady component $(c_{ikpq})_0 \,\varepsilon_{pq}$. Differentiating σ_{ik} with respect to ε_{ik} then gives

$$c_{ikpq} = (c_{ikpq})_0 + \tilde{c}_{ikpq}$$

where

$$\tilde{c}_{ikpq} \equiv \left(\frac{\partial \tilde{\sigma}_{ik}}{\partial \varepsilon_{pq}}\right) = -\sum_r \left(\frac{\partial \ln F_r}{\partial \varepsilon_{ik}}\right)\left(\frac{\partial \ln F_r}{\partial \varepsilon_{pq}}\right)\frac{d\tilde{M}_r}{dH}H^2 \tag{4.26}$$

The difference of sign between (4.24) and (4.26) merely reflects the reciprocal relation between the s and c tensors.

The oscillatory field dependence of the elastic properties manifests itself directly as oscillations in v, the velocity of sound or ultrasound (in practice usually the latter), which are most simply expressed in terms of the $\tilde{c}$

4.3.   Oscillatory mechanical effects

rather than the $\tilde{s}$ tensor. Thus in general* we can write

$$v = (c/\rho)^{1/2} \qquad (4.27)$$

where c is a linear combination of appropriate components of the tensor c_{ikpq} (depending on the direction of propagation in the crystal and the direction of particle motion in the wave; see Appendix 10). The amplitude of the oscillatory variation of ρ is of the same order of magnitude as $\tilde{\varepsilon}$ and, as we shall see directly, this is at least 10^3 times smaller than the amplitude of the oscillations in c, so we can safely ignore it and find for the oscillatory variation

$$\tilde{v}/v = \tfrac{1}{2}\tilde{c}/c \qquad (4.28)$$

The order of magnitude of the amplitude of oscillations in c may be estimated roughly by ignoring the tensor nature of s and c and simply identifying c or $1/s$ with Young's modulus Y; we then have from (4.24) or (4.26), together with (4.21)

$$|\tilde{v}|/v = \tfrac{1}{2}|\tilde{c}|/c = \tfrac{1}{2}|\tilde{s}|/s = \tfrac{1}{2}Y(\partial \ln F/\partial\sigma)(2\pi F/H)|\tilde{\varepsilon}|$$

or

$$= \tfrac{1}{2}(\partial \ln F/\partial\varepsilon)(2\pi F/H)|\tilde{\varepsilon}| \qquad (4.29)$$

and we obtain the estimates of $\tilde{v}/v$ shown in table 4.1. The experimental results of Mavroides et al. (1962) on bismuth and of Alers and Swim (1963) on the gold necks are reasonably consistent with the estimates of table 4.1, though in these pioneer experiments not enough information about the magnetic properties was available to make a detailed comparison possible.[†] Since with good technique v can be measured to 1 part in 10^7 or better, the oscillations in v (in the favourable conditions assumed in our estimates) are typically of amplitude a few thousand times the limit of detectability. Thus, as with the magnetostriction oscillations, the sound velocity oscillations are not difficult to observe, but the margin available for coping with unfavourable conditions is less than for the dHvA effect proper. We have already commented that oscillatory magnetostriction tends to emphasize 'low' frequencies more than does the dHvA effect. This is even

---

* We could also write $v = (1/s\rho)^{1/2}$, but s would in general be a rather unwieldly function of the individual components of the tensor s_{ikpq}, since each component c_{ikpq} is the reciprocal of a linear combination of components of s_{ikpq}.

[†] The value of $\tilde{v}/v$ for the Au necks found by Alers and Swim was in fact about 30 times smaller than our estimate of 5×10^{-4} but their Dingle damping was rather more severe than we have assumed, their temperature of measurement was 4.2 K rather than the 1 K we have assumed and their field orientation was unfavourable, so the discrepancy is not really significant.

more true for the sound velocity oscillations because of the extra derivative of F in (4.24) or (4.26) as compared with (4.20).

4.3.3. Extraction of strain dependence of the Fermi surface from oscillatory data

The main interest of studying magnetostriction and sound velocity oscillations is, of course, to extract from the amplitudes the strain dependence of the Fermi surface in order to compare it with the predictions of band theory.

The first step is to Fourier analyze the oscillations and pick out the amplitude of one particular dHvA frequency. Next the amplitude of $\tilde{M}$ (or equivalently of $d\tilde{M}/dH$) must be determined or otherwise dealt with, so that the stress or strain derivatives of $\ln F$ in (4.20) or (4.26) can be determined. In some of the earlier experiments (e.g. Slavin 1973) the absolute amplitude of $\tilde{M}$ was estimated by direct calculation from the LK formula using parameters determined in earlier dHvA experiments and determining the Dingle factor from the field dependence of the oscillations being studied. Unfortunately, it is only for a few metals that two of the relevant factors, the curvature factor $(|\partial^2 A/\partial k_H^2|)^{-1/2}$ and the spin-splitting factor $\cos[\frac{1}{2}\pi g(m/m_0)]$, are reliably enough known to make this an accurate approach.

In more recent experiments various ingenious arrangements have been used to eliminate $\tilde{M}$ by making *simultaneous* measurements of dHvA oscillations (i.e. effectively of $\tilde{M}$ or $d\tilde{M}/dH$) and of the magnetostriction or sound velocity oscillations. This also makes possible direct determination of the sign of $d\ln F/d\sigma$. A rather elegant solution of the problem, pioneered by Griessen and Sorbello (1972) (see also Griessen, Lee and Stanley 1977 for a more sophisticated version) was to measure the dHvA torque by the capacitor technique described earlier (§3.3.2) and simultaneously to measure the magnetostriction through the changes of another narrow gap capacitor. By bonding a quartz transducer to the top of the sample, the arrangement of Griessen *et al.* (1977) could also be modified to measure sound velocity oscillations instead of magnetostriction oscillations, simultaneously with torque oscillations. We shall not go into the rather complicated details of the geometry of the apparatus which made these combinations possible, but the important point is that since the oscillatory torque is related to $\tilde{M}$ by $(1/F)(dF/d\theta)$, which is accurately known (or can if desired be determined in the same experiment) it is possible to extract stress or strain derivatives of F from such simultaneous measurements. Yet another approach to the problem of how to eliminate the amplitude of $\tilde{M}$ from the formulae is to measure the magnetostriction and sound

4.3. Oscillatory mechanical effects

velocity oscillations on the same sample and to divide one amplitude by the other.

We must now consider a little more concretely just which aspects of the strain dependence are determined by the various experiments. The magnetostrictive change of length $\widetilde{\Delta l}/l$ along a direction (λ, μ, ν) with respect to the axes of reference is related to the strain tensor $\tilde{\varepsilon}_{ik}$ by

$$\widetilde{\Delta l}/l = \tilde{\varepsilon}_1 \lambda^2 + \tilde{\varepsilon}_2 \mu^2 + \tilde{\varepsilon}_3 \nu^2 + 2\tilde{\varepsilon}_4 \mu\nu + 2\tilde{\varepsilon}_5 \nu\lambda + 2\tilde{\varepsilon}_6 \lambda\mu \qquad (4.30)$$

where we have used the conventional abbreviation of a single suffix, 1–6, to replace the double suffixes ik. In what follows we shall identify the axes of reference with reference to the crystal axes, e.g. for a cubic crystal the suffices 1, 2, 3 will refer to the cube axes. We notice that if we measure $\widetilde{\Delta l}/l$ for three mutually perpendicular directions $\sum \widetilde{\Delta l}/l$ reduces to $\varepsilon_1 + \varepsilon_2 + \varepsilon_3$ or $\Delta \tilde{V}/V$ as it should (this follows from relations such as $\lambda_1^2 + \lambda_2^2 + \lambda_3^2 = 1$ and $\mu_1 \nu_1 + \mu_2 \nu_2 + \mu_3 \nu_3 = 0$). In principle, by measuring the oscillatory length changes along six different directions in the sample (but, of course, keeping the field always in the same crystallographic direction), we can determine all the separate components of oscillatory strain $\tilde{\varepsilon}_{ik}$ associated with a particular field direction. For instance, for a cubic crystal it would be sufficient to measure the length changes along each cube edge and face diagonal direction. From the separate strains $\tilde{\varepsilon}_a$, $\partial \ln F/\partial \sigma_a (a = 1\text{--}6)$ would then be given by (4.20) assuming that $\tilde{M}$ had been dealt with by one of the approaches discussed above.

To go from the stress dependence to the strain dependence, we use the relation

$$\frac{\partial \ln F}{\partial \sigma_{ik}} = \frac{\partial \varepsilon_{pq}}{\partial \sigma_{ik}} \frac{\partial \ln F}{\partial \varepsilon_{pq}} = s_{ikpq} \frac{\partial \ln F}{\partial \varepsilon_{pq}} \qquad (4.31)$$

in which the strict summation convention is implied, or with the abbreviated suffix notation (which does not lend itself to the summation convention)

$$\frac{\partial \ln F}{\partial \sigma_a} = \sum_{b=1}^{3} s_{ab} \frac{\partial \ln F}{\partial \varepsilon_b} + 2 \sum_{b'=4}^{6} s_{ab'} \frac{\partial \ln F}{\partial \varepsilon_{b'}} \qquad (4.32)$$

$(a = 1\text{--}6)$

It is important to note that for an arbitrary direction of H (i.e. an arbitrary extremal section of the Fermi surface), *all* six components of $\partial \ln F/\partial \sigma_a$ or of $\partial \ln F/\partial \varepsilon_a$ are needed to specify the stress or strain dependence of F, even though the crystal symmetry may be quite high; in a sense the presence of a magnetic field in an arbitrary direction has destroyed the symmetry of

147

the crystal. It is only for a few special directions of H that some of the derivatives of F vanish and others become equal.

Once the six individual $\partial \ln F / \partial \sigma_a$ (corresponding to the six $\tilde{\varepsilon}_a$) have been extracted from suitable magnetostriction measurements, the six individual $\partial \ln F / \partial \varepsilon_a$ can be determined by solving the equations (4.32). These equations are in practice considerably simplified by the symmetry of the s tensor $(s_{ab} = s_{ba})$ and by the restrictions of crystal symmetry (which of course apply to the unperturbed tensor s that enters in (4.32)). For instance, for cubic symmetry if the 1, 2, 3 axes coincide with the cube axes, the only non-vanishing components of s are

$$s_{11} = s_{22} = s_{33}, \quad s_{12} = s_{23} = s_{31} (= s_{21} = s_{32} = s_{13}),$$

$$s_{44} = s_{55} = s_{66} \tag{4.33}$$

Turning now to the oscillations of sound velocity, we see from (4.26) that the oscillatory components $\tilde{c}_{ab}$ of the elastic moduli c_{ab} are rather more directly linked to the strain derivatives $\partial \ln F / \partial \varepsilon_a$ than are the oscillatory strains $\tilde{\varepsilon}_a$ which determine the oscillatory magnetostriction. However, for an arbitrary direction of propagation (and for an arbitrary direction of H), the c which enters into the sound velocity v (see (4.27)) is a complicated linear function of the moduli c_{ab} even in the absence of a magnetic field and still more complicated when the oscillatory behaviour is considered. This problem is discussed in Appendix 10 but we need not worry about the general case if our main concern is only to extract the strain derivatives $\partial F / \partial \varepsilon_a$. For this purpose we need only consider propagation along symmetry axes. Thus, for propagation in a cubic crystal along one of the cubic axes, the oscillations $\tilde{v}$ are given (for an arbitrary direction of H and one particular oscillatory frequency F) by

$$\frac{\tilde{v}}{v} = \frac{1}{2} \frac{\tilde{c}_{aa}}{c_{aa}} = \frac{1}{2} \left(\frac{\partial \ln F}{\partial \varepsilon_a} \right)^2 \frac{d\tilde{M}}{dH} \frac{H^2}{c_{aa}} \tag{4.34}$$

where if $a = 1, 2$ or 3 the waves are longitudinal and propagation is along the 1, 2 or 3 axis respectively, while if $a = 4, 5$ or 6 the waves are transverse, with propagation and vibration directions along 2, 3 (or *vice versa*) for $a = 4$, and correspondingly for $a = 5$ or 6. In (4.34) the unperturbed moduli conform to cubic symmetry so that

$$c_{11} = c_{22} = c_{33}, \quad c_{44} = c_{55} = c_{66}$$

but as far as the oscillatory contributions are concerned, the cubic symmetry is destroyed by the presence of a magnetic field in an arbitrary direction and in general

$$\tilde{c}_{11} \neq \tilde{c}_{22} \neq \tilde{c}_{33}, \qquad \tilde{c}_{44} \neq \tilde{c}_{55} \neq \tilde{c}_{66}.$$

It can be seen from (4.34) that if $d\tilde{M}/dH$ is known, measurements of sound velocity for longitudinal and transverse waves along each cubic axis give the six strain derivatives of F individually. It should be noticed, however, that since it is the square of a derivative that enters in (4.34) it is impossible to determine the sign of $\partial \ln F/\partial \varepsilon_a$ (the oscillations of v will always be in phase with those of $d\tilde{M}/dH$). Apart from this difficulty of sign, the whole of the strain dependence of the Fermi surface can in principle be determined by carrying out the basic measurements of oscillatory sound velocity for all directions of H relative to the crystal axes.

We now return briefly to the possibility of eliminating $\tilde{M}$ by combining measurements of oscillatory magnetostriction and sound velocity. This inevitably gives somewhat complicated combinations of the $(\partial \ln F/\partial \varepsilon_a)$. For instance the ratio of oscillatory longitudinal sound velocity $\tilde{v}_{\mathrm{L}}$ to the magnetostriction amplitude $\widetilde{\Delta l}$ along the [100] axis of a cubic crystal is given by

$$\frac{(\tilde{v}_{\mathrm{L}}/v_{\mathrm{L}})_{100}}{(\widetilde{\Delta l}/l)_{100}} = \frac{(\pi F/c_{11}H)(\partial \ln F/\partial \varepsilon_1)^2}{s_{11}(\partial \ln F/\partial \varepsilon_1) + s_{12}(\partial \ln F/\partial \varepsilon_2 + \partial \ln F/\partial \varepsilon_3)} \qquad (4.35)$$

and similar equations in which the suffices of ε are cyclically changed (but those of c_{11}, s_{11} and s_{12} left unchanged) give the ratios for measurement along the [010] and [001] axes (the field remaining of course in the same direction throughout). These three simultaneous quadratic equations can be solved to give the separate $\partial \ln F/\partial \varepsilon_a$ ($a = 1, 2, 3$). For transverse wave velocity amplitudes c_{11} in (4.35) is replaced by c_{44} and in the numerators of the above equations the $(\partial \ln F/\partial \varepsilon_a)^2$ ($a = 1, 2, 3$) are replaced by $(\partial \ln F/\partial \varepsilon_{a'})^2$ ($a' = 4, 5, 6$) so that these derivatives too can be determined. Another possibility is to measure the magnetostriction transverse to the direction of the sound velocity. For example the ratio $(\tilde{v}_{\mathrm{L}}/v_{\mathrm{L}})_{100}/(\widetilde{\Delta l}/l)_{001}$, which is given by a formula similar to (4.35) (but with the suffices of ε in the denominator changed from 1, 2, 3 to 3, 1, 2) could be measured to give another relation between the strain derivatives. More complicated combinations of the derivatives are obtained if other crystal directions are used and these could also be useful in providing independent checks on the reliability of the measurements.

It should be noticed that we have so far supposed the magnetic field to be along an arbitrary direction; this is of course because we have regarded the main aim of the experiment as being to determine the strain dependence of the Fermi surface and this can only be fully revealed if the strain dependence of F is determined for a range of field directions. In fact, however, the more limited information obtained for the field only along special directions, such as symmetry axes, can still be very useful if interpreted in conjunction with a parameterised theory of the strain dependence, and this approach has indeed been successfully exploited (see chapter 5). For such special directions the formulae are often appreciably simplified; thus in (4.34), if the field is along [100] (i.e. along axis 1) we have for an orbit which has square symmetry, such as the $\langle 100 \rangle$ belly or rosette

$$\frac{\partial \ln F}{\partial \varepsilon_2} = \frac{\partial \ln F}{\partial \varepsilon_3} \quad \text{and} \quad \frac{\partial \ln F}{\partial \varepsilon_a} = 0 \quad (a = 4, 5, 6)$$

The second relation implies that there is no oscillatory variation of the transverse sound velocity if the field is along a cube axis, and this is confirmed experimentally. This vanishing of the shear derivatives follows because positive and negative shears ε_4 or ε_5 or ε_6 must have equivalent effects on an area of cross section normal to [100] for a cubically symmetrical surface.

Finally we must mention that we have so far ignored a complicating effect in the interpretation of the amplitude of sound wave oscillations. This is that at the high frequencies of the ultrasonic waves used in the sound velocity experiments, the sound wavelength is long compared to the classical skin depth of eddy currents, and eddy currents are set up to maintain constancy of flux through any loop tied to the particles of the metal. This implies that strain and B are linked, so that in differentiating with respect to strain (as in (4.26)) extra terms arise from the associated modification of the local value of B. As first shown by Alpher and Rubin (1954), this modifies the velocity of sound in a magnetic field and a more detailed calculation by Testardi and Condon (1970) (see also de Wilde 1978) shows that in general both the steady and oscillatory components of the sound velocity are modified. We shall not discuss this complication further, beyond mentioning that for certain special geometries this Alpher–Rubin effect can be avoided altogether (for longitudinal waves propagating along the field direction and for shear waves propagating normal to the field). In general, however, the effect can be quite large and must be properly allowed for in any quantitative interpretation of the sound velocity oscillations.

4.4. Oscillations of the Fermi energy

As discussed in §2.4, oscillations of Fermi energy are to be expected with the same periodicity as that of the dHvA effect but of very small amplitude: microvolts or less for most metals, but perhaps approaching millivolts for semimetals such as bismuth and antimony and for the 2-D system of the silicon inversion layer. This effect was first predicted by Kaganov, Lifshitz and Sinelnikov (1957) who suggested it might be observed through the corresponding oscillations in contact potential. Soon afterwards Kulik and Gogadze (1963) suggested a practical way of observing oscillations in the contact potential. The idea was to make the metal crystal one plate of a capacitor and then to sweep the magnetic field sufficiently rapidly to make the time for one oscillation much shorter than the leakage time of the capacitor. In these conditions the oscillatory changes in contact potential would appear as an oscillating voltage across the capacitor. The method was tried by various authors (Verkin, Pelikh and Eremenko (1964), Pelikh (1966), Whitten and Piccini (1966), Piccini and Whitten (1966)) and oscillations were indeed observed, though sometimes of an amplitude much larger than the theoretically expected one.

A somewhat similar method of detection was developed by Caplin and Shoenberg (1965) in which periodic field modulation was used instead of a rapid field sweep. The idea was essentially analogous to replacing the pulsed field method by the field modulation method in detecting the dHvA effect, except that here it was the potential across the capacitor that was studied rather than the e.m.f. induced in a pick-up coil. If the potential V across the capacitor (i.e. the oscillatory part of contact potential) is given by

$$V = V_0 \sin(2\pi F/H + \phi) \tag{4.36}$$

and if a modulation $h \cos \omega t$ is added to H it is easily shown that a periodic voltage v given by

$$v = 2V_0 \sum_{k=1}^{\infty} J_k\left(\frac{2\pi h F}{H^2}\right) \cos k\omega t \sin\left(\frac{2\pi F}{H} + \phi - \frac{k\pi}{2}\right) \tag{4.37}$$

will be superimposed on V. The variation of the amplitude of v as H is slowly swept should then show up the looked for oscillatory effect. The maximum value of J_2 is 0.486, so for 2ω detection the optimum amplitude of $\cos 2\omega t$, if h is correctly adjusted, should be very nearly equal to $V_0 \sin((2\pi F/H) + \phi)$ i.e. to the potential V.

In the practical realization of the method a cylindrical capacitor was constructed in which the inner plate was a cylindrical lead crystal (about 5 mm long and 2 mm in diameter) on which a thin silicon monoxide layer and an aluminium film were successively evaporated. The capacity between the lead crystal and the aluminium film was of order 3000 pF and the leakage resistance was of order 0.3 MΩ, giving a leakage time of order 1 ms. In order that leakage should be negligible during the period of the field modulation, the frequency ($\omega/2\pi$) was chosen to be 16 kHz. Very clear dHvA like oscillations were seen, with the field periodicities expected for the

151

crystal orientation used (H along $\langle 100 \rangle$), but the amplitudes were suspiciously high. Thus V_0 was observed to be $0.3\,\mu V$ and $0.8\,\mu V$ for the observed β and γ oscillations respectively, while the predicted values should not have exceeded a few times $0.01\,\mu V$ at best.

The probable cause of the discrepancy was subsequently suggested by Professor M. Peter during a visit to Cambridge in 1967 and confirmed experimentally by Randles (see Peter, Randles and Shoenberg 1970). What Peter pointed out was that quite an appreciable Hall voltage is generated between the axis and the periphery of the crystal through the action of the main magnetic field on the eddy currents induced by the time variation of the field. If it is assumed that the modulation frequency ω is too low to cause appreciable skin effect, the Hall potential, V_H is easily shown to be

$$V_H = \tfrac{1}{4}R\sigma a^2 B \, \mathrm{d}B/\mathrm{d}t \tag{4.38}$$

where R is the Hall constant, σ the conductivity and a the radius of the lead crystal cylinder. This Hall potential must be added to any genuine contact potential V such as (4.36) and it turns out that it can easily contain a component which oscillates with field much more strongly than does V itself. Usually the main de Haas–van Alphen like contribution to V_H at frequency 2ω comes from the dHvA oscillations of $\mathrm{d}B/\mathrm{d}t$ and this contribution v_H can be shown to be

$$v_H = R\sigma a^2 \omega H J_2(2\pi Fh/H^2) 4\pi \tilde{M} \sin 2\omega t \tag{4.39}$$

In the conditions of the Caplin and Shoenberg experiment this would have amounted to something like 20 times the size of the genuine effect and various tests by Randles confirmed that this eddy current effect was indeed the dominant effect observed. The same eddy current effect was probably also responsible for the large observed effects in most of the experiments based on the impulsive field method, though without detailed knowledge of the geometry of the experiments only the order of magnitude of the eddy current effect can be estimated. Only for bismuth (Verkin *et al.* 1964) and perhaps for antimony (Pelikh 1966) was the observed effect of the order of magnitude to be expected from the oscillations of the Fermi energy. Even here, however, the eddy current effect should have been of a comparable order of magnitude, though coming, in this case, more from the strong dHvA like oscillations in the Hall coefficient R than from the oscillations in $\mathrm{d}B/\mathrm{d}t$.

To summarize, the situation is rather unsatisfactory in as far as there has not yet been any really convincing direct experimental demonstration of the Fermi energy oscillations. With some modification of the geometry of the capacitor arrangement and with careful attention to the phase and frequency dependence of the 2ω signal output, the Caplin and Shoenberg method might yet be made to work. As mentioned in §2.3.4 it might be applied to the 2-D system of a silicon inversion layer, through until this system is better understood it is not obvious what the result would be.

As was discussed in §2.4, the oscillations of Fermi energy are usually too feeble to have any appreciable frequency modulation effects, but Bi provides a striking exception. For Bi the oscillations of Fermi energy associated with the low frequency electron dHvA oscillations are sufficiently large to cause frequency modulation of the high frequency hole oscillations. Frequency modulation of as much as 30% has been observed by Brandt and Lyubitina (1964) in the dHvA and by Volski (1964b) in the SdH effect. The line shape (i.e. harmonic content) of the low frequency oscillations near the quantum limit should also be modified by 'self' frequency modulation, but this effect has not been studied in detail.

4.5. Shubnikov–de Haas effect

Oscillations in the field dependence of electrical resistivity were first observed by Shubnikov and de Haas (1930) in Bi and, as we saw in chapter 1, it was this discovery which led to the discovery of the de Haas–van Alphen oscillations in magnetic properties. It turns out, however, that the effect is not marked except in semimetals and semiconductors and also in conditions of magnetic breakdown (see chapter 7). Usually the effect is feeble and not easy to observe and it has in fact been observed only in rather few metals. The theory of the effect (Adams and Holstein 1959) is quite complicated, since it involves the detailed problem of electron scattering in a magnetic field. Fortunately however the effect can be qualitatively understood on the basis of a simple argument due to Pippard (1965b). He pointed out that the probability of scattering is proportional to the number of states into which the electrons can be scattered, and so this probability, which determines the electronic relaxation time τ and the resistivity, will oscillate in sympathy with the oscillation of the density of states $\mathscr{D}(\zeta)$ at the Fermi energy, discussed in §2.5.

If there is only one sheet of FS concerned (as with monovalent metals) we might expect that the order of magnitude of $|\tilde{\sigma}|/\sigma$ at $T = 0$, i.e. of the relative amplitude of the SdH oscillations would be given by $|\widetilde{\mathscr{D}(\zeta)}|/\mathscr{D}_0(\zeta)$ where $\mathscr{D}_0$ is the steady density of states. For a finite temperature T, the amplitude would be multiplied by R_T where R_T is the temperature reduction factor of (2.128). The real situation for a polyvalent metal is however usually more complicated, since the total conductivity comes from electrons on a number of different sheets of FS, whose contributions to the total conductivity depend on other factors as well as their separate densities of states. Nevertheless some rough idea of the order of magnitude of the Shubnikov–de Haas oscillation amplitude can be obtained by ignoring this complication and identifying $|\tilde{\sigma}|/\sigma$ with $R_T|\widetilde{\mathscr{D}(\zeta)}|/\mathscr{D}_0(\zeta)$, where $\mathscr{D}_0(\zeta)$ is the *total* density of states of all sheets of the FS, while the oscillatory part refers to one particular sheet; the value of $\mathscr{D}_0(\zeta)$ can of course be estimated from the measured value of the electronic specific heat.

Another rough way of estimating the order of magnitude of $|\widetilde{\mathscr{D}(\zeta)}|/\mathscr{D}_0(\zeta)$ is to consider the special case of a spherical FS, for which $\mathscr{D}_0(\zeta)$ can be given explicitly as

$$\mathscr{D}_0(\zeta) = Vmk_F/\pi^2\hbar^2 = (Vm/\pi^2\hbar^3)(2m\zeta)^{1/2} \tag{4.40}$$

and, of course, $|A''| = 2\pi$. For the fundamental oscillations ($p = 1$ in (2.167)), we then find

$$\frac{1}{R_D R_s}\frac{|\widetilde{\mathscr{D}(\zeta)}|}{\mathscr{D}_0(\zeta)} = \left(\frac{eH}{chk_F^2}\right)^{1/2} = \left(\frac{\pi eH}{chA}\right)^{1/2} = \left(\frac{H}{2F}\right)^{1/2} \tag{4.41}$$

This result was first given by Pippard (1965b). If we apply it to Cu, we find that at 50 kG, where F/H, the number of Landau tubes inside the Fermi surface, would be about 10^4, the conductivity oscillations should be rather less than 1% even if the reduction factors R_D and R_s are ignored. For typical conditions (say $T = 1\,\text{K}, x = 0.5\,\text{K}, m/m_0 = 1.4, R_s = 0.5$) the factors R_D and R_s together with R_T (which would damp $\tilde{\sigma}/\sigma$, though not $\widetilde{\mathscr{D}(\zeta)}/\mathscr{D}_0$) might reduce this estimate by a factor of about 100, so we should not expect oscillations of more than 10^{-4} in the conductivity. For some of the oscillations in polyvalent metals, F/H would be a good deal smaller and rather larger effects would be expected. In particular, for Bi (and impurity semiconductors), where the quantum limit can be reached in relatively modest fields, F/H may be below 10 and much larger effects might be expected (though of course since the FS is not spherical, (4.41) is a gross oversimplification).

The detailed theory of Adams and Holstein (1959) which is also based on an isotropic model (i.e. assumes spherical energy surfaces) leads to results which are qualitatively similar to those presented above. For a particular scattering mechanism, that of phonon scattering, they find (in our notation)

$$\frac{1}{R_T}\frac{\tilde{\sigma}}{\sigma} = \frac{5}{2}\frac{\widetilde{\mathscr{D}(\zeta)}}{\mathscr{D}_0(\zeta)} + \frac{3}{2}\left[\frac{\widetilde{\mathscr{D}(\zeta)}}{\mathscr{D}_0(\zeta)}\right]^2 \tag{4.42}$$

with $\widetilde{\mathscr{D}(\zeta)}$ essentially the same as in our treatment. The first term comes from scattering of electrons in the last occupied Landau tube to states in other Landau tubes, while the second term comes from scattering to states within the same tube; the physical origin of the numerical factors $\frac{5}{2}$ and $\frac{3}{2}$ is

154

not obvious. It should be noted that in its Fourier analyzed form the second term gives harmonics of phases different from those given by the first term. Since, as we saw above, the oscillations in density of states are generally feeble, the second term is usually quite negligible, but for low values of F/H, which may be relevant in Bi and semiconductors, it can become comparable. An extremely different scattering mechanism, that by ionized impurities, leads to a result of the same order of magnitude as (4.42) and Adams and Holstein conclude that the nature of the scattering mechanism does not matter too much.

4.5.1. Experimental methods and results

The obvious method of measuring the potential drop across a single crystal wire carrying a steady current is good enough if the SdH oscillations are not too small and this is the method that was used in the early experiments on Bi. Much greater sensitivity can, however, be achieved by field modulation as described earlier in other contexts. Indeed, the relevant theory is essentially given by (4.37), if $i|\tilde{\sigma}|/\sigma^2$ is substituted for V_0 where i is the current in the sample; with phase sensitive detection it is only one of the $\cos n\omega t$ terms which is amplified, while the much larger steady component is eliminated. It is also possible to use the pulsed field method, but in that case the time-frequency of the oscillations varies (falling to zero at the peak of the pulse, where $dH/dt = 0$) and so phase sensitive detection cannot be used.

A possible complication of a method in which the field varies rapidly (i.e. in both the field modulation and pulsed field methods) is that the voltage across the sample may contain a small induced voltage contribution proportional to dB/dt. The magnitude of such a contribution will depend on the exact geometry of the sample and its leads but, if it is appreciable, it too will oscillate with field because of the dHvA oscillations of M. This complication may make it difficult to be certain that any feeble oscillations of v with field are genuine SdH oscillations rather than a by-product of the dHvA effect.

Another technique for observing the SdH effect is to set up as if to observe the dHvA effect by the field modulation technique (i.e. as described in §3.4.2.1, to use a pick-up coil round the sample), but to raise the modulating frequency to make eddy current effect appreciable. If the phase of the PSD is set in quadrature to the normal setting for detecting the dHvA effect at low ω, the amplitude of the output signal contains comparable contributions from the resistivity and susceptibility oscillations and the SdH oscillations can be directly compared with the dHvA ones (the normal phase setting at low ω gives the dHvA effect alone). This method was used

155

by Lonzarich and Holtham (1975) (see also Knecht, Lonzarich, Perz and Shoenberg 1977) to study the rather large resistivity oscillations of Al in conditions of magnetic breakdown, but it is unlikely to find much general application.

It is convenient to mention here that, just as some care is needed to make sure that the dHvA effect induced e.m.f.s do not mask small genuine SdH oscillatory potentials, so equally, care is needed to make sure that in looking for dHvA oscillations by field varying techniques, SdH oscillations do not enter through eddy current effects. Thus in the pulsed field method, where the phase setting criterion of the field modulation technique is not available, the observed oscillations of e.m.f. in the pick-up coil can be due more to SdH oscillations in the induced eddy currents than to genuine dHvA oscillations in magnetization, It is probable that the oscillations observed in InSb by Bliek and Landwehr (1969) using the pulsed field method were predominantly SdH rather than dHvA oscillations and indeed the rather feeble genuine dHvA oscillations have been observed only by the Foner method (Wampler and Springford 1972, Brignall 1974).

Experimental work on resistivity oscillations other than caused by magnetic breakdown (chapter 7) has been mostly on semimetals and impurity semiconductors (for a review see Kahn and Frederikse 1959) and mostly concerned with frequencies rather than amplitudes; the frequencies have always proved to agree well with dHvA effect determinations. There have been a few isolated observations of relatively low frequency oscillations in Zn and Sn (Alers 1956, 1957*), Ga (Yahia and Marcus 1959) and Pb (Tobin, Sellmyer and Averbach 1969), but attempts at systematic study of the higher frequency oscillations associated with major pieces of the Fermi surface have been only partially successful. Thus Grassie (1963), using the pulsed field method did on isolated occasions see high frequency SdH oscillations in Sn and Pb with $|\tilde{\sigma}|/\sigma$ of order $10^{-4}-10^{-3}$, but because of technical difficulties (spurious signals overlaying the genuine effect) was unable to make any systematic study. Dean (1976) attempted a study of the effect in Cu by the field modulation method and may perhaps have seen oscillations with $|\tilde{\sigma}|/\sigma$ of order 10^{-3}, but he was unable to prove convincingly that he had eliminated the possibility of spurious effects (such as those discussed above due to dB/dt). In these various studies the amplitudes were only very roughly determined, if at all, and though the observed orders of magnitude were reasonably consistent with theory, they provide little basis for any detailed comparison. Here again, then, is a

---

* Alers also observed low frequency oscillations in the field dependence of the thermal resistance of Zn and Sn.

challenge for the future. With refinements of experimental technique the weak high frequency SdH oscillations should be amenable to detailed study and such studies–in the first place on the relatively simple monovalent metals–might well throw useful light on the theory of electron scattering in a magnetic field.

As already mentioned in §2.3.4, a strong SdH effect was observed by Fowler *et al.* (1966) in the 2-D system of a Si inversion layer (the so called MOSFET) in which the number of electrons can be varied by changing the gate voltage (V_g). In these experiments and many others since (e.g. Kawaji and Wakabayashi 1976, Pepper 1978), a Corbino disc geometry was used so that the conductivity σ_{11} was directly measured without the intervention of the Hall effect. It is however rather more convenient to discuss more recent experiments using a four electrode arrangement so that the resistance ρ_{11} and the Hall resistance ρ_{12} are directly measured (note that because of the 2-D geometry ρ and σ are measured in ohm and ohm^{-1} rather than ohm cm and ohm^{-1} cm^{-1} as for 3-D). The connection between ρ and σ is given by

$$\rho_{11} = \sigma_{11}/(\sigma_{11}^2 + \sigma_{12}^2) \qquad \rho_{12} = -\sigma_{12}/(\sigma_{11}^2 + \sigma_{12}^2)$$

so provided σ_{12} remains finite, a vanishing of σ_{11} somewhat paradoxically implies that ρ_{11} also vanishes and that $\rho_{12} = -1/\sigma_{12}$. Two such experiments are illustrated in fig. 4.3: (*a*) in which the Fermi energy is moved through the Landau levels by varying the gate voltage V_g of a MOSFET at a fixed high magnetic field (Wakabayashi, Myron and Pepper 1982) and (*b*) for a GaAs–Al$_x$Ga$_{1-x}$As heterostructure in which $n \sim 5 \times 10^{11}$ cm^{-2} (Ebert, von Klitzing, Probst and Ploog 1982) and the Landau levels are moved through the Fermi energy by varying H.

The striking features of both these experiments are the vanishing of ρ_{11} over appreciable ranges of either V_g or H and the corresponding plateaus in the Hall resistance ρ_{12}. These features can be most simply understood by supposing that for these samples it is ζ rather than n which remains constant as H is varied, though this is probably a considerable oversimplification, and probably the interpretations for the MOSFET (see for instance Paalanen, Tsui and Gossard 1982) and the heterostructure (see for instance Baraff and Tsui 1981) are somewhat different in detail, and indeed both interpretations are as yet somewhat incomplete.

The basic point is that when ζ falls between two Landau levels all the levels below ζ are completely full and the density of states at ζ is zero. Thus as H or n is varied, the density of states at the Fermi energy stays zero, except at the 'singular' points when a Landau level passes through ζ. Correspondingly the resistance ρ_{11} should be zero except at these singular points–roughly as in fig. 4.3–apart from the broadening due to various

Other oscillatory effects

Fig. 4.3. Shubnikov–de Haas effect in two-dimensional systems. (a) Si inversion layer (MOSFET) showing variation of ρ_{11} (resistance) and ρ_{12} (Hall resistance) at 1.4 K for varying gate voltage (i.e. varying electron number) in a fixed field of 250 kG (after Wakabayashi, Myron and Pepper 1982). There are four peaks for each Landau quantum number (only number 1 and three peaks of number 0 are shown). The 'minor' splitting arises from valley degeneracy and alternate pairs of peaks correspond to opposite spins. The zeros of resistance correspond to the Fermi level staying between neighbouring sub-levels, i.e. complete occupation of an

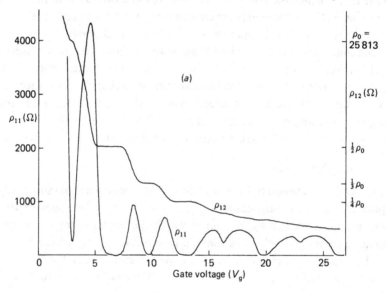

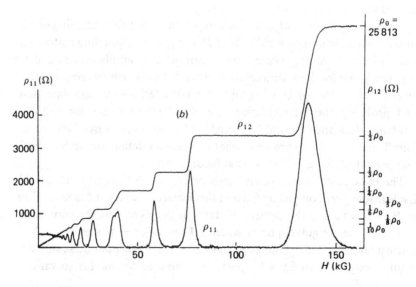

158

4.5. SdH effect

complications such as broadening of Landau levels by electron scattering. The plateaus of the Hall resistance ρ_{12} also find a simple interpretation which proves to be particularly interesting. As in well known,

$$\rho_{12} = -1/\sigma_{12} = H/nec \qquad (4.43)$$

and if the r Landau levels below ζ are exactly full, the electron density must be given by

$$n \equiv N/S = rD/S = reH/\pi chp \qquad (4.44)$$

where N is the total number of electrons, S the sample area, D the degeneracy of the Landau levels as given by (2.84) and p is 2 or 1 according as the spin sub-levels are resolved or treated together. This result is of course also equivalent to (2.62).* For both our examples the spin peaks are in fact well resolved so $p = 2$ except at low fields and the Hall resistance becomes

$$\rho_{12} = 2\pi\hbar/re^2 = h/re^2 = (25813/r)\text{ohm} \qquad (4.45)$$

This then not only explains why ρ_{12} remains constant over the regions where ρ_{11} vanishes, but shows that we have a 'quantum Hall effect'. The remarkable result that the plateaus directly measure integral sub-multiples of the fundamental constant h/e^2 was first given by von Klitzing, Dorda and Pepper (1980). Closer examination of the method of measurement and the above interpretation shows that it is surprisingly free from the need of any corrections and so provides a new type of high precision standard resistance. Alternatively, measurements of the plateau against existing

Caption for Fig. 4.3. (*cont.*)
integral number r of sub-levels and the Hall resistance at each plateau is h/re^2 (or $(25813/r)\Omega$). The measuring current was $1\,\mu$A. (*b*) GaAs – $Al_xGa_{1-x}As$ heterostructure with electron density $\sim 5 \times 10^{11}\,\text{cm}^{-2}$ showing variation of ρ_{11} and ρ_{12} with H at 0.35 K (after Ebert *et al.* 1982). The peaks at higher fields are alternatively of opposite spins, but the spin splitting falls off as H is lowered and becomes negligible below 30 kG. The interpretation of the zeros and plateaus is as in (*a*). Note that the sensitivity for ρ_{11} is halved beyond about 100 kG so the highest peak of ρ_{11} is about 8800 Ω. The measuring current was $20\,\mu$A. These diagrams were kindly provided by Dr. Wakabayashi and Professor von Klitzing; they refer, however, to experimental conditions somewhat different from those described in the published papers.

---

* The saw-tooth variation of the number of mobile electrons with H implied by (4.44) or (2.62) presumably comes about from the underlying structure of the device, but the mechanism is not fully understood.

resistance standards can be regarded as a new method of determining h/e^2 with high precision. At the time of writing, determinations in different laboratories using quite different 2-D systems agree to something like 1 in 10^7 and the precision is likely to be still further increased.

4.6. Oscillations in ultrasonic attenuation; giant quantum oscillations

When ultrasonic waves are propagated in the presence of a magnetic field, not only the wave velocity (see §4.3.2), but also the attenuation oscillates as the field is varied. In 'ideal' conditions (low temperature, little electron scattering and little phase smearing) the oscillations in attenuation assume a very striking spike-like form (see fig. 4.4) and were called 'giant quantum oscillations' (GQO) by Gurevich, Skobov and Firsov (1961), who first predicted the effect. In less ideal conditions, the spikes are smeared out and become almost sinusoidal oscillations which decay as the field is lowered; although the name is no longer really appropriate when the oscillations are almost sinusoidal, they are often still referred to as GQO, because it is basically the same effect as in ideal conditions.

Qualitatively the GQO can be understood as a kind of 'surf riding' effect. The simplest situation is when the wave is propagated along the field direction. As the field is varied there will be a moment just before a Landau tube passes through the Fermi surface when the small component v_H of the Fermi velocity v_F in the wave (and field) direction passes through exact equality with the sound velocity s ($\ll v_F$). At this moment the electron motion keeps exactly in step with the sound wave and is able to extract

Fig. 4.4. Giant quantum oscillations of absorption of longitudinal ultrasonic waves ($v = 100$ MHz) along H in Bi; H along binary axis, $T = 1.2$ K (Fujimori 1969; the original of this diagram was kindly provided by Professor Mase). The division of the highest peak into two separate spin peaks is schematically indicated by the broken lines.

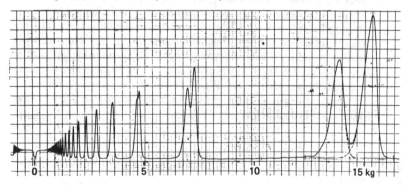

160

much more energy from the wave than at other fields. For this particular field, and similarly for the passage of each successive Landau tube through the Fermi surface, there is a δ-function-like spike of attenuation, but very little attenuation in between. Clearly the periodicity of the GQO is very nearly identical with that of the dHvA oscillations but their novel feature is that they occur for one particular value of κ (the component of k along H) and so they are not reduced in amplitude (as are the dHvA oscillations) by superimposition of oscillations for different κs. Thus, even after the other smearing mechanisms (finite temperature, electron scattering, sample inhomogeneity) are taken into account, the GQO retain a higher harmonic content than the oscillations of dM/dH or of the density of states.

We shall not go into the detailed theory of the attenuation mechanism, which is discussed in a review by Shapira (1968), but present instead a 'phenomenological' kind of treatment, which in fact gives essentially the same answers as the detailed theory, but leaves the average attenuation coefficient as an empirical parameter. Our treatment, however, for the first time takes systematic account of phase smearing due to sample inhomogeneity and this, as we shall see later, helps to clear up some features of the experimental results which were previously not properly understood.*

4.6.1. Conditions for occurrence of the giant quantum oscillations

The 'surf riding' condition may be more precisely formulated by treating the ultrasonic wave as a stream of phonons of energy $\hbar\omega$ and momentum $\hbar q$ (where, of course, ω/q is the wave velocity s) and requiring momentum and energy to be conserved when a phonon is absorbed by an electron. Conservation of momentum requires that after absorption of a phonon, the electron's momentum component along H changes from $\hbar\kappa$ to $\hbar(\kappa + q)$ if the direction of the wave lies along the field. If the direction of the wave is at angle θ to the field, q must be replaced by $q \cos\theta$ in all that follows, but the results are qualitatively similar provided $\cos\theta$ is not so small as to upset some of the inequality conditions that apply for $\theta = 0$. Because in practical conditions the phonon energy is much less than the energy difference between neighbouring Landau tubes (i.e. $\omega \ll \omega_c$ for fields at which the oscillations are observable), the electron stays on the same Landau tube and conservation of energy requires that

$$\varepsilon(a, \kappa) + \hbar\omega = \varepsilon(a, \kappa + q) \tag{4.46}$$

* Bellessa (1973) also took account of phase smearing to interpret his high field data on Hg, but only in terms of a rather special model.

Other oscillatory effects

where we use the notation of chapter 2 to specify the energy in terms of the area of cross section a of a surface of constant energy ε at the plane $k_H = \kappa$. If we expand ε in powers of the small change q, this becomes

$$\hbar\omega = q(\partial\varepsilon/\partial\kappa)_a + \tfrac{1}{2}q^2(\partial^2\varepsilon/\partial\kappa^2)_a \tag{4.47}$$

and since $(\partial\varepsilon/\partial\kappa)_a$ is just $\hbar v_H$, where v_H is the electron velocity in the field direction, and $\omega/q = s$, (4.47) reduces to just the 'surf riding' condition

$$s = v_H \tag{4.48}$$

if the second term on the right of (4.47) is ignored (and this will be justified directly).

The value of κ for which the energy conservation condition is met, can be derived from (4.47) by remembering that $v_H \ll v_F$, which implies that κ is small on the scale of FS dimensions, if κ is measured, as in §2.3.6, from the extremal section of the FS. Thus, using (2.21)

$$(\partial\varepsilon/\partial\kappa)_a = -(\partial a/\partial\kappa)_\varepsilon/(\partial a/\partial\varepsilon)_\kappa = -\kappa A'' \hbar^2/2\pi m \tag{4.49}$$

where A'' is defined as $\partial^2 \mathscr{A}/\partial\kappa^2$ rather than its modulus as in (2.112). One more differentiation gives

$$\partial^2\varepsilon/\partial\kappa^2 = -A'' \hbar^2/2\pi m \tag{4.50}$$

and substituting into (4.47) we find

$$\kappa \equiv \kappa_0 = -\frac{ms}{\hbar}\frac{2\pi}{A''} - \frac{q}{2} \tag{4.51}$$

For a spherical FS $A'' = -2\pi$ and (4.51) agrees exactly with the result for free electrons, which can be obtained in a more elementary way by considering only the κ dependent part $\hbar^2\kappa^2/2m$ of the electron energy.

If we put in some typical numbers, say

$$\omega = 2\pi \times 10^8\,\text{s}^{-1}, \quad s = 3 \times 10^5\,\text{cm s}^{-1}, \quad q = 2 \times 10^3\,\text{cm}^{-1}$$

we find for free electrons ($m/\hbar \sim 1\,\text{s cm}^{-2}$),

$$\kappa \sim (3 \times 10^5 - 10^3)\,\text{cm}^{-1}$$

This confirms both that κ is much less than k_F ($\sim 10^8\,\text{cm}^{-1}$) and that the second term in (4.51) (and so in (4.47)) is ordinarily negligible compared with the first. These remarks are usually still valid for real Fermi surfaces, because the product mA'' does not differ from the free electron value as much as do m and A'' separately.

We have so far been a little vague as to precisely which electrons are able to absorb the phonons. At $T = 0$ and in the absence of any level broadening

162

(as has been implicitly assumed so far), only such states as are at or within the FS are occupied and since, after absorption of a phonon, an electron must go into an empty state, only electrons just within the FS are able to absorb. The situation is illustrated schematically in fig. 4.5. The change ΔH of H, which carries the Landau tube from the position (a) where the absorption of a phonon becomes possible to (b) beyond which it ceases to be possible, is just proportional to the difference ΔA of the cross-sectional areas indicated. The energy range $\Delta \varepsilon$ below the Fermi energy ζ for which absorption can occur is also proportional to this difference and it is easy to see that for spherical energy surfaces with the numbers given above

$$\frac{\Delta H}{H} = \frac{\Delta A}{A} = \frac{\Delta \varepsilon}{\zeta} = \frac{2\kappa q}{k_F^2} \sim 10^{-7}$$

Fig. 4.5. Illustrating conditions for absorption of a phonon at $T = 0$ with no Landau level broadening for the free electron model. Only electrons with $\kappa = \kappa_0$ (as given by (4.51)) can absorb. The positions of a Landau tube (a cylinder of cross section area $A - \Delta A$ at (a) and A at (b)) at fields $H - \Delta H$ and H are determined by the intersections of the plane $\kappa = \kappa_0$ with the spherical surfaces of constant energy $\zeta - \Delta \varepsilon$ and ζ. The Landau tube must expand at least to (a) before absorption is possible (for lower H, $\kappa_0 + q$ would be an occupied state) and absorption is no longer possible beyond (b), since there would then be no electrons on the tube at $\kappa = \kappa_0$. The difference in radius of the two constant energy surfaces is given to a good approximation by $2k_F \Delta k = (\kappa_0 + q)^2 - \kappa_0^2 = 2\kappa_0 q$. The sizes of κ_0 and q in relation to k_F are greatly exaggerated in the diagram in order to show the various intersections more clearly.

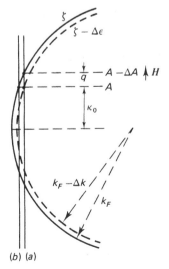

(b) (a)

This is usually a minute fraction of the $\Delta H/H$ corresponding to the separation of peaks, which is H/F, and it is quite a legitimate approximation to treat the spikes in the field dependence of attenuation as a series of δ-functions. As we shall see below, the smearing effects associated with finite temperature, finite Landau level broadening and sample inhomogeneity introduce broadening of the spikes far larger than the 10^{-7} of the ideal situation. For most purposes indeed, we may ignore the small finite value of κ_0 altogether and suppose that absorption occurs at $\kappa = 0$, i.e. when the Landau tube just parts company with the FS, i.e. when $F/H = r + \frac{1}{2}$.

4.6.2. Field dependence of the oscillations

On the basis of the above discussion, we may describe the field dependence of the attenuation Γ in 'ideal' conditions ($T = 0$, no electron scattering, perfect sample) as

$$\Gamma = 2\pi\Gamma_0 \sum_{r=0}^{\infty} \delta(\psi - \psi_r)$$

where   (4.52)

$$\psi = 2\pi F/H \quad \text{and} \quad \psi_r = 2\pi(r + \tfrac{1}{2})$$

and the normalization factor 2π has been introduced so that Γ_0 should be the mean attenuation associated with absorption by the electrons (see fig. 4.6).

We must now consider how this is modified in practical conditions, i.e. at finite T, for finite scattering time τ, and for sample inhomogeneity, and we

Fig. 4.6. Field dependence of GQO at $T = 0$ with no Landau level broadening. Each vertical line is a schematic representation of a δ-function peak multiplied by $2\pi\Gamma_0$ and the area under each peak is therefore $2\pi\Gamma_0$. Since the peaks are 2π apart, the average of Γ (i.e. $\int\Gamma d\psi/\int d\psi$) is just Γ_0.

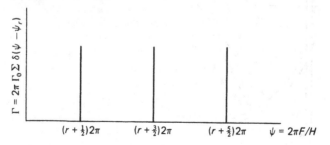

must also allow for the effect of electron spin. As in §2.3.7 we can introduce a phase distribution function $D(z)$ for each effect, where $z = \varphi/\lambda$ and $\phi = \psi - \psi_r$, and smear the ψ in the oscillations (4.52) over this distribution. In this section it is convenient to normalize the distribution functions so that $\int D(z)\,dz = 1$ over the permissible range of z and to omit the harmonic number p from the definition of λ; when later, we wish to consider the damping of the pth harmonic in a Fourier analysis, we shall have to replace λ by $p\lambda$. For finite temperature, the appropriate function is, as shown in §2.3.7.1,

$$D_1(z) = 1/2(1 + \cosh z) \tag{4.53}$$

with

$$\lambda_1 = 2\pi kT/\beta H \tag{4.54}$$

For a Lorentzian distribution of inhomogeneity of F, the distribution function is, as in §2.3.7.3,

$$D_2(z) = 1/\pi(1 + z^2) \tag{4.55}$$

with

$$\lambda_2 = 2\pi(\Delta F)_0/H = 2\pi^2 kx_0/\beta H \tag{4.56}$$

if $(\Delta F)_0$ characterizes the Lorentzian spread of F, or x_0 is the equivalent Dingle temperature in the dHvA effect, due to spread of F.

However, the damping effect of electron scattering is more complicated than in the dHvA effect. Just as before, the effect of electron scattering is to broaden the Landau levels and this introduces a phase smearing of the same kind as (4.55) with

$$\lambda_3 = \pi\hbar/\beta H\tau = 2\pi^2 k(x - x_0)/\beta H \tag{4.57}$$

where x is the total Dingle temperature as observed in the dHvA effect. But there is another consequence of the level broadening, which can sometimes be the dominant source of broadening of the 'ideal' δ-function peaks. This is that the energy conservation condition (4.46) can be relaxed to an extent of order of the broadening of the Landau level, i.e. $\hbar/\tau$. This means that instead of a unique (and small) value κ_0 of κ at which the electron can absorb a phonon, there is a considerable spread of permissible values of for absorption and, as is shown in Appendix 11, this is equivalent to introducing a phase distribution function

$$D_4(z) = 1/\pi z^{1/2}(1 + z) \tag{4.58}$$

with

$$\lambda_4 = \left(\frac{2\pi}{A''}\right)\frac{\pi m}{\beta H q^2 \tau^2} \tag{4.59}$$

Here once again A'' denotes $d^2A/d\kappa^2$ rather than $|d^2A/d\kappa^2|$ and it should be noted that z must be positive, going from 0 to a high value which can be taken as ∞. Correspondingly, the phase smearing is quite asymmetrical, with the smearing variable ϕ having the same sign as A''. It is easy to show that the $|\lambda_4|$ of (4.59) is normally much bigger than the λ_3 of (4.57); for a spherical FS the ratio is in fact

$$|\lambda_4|/\lambda_3 = k_F l/q^2 l^2 \tag{4.60}$$

where l is the mean free path given by $v_F\tau$. Taking $l \sim 10^{-2}$ cm, $k_F \sim 10^8$ cm^{-1}, $q \sim 2 \times 10^3$ cm^{-1}, we find typically

$$|\lambda_4|/\lambda_3 \sim 2.5 \times 10^3 \tag{4.61}$$

Thus the 'direct' broadening effect of λ_3 should be completely negligible compared with the 'indirect' λ_4 effect.

The phase smearing effect of any one of the mechanisms acting alone is obtained very simply by convolution of the phase distribution function with the unsmeared variation (4.52), giving

$$\Gamma = 2\pi\Gamma_0 \int \sum_{r=0}^{\infty} \delta(\psi - \psi_r + \varphi)D(\varphi/\lambda)\mathrm{d}(\varphi/\lambda)$$

or

$$\Gamma = \frac{2\pi\Gamma_0}{\lambda} \sum_{r=0}^{\infty} D\left(\frac{\psi_r - \psi}{\lambda}\right) \tag{4.62}$$

If $\lambda \ll 2\pi$, i.e. if in a range of ψ away from ψ_r of much less than one period of oscillation D becomes negligible, then (4.62) simply represents a series of peaks each of which has the line shape of the function D.* When, however, various smearing effects are simultaneously present with comparable λs, the result becomes more complicated since the various effects must all be convoluted together, leading to rather cumbersome expressions which we shall not write down here. The detailed interpretation of observed line shapes is correspondingly complicated and is essentially a matter of trial and error. The line shapes of the oscillations for each one of the effects

---

* If D is a symmetrical function (as for D_1, D_2 and D_3) it is immaterial whether we put $\psi - \psi_r$ or $\psi_r - \psi$ in (4.62), but for D_4, which is asymmetrical, it is important to remember that the function exists only on the side of ψ_r such that $\psi_r - \psi$ has the same sign as λ_4.

separately are illustrated in fig. 4.7, together with an example of the line shape for several comparable smearing effects acting together.

The other limiting case, $\lambda \gg 2\pi$, when the smearing spans a range of several periods of oscillation, is rather easier to discuss, because it then becomes useful to Fourier analyze the series of δ-functions (4.52) and, as in §2.3.7, to take account of phase smearing by calculating appropriate amplitude reduction factors R_T, etc. given by the Fourier transforms of the various distribution functions. In this limit the reduction factors for the higher harmonics become so severe that only the fundamental and perhaps a few harmonics need be considered (in contrast to the case of $\lambda \ll 2\pi$, when so many harmonics would be significant that the Fourier analysis approach would not be very helpful). The Fourier analysis of (4.52) gives

$$\Gamma = \Gamma_0 [1 + 2 \sum_{p=1}^{\infty} (-1)^p \cos 2\pi p\psi] \tag{4.63}$$

and we can immediately write down the final result after the phase smearing from all the various mechanisms has been introduced.

We find that the contribution Γ_p of the pth harmonic of the oscillatory behaviour is

$$\Gamma_p = 2\Gamma_0 (-1)^p \left(\frac{\pi p \lambda_1}{\sinh \pi p \lambda_1} \right) (e^{-p\lambda_2})(e^{-p\lambda_3})$$

$$\times (\pi p |\lambda_4|)^{-1/2} \cos\left(2\pi p\psi \pm \frac{\pi}{4} \right) \tag{4.64}$$

Here each of the reduction factors from the various mechanisms appears bracketed and though no assumption has been made about the magnitudes of the first three reduction factors, the last reduction factor, i.e. coming from the Fourier transform of $D_4(z)$, is valid only in the limit $|\lambda_4| \gg 2\pi$ (a general expression for this reduction factor is given in Appendix 11). The phase shift of $\frac{1}{4}\pi$ in (4.64) comes about because in the limit $|\lambda_4| \gg 2\pi$ the Fourier transform of $D_4(z)$ has equal real and imaginary components.

It can be seen that as the field is lowered, although $|\lambda_4|$ stays much larger than λ_3, eventually $e^{-p\lambda_3}$ becomes smaller than $(\pi p |\lambda_4|)^{-1/2}$. However for $|\lambda_4|/\lambda_3 \sim 2500$ this happens only for $|\lambda_4| > 12\,000$ (for $p = 1$) and by then the amplitude of the oscillatory attenuation is certainly less than 1% of the mean value Γ_0 and so would be only barely observable. In any case there are probably other mechanisms causing feeble oscillatory attenuation besides the one responsible for the giant quantum oscillations, so that (4.64) is anyhow unlikely to be exactly valid in the limit where the Dingle factor $e^{-p\lambda_3}$ is relevant. If we put in the explicit meanings of the λs and of ψ, (4.64)

Other oscillatory effects

Fig. 4.7. Phase distribution functions which determine GQO line shape. (a)$1/2(1 + \cosh z)$: effect of temperature. (b) $1/\pi(1 + z^2)$: effect of phase smearing. (c) $1/\pi z^{1/2}(1 + z)$: effect of electron scattering. (d) convolution of (a) and (b) with $\lambda_1 = 0.0233$ ($T = 1.34\,\text{K}$, $m/m_0 = 0.0513$, $H = 138\,\text{kG}$) and $\lambda_2 = 0.0110$. The broken curve shows the line shape before convolution with (b), to illustrate how the convolution broadens the curve and reduces its height. The value of λ_2 has been chosen to match one of the Ga spin-split peaks of fig. 4.9, though exact comparison is complicated by the non-linear scale of fig. 4.9. (e) convolution of (a) with (b) with (c) for $\lambda_1 = 0.034$, ($T = 1.2\,\text{K}$, $m/m_0 = 0.0091$, $H = 15.1\,\text{kG}$) $\lambda_2 = 0.044$, $|\lambda_4| = 0.16$. These values have been chosen in an attempt to match the last peak of fig. 4.4. Note that the ordinates in (a), (b) and (c) are equivalent to $\lambda\Gamma/2\pi\Gamma_0$, but in (d) and (e) to $\Gamma/2\pi\Gamma_0$; the sign of $\phi = \psi - \psi_r$ in (e) corresponds to a negative A'' (e.g. as in Bi) and ϕ increases with $1/H$. The

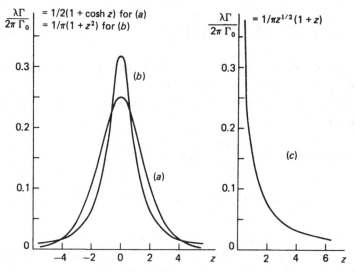

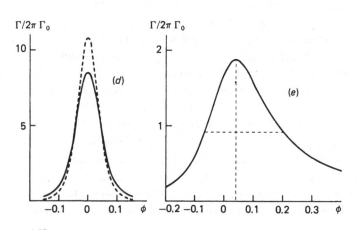

168

becomes

$$\Gamma_p = \sqrt{2}\Gamma_0(-1)^p \frac{2\pi^2 pkT/\beta H}{\sinh 2\pi^2 pkT/\beta H}$$

$$\times \exp(-2\pi^2 pkx/\beta H)\frac{(\beta H)^{1/2}q\tau}{\pi^{3/2}(pm/|A''|)^{1/2}}\cos\left(\frac{2\pi pF}{H} - \frac{\pi}{4}\right)$$

(4.65)

For free electrons, $|A''| = 2\pi$ and we can put $\tau = l/v$, $\frac{1}{2}mv^2 = \zeta$, and $\zeta/\beta = F$, so (4.65) reduces to

$$\Gamma_p = \frac{\sqrt{2}\Gamma_0}{\pi}(-1)^p \frac{2\pi^2 pkT/\beta H}{\sinh 2\pi^2 pkT/\beta H}$$

$$\times \exp(-2\pi^2 pkx/\beta H)ql\left(\frac{H}{pF}\right)^{1/2}\cos\left(\frac{2\pi pF}{H} - \frac{\pi}{4}\right)$$

(4.66)

Apart from inclusion of the Dingle factor and the $\sqrt{2}$ factor, (4.66) agrees with a formula derived by Skobov (1961) using more sophisticated techniques. Professor Skobov has been kind enough to confirm that he had inadvertently omitted a factor $\sqrt{2}$ in his paper.

So far we have ignored spin; if we are in the 'genuine' giant oscillation regime, i.e. all the λ s $\ll 2\pi$, its effect is to split each peak into two. Thus (4.52) is replaced by

$$\Gamma = \pi\Gamma_0\left[\sum_{r=0}^{\infty}\delta(\psi - \psi_{1r}) + \sum_{r=0}^{\infty}\delta(\psi - \psi_{2r})\right]$$

(4.67)

where

$$\psi_{1r} = (r + \tfrac{1}{2} + \tfrac{1}{4}g)2\pi, \qquad \psi_{2r} = (r + \tfrac{1}{2} - \tfrac{1}{4}g)2\pi$$

(4.68)

After phase smearing the two separate series of broadened peaks must be superimposed. In the limit of considerable broadening (the λ s $\gg 2\pi$), when the Fourier approach is useful, the superposition simply introduces the

Caption for Fig. 4.7. (cont.)
convolutions which were kindly computed by Mr C. M. M. Nex have the form:

$$F(\phi) = \frac{1}{2\pi\lambda_1\lambda_2}\int_{-\infty}^{\infty}d\theta/\left[1 + \cosh\left(\frac{\phi - \theta}{\lambda_1}\right)\right]\left(1 + \frac{\theta^2}{\lambda_2}\right) \text{ for } (d)$$

and

$$G(\phi) = \frac{1}{\pi|\lambda_4|}\int_0^{\infty}F(\phi - \theta)d\theta/(\theta/|\lambda_4|)^{1/2}[1 + (\theta/|\lambda_4|)] \text{ for } (e).$$

extra reduction factor

$$R_s = \cos\left(p\frac{\pi}{2}g\frac{m}{m_0}\right) \tag{4.69}$$

into (4.64) to (4.66), as in the dHvA effect.

The GQO in their characteristic spiky form were observed experimentally in Zn by Korolyuk and Prushchak (1961) very soon after their theoretical prediction by Gurevich *et al.* Because the λ_4 is so enormously greater than λ_3 (see (4.61)), it is only in extremely pure samples that the oscillations can be seen as well separated spikes at high fields. Typically, if impurity scattering produces a Dingle temperature of more than something like 10^{-2} K, the spikes are broadened out over several oscillations, even at the highest fields, so that only the smoothed out oscillations given by (4.65) can be observed. The spikes have in fact been observed in such metals as Zn, Bi (Mase, Fujimori and Mori 1966), Ga (Shapira and Neuringer 1967) and Hg (Bellessa 1973), which can be made exceptionally pure and moreover have oscillations characterized by very low cyclotron mass (i.e. high β) so that all the broadening effects can be made reasonably small at accessible fields.

The general form of the oscillations is particularly well illustrated in Bi (fig. 4.4) where they can be followed all the way from the quantum limit near 15 kG down to a few hundred G where they are practically sinusoidal before they fade out. According to (4.62) the heights of the peaks in the high field region, where the different orders r do not overlap appreciably, should vary linearly with H, since all the λs of the various broadening effects vary as $1/H$. Comparisons with experiment are complicated by the spin-splitting: for the last peak of all, the two spin peaks are well separated and can be considered separately, but as the order increases the spin peaks rapidly coalesce and by the fifth peak it is reasonable to regard the peak as of a height which is the sum of the heights of the two separate spins. On this basis the linear variation with H is roughly verified by noting that a straight line joining the origin to the half height of the fifth peak passes through the average height of the well resolved peaks at the quantum limit. The fact that the two spins have appreciably unequal peaks may perhaps indicate that Γ_0 is different for the two spin states in the absence of a field.

The theory of broadening of the peaks by electron scattering predicts that they should be broadened asymmetrically, more on the low field than on the high field side of Bi, which has A'' negative (convex FS), and this asymmetry is evident in the quantum limit peaks, where the spins are well separated. The experiments of Bellessa (1973) on Hg confirm that the asymmetry should be reversed if A'' is positive rather than negative and as

170

can be seen from fig. 4.8 the broadening for the oscillations in Hg, for which it is known that the relevant A'' is positive, is indeed on the high field side in contrast to the Bi peaks.

Although the theory gives a good qualitative account of the experiments, attempts to make a quantitative fit are only partially successful. Several features of the data can be compared with the theory; these are:

(1) The breadth of a peak at half height, which is about 0.26 in phase (0.69 kG in H) for the last peak for Bi in fig. 4.4 (i.e. for a single spin).
(2) The asymmetry of a peak, which can be specified by the ratio ρ of the width at half height on one side of the peak maximum to the width on the other side. For the last peak in fig. 4.4, ρ is about 2.1.
(3) The ratio of the peak height to the mean level of the oscillations at low fields. This information is available only for Bi and suggests that for the last peak of fig. 4.4, $\Gamma/2\pi\Gamma_0 \sim 2.7$, with an uncertainty of order 10% (or possibly more if the recording is not exactly linear on a decibel scale).
(4) The decay of amplitude at low fields, where (4.65) applies should determine the Dingle x due to phase smearing. This information again is available only for Bi. A plot of $\ln(|\tilde{\Gamma}|H^{-1/2}\sinh z/z)$ against $1/H$ (with $z = 2\pi^2 kT/\beta H$) is indeed roughly linear (within the considerable uncertainty of the data in this region) and indicates $x \sim 1$ K.

Let us first consider (1) and (2) for the last peak of Bi. The value of λ_1 is specified by (4.54) as 0.034 and this by itself would give a phase width $\Delta\phi$ at half height of 0.12 (0.34 kG) compared with the observed phase width of 0.26 (0.69 kG). It is clear therefore that extra broadening must come from the λ_2 and λ_4 mechanisms, but here we run into difficulties. If we take λ_2 as 0.09, corresponding to $x = 1$ K (see (4) above) the convolution shows that practically all the extra broadening is accounted for without introducing any asymmetry of the kind caused by λ_4, so that the observed asymmetry of $\rho \sim 2.1$ is not explained. If however we try to explain the observed asymmetry by introducing an appreciable λ_4 (something like 0.3 is

Fig. 4.8. GQO in Hg (Bellessa 1973); $v = 20$ MHz, T = 0.45 K, longitudinal waves along H in the [100] direction. Note the slight asymmetry of the peaks (more broadening towards higher fields). F/H is about 12 at 67 kG.

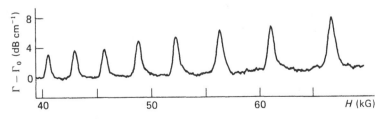

required), the total width at half height becomes appreciably greater than observed. A convolution assuming $\lambda_1 = 0.034$, $\lambda_2 = 0.044$ and $\lambda_4 = -0.16$ provides a rough compromise (see fig. 4.7e). The half height width $\Delta\phi = 0.22$ agrees reasonably with the observed 0.26 and the height ($\Gamma/2\pi\Gamma_0 = 1.0$) is not too far off the rather uncertain observed 2.7. However, the λ_2 value implies an x of only 0.5 K compared with the roughly estimated 1 K, the asymmetry ρ is only 1.5 compared with the observed 2.1, and moreover the attenuation decays too gradually as ϕ increases.

These quantitative discrepancies may be a consequence of oversimplified assumptions in the theory or perhaps come from some unsuspected errors in the experiment (such as insufficient allowance for recorder time constant or deflections being slightly off perpendicular to time axis), but leaving these difficulties to one side, one important conclusion stands out. This is that a Dingle temperature of anything like 1 K or even 0.1 K cannot be accounted for by the electron scattering implied in the estimate of λ_4 required to produce the observed asymmetry. Even if λ_4 were as high as 0.4, the corresponding value of x without any phase smearing (i.e. for $\lambda_2 = 0$) would be only something like 0.02 K in contrast to the much higher value indicated by the decay of the low field oscillations and indeed in contrast to the usual Dingle temperatures of between 0.1 and 1 K found in the dHvA effect. This strongly suggests that phase smearing provides the

Fig. 4.9. Spin-split GQO peak in Ga (Shapira and Neuringer 1967); $v = 20\,\text{MHz}$, $T = 1.34\,\text{K}$, longitudinal waves along $\boldsymbol{H}$ and along the b-axis, F/H is about 2.5.

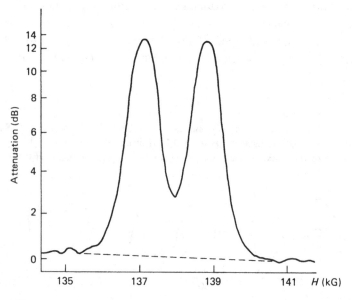

dominant contribution to the Dingle temperatures in pure metals. The evidence from the Hg and Ga data, though more limited, also supports this suggestion, and indeed the asymmetry in Hg is rather smaller than in Bi, while in Ga it is almost completely absent, so it is even more essential to introduce the λ_2 phase smearing contribution. We shall comment further on the value of GQO as a clue to understanding the mechanism of Dingle temperatures in chapter 8. The great sensitivity of the GQO to impurity scattering and the possibility of using the asymmetry to separate the broadening contributions of scattering and phase smearing suggests that the GQO might be a useful tool for studying alloying effects at very low concentration; since the breadth goes as $1/\tau^2$ (see (4.58) and (4.59)) the effect of alloying should be quite dramatic. This potentiality has not so far been exploited.

Finally we should mention that the form of the thermal broadening is particularly well confirmed by the line shapes of the GQO in Ga (Shapira and Neuringer 1967). Their samples were so pure that the electron scattering asymmetry is no longer appreciable (fig. 4.9) and the breadth of the line at half height is only 20% greater than the thermal breadth predicted by (4.7). The extra width can be interpreted as due to phase smearing corresponding to a Dingle temperature of order 0.2 K, and if the appropriate convolution is made (see fig. 4.7*d*) the line shape produced agrees very well with that observed (fig. 4.9).

4.7. Oscillations of other physical properties

4.7.1. Optical properties

The reflectivity of metals may oscillate with magnetic field through two rather different mechanisms. The first is simply a manifestation of the Shubnikov–de Haas effect in as far as the optical behaviour at long wavelengths depends on the electrical conductivity. This effect was first observed by Dresselhaus and Mavroides (1964) in Sb. The second mechanism is rather more complicated and involves transitions across the energy gap between a valence and a conduction band. Thus the reflectivity is modified whenever the photon energy just matches the separation between Landau levels in the two bands. The effect has been most studied in Bi where a conveniently small energy gap occurs, and the appropriate photon energies fall in the infrared. From details of how the oscillations depend of frequency, crystal orientation and other factors, valuable evidence about the band structure can be obtained, but the interpretation is complicated and will not be discussed here. The first observation of this type of oscillation was by Brown, Mavroides and Lax (1963) and more recent detailed studies have been made by Maltz and Dresselhaus (1970) and Vecchi and Dresselhaus (1974).

4.7.2. Miscellaneous transport properties

Oscillations have been observed not only in the electrical conductivity (see §4.5), but in most of the other transport properties, such as Hall effect. thermal conductivity and thermoelectric effects (see for instance Grenier, Reynolds and Zebouni (1963); Fletcher (1981)). The periodicity of the oscillations is always identical with that of the SdH and dHvA effects but the theory of the amplitudes is complicated and has not been fully worked out.

4.7.3. The Knight shift in nuclear magnetic resonance

The slight change in the frequency of nuclear magnetic resonance in a metal as compared with an insulator (the 'Knight shift') shows oscillations as the magnetic field varies (see for instance Goodrich, Khan and Reynolds 1971). Actually what is observed for given resonant frequency v is the small difference ΔH of the magnetic field at resonance between a metal crystal and an insulator containing the nuclear species being studied. The value of $\Delta H/H$, typically of order a few times 10^{-3}, is found to show oscillations as v (and consequently H) varies, whose amplitude is typically of order a few times 10^{-5}. Qualitatively such an effect is to be expected since the magnetic field seen by the nucleus must oscillate as the magnetization of the metal oscillates, and this would suggest that the amplitude of the oscillations in $\Delta H/H$ should be of order $4\pi|\tilde{M}|/H$, which indeed is not too far wrong. However the detailed theory of the screening effect of the electrons in the atom containing the resonant nucleus and of the other electrons is complicated and the interpretation of the experimentally observed amplitudes is not straightforward.

4.8. Magnetic oscillations of other kinds

For the sake of completeness we shall mention here a number of other oscillatory effects whose periodicities are *not* linked with that of the dHvA effect. The first four effects are discussed more fully in a review by Chambers (1969):

4.8.1. Azbel–Kaner cyclotron resonance

Here the microwave impedance is modified whenever the microwave frequency ω is an integral multiple of the cyclotron frequency ω_c, which is of course proportional to H. This follows either from thinking of the classical orbit of the electron re-entering the skin depth region when the

174

alternating field there has undergone an integral number of oscillations or from thinking of a photon $\hbar\omega$ being absorbed in a transition from one Landau level to another. Thus the microwave impedance shows oscillations with a period in $1/H$ given by

$$\Delta(1/H) = e/\omega mc \qquad (4.71)$$

and the effect provides a method of determining the cyclotron mass m.

4.8.2. Magnetoacoustic effect ('geometric resonance')

Here the attenuation of an ultrasonic wave varies in oscillatory fashion with field, but by a mechanism entirely different from that of the GQO discussed in §4.6. Here it is a question of the electron orbit size spanning an odd number of half wavelengths of the sound wave so that the electrons can absorb energy from the electric field associated with the sound wave. The periodicity is given by

$$\Delta(1/H) = e\lambda/hcp \qquad (4.72)$$

where λ is the ultrasonic wavelength and p is an appropriate 'caliper' dimension of the FS. This periodicity is of course quite different from that of the GQO, which is essentially that of the dHvA effect and quite independent of the sound wavelength. Measurement of the caliper dimension p by this magnetoacoustic effect provides an alternative method for determining the FS.

4.8.3. The r.f. size effect (Gantmakher effect)

This is somewhat similar to the magnetoacoustic effect, except that the electron orbit size is matched to the thickness d of a plate-like sample rather than to a sound wavelength. If the electron mean free path is long compared to d the r.f. impedance of the sample shows anomalies whenever d is an exact multiple of the electron orbit size and the oscillations are now periodic in H rather than $1/H$. The period is given by

$$\Delta H = pch/ed \qquad (4.73)$$

where p is, as in (4.72), an appropriate caliper dimension of the FS. This not only provides another method of determining the FS, but information can also be deduced about electron velocities from the fine structure of the effect (Gantmakher, Lebech and Bak 1979).

4.8.4. The d.c. size effect (Sondheimer effect)

If the mean free path is long compared to the thickness d of a plate-like sample, the resistivity of the sample in the direction normal to the plate shows oscillations as H varies. For an electron of given k_H, oscillations occur when the time to cross the thickness d matches the time for an integral number of cyclotron orbits, and this leads to a periodicity ΔH given by

$$\Delta H = (c\hbar/ed)(\partial a/\partial k_H)_\varepsilon \qquad (4.74)$$

However since this ΔH is a function of the k_H of the electron, the oscillations from electrons of differing k_H tend to interfere destructively. Usually only a feeble oscillation survives, with a ΔH corresponding to the limiting k_H of the Fermi surface, for which (if the FS is closed),

$$(\partial a/\partial k_H)_\varepsilon = 2\pi K^{-1/2} \qquad (4.75)$$

where K is the Gaussian curvature of the FS at the limiting point.

4.8.5. Magnetic surface states (see for instance Khaikin 1979, Doezema and Koch 1975)

In a very pure metal at very low magnetic fields, typically a few G, the radius of curvature of an electron orbit may be as large as a few mm and some electrons are able to travel in a 'skipping' orbit close to a plane surface of the sample, with specular reflection each time the orbit meets the surface. This kind of periodic motion is quantized and the permitted states correspond to a pattern of magnetic field dependent energy levels, arranged rather like those of the hydrogen atom. The separations between these levels are comparable to those of a microwave frequency quantum $\hbar\omega$ and consequently the microwave impedance of the sample at frequency ω shows absorption peaks whenever $\hbar\omega$ matches particular level separations as the field is varied. The result is a complicated pattern of rather irregular looking oscillations in the impedance as a function of field, from which, however, much useful information can be derived about various aspects of electronic structure, in particular electron velocities and relaxation times.

4.8.6. Magneto-phonon effect

Marked oscillations in the resistivity of semimetals and non-degenerate semiconductors occur when the energy of an optical mode phonon matches the separation between Landau levels, i.e. is an integral multiple of $\hbar\omega_c$. These magneto-phonon oscillations can be thought of as resonances in the inelastic scattering of electrons by optical phonons and were first predicted by Gurevich and Firsov (1961); for a general review see Harper, Hodby and Stradling (1973). Superficially the oscillations look rather like the Shubnikov–de Haas effect, but they are really quite different. Just as in the SdH effect they are periodic in $1/H$, but the period is given by

$$\Delta(1/H) = e/\omega_0 mc \qquad (4.76)$$

where ω_0 is the frequency of the optical phonon, rather than by the extremal area of the Fermi surface. Secondly the oscillations require optical phonons to be appreciably excited, which implies a temperature T such that

$$kT \gtrsim \hbar\omega_0 \tag{4.77}$$

Typically T is of order 300 K and the oscillations fade out completely at temperatures low enough for the de Haas–van Alphen and Shubnikov–de Haas effects to appear.

5

Fermi surfaces and cyclotron masses

5.1. Introduction

During the last 25 years de Haas–van Alphen studies have led to a spectacular advance in our knowledge of the Fermi surface of metals and (to a lesser extent) of the differential properties of the surfaces of constant energy in the vicinity of the FS. This progress has been made possible partly by parallel developments in the theoretical understanding of band structures, but perhaps more significantly by advances in technology. These advances, in the production of high magnetic fields and low temperatures, in electronic techniques and data processing and in the growing of purer and more perfect single crystals, are still continuing and hopefully will continue to be exploited to extend our knowledge still further.

By far the greatest effort has gone into measurements of dHvA frequencies F with a view to the determination of the FS of metals through the Onsager relation and by now the FS of nearly all the metallic elements and of many intermetallic compounds have in fact been determined. The level of determination achieved however, varies both in the degree of certainty with which the qualitative nature of the surface has been established and in the precision of the quantitative specification of the surface. At best, the qualitative nature of the surface (i.e. the number and shapes of the separate sheets) is reliably known and the dimensions of the various sheets determined with a precision of order 1 in 10^3. But this is true of only a few metals, and for many others there are still doubts about qualitative details and FS dimensions have been determined to only a few %.

The cyclotron mass can be determined from the temperature dependence of the dHvA amplitude or from cyclotron resonance and, as will be explained in §5.2.2, a detailed knowledge of the orientation dependence of cyclotron mass leads to a specification of the differential properties of the surfaces of constant energy in the vicinity of the FS. This in turn leads to a determination of the density of states and a specification of the electron velocity at all points of the FS. Up to now, however, this programme has been carried through for only a few metals and much remains to be done. As has already been said in §2.6, the observed cyclotron masses (and so also

178

the deduced electron velocities) are appreciably enhanced by many-body interactions (particularly the electron–phonon interaction) as compared with the predictions of independent particle band structure calculations. These enhancements are directly relevant to the interpretation of the superconducting behaviour of metals.

In this chapter we shall first outline the general principles involved in going from the experimental data on frequencies and masses to specifying the details of the FS and their differential properties. This will be followed by a survey of what has been achieved for a few selected metals. We shall not attempt any complete coverage in this respect, since this would require a book in itself and indeed a book (Cracknell and Wong 1973) already exists, as well as more recent reviews emphasising certain aspects (Gold (1974) and Lonzarich (1980) deal with ferromagnetic metals, Young (1977) gives a general review with emphasis on the rare earth metals, Edelman (1976) deals with Bi and Sellmyer (1978) deals with alloys and compounds). These sources between them provide a reasonably up-to-date bibliography of the very extensive literature. Our aim will rather be to discuss simple examples which illustrate and emphasise the general principles of what can be done. Some features of the examples chosen will also be useful later in discussing other aspects of the dHvA effect which require for their interpretation a concrete picture of the FS concerned. We shall also review briefly results on the modifications of Fermi surfaces caused by strains and by dilute alloying (i.e. addition of foreign atoms).

5.2. General principles

5.2.1. Determination of a Fermi surface from de Haas–van Alphen frequencies

Measurement of the dHvA frequency F as a function of the magnetic field direction specifies the extremal areas A of the FS normal to all directions relative to the crystal axes. This provides a great deal of geometrical information about the FS, but it is only under rather restricted conditions that the actual shape and size of the Fermi surface (i.e. a specification of $|k|$ as a function of the direction of k) can be uniquely deduced from this geometrical information. The requirements are that the surface should have a centre of symmetry and that a radius vector from the centre of symmetry should intersect the surface in only two symmetrical points. For a surface of this kind Lifshitz and Pogorelov (1954) deduced an elegant theorem which gives a recipe for calculating the radius vector in terms of the areas of plane sections through the centre of symmetry (which in this case are indeed extremal areas). Unfortunately, however, many FS do not satisfy the conditions necessary for the theorem to be valid, and even if they do, the inversion from areas to radii is more simply achieved by other means. In practice the unravelling of the frequency (F) data to determine a FS has usually been somewhat of an empirical procedure, based on some preliminary idea from band structure theory of what kind of FS might be expected. Without some such guidance it is almost impossible to make much of an F-spectrum as complicated as that of fig. 5.18 or 5.23. Additional guidance has sometimes also been available from other phenomena (anomalous skin effect, magnetoresistance, magnetoacoustic effect etc.; for a brief review see Shoenberg (1969b)).

A few general points will now be mentioned which provide interpretative clues from the F-spectrum itself and help to distinguish between different possible FS models. Where there are several branches fairly close together in the spectrum (e.g. the ρ branches in fig. 5.23) it is likely that they come from a group of symmetry-related pieces of FS (cf. the ellipsoids in fig. 5.24). If the oscillations can be followed over the whole range of orientations in different planes of rotation (as with ρ) it is likely that the individual pieces are closed surfaces, e.g. ellipsoids. However this is not inevitable, as can be seen from the example of fig. 5.18 for Al, where, as we shall see in §5.5.1, the continuous branches of the spectrum prove to correspond to a rather complicated tubular surface. Even when a simple closed surface is reliably indicated, the F-data do not uniquely specify the shape unless it is known to be centro-symmetric. For instance dHvA data alone cannot distinguish between the pair of non-centro-symmetric surfaces (a) sketched in fig. 5.1 and the single centro-symmetric surface (b) which has the same extremal areas.

If the oscillations on a particular branch cease abruptly at certain critical orientations (e.g. ξ in fig. 5.23) it is an indication that they correspond to an extremal section of the FS which exists over only a limited range of orientations and this usually implies that the relevant sheet of FS is somewhat complicated (cf. the 'jack' in fig. 5.24).

Once a fairly detailed qualitative model of the various sheets of a Fermi surface has emerged, and has been checked against the predictions it makes of how F should vary with orientation in various planes of rotation, the

Fig. 5.1. Schematic diagram showing normal sections of (a) two symmetry related non-centro-symmetrical cigar-like surfaces (these could be 2 out of 6 at the corners of a hexagon) and (b) a single cigar-like surface with central symmetry, and extremal areas identical with those of (a).

a   b   a

180

model has to be specified in some more quantitative way and its parameters determined. The most satisfactory situation is if the surface can be specified in terms of some mathematical representation which takes into account the crystal symmetry and involves only a few parameters to be determined empirically by fitting the representation to the F data. Such representations have been successfully achieved for only a few metals (e.g. in terms of cubic harmonics for the alkalis, Fourier series for the noble metals and ellipsoids for bismuth) and have the merit that they provide an objective and simple description of the surface, independent of any theoretical treatment of the band structure. In recent years, however, band structure calculations have become more and more dependable and an alternative approach (and indeed sometimes the only practicable approach) is to compare the F data with the predictions of a parametrized band structure calculation in which the parameters (such as the phase shifts and Fermi energy in a Korringa–Kohn–Rostoker (KKR) calculation, or where appropriate, a set of pseudopotential coefficients) are adjusted to give agreement between theory and experiment. This approach involves more complicated computations, since for a band structure calculation there is a much more devious path from assumed values of the parameters to the F spectrum they imply than there is for a direct mathematical representation of the surface. However, with modern computing methods this is no longer the obstacle it used to be and of course the parametrized band structure approach has the merit that the parameters determined have some physical meaning, rather than being purely empirical, as in the direct approach.

One important aspect of using a parametrized representation of the surface, whether it be a parametrized band structure or a direct mathematical formula, is that provided the number of parameters is small, their values are fully determined by F data for angular traverses in only one or two zones (e.g. the (100) and (110) zones for a cubic crystal). Indeed, if there are only three parameters, only the F values at three distinct orientations (e.g. $\langle 100 \rangle$, $\langle 110 \rangle$ and $\langle 111 \rangle$) are needed to determine the parameters, and F values at other orientations provide some check on the reliability of the representation. This involves considerably less data taking than would application of the Lifshitz–Pogorelov theorem, which requires F to be known at *all* orientations (or at least at sufficient orientations to permit adequate interpolation).

This feature of parametrization, though evidently advantageous, also carries some risk, especially if more than very few parameters are fitted to data in a single crystallographic zone. The risk is that the representation may prove less successful for orientations other than those in the particular zone studied. Ideally, checks should be made over a number of zones, but if very high accuracy is to be achieved it must be remembered that the accuracy of orientation determination may prove as much of a limiting factor as the accuracy of measurement of F. To give a concrete example, for the noble metal bellies the orientation at a non-symmetry direction must be known to better than $0.1°$ to match an accuracy of measurement of F of 1 part in 10^4. It is only at orientations where F has an absolute maximum or minimum (usually symmetry points) that precise determination of orientation becomes relatively less important.

An ingenious method of improving the precision of determining non-symmetry orientations has been developed by Mueller, Windmiller and Ketterson (1970). The method exploits the fact that if an arbitrary zone is traversed there will be several pairs of orientations on the traverse which are exactly equivalent because of crystal symmetry and so must have the same F. With most sample rotation mechanisms the field effectively rotates about a cone (of semi-angle close to $90°$) with respect to the crystal and, if the orientation of the axis and semi-angle of the cone are precisely specified, so too are the orientations of the pairs of crystallographically equivalent positions in the traverse. Thus, provided the cone axis orientation and semi-angle are approximately known to start with, they (and the orientations of all points on the traverse) can be much more precisely established by a process of trial and error (with the cone axis orientation and the cone semi-angle as adjustable variables) by successive trials in which the predicted positions of equivalent pairs 'home in' on the positions actually observed. By measuring F over a number of non-symmetry zone traverses whose orientations are sufficiently precisely determined by this technique, it is possible to obtain a more reliable fit to a parametrized representation than can be obtained by using only symmetry zones.

Improvements in the precision of determining Fermi surfaces have proved important in two respects. First they have made possible ever more significant comparisons with band structure theory and secondly the theoretrical calculations of various electronic effects in metals require precise specification of the FS, so that improved precision permits more stringent checks of the theory of the effect. There are however some basic limitations (other than technical ones) to the possibility of increasing precision indefinitely.

First there are some inherent approximations in the LK formula, so that description of the oscillations as proportional to $\sin(2\pi F/H + \phi)$, with constant F and constant ϕ, is not strictly exact except in the limit of high F/H and at very low temperatures. If F is to be measured to better than 1 part in 10^4 it is important to make sure the oscillations are followed down to low enough fields and that the constancy of F and ϕ is checked over the field range used. The FS is also in principle slightly temperature dependent because of the effect of lattice vibrations in slightly modifying the band structure through electron–phonon interactions, and because of the slight temperature dependence of the Fermi energy. Typically the changes of F are only parts in 10^4 over the range of temperature for which oscillations can be observed, but, as will be discussed in §5.5, much larger effects can be observed in special cases. In ferromagnetic metals the FS can be appreciably modified through spin-orbit coupling by the direction of the ferromagnetic saturation magnetization and so cannot be uniquely specified to better than parts in 10^3 (and for some parts of the FS of iron only to parts in 10^2). There may also be another extremely slight temperature dependence of F in ferromagnetics arising from interaction with the magnetic excitations and study of this effect can provide valuable evidence about the basic theory of ferromagnetism (Lonzarich 1980). Some procedures for precision determination of F will be described later (§9.3) in connection with absolute phase determination.

Finally we must mention two complications coming from effects which we have not yet properly considered; magnetic interaction and magnetic breakdown. As we shall see later, both these effects can lead to the appearance of combination and difference frequencies, while harmonics of the 'basic' frequencies can be enhanced by magnetic interaction as compared with the predictions of the LK theory. Unless such 'non-basic' branches of the F spectrum are recognised for what they are, the raw data of the F spectrum may suggest all sorts of extra sheets of FS which are really entirely spurious; some of these non-basic branches in the spectrum of W are indicated in fig. 5.23. Another complication of magnetic breakdown is that as the magnetic field is increased, originally separate sheets of the FS effectively become connected and it is as if the FS has a different form at low and high fields (with an even more complicated situation in the transition range!). In what follows we shall do little more than mention such complications where it is essential, but leave a more detailed disussion until later (chapters 6 and 7).

5.2.2. Determination of differential properties from cyclotron masses

As was first pointed out by Lifshitz and Pogorelov (1954) knowledge of the cyclotron mass for all field orientations can be combined with knowledge of the Fermi surface to derive differential properties of the band structure at the Fermi energy ζ. The mass m directly gives $(dA/d\varepsilon)_{\varepsilon=\zeta}$ and since the extremal area A of cross-section of the surface of energy ζ is known, the extremal area $A + \Delta A$ of an immediately neighbouring surface of energy

$\zeta + \Delta\varepsilon$ is given as

$$A + \Delta A = A + (dA/d\varepsilon)\Delta\varepsilon \qquad (5.1)$$

Thus, if $(dA/d\varepsilon)$ is known for all directions, so too is $A + \Delta A$, and whatever method was used to determine the FS from the values of A (i.e. of F), can now be used to determine the surface of constant energy $\zeta + \Delta\varepsilon$. Once this determination has been made (for a value of $\Delta\varepsilon$ sufficiently small for (5.1) to be valid, yet large enough to permit reliable computation) the problem is essentially solved. Thus the volume of k-space between the surfaces of energy ζ and $\zeta + \Delta\varepsilon$ immediately gives the density of states, while the distance Δk between the surfaces along the normal at any point immediately gives $\mathrm{grad}_k\varepsilon$ and so the electron velocity (see (2.6)). It should be noticed that as regards the density of states this procedure is simply equivalent to determining the appropriate average of m which should be inserted into the formula $(2.165a)$ for the density of states in the special case of spherical surfaces of constant energy.

Although this programme is simple in principle, it has not yet been much exploited. This is partly because of computing difficulties and partly because the acquisition of sufficiently accurate data on masses involves considerably more work than on frequencies. The experimental procedure for determining m is based on the formula (2.128) (or its approximate form (2.129)) for R_T, the theoretical reduction factor of the dHvA amplitude $A(T)$ from its value at $T = 0$. If, as is usually appropriate, the fundamental amplitude ($p = 1$) is measured, a plot for given H of $\ln(A/T)$ against T should, in the limit of high T, be linear with a slope given by $-1.47 \times 10^5 \, (m/m_0)/H$. At low temperatures, however, the difference between $\sinh z$ and $\frac{1}{2}e^z$ (with $z = 1.47 \times 10^5 \, (m/m_0)T/H$) may become appreciable and the experimental points will then fall above the linear extrapolation of the high T points. A convenient way of allowing for this to sufficient approximation is to estimate a provisional value of (m/m_0) from the high T points and then subtract e^{-2z} from each point on the logarithmic plot for which e^{-2z} is appreciable on the scale of the plot. It is important to remember that the theoretical formula (2.128) is valid only if magnetic interaction (see chapter 6) is negligible or properly allowed for, but with this proviso and with sufficient care in the control and measurement of temperature (and of course of A itself and H), excellent linear plots can be obtained, thus confirming the validity of the theory, and (m/m_0) can be determined with a precision of a few parts in 10^3.

An alternative possibility is to use cyclotron resonance to determine the cyclotron mass, but this has also been little exploited so far. Not only is the experimental technique rather more difficult than that of de Haas–van Alphen experiments, but there are difficulties of interpretation and the mass measured involves a rather different integration over k_H and so is not exactly the same as the 'extremal' mass that is needed for the calculation of velocities. In fact at the level of precision so far achieved (of order 1% by cyclotron resonance and rather better by dHvA amplitudes) no significant difference has been observed between masses determined by the two methods.

5.3. Results for some metals

5.3.1. The alkalis

As might be expected, the monovalent metals have the simplest Fermi surfaces and of the monovalent metals the alkali FS are the simplest of all, departing only slightly from the ideal spherical surface of the free-electron model. Because the departures are so slight, the frequency F varies only little with orientation and a sensitive method of measuring the variation is to observe the oscillations as the sample is rotated in a constant field H. Since F/H is typically of order a few thousand (3600 for K at 5×10^4G) the passage of one oscillation as the sample is rotated corresponds to a change of F of only one part in a few thousand and thus the difference ΔF from a reference value F at any particular orientation can be mapped out with considerable precision. The reference value F can of course be determined by the usual method of counting oscillations as the field is swept. Fig. 5.2*a* shows a rotation curve for K obtained by Shoenberg and Stiles (1964) in the original application of this method and the corresponding changes ΔF are shown in fig. 5.3. A more recent rotation curve (Templeton 1981) is shown in fig. 5.2*b* to illustrate the considerable improvements of technique since the early experiments. Particularly striking is the constancy of amplitude with rotation; the significance of this constancy will be pointed out a little later (p. 196).

Since the FS of the alkalis are closed and have cubic symmetry, they can be conveniently represented by expansions in cubic harmonics. In particular the orientation dependence of the central areas of cross-section A can be expressed as

$$\Delta A/A_0 \equiv (A - A_0)/A_0 = c_4 K_4 + c_6 K_6 + c_8 K_8 + c_{10} K_{10} \quad (5.2)$$

where the Ks are cubic harmonics*, the cs are coefficients, whose values are

---

* Beyond K_{10} this notation becomes inadequate and an extra index is required, but this extra index is omitted here for simplicity. Explicit definitions of the first few harmonics may be found in Shoenberg and Stiles (1964) and in somewhat different form and extended to more harmonics in Lee and Falicov (1968); see also Gaertner and Templeton (1977). As has been pointed out by Mueller (1966) and Mueller and Priestley (1966) this type of expansion is the special form for cubic symmetry of a more general expansion in spherical harmonics; see also Foldy (1968) who relates the Mueller expansion to the Lifshitz–Pogorelev theorem.

Fig. 5.2. dHvA oscillations for rotation of a K crystal at constant $H \sim$ 50 kG, $T \sim 1.2$ K. (a) Shoenberg and Stiles (1964). The numbers identify successive increases and decreases of π in phase (the total phase $2\pi F/H$ is of order 7400). The rotation is in a (110) zone. (b) Templeton (1981). This rotation diagram over 360° illustrates the improvement in experimental technique since 1964. Rotation is not in an exact symmetry zone but not far off (110).

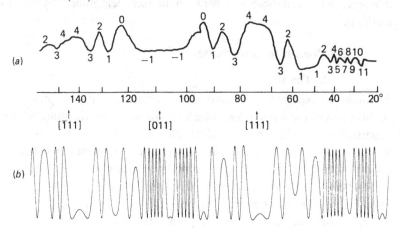

Fig. 5.3. Orientation dependence of dHvA phase (i.e. of F or A) for K in a (110) zone. The points marked ○ are taken from the phase changes shown in fig. 5.2a, those marked + are from a rotation experiment on a different crystal. $\Delta A/A$ is about 1 part in 7400 for each π of phase change. The full curve is based on (5.2) with $c_4 = 7.8 \times 10^{-4}$, $c_6 = 3.5 \times 10^{-4}$, $c_8 = 4.5 \times 10^{-4}$, $c_{10} = 0$.

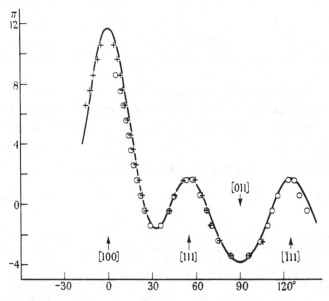

to be determined by fitting the formula to experimental data, and A_0 is the average of A over the FS. The merit of this type of representation is that it leads immediately to an explicit expression for the radius k of the surface, which becomes particularly simple when the relative variations in k are small.

In general it can be shown that

$$\frac{\Delta(k^2)}{k_0^2} \equiv \frac{k^2 - k_0^2}{k_0^2} = \frac{c_4 K_4}{P_4(0)} + \frac{c_6 K_6}{P_6(0)} + \frac{c_8 K_8}{P_8(0)} + \frac{c_{10} K_{10}}{P_{10}(0)} + \cdots \qquad (5.3)$$

where $P_l(0)$ is the value of the lth Legendre polynomial at the origin and $\pi k_0^2 = A_0$ (i.e. k_0 is the radius of the sphere whose diametral cross-section has area A_0). If it can be assumed that $(k - k_0) \ll k_0$, the l.h.s. of (5.3) can be replaced by $2(k - k_0)/k_0$ and putting in the numerical values of $P_l(0)$, we have an explicit expansion in cubic harmonics for the radius k:

$$\frac{\Delta k}{k_0} = \frac{k - k_0}{k_0} = \frac{4}{3} c_4 K_4 - \frac{8}{5} c_6 K_6 + \frac{64}{35} c_8 K_8 - \frac{128}{63} c_{10} K_{10} + \cdots$$

$$(5.4)$$

and the problem of inversion from areas to radii is solved once the cs have been determined. It can also be shown that the values of k_0 and A_0 are to a good approximation those of a sphere of the same volume as that of the actual surface. If, however, $\Delta k/k_0$ becomes appreciable, the approximation leading to (5.4) becomes inadequate and (5.3) must be used as it stands; the value of k/k_0 is then given by the square root of [1 plus the r.h.s. of (5.3)]. The largest errors in k caused by using (5.4) rather than (5.3) are of order $4 \times 10^{-7} k_0$ for Na, $10^{-6} k_0$ for K, $5 \times 10^{-5} k_0$ for Rb and $4 \times 10^{-3} k_0$ for Cs. The best experimental accuracy so far achieved is of order $10^{-5} k_0$ for Na and K, $4 \times 10^{-5} k_0$ for Rb and $3 \times 10^{-4} k_0$ for Cs, so it is only for Cs that the simplifying approximation of (5.4) is appreciably inadequate.

In fitting experimental data such as shown in fig. 5.3 to (5.2), the origin of ΔF or ΔA must be treated as adjustable, i.e. effectively the experimental curve of ΔF against orientation has to be slid up and down until it fits best with the cubic harmonic expression. The origin for the best fit is then such that ΔA is $A - A_0$, where A_0 is the average of A over all orientations, since all the cubic harmonics K_4 etc. are defined to have zero averages. Once the coefficients c_4 etc. in (5.2) have been determined, the value of A_0 (and hence of k_0) can be determined if F (and hence A) is measured absolutely at any particular orientation (preferably a symmetry orientation, where errors in orientation matter least). Thus A_0 is then simply A divided by [1 plus the r.h.s. of (5.2)]. It is not difficult to show that the volume of the FS is very close to $\frac{4}{3} \pi k_0^3$, i.e. to that of a sphere of radius k_0. In fact the volume differs

187

from $\frac{4}{3}\pi k_0^3$ by an amount involving the squares of the cs, which even for Cs amounts to only a few parts in 10^4.

In the early experiments of Shoenberg and Stiles the FS of K and Rb could be represented within experimental uncertainties by using only the three harmonics K_4, K_6 and K_8, though a small addition of K_{10} marginally improved the fit for Rb. However, the later work of Lee and Falicov (1968) and Templeton (1981) on K and that of Gaertner and Templeton (1977) on Rb and Cs showed that to match the precision of the experimental data, about seven harmonics were required for K, nine for Rb and eleven for Cs. The least square deviations between the data and its representation no longer improved if more harmonics were included. The most accurate determinations so far have been those of Gaertner and Templeton (1977) and Templeton (1981) who made use of the 'arbitrary' traverse method of Mueller et al. (1970) (see §5.2.1) to cover a wide spread of non-symmetry orientations, and also measured fields very precisely by n.m.r. and used sophisticated computing methods to analyse the large mass of data efficiently and objectively. The precision of the results for Na too could probably be considerably improved by these techniques. Fig. 5.4a shows a

Fig. 5.4. (a) Contours of $\Delta A/A_0$ for K on a stereographic projection, computed from a seven term cubic harmonic expression (Templeton 1981); the three term fit (Shoenberg and Stiles 1964) is shown dotted where it can be distinguished from the seven term fit. The contours are marked in units of 10^{-4}. (b) Contours of $\{\Delta(\mathrm{d}\ln A/\mathrm{d}p)/(\text{mean value of } \mathrm{d}\ln A/\mathrm{d}p)\}$ for K (Templeton 1981); units of 10^{-3}. Note: Strictly speaking only one sixth of either diagram is needed to specify all the information, but a much better visual impression of how the deviation varies over the sphere is obtained by including the redundant five sixths. The plots were kindly provided by Dr I. M. Templeton.

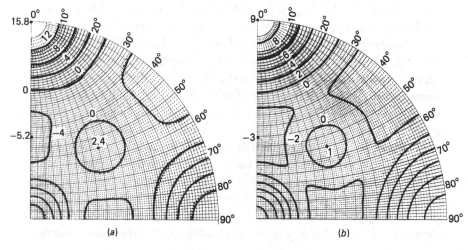

5.3. Results for some metals

contour diagram of the area differences $\Delta A/A_0$ for K plotted on a stereographic projection and it can be seen that the early results of Shoenberg and Stiles agree quite well with the much more precise results of Templeton (1981). Fig. 5.4*b* is discussed on p. 238. The contours of $\Delta k/k_0$ for the various alkalis as deduced from the most recent data are shown in fig. 5.5.

The FS volumes as determined by precise absolute measurements of F at particular orientations together with the known values of the cubic harmonic coefficients prove to be, as they should, just those corresponding to one electron per atom. For K, Rb and Cs the precision of the dHvA volume determination is better than 1 part in 10^3 (the main error comes from the determination of the cs rather than from that in the absolute values of F); this is somewhat better than the precision of the theoretical estimate–of order a few parts in 10^3–which is limited by the error in determining the lattice constant. For Na and Li the FS volume has not yet been as precisely determined by the dHvA method, but there is no significant discrepancy with the theoretical estimate.

The results for Li shown in fig. 5.5*a* are based on an indirect method and are much less accurate than any of the others. The difficulty with Li is that a good single crystal in the body-centred cubic (BCC) phase at room temperature is completely broken up by the martensitic transformation it undergoes as it is cooled through 75 K. There is a similar difficulty with Na but the martensitic transformation occurs at a lower temperature (30 K) and a sufficiently perfect crystal can, with care, be cooled through the transition temperature without transforming. Although observations on a good single crystal are essential for an accurate and reliable FS determination, it is possible to obtain some useful semi-quantitative information by an ingenious indirect method due to Randles and Springford (1976). They found that colloidal preparations of both Li and Na showed feeble dHvA oscillations at high fields (85–100 kG) and low temperatures (1.2 K), and this seems to imply that for lack of a nucleus, the martensitic transformation does not occur in many of the very small particles (typically 5 μm diameter for Li and 10 μm diameter for Na) in the samples. The feebleness of the oscillations is of course because of destructive interference between the signals from the randomly oriented individual particles (see §8.5.4 for further discussion) and the fact that the amplitude was something like 20 times weaker for colloidal Li than for a comparable volume of colloidal Na, immediately suggests that the FS of Li is much more distorted than that of Na.

This qualitative inference is supported by the relative scales of the beat structures in the oscillations (fig. 5.6) and the distortion shown in fig. 5.5*a*

189

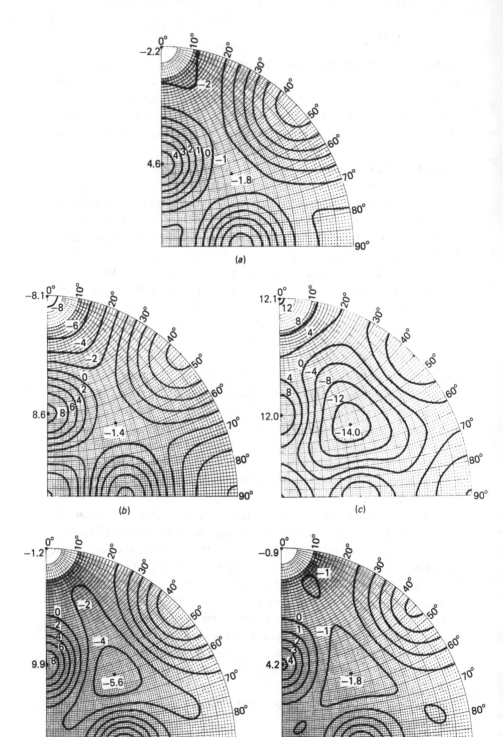

5.3. Results for some metals

Fig. 5.5. Fermi surfaces of the alkali metals shown as contours of $\Delta k/k_0$, as computed from cubic harmonic fits (Dr I. M. Templeton kindly provided these plots). (a) Li (Randles and Springford (RS) 1976); contour units of 10^{-2}, (b) Na(data of Lee 1966, fitted by RS); units of 10^{-4}, (c) K (Templeton 1981) units of 10^{-4}, (d) Rb (Gaertner and Templeton (GT) 1977) units of 10^{-3}, (e) Cs (GT); units of 10^{-2}. For (d) and (e), the published form of the contours included a zero offset and was expressed in terms of Δk^2; here there is no zero offset and the results have been recomputed to give $\Delta k/k_0$. It should be noted that (a) is little more than schematic and is mainly intended to give a rough idea. *Note in proof*: (a) was inadvertently computed for model 1 of RS, which exaggerates the values of $\Delta k/k_0$ by a factor of about 2.6 as compared with the RS data, and even more as compared with recent position annihilation data (see p. 566).

was estimated by comparing the observed beat structure with the beats predicted by averaging the oscillations over all orientations on the basis of various theoretical models of the FS. It turns out that the scale of the beat structure (i.e. the average frequency of the beats) is rather insensitive to the details of the distortion of the FS but depends mainly on the scale of the distortion as specified by a parameter δ, where

$$\delta = [k(110) - k(100)]/k_0 \qquad (5.5)$$

and the coefficients of the expansion in cubic harmonics for each model are taken to be proportional to δ. The best fit to observation (based on an average of the predictions of four different models) gave $\delta = 0.026 \pm 0.009$, and this is, as expected, a considerably larger distortion than that of the FS of Na, for which $\delta = 0.0017$. Evidently this ingenious method can at best be only semi-quantitative and in particular it does not really tell much about the detailed shape of the FS. Moreover the quantitative analysis, by which the scale of the departures from a sphere is deduced, is based on various simplifying assumptions, so that the actual value of δ may be more in error than the indicated $\pm$ implies. Fortunately a check on the reliability of the method is available by applying it to a colloid of Na and comparing the resulting estimate of δ from the beats with that obtained from the FS as established by single crystal data. This check is quite encouraging in giving $\delta = 0.0018 \pm 0.0003$, which agrees well with the single crystal value 0.0017. The check is also demonstrated by the fair agreement between the observed pattern of beats and that computed assuming the known FS (fig. 5.6). However in this check it is assumed that the geometry of the FS is as determined from single crystal data and that only the scaling factor is to be determined. This is in contrast to the calculation for Li for which the geometry of the FS is little better than guessed.

Cyclotron masses for the alkalis have been determined both from the

191

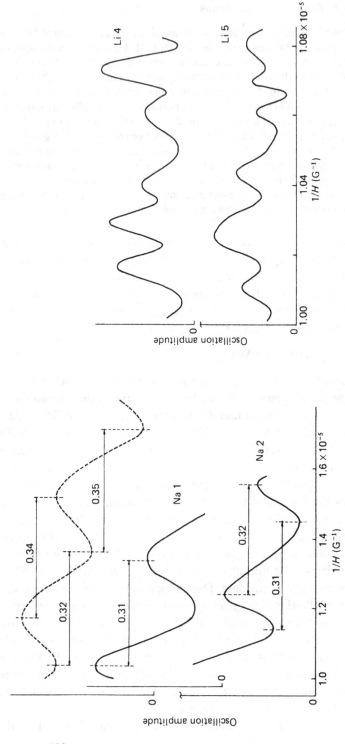

Fig. 5.6. Beats in dHvA oscillations of Li and Na colloids (RS 1976). The beat patterns vary in detail but not, in general, in scale as between different samples of each metal; the beat 'frequency' is about 15 times higher for Li than for Na. The broken curve for Na is theoretical, based on the known FS and averaging the oscillations over all orientations. The amplitude units are arbitrary but different for Li and Na.

192

5.3. Results for some metals

temperature dependence of dHvA amplitude at a few orientations and from cyclotron resonance in polycrystalline samples, and the results are summarised in table 5.1. Within the rather poor precision so far achieved there is no significant orientation dependence of mass for Na and K, but for Rb and Cs the mass varies more with orientation than does the radius of the FS. The cyclotron resonance masses are, as they should be, reasonably close to the average dHvA values. Although the orientation dependence data are too unsystematic and too imprecise to permit any worthwhile calculation of the electron velocity distribution over the FS on the lines described in §5.2.2, the average mass m may be used to estimate the density of states and hence the electronic specific heat by treating the constant energy surfaces as spherical. In fact it is easy to show that the ratio of the electronic specific heat γT to $\gamma_s T$, the value for a free electron model of the same electron density, should be simply m/m_0. As can be seen from table 5.1, for Na and K the electronic specific heat estimated in this way agrees quite reasonably with that measured calorimetrically, but there appear to be significant discrepancies for Li, Rb and Cs. The discrepancy for Li may well be a consequence of the fact that the calorimetric sample contained a mixture of BCC and HCP phases, but the origin of the discrepancy in Rb and Cs is not yet understood. It may be that in deriving the electronic contribution to the

Table 5.1. *Cyclotron masses of alkali metals compared with electronic specific heats*

	m/m_0		γ/γ_s(calorimetric)[g]
	dHvA	cyclotron resonance	
Li	1.86 ± 0.10[a]	—	2.20
Na		1.24[d]	1.26
K	1.22[b]	1.217 ± 0.002[e]	1.24
Rb	$1.19 - 1.25$[b]	1.20[f]	1.37
Cs	$1.35 - 1.5$[b,c]	1.44[f]	1.80

Notes: For spherical surfaces of constant energy we should have $m/m_0 = \gamma/\gamma_s$ where γT is the calorimetrically measured value of electronic specific heat and $\gamma_s T$ is the value for a free-electron gas of the same density ($\gamma_s = 1360 \, V^{2/3}$ erg K^{-2} for atomic volume V). The accuracy of determination of both m/m_0 and γ/γ_s is usually about ± 0.02, apart from the exceptions noted. The cyclotron resonance data are for polycrystals.

References (a) Randles and Springford (1976), (b) Knecht (1975), (c) Gaertner and Templeton (1977), (d) Grimes and Kip 1963, (e) W. M. Walsh quoted by Allen, Rupp and Schmidt (1973), (f) Grimes, Adams and Schmidt (1967), (g) Martin (1970) (contains references to earlier work).

observed specific heat some anomalous feature of the lattice specific heat has not been adequately allowed for, and specific heat measurements at still lower temperatures might clarify the issue. However the interesting possibility that something is wrong with our basic ideas must be kept in mind and it would be well worth while to make more systematic and more accurate dHvA measurements as well as extending the specific heat measurements to lower temperatures.

The Fermi surfaces of the alkalis have been studied by various other techniques (reviewed by Shoenberg 1969b) besides that of the dHvA effect, but so far, apart from one exception, none of them proved delicate enough to reveal, let alone measure, the small departures of the FS from spherical form. The one exception is the positron annihilation technique, which has the great asset that it does not require low temperatures and so avoids the difficulty of the martensitic transformation in Li. The most recent study by Paciga and Williams (1971) indicated an anisotropy δ (see (5.5)) of 0.029, appreciably smaller than a value of about 0.05 found earlier by Donaghy and Stewart (1967), but in fair agreement with the value found from the dHvA oscillations of colloids.

First principles band structure calculations, though correctly describing the general trend of a considerably distorted Fermi surface for Li followed by an almost perfect sphere for Na and then successively increasing distortion for K, Rb and Cs, have not yet achieved an agreement with the experimental results to within anything like the experimental accuracy (for references see Lee (1969a), Randles and Springford (1976), Gaertner and Templeton (1977), MacDonald (1980)). This is no doubt because the calculations of the slight departures are sensitive to the many approximations involved and hopefully the existence of accurate experimental data may serve as a guide to improving the approximations in the future.

Much greater success has been achieved by parametrized band structure calculations of varying degrees of sophistication, in which the parameters are adjusted to make the calculated Fermi surface agree with experiment. The simplest approach of this kind (first used by Ashcroft (1965)) is to treat the electrons as if they were nearly free, but perturbed by a relatively feeble local pseudopotential specified by a small number of Fourier coefficients $V_{110}, V_{200}, V_{211}$ etc. Lee (1966) was able to fit his experimentally determined Na Fermi surface to well within the experimental accuracy by using only the single coefficient V_{110} and choosing its magnitude to be

$$|V_{110}| = 0.23 \, \text{eV}$$

(the sign is not determined in this calculation). This agrees quite reasonably

5.3. Results for some metals

with Heine and Abarenkov's (1964) 'model potential' calculation which gives $V_{110} = +0.25 \pm 0.10\,\mathrm{eV}$. However this relatively simple approach is less successful for the other alkalis. For Li this is because there are no p-electrons in the ionic core, so that the cancellation which keeps the effective potential small no longer applies to the p-states. For K, Rb and Cs the d-like bands come increasingly closer to the Fermi energy and the use of a local pseudopotential is no longer a good approximation. The non-locality was first taken into account by Lee and Falicov (1968) for K by adding angular momentum dependent square-potential wells within the ion cores and a rather better fit to the Fermi surface was obtained. However the most successful parametric calculation (Lee 1969a) was based on an APW band structure calculation with three adjustable phase shifts (for $l = 0$, 1 and 2) and an adjustable Fermi energy. With only these four adjustable parameters Lee was able to fit the reliably measured FS to within the experimental accuracy. It is impressive that so few parameters are needed to describe surfaces which need as many as ten cubic harmonic coefficients for a specification of similar precision.

Although conventional band structure theory seems to account quite successfully for the dHvA data, Overhauser (1964, 1968 and a review with full references in 1978) has proposed a radically different electronic structure for K, the compatibility of which with the experimental evidence is still a controversial issue. His original (1964) proposal, on the evidence of optical data, was that the ground state of K is one with a spin density wave (SDW), but later he suggested that a charge density wave (CDW) was more likely. Either kind of wave implies a Fermi surface instability such that in the ground state the nearly spherical FS is distorted into a lemon-like shape with the axis of the lemon along the wave vector and with an anisotropy of linear dimensions of a few %.

To account for the absence of any such anisotropy in the FS deduced from the dHvA data, Overhauser at first suggested that the CDW vector aligns itself along the high magnetic field of the dHvA experiments, which would imply that the area corresponding to the dHvA frequency was always that of the cross-section normal to the long axis of the lemon and, as observed, varying only slightly with crystal orientation. The FS deduced from such data would then be some kind of 'pseudo'-surface with a volume slightly, but appreciably, smaller than the true volume expected for one electron per atom. However, as mentioned earlier, the most accurate determinations do not show any volume discrepancy of the required magnitude.

More recently Giuliani and Overhauser (1980) have suggested that the CDWs have preferred crystallographic directions (close to $\langle 110 \rangle$) so that

195

the sample breaks up into domains; this would once again imply a drastic reduction of the lemon anisotropy and so might help to explain what Overhauser in his review calls 'the de Haas–van Alphen difficulty'. But, as emphasized by O'Shea and Springford (1981), this domain hypothesis would imply considerable phase smearing (i.e. interference between oscillations from different domains) and variation of amplitude with orientation in contrast to the strikingly constant amplitude actually observed with good samples. Moreover the interference would reduce the absolute amplitude to an extent which would be difficult to reconcile with the experiments described in chapter 9 from which a g value is deduced in good agreement with determinations by quite different methods.

Perhaps some convincing way can yet be found of reconciling the CDW hypothesis with the dHvA results, but until this is done the hypothesis must be regarded as unproven, even though it does seem to be supported by other evidence. Particularly striking is the explanation by Huberman and Overhauser (1981), by means of the CDW hypothesis, of the irregular oscillations observed by Coulter and Datars (1980) in the eddy current torque acting on a sphere of K slowly rotated in a magnetic field. A possibility might be that the CDW occurs in the large samples of Coulter and Datars, but not in the small samples used in dHvA studies. A key experiment would be to study the dHvA anisotropy on a large sample which showed the eddy current torque oscillations.

5.3.2. The noble metals

The next simplest Fermi surfaces are those of the monovalent noble metals, Cu, Ag and Au which are, because of the proximity of the d-band to the Fermi level, just complicated enough to show new features of interest. As mentioned in chapter 1, the dHvA effect in the noble metals was not observed until after Pippard (1957a) had already studied the FS of Cu by the anomalous skin effect technique. His results suggested that the FS was probably sufficiently distorted from a sphere to make contact with the $\langle 111 \rangle$ faces of the Brillouin zone (fig. 5.7(a)) and this would imply a multiply connected open surface with qualitatively new features. Most of these new features were indeed soon confirmed by the dHvA experiments, not only in Cu, but also Ag and Au.

The most striking new feature predicted by Pippard's model is that in the extended zone representation there should be 'necks' joining the surfaces within each separate zone. Thus a field along $\langle 111 \rangle$ should 'see' not only a large extremal area round the 'belly' of the surface at B but also a small minimum area round the neck at N. Correspondingly, there should be dHvA oscillations of much lower frequency, associated with the neck, and coexisting with the high frequency oscillations associated with the belly. Such neck oscillations were first observed in Au, with a frequency about 30 times lower than the belly frequency and this immediately indicated that the neck radius must be a little over five times smaller than that of the belly. A neck of comparable size was soon afterwards found in Cu and a rather smaller neck in Ag. The manner in which the frequency increases as the field is tilted away from ⟨111⟩ indicates that the shape of the neck is approximately hyperboloidal.

The other main qualitative consequences of a multiply connected surface are that there should be:

(1) a 'dog's bone' orbit (D) normal to ⟨110⟩ (fig. 5.7c and e), of area about 0.4 that of the belly; for this orientation, however, the central belly orbit is broken by the necks and so no belly frequency should be observed together with the dog's bone.

(2) a 'four-cornered rosette' orbit* (4-R) normal to ⟨100⟩ (fig. 5.7b and f) of area comparable to that of the dog's bone; the rosette oscillations should coexist with belly oscillations.

(3) A 'six-cornered rosette' (6-R) normal to ⟨111⟩ (fig. 5.7d and g) of area nearly twice that of the belly; the corresponding oscillations should coexist with both neck and belly oscillations.

Experimental observation of the dog's bone and four-cornered rosette oscillations at ⟨110⟩ and ⟨100⟩ respectively (and for a small range of neighbouring orientations), together with the absence of belly oscillations around ⟨110⟩, provided convincing confirmation that the low frequency oscillations at ⟨111⟩ were indeed due to necks of the kind envisaged in Pippard's model. The six-cornered rosette oscillations have so far been observed only in Ag, and only somewhat indirectly. As discussed in §6.7.1.3, the difficulty in detecting them is that they are overshadowed by the strong second harmonic of the belly oscillations due to magnetic interaction (the frequency of this harmonic is only 5% higher than that expected for the six-cornered rosette).

Once the qualitative nature of the model could be regarded as completely reliable, the next step was to determine the FS as precisely as possible from the detailed orientation dependence of the frequencies. The early experiments using the pulsed field technique showed that the belly frequencies were indeed close to the predicted value for a free electron sphere, as was to be expected, and that they did not vary more than a few per cent over the

---

* To help visualization, a crude model consisting of spheres joined by short necks is often used to represent the multiply connected Fermi surface. In this model the orbit concerned does have some resemblance to a four-cornered rosette, but for the real FS the orbit turns out to be almost square rather than rosette-like (except for Ag–see fig. 5.7h). However, even though it is not very appropriate, the name has stuck.

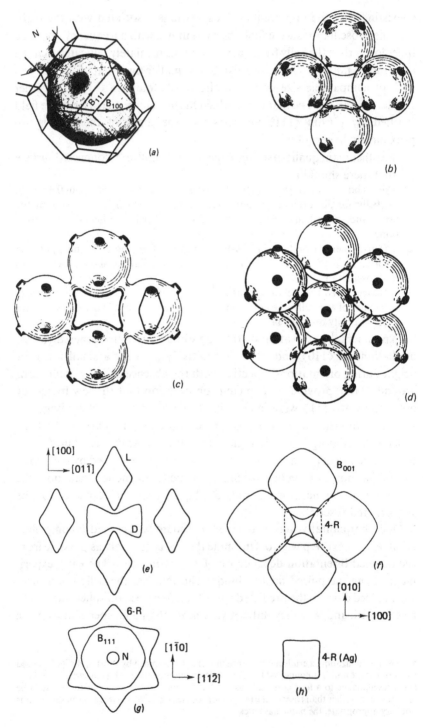

5.3. Results for some metals

Fig. 5.7. Fermi surface of the noble metals. (a) Schematic; showing the FS within a single zone joined by a neck (N) to the surface in a neighbouring zone; the belly orbits normal to $\langle 111 \rangle$ and $\langle 100 \rangle$ are also indicated. (b), (c), (d) Schematic sketches of the FS in the extended zone scheme looking along $\langle 100 \rangle$, $\langle 110 \rangle$ and $\langle 111 \rangle$ respectively and showing the location of four-cornered rosette in (b), of dog's bone and lemon in (c) and of six-cornered rosette in (d); in (b) and (d) the hidden parts of the orbits are shown broken. (e), (f), (g) are accurate scale drawings of the various extremal orbits specifically for Cu; it should be noted that in (e) the lemon (L) and dog's bone (D) are in different planes, in (f) the 4-R and each pair of B_{001} orbits, likewise, while in (g) 6-R and N are in the same plane but B_{111} is in a different plane. The scale of (e), (f) and (g) is such that the biggest radius of B_{001} is 1.06 times k_s, the radius of a free electron sphere of the same electron density. (h) the 4-R orbit in Ag to illustrate the slight resemblance to a rosette. The sketches of (b), (c) and (d) are based on those of Lee (1969b); (e)–(h) were kindly computed by Mr C. M. M. Nex.

ranges of orientation for which they could be observed. In order to measure this slight orientation dependence of F sufficiently precisely, a method more sensitive than direct counting of oscillations was devised, in which the oscillations from the sample were made to beat against oscillations from a fixed reference sample (see §3.4.1). It was from data obtained in this way (Shoenberg 1962) that Roaf (1962) produced the first mathematical representation of the Fermi surfaces of the noble metals, in which the radii could be regarded as reliable to 1%.

However, with the advent of superconducting magnets, this indirect beat method has been superseded by the rather more precise technique of counting oscillations as the sample is rotated in a constant field – the same technique that was used for the alkalis. Oscillations of this kind, obtained by Halse (1969) for rotation of a Cu crystal in a (110) zone, are shown in fig. 5.8. Leaving aside, for the moment, some of the interesting details revealed by this rotation curve, it is immediately evident that the belly in Cu is a good deal more anisotropic than the FS of K, as evidenced by the much larger number of oscillations for rotation over a comparable angular range. The orientation dependence of $F - F_{100}$ as deduced from the prominent oscillations in fig. 5.8 is shown as B_1 in fig. 5.9 and it can be seen that, in agreement with the evidence of the direct frequency measurements, the variation amounts to only a per cent or so, which is indeed about ten times more than the variation in K (see figs. 5.2 and 5.3). The curve B_2 of fig. 5.9 will be discussed later.

The simplest analytical representation of the FS which takes account of the cubic symmetry and the multiple connectivity of the surface is a Fourier

199

expansion of the form (Roaf 1962, Halse 1969)

$$
\begin{aligned}
C_0 = 3 &- \sum \cos\tfrac{1}{2}ak_y \cos\tfrac{1}{2}ak_z + C_{200}(3 - \sum \cos ak_x) \\
&+ C_{211}(3 - \sum \cos ak_x \cos\tfrac{1}{2}ak_y \cos\tfrac{1}{2}ak_z) \\
&+ C_{220}(3 - \sum \cos ak_y \cos ak_z) \\
&+ C_{310}(6 - \sum \cos\tfrac{3}{2}ak_y \cos\tfrac{1}{2}ak_z - \sum \cos\tfrac{3}{2}ak_z \cos\tfrac{1}{2}ak_y) \\
&+ C_{222}(1 - \cos ak_x \cos ak_y \cos ak_z) \\
&+ C_{321}(6 - \sum \cos\tfrac{3}{2}ak_x \cos ak_y \cos\tfrac{1}{2}ak_z \\
&\qquad\quad - \sum \cos\tfrac{3}{2}ak_z \cos ak_y \cos\tfrac{1}{2}ak_x) + \cdots
\end{aligned}
\tag{5.7}
$$

Fig. 5.8. dHvA oscillations for rotation of a Cu crystal in a constant field (49 kG) from ⟨100⟩ in a (110) zone (Halse 1969). One oscillation corresponds to $\Delta F/F \sim 8 \times 10^{-5}$. The features marked are A: 4-R superimposed on B_{100} oscillation near ⟨100⟩ (although 4-R has the lower F, it varies about five times as fast with angle), B: the central belly area (B_1) goes through a minimum (stationary phase); the faster oscillations are from the non-central belly (B_2), C: B_2 passes through zero because of vanishing spin-splitting factor, D: large amplitude at merging of central and non-central orbits, E: B_1 passes through zero because of vanishing spin-splitting factor, F: B_1 disappears because of intervention of necks, G: chart recorder sensitivity doubled. The minimum just beyond 25° may be a beat minimum associated with sample inhomogeneity. Some further discussion of the region between C and E is given in Appendix 5.

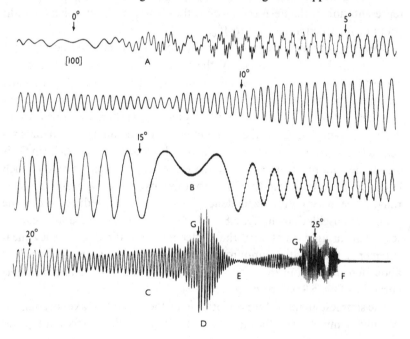

where a is the lattice constant and the Cs are adjustable parameters to be
fitted to the data. The values of the parameters are chosen by a process of
trial and error in which extremal areas for particular orientations are
computed for trial values of the parameters and compared with the
experimentally measured areas. Once agreement has been obtained, the
areas at other orientations provided a check of the quality of the fit. In
principle, to fit seven parameters (such as appear in (5.7)), seven experimental features have to be fitted. In the early work of Roaf and Halse, the
absolute accuracy of the measured area was only of order 1%, so only *ratios*
of areas (i.e. of F values) were used, which were determined much more
precisely, and the absolute scale was established by computing the *volume*
of the surface within the fundamental zone and requiring this to be exactly
the volume needed for one electron per atom. In more recent work

Fig. 5.9. Variation of $\Delta F/F_s$ in (110) zone of Cu as deduced from fig. 5.8
(Halse 1969); ΔF is $F - F_{100}$ and F_s is for the free electron sphere model.
The points are experimental and the curves are computed from the
Fourier series description of the FS; B_1 and B_2 refer to the central and
non-central extremal orbits.

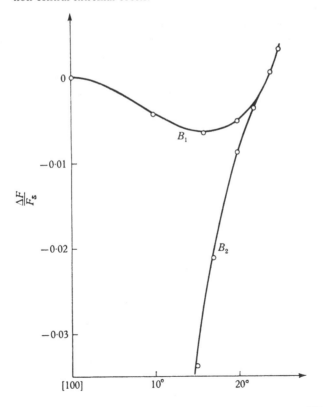

(Coleridge and Templeton 1972, Bosacchi and Franzosi 1976) measurement of field by n.m.r. enabled the absolute accuracy to match the relative accuracy and this has made possible a fit based on areas alone, probably the most reliable fit to a Roaf-like formula available at the time of writing. The values of the Cs are given in table 5.2. The computed volume of the FS finally arrived at is found, as it should, to agree to better than 0.1% with the volume corresponding to 1 electron per atom and this provides a valuable additional check on the reliability of the representation.

As can be seen from the B_1 curve of fig. 5.9, the orientation dependence of $(F - F_{100})/F_s$ (where F_s is the free electron sphere value) as computed from Halse's Cu7 representation (involving seven coefficients) agrees well with the experimental values obtained by counting oscillations on a rotation curve. This, together with agreement of similar quality over other angular ranges in various zones and for various types of orbit, is already good evidence for the reliability of the representation, but even more impressive confirmation comes from one of the other interesting features in fig. 5.8 (further points of interest are mentioned in the caption). This is the presence of rapid subsidiary oscillations which set in at about 15° from $\langle 100 \rangle$ and then gradually slow down until they disappear at about 23°.

These subsidiary oscillations were first observed by Joseph and Thorsen (1964*b*) and correctly interpreted as coming from non-central extremal areas of the FS. These non-central extrema are minima lying symmetrically above and below the central maximum (see fig. 5.10); their existence is of course a consequence of the distortions caused by the necks and they occur only over certain limited ranges of orientation. As the orientation is varied,

Table 5.2. *Values of Cs in Fourier representation of the FS for the noble metals* (Coleridge and Templeton 1972)

	C_0	C_{200}	C_{211}	C_{220}	C_{310}	10^4 $\times$ r.m.s.
Cu(5VI)	1.691314	0.006574	−0.426081	−0.018050	−0.036283	4
Ag(5IV)	−0.898274	−0.120728	−0.904789	−0.140823	−0.092808	3
Au(5VI)	−2.263657	−0.167472	−1.263244	−0.102181	−0.120280	6

Note: The designation of the formulae (Cu5VI etc.) is that of the authors. The column r.m.s. gives the r.m.s. of the differences between F/F_s as calculated from these formula and the observed values. Bosacchi and Franzosi (BF) (1976) using the same experimental data showed that the r.m.s. errors can be considerably reduced by fitting a seven rather than a five term formula; the fitting error then becomes comparable to the experimental error. It should be noted that BF do not quote Halse's formula (5.7) quite correctly. Improved seven term formulae have recently been given by Coleridge and Templeton (1982).

the non-central extremal areas vary more rapidly than the central ones and so give rise to the more rapid oscillations. The points on the curve B_2 in fig. 5.9 are derived by counting these non-central oscillations and the significant feature is that they agree so well with calculations based on a mathematical representation of the FS which was arrived at from entirely independent data. Not only do the computed plots of area as a function of κ show minima for non-zero values of κ, at orientations where the subsidiary oscillations are observed, but these non-central minimum areas agree almost perfectly with the experimental points. This remarkable agreement and similar agreement over other angular ranges for which non-central extrema occur (for details see Halse 1969) is among the strongest evidence for the meaningfulness of the Fermi surface as an extremely precise entity.

One final piece of evidence of a similar kind was the prediction by Halse of a 'lemon'-shaped extremal section (fig. 5.7c and e) normal to $\langle 110 \rangle$. The computed area as a function of κ for Cu in this direction shows a minimum at $\kappa/\kappa_s = 0.86$, and this minimum area should correspond to a frequency slightly lower than that of the dog's bone. These 'lemon' oscillations were expected to be feeble so they were specially looked for at a very low temperature (0.4 K) and high field (90 kG) and they did indeed show up as a beat of exactly the expected frequency in the stronger dog's bone oscillations. Halse has predicted that the lemon should occur also in Au, but not in Ag: these predictions have not, however, yet been verified.

Fig. 5.10. Belly areas in Cu computed as a function of $|\kappa|/k_s$ normal to various directions in a (110) zone. Between about 16° and 23° from $\langle 100 \rangle$ a minimum occurs close to the end of the range of κ for which a closed orbit occurs, as well as the maximum at $\kappa = 0$ (i.e. the central area). These non-central minima give rise to the B_2 curve of fig. 5.9.

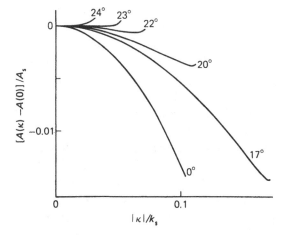

203

As with the alkalis, the alternative method of specifying the FS, by a suitable choice of phase-shift parameters (and of Fermi energy) in a band structure calculation, has also proved remarkably successful. This approach was pioneered by Lee (1969*b*) who with the use of only four parameters (three phase shifts and the Fermi energy) was able to fit Halse's experimental data for Cu about as well as Halse's own fit with a seven-coefficient Fourier formula; a still better fit was obtained by Nowak and Lee (1972) to the more accurate results of Coleridge and Templeton (1972). The advantages of the band structure representation as compared with the Fourier representation of having fewer parameters and parameters of some physical significance, are of course to some extent offset by the greater complexity of the computations necessary to arrive at any value for an area or a radius of the FS. Probably the most reliable values of the fitting parameters in a parametrized KKR band structure calculation for Cu, Ag and Au are those of Coleridge quoted (together with references to other values) in Table VI of Bibby, Coleridge, Cooper, Nex and Shoenberg (1979). There have been many 'first principles' band structure calculations, particularly for Cu (for a review see Dimmock 1971) but although they predict Fermi surfaces of the right general shape the detailed agreement with the experimentally determined surfaces is at best of order 1% in radius and it is difficult to pin down just which of the many assumptions and approximations need modification to give better agreement.

The first significant application of the method outlined in §5.2.2 to obtain the distribution of electron velocities over the FS from the orientation dependence of cyclotron mass was achieved by Halse (1969) for the noble metals. However, his analysis was based on rather sparse and not very accurate data and much improved results have since been obtained by Lengeler, Wampler, Bourassa, Mika, Wingerath and Uelhoff (1977) from careful measurements of the temperature dependence of oscillation amplitudes at many orientations. The variation of cyclotron mass with orientation for Cu is shown in fig. 5.11; the agreement with the masses obtained from cyclotron resonance (Koch, Stradling and Kip 1964) is only to within a few per cent, but the discrepancies are probably within the uncertainties of the cyclotron resonance data. From the cyclotron masses Halse-like surfaces Cu5+ and Cu5− were constructed in such a way that the extremal areas of these two surfaces for any orientation should differ from the extremal area of the Halse-like five-coefficient Fermi surface Cu5 by $\pm(dA/d\varepsilon)\Delta\varepsilon$, where $dA/d\varepsilon$ was deduced from the observed cyclotron mass and $\Delta\varepsilon$ was given the small value of 6×10^{-4} times the Fermi energy of the free electron sphere (Halse had used a value of $\Delta\varepsilon$ ten times larger, which may have led to appreciable errors). From the Cu5+ and Cu5−

5.3. Results for some metals

formulae the area differences and hence the cyclotron masses could be computed at all orientations, and these computed masses (indicated by the continuous curve of fig. 5.11) can be seen to fit all the sixteen data points extremely well even though essentially only five data points determine the coefficients of the formulae. (In fact a least squares procedure was used in which all the data points were used in the fitting rather than five particular points.)

The Fermi velocity at any point on the FS now follows from the computed distance along the normal between the ± surfaces, and the

Fig. 5.11. Variation of cyclotron mass of Cu with orientation in (100) and (110) zones (Lengeler *et al.* 1977). The full curves are fitted to the data by means of five adjustable parameters and it can be seen that the authors' 17 experimental points ($\bigcirc$) fall very well on the fitted curves. The other points (∇ Coleridge and Watts 1971*a* from dHvA effect and $\square$ Koch *et al.* 1964 from cyclotron resonance) are included to give some idea of the consistency between different experiments.

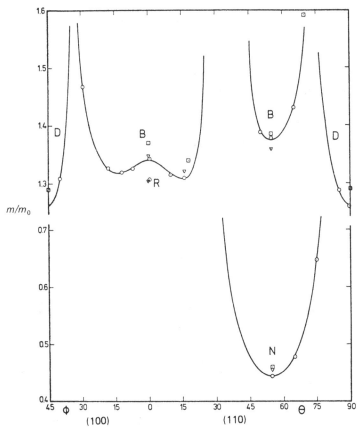

resulting orientation dependence of v/v_s (where v_s is the velocity for a free electron sphere) is shown in fig. 5.12 for Cu, Ag and Au. Another method of determining Fermi velocities is that based on microwave resonance spectra of magnetic field induced surface states and it can be seen that the results for Cu (Doezema and Koch 1972) agree reasonably well with those from the dHvA effect though the agreement for Ag (Deimel and Doezema 1974) is less satisfactory. This alternative method, however, requires extremely perfect samples and for this reason has not yet been applied to many metals. Moreover to obtain the velocity at a particular point on the FS the radius of curvature of the surface at that point must be accurately known, so that the final result is rather sensitive to the assumed specification of the FS.

The computed volume in k-space between the $\pm$ surfaces determines the density of states at the Fermi energy and hence the electronic specific heat. As can be seen from table 5.3, the results obtained are in excellent agreement with the directly measured electronic specific heats, thus confirming again both the accuracy of the specification of the constant energy surfaces around the FS and the reliability of the theory.

The ratio of the experimentally measured velocity at any point of the FS to the velocity given by a band structure calculation in which many-body effects are ignored, measures the reciprocal of the electron–phonon 'enhancement factor' $1 + \lambda(\boldsymbol{k})$ (as discussed in §2.6, electron–electron enhancement contributes only slightly). On the basis of Lee's (1970) band structure calculation of v, $\lambda(\boldsymbol{k})$ is found to be appreciably anisotropic (fig. 5.13). The average value $\bar{\lambda}$ over the FS is found to be 0.11 as compared with theoretical estimates ranging from 0.12 to 0.15. In view of the uncertainties of the velocities calculated from band structure and of the

Table 5.3. *Comparison of electronic specific heats of the noble metals deduced from dHvA data and calorimetrically*

	γ/γ_s	
	dHvA (*a*)	Calorimetric (*b*)
Cu	1.382 ± 0.010	1.383 ± 0.002
Ag	1.008 ± 0.007	1.004 ± 0.002
Au	1.074 ± 0.007	1.083 ± 0.002

Notes: As explained in the note to table 5.1, dHvA estimates of γ/γ_s are essentially equivalent to estimates of m/m_0 where m is an appropriate average of the cyclotron mass over the FS. Examination of fig. 5.11 shows that for Cu this average is dominated by the belly masses and this is also true for Ag and Au.

References (*a*) Lengeler *et al.* (1977) and (*b*) Martin (1973).

5.3. Results for some metals

Fig. 5.12. Variation of electron velocity with orientation for the noble metals (Lengeler *et al.* 1977); v is the velocity at a point on the FS specified by the angle from $\langle 100 \rangle$ in the (100) and (110) zones, and v_s is the velocity for the free electron sphere of the same electron density ($= 1.58 \times 10^8\,\mathrm{cm\,s^{-1}}$ for Cu, 1.40 for Ag and Au). The chain curves for Cu and Ag are those of Doezema and Koch (1972) and Deimel and Doezema (1974) respectively, based on surface state resonances.

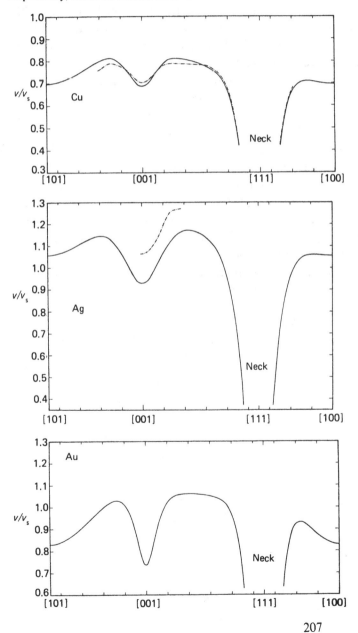

theoretical estimates of $\bar{\lambda}$ (typically ± 0.02), this is satisfactory agreement, though it would of course be more convincing if the anisotropy of λ were also theoretically calculated.

Finally we should mention that other Fermiological techniques produce results which are, with few exceptions, entirely consistent with those derived from the dHvA effect. Thus 'caliper' measurements based on the magneto-acoustic 'geometrical' resonances (e.g. Bohm and Easterling 1962) and on the Gantmakher radio frequency effect (Gasparov and Harutunian 1976) indicate the same shape and size of Fermi surface, though the former results have a precision of only 1% or so in linear dimensions, rather than the 0.1% of the dHvA method, and the latter achieve a precision of order 0.2% only by the use of extremely pure samples and elaborate orientating techniques. Magnetoresistance measurements confirm the topology of the surface and give about the right sizes for the necks (Priestley 1960). There are, however, some puzzling aspects of the electrical and thermal Hall effect measurements, which should be related by the Wiedemann–Franz law, and should essentially give the difference of the volumes of the FS for which the sections normal to the field are closed or open. Although for the field along $\langle 100 \rangle$ and $\langle 110 \rangle$ the electrical (Kunzler and Klauder 1961) and thermal (Lipson 1966) results are reasonably consistent with each other and with predictions based on the FS neck diameters, for the field along $\langle 111 \rangle$ they are wildly discordant with each other and with the FS prediction. The cause of these discrepancies is not clear but it should be said that the experiments

Fig. 5.13. Anisotropy of the phonon enhancement factor $1 + \lambda(k)$ in Cu (Lengeler *et al.* 1977). The factor $1 + \lambda(k)$ is obtained from the ratio of the value of v at any orientation as obtained by Lee (1970) from a band structure calculation in which electron–phonon interaction is omitted, to the experimental value of Lengeler *et al.*

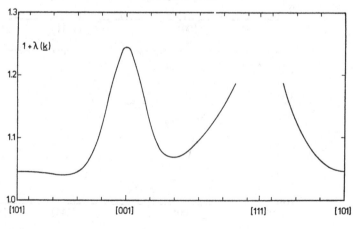

are difficult and some of the underlying assumptions behind the predictions may not have been fulfilled in the experiments. Thus the field may not have been high enough to achieve true saturation of the gradient of the field dependence of the observed potential or temperature difference, the Wiedemann–Franz law may not be reliable in a magnetic field, and the results are extremely sensitive to orientation around $\langle 111 \rangle$. A worrying feature, too, is that Lipson's results at $\langle 111 \rangle$ appeared to be strongly sample dependent. It would be well worth while to repeat both types of experiment using improved techniques and ideally measuring the electrical and thermal Hall effects simultaneously.

Positron annihilation experiments which, as mentioned earlier, have the advantage that they do not require low temperatures and moreover are not sensitive to electron scattering, confirm the form of the FS of the noble metals as determined by the dHvA effect, though the precision is only of order 1% at best. The interpretation of positron annihilation data is however rather subtle (see Berko 1978 for a good general review), and it is only in relatively simple situations (such as the alkalis or the noble metals) that it is possible to go unambiguously from the data to a Fermi surface. In more complicated situations it is rather the other way round: if the FS is known it is possible to compare the observed data with the predictions according to band structure calculations, and in this way to test the reliability of the calculations. Where positron annihilation really comes into its own is in making it possible to determine the FS of disordered alloys, in which the dHvA oscillations are too feeble to show up at all. One example of its value is in showing that the necks of the Cu surface continue to exist even when the electron per atom ratio is increased by as much as 30% (e.g. for addition of 30% Zn); the neck diameter for such alloys is roughly double that of pure Cu (Berko 1979).

Although it was anomalous skin effect studies which led Pippard (1957*a*) to suggest what has proved to be the correct qualitative form of the FS of Cu, the rather complicated geometrical information provided by the effect (an average round a zone of a radius of curvature) cannot be used to go unambiguously from the experimental data to a Fermi surface. Somewhat as with positron annihilation, it is rather the other way round. Once the FS is known, it is possible to compute just what the anomalous skin effect should be for each orientation of the sample surface. Such computations were first made by Roaf (1962) and refined by Halse (1969), using his rather more accurately specified FS; as can be seen from fig. 5.14 the computed surface resistances for Cu and Ag do not differ from the experimental values by much more than the experimental uncertainties. Such slight discrepancies as there are, could probably be removed by very slight adjustments to the FS, though they may partly be due to unsuspected errors in the experiments.

5.3.3. 'Simple' polyvalent metals

As outlined in the historical introduction (§1.2) the 'nearly free electron' model (NFE) proves to give a remarkably close guide to the Fermi surfaces of many polyvalent metals which 'simple' in the sense that the d-bands are not too close to the Fermi level. In its crudest form, the 'free electron' (FE) model, this is just a sphere in extended k-space, of the volume necessary to hold the correct number of valence electrons per atom. If the sections of this sphere falling in each of the various zones are remapped in the periodically repeated fundamental zone, we find a number of separate

Fig. 5.14. Anomalous skin effect in (a) Cu and (b) Ag. The full curves are the values of surface resistance computed by Halse (1969) from his seven term Fourier formulae for the Fermi surfaces. The points are experimental (Pippard 1957a, Morton 1960). One unit of the ordinate represents the surface resistance for a spherical FS.

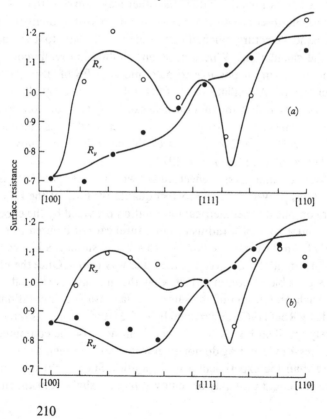

sheets of FS such as shown in fig. 5.15. The effect of the lattice potential, or rather of the equivalent weak pseudopotential, is to distort the surface of the original sphere wherever it crosses zone boundaries and this causes some modifications in the details of the remapped sheets. This type of model was originally suggested by Gold (1958) to interpret his dHvA experiments on Pb and it was later shown by Harrison (1960) to have a fairly wide validity for many other metals. The theoretical explanation of why perturbation by only a weak pseudopotential is adequate to describe the FS came soon afterwards (see review by Heine *et al.* 1970). Al provides one of the simplest examples of how the dHvA data can be analysed to determine values of the pseudopotential coefficients as well as the detailed form of the Fermi surface. We shall also briefly discuss the case of Pb, which is rather more complicated, because spin-orbit coupling has to be taken into account, and the series of divalent hexagonal metals in which interesting complications are introduced by magnetic breakdown.

5.3.3.1. Aluminium

The special feature of Al is that the three-electron sphere passes very close to the corners W of the zone, so that quite a small perturbing potential can modify the detailed form of the sharp points of the second zone free-electron hole surface and the detailed form of the rather thin arms of the third zone 'monster' (fig. 5.15). Using only two pseudopotential coefficients V_{200} and V_{111} and a 4-OPW calculation, Ashcroft (1963) showed that

Fig. 5.15. Free electron FS for f.c.c. metals of valence 3 and 4 (Harrison 1960), Note that the third band surfaces are displaced by half a reciprocal lattice vector in the direction ΓX to exhibit the form of the 'monster' more clearly and that the fourth band surfaces are similarly displaced in the direction ΓL.

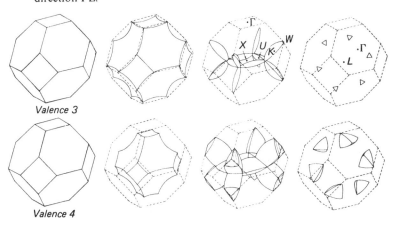

Valence 3

Valence 4

different topological situations arise for different ranges of these two parameters. Thus in region A of the $V_{200} \mid V_{111} \mid$ plane (fig. 5.16) the monster has the same connectivity as for $V_{200} = V_{111} = 0$ (the strictly FE case), in C_1 the monster breaks into 'rings of four' (fig. 5.17), while in D the monster is

Fig. 5.16. The $V_{200} \mid V_{111} \mid$ plane is divided into regions A, B, C_1, C_2 and D, each of which corresponds to a characteristic topology of the FS of Al (Ashcroft 1963). The units for both axes are Rydbergs.

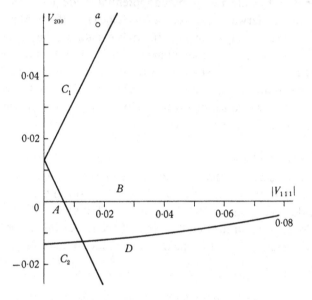

Fig. 5.17. Sketch of the 'rings of four' into which the multiply-connected monster of fig. 5.15 breaks up if V_{200} and $\mid V_{111} \mid$ for Al have the values indicated by (*a*) in fig. 5.16 (Volski 1964*a*).

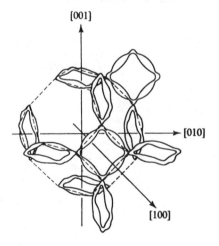

5.3. Results for some metals

completely broken up into separate 'sausages' and thin tunnels form between the second zone hole surfaces, and so on.

For each of these topological situations there are several very small extremal areas, either close to where the arms of the monster join or along the tunnels joining the second zone surfaces, and these give rise to a complicated pattern for the orientation dependence of the corresponding very low dHvA frequencies ($F \sim 3 \times 10^5$ G), which depends on the particular type of topology. It proved that only values of V_{200} and $|V_{111}|$ in the region C_1 reproduced the experimentally observed pattern and the point a gave an almost perfect quantitative fit (see fig. 5.18) not only to the

Fig. 5.18. Variation of medium and low frequency periods for Al (a) in the (100) zone (b) in the (110) zone. The curves are computed from Ashcroft's model; the points are experimental (Larson and Gordon 1967).

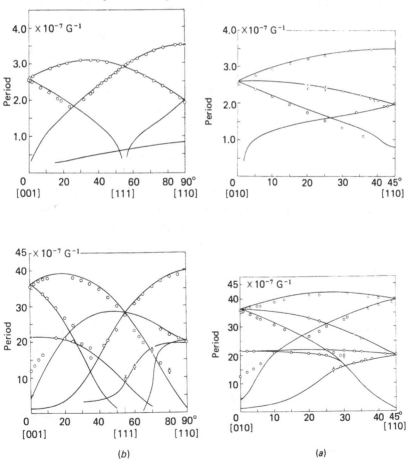

(b)   (a)

213

observed very low frequencies, but also to the medium frequencies ($F \sim 3 \times 10^6$ G) corresponding to the fatter parts of the arms and the high frequencies ($F \sim 4 \times 10^8$ G) corresponding to the second zone surface. In developing his theoretical model in 1963,. Ashcroft had available only somewhat fragmentary dHvA data: early results of Shoenberg (1952a), more extensive ones of Gunnersen (1957), preliminary results of what eventually proved to be the most thorough survey until a good deal later, by Larson and Gordon (1967) and some high frequency results by Priestley (1962). By a curious coincidence Volski in Moscow was also making a detailed survey of the low and medium frequency oscillations but did not learn of Ashcroft's calculations until his own measurements were complete. The reliability of both theory and experiment was convincingly demonstrated by the almost perfect agreement that Volski (1964a) found between his observed frequencies and the calculated values sent to him by Ashcroft.

Of course, the higher frequencies of the second zone surface are less sensitive than the very low ones to the exact choice of the Vs, but, nevertheless, it was found that the choice indicated by fitting the very low frequencies was appreciably better than the strictly FE model. Thus, paradoxically, it is the very low frequencies, associated with the very fine detail of the FS, which provide the key to an accurate determination of the whole surface–a case of the tail wagging the dog.

Subsequent studies (Anderson and Lane 1970, Coleridge and Holtham 1977) have considerably improved the accuracy of the frequency measurements and cleared up a number of puzzles, in terms of magnetic interaction and magnetic breakdown effects, but the Ashcroft model still stands as a valid description of the FS of Al. His values of the Vs and the Fermi energy have changed only slightly, as much through use of a more accurate value of the lattice constant as because of slight changes of the F values to be fitted (P.M. Holtham 1979, private communication). The physical meaningfulness of the pseudopotential coefficients deduced from the dHvA frequencies is supported in several ways. First they are in agreement with estimates based on fundamental band theory within the considerable uncertainties of the latter (Animalu and Heine 1965). Secondly, the values of the pseudopotentials may be used to calculate the resistivity of liquid aluminium and good agreement with experiment was obtained by Ashcroft and Guild (1965). Further support is provided by experiments on the stress dependence of the FS which will be discussed in §5.9.

Finally it should be mentioned that the FS of In is very similar to that of Al (Hughes and Shepherd 1969, van Weeren and Anderson 1973) and much of what has been said of Al applies at least qualitatively to In also. The FS of In will be briefly discussed in §5.6.

214

5.3.3.2. Lead

As already mentioned, Pb is more complicated than Al because spin-orbit coupling can no longer be neglected. Anderson and Gold (1965) analysed their very detailed dHvA data in terms of 4 OPWs involving pseudo-potential coefficients V_{200} and V_{111} as in Ashcroft's scheme, but also introduced a spin-orbit coupling parameter λ and treated the Fermi energy ζ as another adjustable parameter. As can be seen from fig. 5.19 they were able to find a set of values of the Vs, λ and ζ which fit the experimental points on the whole very well; a rather poorer overall fit was obtained in an earlier attempt in which the spin-orbit coupling was ignored. Because of the extra complications of the situation, inevitably even more simplifying assumptions have been introduced to make the calculation tractable, and

Fig. 5.19. Variation of periods (note logarithmic scale) for Pb in the (110) zone. The curves are computed from the four parameter model discussed in the text (Anderson and Gold 1965).

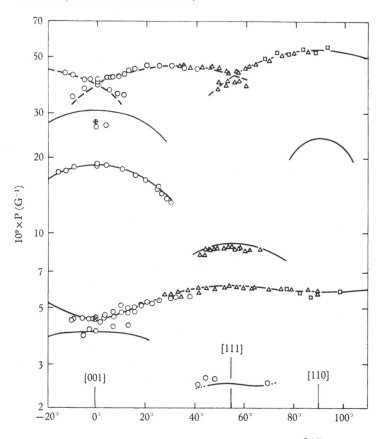

the precise physical significance of the parameters is probably more in doubt than for Al. It may be mentioned, for instance, that the value found for λ is about 50% greater than for the free atom and this excess is perhaps rather more than might be expected; moreover the concept of a local pseudopotential is less likely to be valid. More recently Anderson, O'Sullivan and Schirber (1972) have redetermined the absolute frequencies for the symmetry orientations with a considerably improved accuracy and have shown that a better fit to the data can be obtained by taking non-locality into account.

5.3.3.3. The divalent hexagonal metals

The series Be, Mg, Zn and Cd is of interest both in providing further illustrations of the use of the FE model and in illustrating the complications caused by magnetic breakdown (see chapter 7). The effect of magnetic breakdown is that the FS in magnetic experiments is effectively different according as the magnetic field is 'low' or 'high'. At low fields the 'single-zone' picture $(\text{fig. } 5.20(a)-(f))$ must be used because of the spin-orbit splitting across the plane ALH, and the corresponding free electron FS consists of:

(a) hole pockets in the first zone
(b) a 'monster' of holes in the third zone which is multiply connected at its corners both along the hexagonal axis and in directions perpendicular to it
(c) 'needles' or 'cigars' along the edges of the third zone
(d) a 'lens' at the centre of the third zone
(e) partial discs, which remap to form 'four-winged butterflies' in the third zone
(f) electron segments, which remap to form 'cigars' in the fourth zone.

When the field is high enough $(10^2–10^3 \text{ G for Be and Mg, but } 10^4–10^5 \text{ G for Zn and Cd})$, magnetic breakdown occurs and the electrons are able to cross the small energy gaps in the plane without feeling them, and the double-zone picture becomes appropriate. The effective free electron FS is now considerably modified in the following respects:

(1) The first zone pockets remap on to the monster at the points H, and the monster is no longer connected along the hexagonal axis, though it still joins up laterally with the neighbouring monsters at these points.
(2) The features (e) and (f) combine to give surfaces rather like partially opened clam shells (g).

The needles and the lens are unaffected. For intermediate fields the situation is more complicated and features of both schemes appear.

The detailed dimensions in either scheme depend on the c/a ratio and some features are absent if c/a is too large or too small. Also, of course, the FE model is only a crude approximation to the truth and in the actual metals some of the dimensions are appreciably modified from the FE prediction, or even completely eliminated. With so many different sheets of

5.3. Results for some metals

Fig. 5.20. Free electron FS for a divalent hexagonal metal with the c/a ratio of Mg (after Ketterson & Stark 1967). All the sketches except (g) refer to the single-zone picture: (a) first zone hole pocket centred on H (two complete pockets per zone); (b) second-zone hole monster; (c) third zone cigar centred on K (two per zone); (d) third zone lens centred on Γ; (e) third zone butterfly centred on L (three butterflies per zone); (f) fourth zone cigar centred on L (three per zone). In the double-zone picture one sixth of a pocket (a) is stuck on like a cap to each of the twelve arms of the monster in the top and bottom planes ALH in (b); the end faces of the double zone are at distance HK above and below the faces ALH of the single zone. The upper (or lower) half of each butterfly joins on to the lower (or upper) half of each fourth zone cigar to give the clam (g) of which there are six per zone. To understand the diagrams it is important to realize that only one typical zone point of each kind is shown in (b); thus there is a point M at the centre of each square face of the zone shown and adjoining zones, a point H at each hexagon corner and so on.

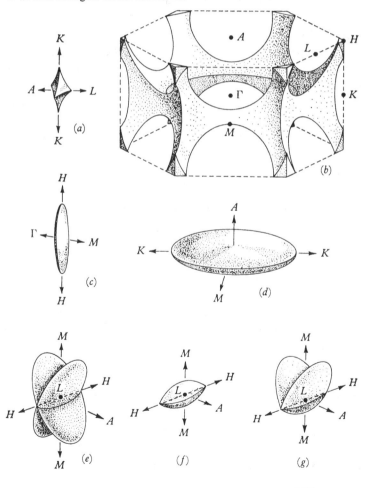

217

complicated shape, there are many (typically six or so) extremal areas for any given field direction and correspondingly the pattern of orientation dependence of dHvA frequencies is extremely complicated and correspondingly difficult to interpret reliably.

The complication of magnetic breakdown can be avoided by experiments at sufficiently low fields and Ketterson and Stark (1967) have shown that by the use of extremely pure samples of Mg it is possible to observe magnetoacoustic oscillations (see §4.8.2) down to fields as low as 100 G where the dHvA oscillations are no longer of appreciable amplitude. These magnetoacoustic oscillations give extremal caliper dimensions of the FS which help to sort out the complicated geometry.

A great deal of work has gone into ever more accurate measurements of the dHvA and magnetoacoustic frequencies, and magnetoresistance measurements have been a considerable help in sorting out the topology of the surface and its modifications by magnetic breakdown. A fairly clear picture of the FS both at low and at high fields has now been established and the departures from the FE model have been successfully interpreted either in terms of pseudopotential schemes or first principles band structure calculations.

We shall not attempt to describe all the observed features, but mention just a few of the simpler results. The third zone cigars and lens are the easiest pieces of the FS to identify and their shapes and sizes have been accurately determined. It is found that their linear dimensions are usually fairly close to those predicted by the FE model. The case of Zn is particularly striking since the FE sphere radius is only $\frac{1}{2}\%$ greater than the hexagon 'radius', so that only the slightest distortion would eliminate the overlap producing the needles. Yet the 'needle' (a more appropriate name here than 'cigar' because of its thinness) has almost exactly the diameter and length of the FE prediction. A comparison to better than 20 or 30% is however difficult because the dimensions are extremely sensitive to the c/a ratio, which is not accurately enough known at low temperatures. The dHvA frequency of the Zn needle ($F = 1.5 \times 10^4$ G) proves to be about the lowest observed for any metal apart from Bi. Because of its extreme sensitivity to the c/a ratio, this frequency is very sensitive to pressure (O'Sullivan and Schirber 1966), as can be seen in fig. 5.31 (see p. 241), increasing by a factor of about 2.1 for 3500 atmospheres, just about the expected amount for the corresponding change (from 1.830 to 1.824) of c/a. For Mg the cigar has almost exactly the rather triangular section of the free electron model, though it is slightly fatter and longer (Stark 1967). For Be the cigar is considerably fatter and rounder and rather longer than the free electron model (Watts 1964), while for Cd the c/a ratio is sufficiently large to eliminate the cigar and none is indeed observed (Tsui and Stark 1966). The dimensions of the lens are also very nearly those of the FE model, except for Be where it is entirely absent (Be has the smallest c/a ratio and should therefore have the smallest lens).

The study of the monster is considerably complicated by magnetic breakdown effects. Not only is there the complication of the changeover from the single- to the double-zone scheme which alters the possible extremal sections, but even in the double-zone scheme there is the further possibility of breakdown at a number of places where the band gap is small, so that all kinds of new orbits become possible for high enough fields. The classic case is that of the 'giant' orbit which was first observed in Mg for the field along the hexagonal axis (Priestley 1963): instead of following an orbit around the outside of the monster, the electron prefers at high fields to go round a large circular orbit made up of segments of the monster and the cigar (see fig. 7.1).

Another interesting feature is shown by Cd (Tsui and Stark 1966) where the monster is broken at each of the 'waists' of fig. 5.20. The six resulting pieces can then be remapped round the zone edge HK to form an undulating cylinder of trifoliate section which, in the periodically repeated scheme, continues indefinitely along the hexagonal axis (fig. 5.21a). The first zone hole pockets are rather large and extend nearly from K to K; at very

Fig. 5.21. FS of Cd (Tsui and Stark 1966). (a) In the single-zone picture the broken monster remaps into the continuous cylinder shown, with each minimum section centred on H and each maximum (trifoliate) section centred on K. (b) In the double-zone picture the hole pocket 'caps' the top and bottom of that part of the cylinder between H and H, giving a closed surface centred on K and reaching nearly to K on either side (in the double zone, of course, the end points K are no longer equivalent to the central point K).

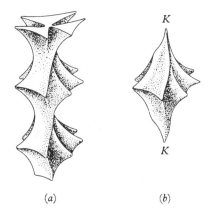

K

K

(a) (b)

high fields where the double-zone picture is appropriate, half of each hole pocket 'caps' the cylinder at top and bottom giving the closed piece shown in fig. 5.21(b).

From an experimental point of view, the simplest FS of the series is that of Be, paradoxically because it departs the most seriously from the free electron model. The relevant spin-orbit splitting is so small that the double-zone scheme applies at any fields big enough to give an appreciable dHvA effect, and moreover the lens and the clams are altogether suppressed. The FS consists only of the first and second zone monster (which is completely contained within the zone and has very thin waists) and the third zone cigars; the monster is sometimes called a 'coronet' (fig. 5.22). Because of this relative simplicity, Watts (1963, 1964) was able to determine the leading dimensions of both cigars and the coronet without having to appeal to any detailed model; a gratifying feature of his analysis is that the volumes of the monster and the cigars are equal within experimental error, as they should be if there are no other sheets of FS. It should be noted that the Be cigar has

Fig. 5.22. FS of Be: (a), (b) Sections through the double zone showing cigar and coronet (Watts 1963): the dimensions (in reciprocal atomic units) are as follows: *cigar ab* = 0.180, *cd* = 0.183, *ef* = 0.96; *coronet gh* = *kl* = 0.23, Γl = Γn = 0.57, *on* = 0.02; *zone* ΓA = 0.464, ΓK = 0.971. (c) Perspective sketch of coronet (Loucks and Cutler 1964).

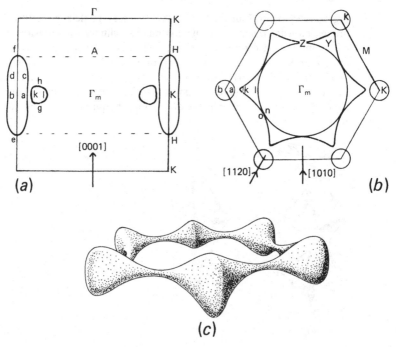

(a)

(b)

(c)

a *minimum* area normal to the hexad axis at its central cross-section and two non-central maxima on either side. This gives rise to beats between the two slightly different frequencies and this is relevant in some experiments to be discussed later (see §§6.7.1.2, 6.8 and 6.9).

As already mentioned, the departures from the exactly free-electron FS prove to be well describable in terms of pseudopotentials, though non-locality and spin-orbit coupling need to be taken into account to some extent, particularly in Cd (Kimball, Stark and Mueller 1967 (Mg), Stark and Falicov 1967b (Zn, Cd), Tripp, Everett, Gordon and Stark 1969 (Be)). The results agree also, within the uncertainties involved, with first principles band structure calculations such as those of Falicov (1962) for Mg and of Loucks and Cutler (1964) for Be.

5.3.4. Transition metals

The Fermi surfaces of the polyvalent transition metals (both those with incomplete *d*-shells and with incomplete *f*-shells) are more complicated than those we have considered so far, basically because the Fermi level falls right in the middle of the *d*-bands and no simple approximation such as the free electron model can be used, even as a rough guide, to the interpretation of the correspondingly complex spectrum of the dHvA frequencies. A further complication is that in some of these metals exchange interaction is strong enough to make them ferromagnetic while in Pt and Pd the electron–electron interaction causes strongly enhanced paramagnetism. In spite of these complexities there has been remarkable progress during the last 15 years or so in unravelling the complex Fermi surfaces of most of the transition metals (for a review see Mackintosh and Andersen 1980). This has come about as much from developments in material science which have provided pure and perfect enough single crystal samples, as from improvements in measuring and computing techniques and from developments of band structure theory which have enabled the other improvements to be exploited in the interpretation of the raw data. In what follows we shall not attempt to do more than outline some of the salient points and present a few pictures to illustrate the complexity of the results. We shall not discuss the rare earth (incomplete *f*-shell) metals beyond mentioning that they have particularly complicated Fermi surfaces, and that information is as yet far from complete; for a detailed recent review see Young (1977).

5.3.4.1. Non-ferromagnetic transition metals with unfilled *d*-shells

One important general feature of the theoretical band structures (Lomer 1962) which is supported by the experimental results, is that the band structures are all rather similar in character and it is possible to go from one transition metal to another of the same crystal structure merely by shifting the position of the Fermi level and making minor adjustments to some of the details of the individual bands. A nice application of this 'rigid band' kind of approach was Coleridge's (1966) study of the FS of Rh. At that time no theoretical band structure was available for Rh, but Coleridge was able to get a good idea of the band structure by slightly adapting a theoretical band structure for non-ferromagnetic Ni (itself similar to the band structure of Cu). He was able to reconcile his dHvA spectrum in fair detail with an FS based on this 'do-it-yourself' band structure and later a detailed band structure calculation specifically for Rh (Andersen and Mackintosh 1968) justified this approach. The 'rigid band' idea shows up particularly in the striking similarity of the FS of Nb and Ta, of Mo and W and of Pd and Pt. Each pair has the same crystal structure and effectively the same number of electrons to be fitted into the bands, so if the bands are almost the same, so too should be the Fermi surfaces.

Tungsten provides a nice example of one of the early rather detailed determinations of transition metal FS (Girvan, Gold and Phillips 1968). The complexity of the de Haas–van Alphen spectrum (determined by a sophisticated version of the impulsive field technique) is illustrated by fig. 5.23, in which, as mentioned earlier, in addition to all the fundamental branches in the frequency spectrum, many non-fundamental branches (harmonics and combination frequencies due to magnetic interaction) also occur. A coherent interpretation of the spectrum was achieved mainly by consideration of the FS indicated by first principles band structure calculations, with some slight modifications to fit the dHvA data. An artist's view of the final surface is shown in fig. 5.24, and it was found possible to describe the details of the various sheets analytically by rather complicated formulae containing a considerable number of adjustable parameters. As can be seen from fig. 5.23 this semi-empirical model FS matches the dHvA spectrum extremely well. As already mentioned, the FS of Mo (Sparlin and Marcus 1966, Hoekstra and Stanford 1973) is very similar in general appearance to that of W, though the detailed dimensions of the various pieces are somewhat different.

More recent determinations by the Argonne group (reviewed by Crabtree, Dye, Karim and Ketterson 1979) have been very successful in exploiting the parametrized KKR band structure approach to determine not only the Fermi surfaces but also the electron velocities for Nb, Pd and Pt in terms of a relatively small number of parameters from the dHvA frequencies and the cyclotron masses. For the noble metals only three phase shift parameters and the Fermi energy ζ were sufficient to give a good description of the Fermi surface, but here one or two extra parameters prove necessary. The cyclotron masses can similarly be interpreted by using as fitting parameters the derivatives of the phase shifts with respect to energy. The fits obtained are most impressive and a particularly convincing aspect (as for the noble metals) is the prediction by the final scheme of various features such as the existence of non-central extrema at particular orientations, which were not used at all in setting up the scheme. Another important confirmation of the reliability of the determination is the

Fig. 5.23. dHvA frequency spectrum of W (Girvan *et al.* 1968). The points are experimental and the full curves are computed from the FS model illustrated in fig. 5.24; the Greek letters refer to the extremal orbits indicated. The broken curves indicate some experimental points attributed to harmonics or combination frequencies. Note that the scale of F is logarithmic.

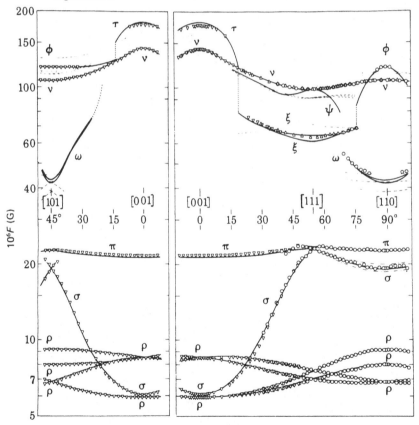

223

excellent agreement (for Nb and Pt about as good as in the noble metals), first between the sum of the volumes of the various sheets of FS and the expectation based on the valency and secondly between the total density of states deduced from the cyclotron masses and the value indicated by the electronic specific heat. For Pd, however, the calorimetrically measured specific heat does not agree well with that deduced from cyclotron masses, possibly because of paramagnon effects (Dye, Campbell, Crabtree, Ketterson, Sandesara and Vuillemin 1981). Recently, Hsiang, Reister, Weinstock, Crabtree and Vuillemin (1981) showed that the calorimetric specific heat

Fig. 5.24. Perspective sketch of FS of W (Girvan *et al.* 1968); the Greek letters indicate the extremal orbits giving rise to the various frequencies of fig. 5.23. For the sake of clarity the 'squareness' of the π orbits has been somewhat exaggerated and only those ellipsoids (carrying the ρ orbits) which lie in the front faces of the zone are shown.

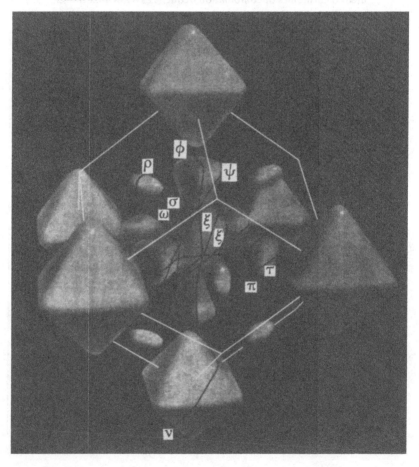

increases with field, but not enough to account plausibly for the discrepancy. It is to be hoped that this parametrization method will before long be applied to all the other transition metals, for which up to now only less meaningful and sometimes less complete specifications of the FS and its differential behaviour have been obtained.

5.3.4.2. Ferromagnetic metals

Observation of the de Haas–van Alphen effect in ferromagnetics, first achieved by Anderson and Gold (1963) in Fe, was somewhat of a *tour de force*, involving as it does the observation of the relatively feeble magnetic oscillations ($|\tilde{M}| \lesssim 1$ G) in the presence of the enormously larger ferromagnetic saturation magnetization ($M_s \sim 10^3$ G). However, provided due attention is paid to the special features associated with the ferromagnetism, it turns out that the oscillations can be studied with only slightly more difficulty than those for any other metal with a complicated FS. The presence of the large ferromagnetic magnetization is not as much of a complication as might be thought, since at the fields (of several times 10^4 G) necessary to produce oscillations, this magnetization is very completely saturated. Thus if the field is varied (as in the impulsive field or in the field modulation techniques) the only significant variation in magnetization comes from the dHvA effect.*

The main novelty of a ferromagnetic is that the phase of the oscillations (apart from a constant) is given by $2\pi F/B$, rather than by $2\pi F/H_e$, where H_e is the applied field and that B differs considerably from H_e in virtue of the ferromagnetic saturation magnetization M_s. For an ellipsoidal sample of demagnetizing coefficient $4\pi n$ we have (since the contribution to B of the much feebler oscillatory magnetization $\tilde{M}$, which is responsible for the magnetic interaction effect discussed in chapter 6, is not usually significant for ferromagnetics),

$$B = H_e + 4\pi(1 - n)M_s \tag{5.9}$$

and this has several important implications. First, the usual procedure for determining F, essentially by plotting the reciprocal of the fields at which oscillations occur against successive integers and determining the slope of the linear plot, is no longer appropriate, since the oscillations are no longer of equal period in $1/H_e$. It is easy to show that if ΔH_e is the field difference between the start and finish of a series of N oscillations starting at a field

---

* However, in the impulsive field technique it is necessary to superimpose an appreciable steady field to saturate the ferromagnetism before the impulse is applied, in order to avoid the huge change of magnetization at low fields, which could swamp the detection circuits.

H_0, then

$$FΔH_e/N = [H_0 + 4π(1 - n)M_s]ΔH_e$$
$$+ [H_0 + 4π(1 - n)M_s]^2 \quad\quad (5.10)$$

Thus a plot either of $ΔH_e/N$ against $ΔH_e$, or of $1/N$ against $1/ΔH_e$, as N varies for fixed H_0, should be linear and both F and $4π(1 - n)M_s$ can be determined from the intercept and slope. In their pioneer experiment on Fe, Anderson and Gold (1963) found (by a somewhat different procedure) that $4πM_s$ did have just the correct value (about 21 kG), thus confirming that the effective field 'seen' by the electron is indeed B; the same was later found for other ferromagnetics.

The other implications of replacing H_e by B arise from the presence of the demagnetizing coefficient $4πn$ in (5.9). It is only for an ellipsoid that the demagnetizing field is strictly uniform inside the sample, so it is important that the sample shape should not depart too much from that of an ideal ellipsoid if phase smearing and consequent loss of amplitude is to be avoided. Again, if the sample is to be rotated in a constant and uniform B (to study anisotropy of F), it is essential that n should not vary significantly with rotation: this implies the use of either a spherical or a disc shaped sample (with the field kept accurately in the plane of the disc).

As already mentioned, the FS are rather complicated and some of the details, particularly in Fe and Co, have not yet been unambiguously sorted out. A first approximation to a description of the FS of a ferromagnetic metal is to start with what would be the calculated band structure of the non-ferromagnetic metal and then introduce a constant exchange splitting between the spin-up (↑) or 'majority' and spin-down (↓) or 'minority' electrons. This leads essentially to separate FS for the ↑ and ↓ electrons, which is part of the reason why the whole picture is complicated but, in addition, the details of the surfaces are appreciably modified by variation of the exchange splitting over the FS and by hybridization between the sp and d states, which modifies the connectivity of the FS in particular regions.

One other complication is that because of spin-orbit coupling the FS depends to some extent (more seriously in Ni than in Fe) on the direction of the saturation magnetization, i.e. of the field. This means that the surface determined from extremal areas normal to different field directions, i.e. in accordance with the usual procedure, is not really exactly a Fermi surface at all, unless theoretical allowance is made for the spin-orbit effect. Little systematic work has yet been done on the cyclotron masses in the ferromagnetic metals, though valuable information about the separate densities of state of the ↑ and ↓ electrons could be obtained from such a

Fig. 5.25. Central (110) section of the FS of Ni (Lonzarich 1980) based mainly on experimental data of R. W. Stark quoted by Wang and Callaway (1977) and (dotted curve) by Tsui (1967). The full and broken lines correspond to majority and minority spin carriers but the FS surface can change character where the full and broken lines would cross if not for spin-orbit coupling (spin-hybridization). Note that around L the FS of the majority carriers has a neck rather similar to that of Cu.

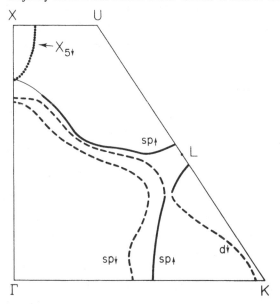

Fig. 5.26. Central (100) and (110) section of the FS of Fe (Lonzarich 1980). The model, based on a combination of experimental data and band theory calculations, is still somewhat conjectural but illustrates the complexity of the FS. Spin hybridization is less important than in Ni and is ignored in this diagram. Note that the minority spin FS is somewhat similar to the FS of W (fig. 5.24).

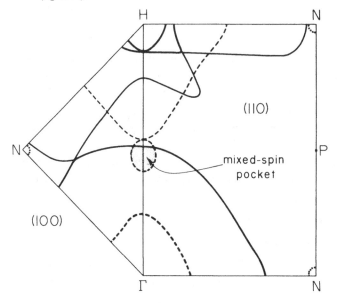

study. A practical difficulty is that some of the cyclotron masses are very high (e.g. $m/m_0 \sim 10$) so that oscillations can only be observed in extreme conditions of high field and low temperatures. Because of the difficulty of temperature control and measurement below 1 K, accurate mass determination is not straightforward. For a more detailed discussion of the ferromagnetic metals see reviews by Gold (1974) and Lonzarich (1980, for Ni and Fe). Some idea of the complexity of the FS of Ni and Fe is given by figs. 5.25 and 5.26.

5.3.5. Bismuth*

Because the FS of Bi is so small, the dHvA effect can be observed in quite small fields (as low as 100 G at 0.3 K) and at fairly high temperatures (up to 20 or 30 K at fields of a few kG). It is in fact the metal in which the effect was first discovered and has probably been more studied ever since than any other metal by nearly all the available Fermiological techniques, as well as theoretically. In a thorough recent review of all this work, Edelman (1976) mentions that the rate of publication on Bi has risen roughly linearly with time from something like ten papers per year in the 1950s to something like 70 per year in the 1970s.

The theory of the dHvA effect was originally based on the assumption that the surfaces of constant energy were ellipsoidal and it is fortunate that the FS of Bi did indeed prove to consist entirely of nearly ellipsoidal sheets. Thus when the first systematic data on the orientation dependence of the oscillations was obtained (Shoenberg 1939) it was possible to interpret it meaningfully in terms of the explicit formula which Landau (see Appendix to Shoenberg 1939) had just then derived, assuming the FS consisted of three ellipsoids with the rhombohedral symmetry of Bi. Landau's three-ellipsoid model has stood the test of time remarkably well though some modifications have proved necessary as more and more refined measurements were made. In due course a single ellipsoid of holes had to be added to the model to account for the higher frequency oscillations found by Brandt (1960) in experiments at very low temperatures, and later experiments by Bhargava (1966) using the field modulation method established the dimensions of the various ellipsoids rather more accurately. In the most recent determination (Edelman 1973) the precision was still further improved and it turned out that there were slight, but significant, departures from the ellipsoidal form. Leaving these slight departures aside for the moment, it is a fair approximation to describe the Fermi surface by

---

* The FS of Sb, which is similar to that of Bi, though larger and more complicated, is briefly discussed in §5.6.

5.3. Results for some metals

three ellipsoids containing electrons and one ellipsoid of revolution containing holes (see figs. 1.4 and 5.27). This model has the merit that all its geometrical properties are readily calculated (see Appendix 1). With respect to the crystal axes (k_x the binary, k_y the bisectrix and k_z the trigonal axis, but note that Edelman's C_1 is our k_y and his C_2 is our k_x) the equations of these ellipsoids are, for the electrons

$$a_1 k_x^2 + a_2 k_y^2 + a_3 k_z^2 + 2a_4 k_y k_z = 1 \qquad (5.11)$$

together with two others obtained by rotation through $\pm 120°$ about the k_z axis (in order to satisfy the rhombohedral symmetry), and for the holes

$$(k_x^2 + k_y^2)/h_1^2 + k_z^2/h_3^2 = 1 \qquad (5.12)$$

The best fit to Edelman's data gives*

$$\left.\begin{array}{ll} a_1 = 3.494 \times 10^{-12}\,\mathrm{cm}^2, & a_2 = 0.0404 \times 10^{-12}\,\mathrm{cm}^2 \\ a_3 = 1.987 \times 10^{-12}\,\mathrm{cm}^2, & a_4 = -0.2206 \times 10^{-12}\,\mathrm{cm}^2 \end{array}\right\} \quad (5.13)$$

Fig. 5.27. Location of FS of Bi in the Brillouin zone (after Brown et al. 1968). Careful consideration of the data shows that the angle ϕ of tilt is definitely positive ($= 6.38°$) as shown, rather than negative. For the sake of clarity the dimensions of the FS ellipsoids (cf. fig. 1.4) are exaggerated by a factor of about 5.6 relative to those of the zone.

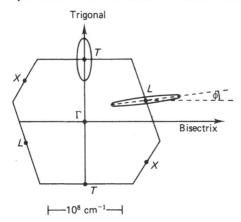

$\longmapsto 10^8\,\mathrm{cm}^{-1} \longmapsto$

* The fit is in fact good to only a few %, but the as are given to four significant figures so that they should be consistent with the lengths of the principal axes as given by the es in (5.14). In particular the value of e_2 in terms of the as depends on the difference of two nearly equal quantities and is therefore extremely sensitive to the values of the as.

FS and cyclotron masses

The significance of the cross-term $2a_4 k_y k_z$ is that the principal axes of the ellipsoid (5.11) are tilted in the k_y, k_z plane from the k_y and k_z axes; the numerical data of (5.13) correspond to a tilt angle of $\theta = 6.38°$ in the sense shown in fig. 5.27 (Brown, Hartman and Koenig 1968). The principal axes have lengths given by

$$e_1 = a_1^{-1/2} = 0.535 \times 10^6 \, \text{cm}^{-1}$$
$$e_2 = (a_2 \cos^2 \theta + a_3 \sin^2 \theta + 2a_4 \cos \theta \sin \theta)^{-1/2}$$
$$= 7.98 \times 10^6 \, \text{cm}^{-1} \tag{5.14}$$
$$e_3 = (a_2 \sin^2 \theta + a_3 \cos^2 \theta - 2a_4 \cos \theta \sin \theta)^{-1/2}$$
$$= 0.705 \times 10^6 \, \text{cm}^{-1}$$

The hole ellipsoid is one of revolution about the k_z axis and its principal axes are

$$h_1 = 1.391 \times 10^6 \, \text{cm}^{-1}, \qquad h_3 = 4.628 \times 10^6 \, \text{cm}^{-1} \tag{5.15}$$

It can be seen that the dimensions of both the electron and hole ellipsoids are indeed small compared to the Brillouin zone dimensions which are of order $10^8 \, \text{cm}^{-1}$ (see fig. 5.27), and calculation shows that the volume v of the three electron ellipsoids is $37.8 \times 10^{18} \, \text{cm}^{-3}$, which is within experimental error equal to that of the hole ellipsoid ($37.5 \times 10^8 \, \text{cm}^{-3}$) as it should be. The corresponding total number of electrons (or holes) per cm^3 is

$$v/4\pi^3 = 3.0 \times 10^{17} \, \text{cm}^{-3} \tag{5.16}$$

This corresponds to only 1.06×10^{-5} electrons (or holes) per Bi atom.

The orientation dependence of the dHvA periods corresponding to this FS is relatively simple (fig. 5.28), with at most only four branches (two of which coalesce in the trigonal-bisectrix plane). In the torque method a single frequency is often dominant, either because another frequency comes near an extremum with respect to orientation (so that its amplitude is small) or because other frequencies (and the corresponding cyclotron masses) are much higher so that the oscillations are considerably damped down. This was a lucky circumstance in the early days, since frequency measurement is particularly easy if a single frequency is dominant (cf. fig. 1.3).

Later Dhillon and Shoenberg (1955) deliberately exploited the idea of choosing an orientation (H at 78° to the trigonal axis in the trigonal-bisectrix plane) such that only one low frequency oscillation survived. This permitted a detailed study of the oscillation line shape (i.e. harmonic content) right up to the quantum limit, which for this orientation is reached

230

5.3. Results for some metals

at a relatively modest field (about 15 kG). More recently Barklie and Shoenberg (1975) again studied a single frequency by setting the field along a binary axis (where two branches of the spectrum cross) and using the more powerful field modulation technique. Such studies at fields approaching the quantum limit are relevant for testing basic features of the theory

Fig. 5.28. Orientation dependence of dHvA periods in Bi (Bhargava 1966): (a) electrons in binary-bisectrix (xy) plane; the plot from 30°–60° would be a mirror image of the 0°–30° plot, (b) electrons in bisectrix-trigonal (yz) plane; the upper curve comes from two equivalent ellipsoids, (c) electrons in binary-trigonal (xz) plane, (d) holes in yz plane (the plot in the xz plane is identical and in the xy plane the period is independent of angle). In all the diagrams the points are experimental and the curves are as calculated from the ellipsoidal FS specified by (5.11)–(5.15) (but with slightly different numerical values). The data were obtained by the field modulation system and some of the tricks described in §3.4.2.1 of setting the pick-up coil axis at an angle to H were used to help separate the various frequencies.

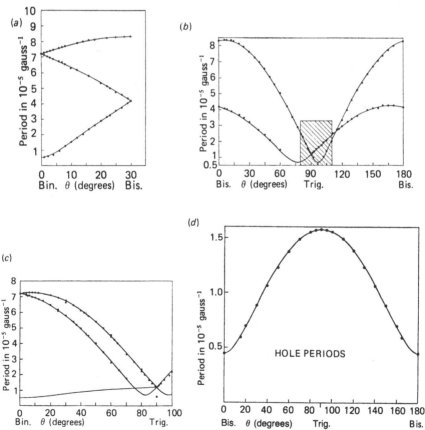

and we shall return to them later in connection with the Dingle temperature (chapter 8) and spin properties (chapter 9).

In the early theoretical work it was assumed not only that the FS was ellipsoidal but that the band was parabolic, i.e. that the surfaces of constant energy ε were similar to (5.11) and (5.12) but with the r.h.s. replaced by ε/ζ where ζ is the Fermi energy (different for the electrons and holes). With this simple 'ellipsoidal parabolic' (EP) model it is easily shown (see §A1.1) that the dHvA frequency is given by

$$F = \zeta/\beta \qquad (5.17)$$

where as usual $\beta = e\hbar/mc$ and m is the cyclotron mass at the Fermi energy. This relation was often taken for granted in the interpretation of early dHvA data, not only for Bi, but with even less validity for other metals, where the Fermi surfaces are sometimes not even approximately ellipsoidal.

The assumption of a parabolic band for the holes in Bi proves to be reasonably valid, but is quite significantly wrong for the electrons. Basically this is because there is only a very small gap ε_g between the top of the full valence band and the bottom of the electron conduction band and simple perturbation theory then suggests that to a first approximation the electron constant energy surfaces (with ε measured from the bottom of the band) should have the form of (5.11) but with the r.h.s. replaced by

$$f(\varepsilon) = \frac{\varepsilon(1 + \varepsilon/\varepsilon_g)}{\zeta(1 + \zeta/\varepsilon_g)} \qquad (5.18)$$

This is the 'Lax' or 'ellipsoidal non-parabolic' (ENP) model (see §A1.3) which although still an oversimplification, provides a convenient explicit form of the $\varepsilon-k$ relation, which can be used for making various semi-quantitative predictions. The main experimental evidence for the Lax model came from the observation of magneto-optical oscillations (see §4.7.1) by Brown, Mavroides and Lax (1963) and experiments by Vecchi and Dresselhaus (1974) suggest that the energy gap has the value

$$\varepsilon_g = 0.0135 \, \text{eV}, \qquad (5.19)$$

though the detailed interpretation of the experiments is complicated and does not fit exactly into the framework of the Lax model; the value of ε_g might well be in error by as much as 10%.

For the Lax model, instead of (5.17) we have (see (A1.34))

$$F = \frac{\zeta}{\beta} \frac{(1 + \zeta/\varepsilon_g)}{(1 + 2\zeta/\varepsilon_g)} \qquad (5.20)$$

5.3. Results for some metals

so ζ can be determined if F, β (i.e. m) and ε_g are known. As pointed out in §A1.2, F/m (or A/m) should be a constant, independent of orientation, for any ellipsoidal model, whatever the form of $f(\varepsilon)$ in (5.18) and this constancy provides a useful test of the validity of the ellipsoidal assumption. In fact, as can be seen from table 5.4, A/m does not vary much with orientation for either electrons or holes, thus confirming that the departures from ellipsoidal shape are not large. If the value

$$A/(m/m_0) = 1.47 \times 10^{14}\,\text{cm}^{-2} \quad \text{or} \quad F/(m/m_0) = 1.54 \times 10^6\,\text{G}$$

and the value of ε_g from (5.19) are substituted into (5.20) we find

$$\zeta(\text{electrons}) = 0.030\,\text{eV} \tag{5.21}$$

This result is insensitive to the assumed value of ε_g and even a 10% change of ε_g only changes ζ by about 1%.

For the holes we assume the band is parabolic, and taking

$$A/(m/m_0) = 0.95 \times 10^{14}\,\text{cm}^{-2} \quad \text{or} \quad F/(m/m_0) = 1.00 \times 10^6\,\text{G},$$

we find from (5.17)

$$\zeta(\text{holes}) = 0.012\,\text{eV} \tag{5.22}$$

Table 5.4. *Cyclotron masses m and Fermi surface extremal areas A of Bi*

Orientation of H	$10^{-12}A\,\text{cm}^{-2}$	$10^2 \times m/m_0$	$10^{-14}A/(m/m_0)\,\text{cm}^{-2}$
Electrons			
Binary	17.7	11.9	1.49
	1.38	0.95	1.45
Bisectrix	1.19	0.82	1.45
	2.37	—	—
Trigonal	8.36	6.3	1.33
Principal ellipsoid axis near trigonal	13.4	8.8	1.52
Holes			
Binary or Bisectrix	20.2	21.2	0.952
Trigonal	6.08	6.39	0.951

Notes: The masses are taken from the first column of Table 6 of Edelman (1976) and are mostly stated to be accurate to $\frac{1}{2}$% or better; however the two entries at and close to the trigonal orientation appear to be subject to larger errors. The areas are based on the ellipsoidal formulae and values given in (5.11)–(5.13) and (5.15); since they represent some smoothing of the original data they are probably only reliable to 2 or 3%.

Considerable effort has gone into theoretical calculations of the band structure of Bi. A number of first principles calculations starting with Mase (1958) have shown that the general picture, derived above on a purely experimental basis, is indeed to be expected theoretically and moreover that the electron ellipsoids are centred on the points L of the Brillouin zone while the hole ellipsoid is centred on T (fig. 5.27). First principles calculations are not precise enough to give more than qualitative agreement with experiment, but various parametrized models – of which the Lax model is perhaps the the simplest – have been developed at various levels of sophistication.

The most recent calculation by McClure (1976) and McClure and Choi (1977), which is essentially a perturbation treatment to high order, involves ten parameters whose values are fitted to the detailed dHvA data, the cyclotron mass data (including limiting point masses, which the Lax model fits only poorly), and also the magneto-optical and surface magnetic resonance data. An interesting feature of the model is that it is able to fit in detail the small deviations of the Fermi surface from the ellipsoidal form found by Edelman. McClure shows also that his model is entirely consistent with a rather different approach developed by Abrikosov and Falkovsky (1962) who started from the fact that the bismuth lattice can be generated by a small stretching along the body diagonal of a face-centred cubic lattice, followed by a relative displacement of neighbouring atoms on this diagonal. Treating this deformation and the spin-orbit coupling as perturbations from the original structure they could predict the detailed band structure around the L and T points. At first it appeared that their treatment gave results disagreeing with those of the more conventional perturbation approach, but McClure shows that this was only because they had left out a higher order term which is not in fact negligible.

5.3.6. Compounds

Extension of de Haas–van Alphen studies and other Fermiological techniques from elements to stoichiometric metallic compounds has greatly expanded the scope of Fermiology and has in a sense added a new dimension to it. This extension has become possible partly through the considerable advances in the techniques of preparing single crystals which are sufficiently stoichiometric and perfect and partly through advances in measuring techniques. The striking progress that has been made since the first observation of dHvA oscillations in BiIn by Thorsen and Berlincourt (1961) and the beginning of systematic studies on several other compounds by the Ottawa group (Beck, Jan, Pearson and Templeton 1963) has been thoroughly reviewed by Sellmyer (1978). The FS of more than 50 compounds have now been either completely or partially determined. Some of these, such as β-brass (e.g. Jan, Pearson and Saito 1967), high field superconducting compounds (e.g. Arko *et al.* 1978) and layer compounds (e.g. Graebner and Robbins 1976, Graebner 1977) are of considerable technical interest and are likely to be studied even more intensively in the future.

Here we shall only mention that in general the FS of metallic compounds tend to be more complicated than those of the metallic elements. Basically this comes about because there can be many more conduction electrons to fit into the electronic structure. For some compounds, where the FE model is reasonably applicable, there may be as many as 14 electrons per unit cell and consequently the free electron sphere intersects several more zones than say in a four-valent metal and the FE Fermi surface becomes extremely complex (compare for instance fig. 5.29 with fig. 5.15). Bearing in mind that the FE model is only an approximation to the real FS the possibilities of misinterpretation of the complicated dHvA spectra are considerable. Where the FE model is not applicable, the situation is even more confusing and considerable appeal to band structure theory is usually essential to obtain any reliable interpretation. The FS of Cu_2Sb as deduced

Fig. 5.29. The free electron FS for hexagonal AuSn which has ten valence electrons per unit cell (after Edwards, Springford and Saito 1969 as shown in Sellmeyer 1978). The subscripts to the labelled symmetry points indicate the zone number (zones 1 and 2 are full) Only the single-zone scheme is shown, though actually the double-zone scheme is more relevant because of magnetic breakdown.

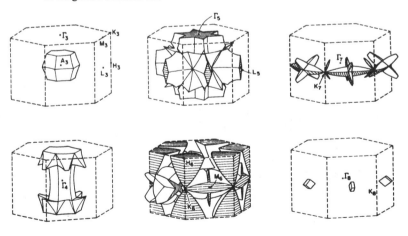

by Jan and Skriver (1977) from a band structure calculation is shown in fig. 5.30 to illustrate a typically complex situation. This FS provides only a semi-quantitative interpretation of the very complicated observed dHvA frequency spectrum.

Fig. 5.30. FS of Cu_2Sb (tetragonal, 14 electrons per unit cell) as calculated by Jan and Skriver (1977): (a) band 25, (b) band 26, (c) band 27, (d) band 28. The capital letters denote symmetry points in the basic zone (indicated in (c) and partially in (d)).

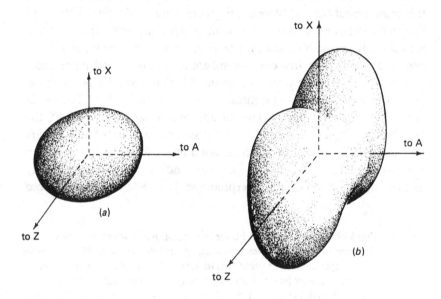

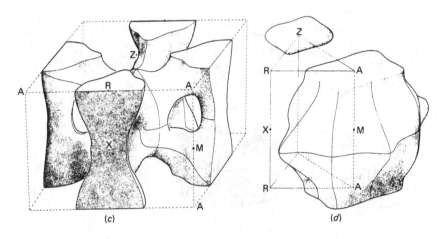

5.4. Strain dependence of the Fermi surface

As we saw in §3.5 and §4.3, information on the strain dependence of the FS can be obtained either directly, by actually applying a stress and observing the change of dHvA frequency, or indirectly from the oscillatory magnetostriction and oscillations in sound velocity. In the former approach the effects of dilatational and shear strains can be determined by combining observations on hydrostatic pressure (which for cubic symmetry produces simply a negative dilatational strain, but for lower symmetries produces a combination of negative dilatation and shear) and on uniaxial stress, whose effect can be resolved into a dilatation and a shear. In the indirect approach the strain dependence can in principle be fully determined independently, but if only limited measurements are made, such as of oscillatory magnetostriction along only one direction in the sample, the information obtained is exactly the same as that from direct application of a uniaxial stress. In this section we shall review briefly some of the results that have been obtained, particularly for the metals whose FS have been discussed earlier in this chapter. We shall see that the experimental results on strain dependence can prove relevant to the understanding of the band structure and also that the potentialities of the available techniques have as yet been only partially exploited. A more detailed discussion can be found in the review by Fawcett *et al.* (1980).

5.4.1. Alkali metals

Because the FS of the alkalis are so close to the corresponding free electron spheres, the most obvious effect of a hydrostatic pressure should be simply to change the volume v of the FS inversely as the sample volume V in real space. Since the dHvA frequency F is proportional to the diametral cross-sectional area of the FS, which varies as the $\frac{2}{3}$ power of the volume, we would expect that to a good approximation

$$\tfrac{3}{2}\,\mathrm{d}\ln F/\mathrm{d}p = \mathrm{d}\ln v/\mathrm{d}p = -\mathrm{d}\ln V/\mathrm{d}p = \mathcal{K} \qquad (5.23)$$

where $\mathcal{K}$ is the compressibility.

Until recently, the experimental evidence on the validity of (5.23) appeared to be rather conflicting, but various previously unsuspected sources of error have since come to light and, as can be seen from table 5.5, there is now no reason to doubt that (5.23) is correct. This disposes of a suggestion (Altounian and Datars 1980) that the high pressure dHvA data for K provide evidence for Overhauser's charge density wave hypothesis (see §5.3.1).

To determine any anisotropy in the pressure coefficient itself requires

237

extremely accurate measurement since the dHvA phase shift for 25 bar is only about 1.6 oscillations, so if the pressure coefficient is only anisotropic to the same extent as the FS, say 1 in 10^3 (as for K), the phase shift would vary with orientation by only about $\pm 0.6°$, which is comparable with the experimental error of the best experiments so far. Fortunately it proves that the anisotropy of the pressure coefficient in K is about six times larger than that of the FS and Templeton (1981) has recently succeeded in establishing the form of the anisotropy quite convincingly. A contour diagram of the anisotropy of $d \ln F/dp$ as fitted to a six term cubic harmonic expansion is shown in fig. 5.4(*b*), and it can be seen that apart from scale it is rather

Table 5.5. *Pressure dependence of F in the alkalis compared with compressibility* $(\mathscr{K})$

Metal	$10^{11} \times \tfrac{3}{2} d \ln F/dp (\text{dyne}^{-1} \text{cm}^2)$		$10^{11} \mathscr{K} (\text{dyne}^{-1} \text{cm}^2)$	
	(*a*)	(*b*)	(*c*)	(*d*)
Na		(1.14)1.25 $\pm$ 0.03		1.35 $\pm$ 0.05
K	2.58 $\pm$ 0.01	(2.30, 2.43)2.59 $\pm$ 0.08	2.73 $\pm$ 0.05	2.70 $\pm$ 0.1
Rb	(3.16)3.21 $\pm$ 0.02		3.27 $\pm$ 0.07	3.45 $\pm$ 0.2
Cs	(4.02)4.17 $\pm$ 0.04		4.33 $\pm$ 0.09	4.65 $\pm$ 0.3
	(4.4) 4.3 $\pm$ 0.1			

Notes: (General) (*a*) dHvA at low pressures (< 25 bar), (*b*) dHvA at high pressures (several kilobar) extrapolated to $p = 0$, (*c*) ultrasonic, so essentially at $p = 0$ (K: Marquardt and Trivisonno 1965, Rb: Gutman and Trivisonno 1967, Cs: Kollarits and Trivisonno 1968), (*d*) from high pressure data extrapolated to $p = 0$ (as quoted by Anderson, Gutman, Packard and Swensen 1969). No errors are quoted by the authors in (*c*); we have somewhat arbitrarily supposed $\pm 2\%$ is appropriate.

Remarks on individual metal entries in (*a*) *and* (*b*) Na: Elliott and Datars (1981). The entry 1.14 is an average between 0 and 4 kilobar; if adjusted by the same factor as appropriate for non-linear compressibility, the $p = 0$ value becomes 1.25. K: (*a*) Glinski and Templeton (1969), Templeton (1981); (*b*) Altounian and Datars (1980) gave 2.30 as an average between 0 and 1 kilobar. Templeton (1981) suggested this should be adjusted to give 2.43 at $p = 0$ to allow for non-linear compressibility and finally Elliott and Datars discovered an error in the pressure scale that had been used, bringing the value to 2.59; Rb: Glinski and Templeton gave 3.16 but made no allowance for anisotropy of the pressure derivative; Jan *et al.* (1980) corrected this to 3.21 as an appropriate average over orientation based on a theoretical estimate of the anisotropy; Cs: Glinski and Templeton gave 4.02 and this was corrected by Jan *et al.* as for Rb; Beardsley and Schirber (1972) gave 4.4 making an allowance for anisotropy, but Jan *et al.* revised this allowance, giving 4.3.

similar to the contour diagram of $\Delta A/A$ (fig. 5.4(a)) i.e. that the pressure coefficients are biggest where the FS bulges most. So far there has been no theoretical estimate of how large the anisotropy of pressure coefficient should be in K, because the size of the anisotropy is comparable to the precision of the necessary band structure calculation. It will be interesting to see what happens for Rb and Cs where band structure calculations predict much larger anisotropies for the pressure dependence than in K (Jan, MacDonald and Skriver 1980). As indicated in table 5.5 these theoretically predicted anisotropies help to bridge the slight apparent discrepancies between compressibilities deduced from dHvA and ultra-sonic data.

5.4.2. Simple polyvalent metals

Stress dependence has been studied on a number of polyvalent metals which are simple in the sense that the NFE model applies. On the whole a reasonably consistent picture has emerged, in which the results can be successfully interpreted by supposing that the pseudopotential coefficients vary with strain in just the way predicted by basic theory.

One of the earliest experiments on pressure dependence was that of Melz (1966) on Al, who used the 'direct' high pressure technique and found that F for the slow β-oscillations at $\langle 100 \rangle$ increased by about 1 in 10^5 per bar and decreased about half as rapidly for the faster γ oscillations at $\langle 110 \rangle$. Melz originally interpreted his results as being in conflict with Ashcroft's model, but it was later pointed out by Sorbello and Griessen (1973) that he had omitted an important term in the expression for the pressure dependence of the FS area as calculated from the pseudopotential model. With the correct treatment, Sorbello and Griessen were able to obtain from Melz's data values for the pressure dependence of the pseudopotential coefficients which were in excellent agreement with theoretical predictions, thus confirming the physical meaningfulness of the coefficients.

The effect of uniaxial stress on the FS of Al has also been studied and indeed this was the first application by Griessen and Sorbello (1972) of their ingenious technique of combining oscillatory magnetostriction and torque measurements in a single experiment to derive stress derivatives of FS areas (see §4.3.3). Once again the reliability of the pseudopotential scheme was demonstrated by the excellent agreement which they obtained between their experimental stress derivatives and those predicted by the scheme.

Results are also available on the effect of both hydrostatic pressure (Anderson, O'Sullivan and Schirber 1972) and uniaxial stress (Joss 1981) on a number of extremal areas of the FS of Pb. Joss has successfully interpreted these data in terms of a local pseudopotential scheme ignoring spin-orbit coupling and possible non-locality. This scheme does not give as good a fit to the zero-pressure dHvA frequencies as does a calculation by Anderson *et al.*, who took account of both spin-orbit coupling and non-locality and used experimental data more accurate than those of Anderson and Gold (1965). Nevertheless, the simpler scheme provides a rather better fit to the observed dilatation and shear derivatives of a large number of extremal areas in terms of a set of strain derivatives of the basic parameters, in reasonable agreement with the predictions of basic theory. The fact that the strain dependence can be interpreted almost equally well by alternative schemes based on rather different physical concepts suggests that perhaps the fitting parameters may have less physical meaning for Pb than for Al.

The stress dependence properties of the FS of indium, which is tetragonal, and of the divalent hexagonal metals have also been extensively studied, by both direct and indirect models. A complication with these non-cubic metals is that hydrostatic pressure is no longer equivalent to a pure negative dilatation, since the axial ratio is also a function of pressure. Indeed, as already mentioned in §5.3.3.3, for the Zn needles, the great sensitivity of frequency to pressure (see fig. 5.31) is mostly due to the change of axial ratio with pressure. However, as with the other 'simple' metals, the observed stress dependence is reasonably consistent with pseudopotential theory, provided non-locality and spin-orbit coupling are taken into account where appropriate. For details see Sorbello and Griessen (1973), Griessen and Sorbello (1974), Gamble and Watts (1973), Watts and Sundström (1979) and Watts and Mayers (1979). The papers by Watts and colleagues emphasise the physical significance of the various contributions to the theoretically calculated stress derivatives.

5.4.3. The noble and transition metals

As we have seen earlier, the noble and transition metals have in common that their Fermi surfaces can be well represented by parametrized band structures in which the adjustable parameters are the phase shifts η and the Fermi energy ζ. Once again the physical meaningfulness of the fitting scheme is demonstrated by the extent to which the strain dependence of the fitted parameters agrees with the predictions of basic theory.

We shall deal first, and in rather more detail, with the noble metals because of the relative simplicity of their Fermi surfaces, though, except for Cu, the strain dependence has as yet been less completely studied than that of some of the more complicated transition metals. So far the most thoroughly studied aspect has been the pressure dependence, and the most accurate results obtained by the liquid helium pressure method (Templeton 1966, 1974) are summarised in table 5.6. Since the FS of the noble metals are recognizable distortions of the free electron sphere model (though of course much more considerable distortions than for the alkalis), it is helpful to express the observed pressure dependence of the dHvA frequencies (i.e. of the FS extremal areas A) by comparison with the pressure dependence of A_s, the area of the diametral plane of the free electron sphere. Thus if the FS changes exactly on the scale of the free electron sphere we should find

$$\mathrm{d}\ln A/\mathrm{d}\ln A_s = 1 \qquad\qquad (5.24)$$

As we have seen, this is a good approximation for the alkali metals (note that (5.24) is equivalent to (5.23) since $\mathrm{d}\ln A_s/\mathrm{d}p$ is $\tfrac{2}{3}\,\mathscr{K}$), but for the noble metals it is only a rough approximation, even for the most free electron-like

Fig. 5.31. Change of Zn needle frequency with pressure and temperature showing that both effects are essentially mediated by variation of the c/a ratio (after O'Sullivan and Schirber 1966); ● effect of pressure measured by the authors based on dHvA effect, △ effect of pressure based on SdH effect (Schirber 1965), ○ effect of temperature based on dHvA effect (Berlincourt and Steele 1954). Note that increasing pressure decreases c/a while increasing temperature increases c/a.

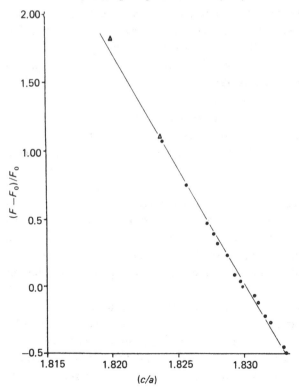

orbits, i.e. the belly orbits, and is quite invalid for the necks. A convincing check of the reliability of the experimental data is that if the FS under pressure is described by a parametrized representation–either a Fourier expansion or a band structure representation–the volume of the FS is found to scale inversely as the sample volume, as of course it should.

The effect of pressure on the FS can be estimated theoretically from a 'first principles' band structure calculation in which the lattice parameter is slightly varied from its zero pressure value (Davis, Faulkner and Joy 1968) and as indicated in table 5.6 for Cu there is reasonable agreement with experiment within the considerable uncertainties of the calculation. As regards the representation of the band structure in terms of the ηs and ζ it can be seen that the observed pressure dependence of FS areas can be fitted very well by a suitable choice of the derivatives $d\eta_1/d \ln A_s$ with a theoretically estimated value of $d\zeta/d \ln A_s$ (Fawcett et al. 1980). However, the chosen values of $d\eta/d \ln A_s$ are essentially empirical parameters and cannot easily be interpreted in more basic terms.

The meaningfulness of the phase shift parameters is more significantly tested in the interpretation of experiments involving shear dependence of the FS, such as experiments on the effect of uniaxial stress or equivalent magnetostriction experiments. The results of such experiments, however, present a rather confused picture as can be seen from table 5.6. The magnetostriction data of Slavin (1973) for the necks gave appreciably higher values of $d \ln A/d \ln A_s$ than those of the earlier 'direct' measurements of Shoenberg and Watts (1967) on the effect of uniaxial tension. This discrepancy led Watts to suggest that the tensions had been systematically overestimated because of friction in the tension-transmitting link and this suggestion appeared to be confirmed by Gamble and Watts (1972), who avoided any frictional effects in measurements of the effect of uniaxial compression (but only on the neck of Au). However recent magnetostriction experiments by Ruesink and Perz (1983) for several orbits (but only for Cu) give values of $d \ln A/d \ln A_s$ which agree better with Shoenberg and Watts than with Slavin. Probabily the Ruesink and Perz experiments which combined magnetostriction and torque measurements should be regarded as more reliable than those of Slavin, who estimated the absolute value of $|\tilde{M}|$ from the LK formula, but evidently the experimental situation is rather unsatisfactory.

The theoretical interpretation of $d \ln A/d \ln A_s$ for uniaxial stress is based on resolving the strain caused by a uniaxial stress into a pure dilation (the effect of which is known from the hydrostatic pressure dependence) and a volume conserving shear strain. For a cubic crystal it can be shown that the ηs are unchanged by a pure shear and so the ηs of the unstrained lattice

242

Table 5.6. *Stress dependence of noble metal FS areas*

Orbit	Hydrostatic			Uniaxial	
			$d \ln A / d \ln A_s$		
Cu	exp.	fitted	band	exp.	calc.
B⟨100⟩	0.94	0.94	1.00	$\begin{cases} 2.4 \pm 0.5^a \\ 2.2 \pm 0.1^d \end{cases}$	3.0
B⟨111⟩	0.90	0.90	0.94	$\begin{cases} 0.6 \pm 0.2^a \\ 0.72 \pm 0.04^d \end{cases}$	1.0
R⟨100⟩	0.94	0.95		$\begin{cases} -2.1 \pm 0.8^a \\ -1.5 \pm 0.3^d \end{cases}$	-1.7
D⟨110⟩	0.86	0.86	0.85	-13.9 ± 0.4^d	-13.3
N⟨111⟩	4.2	4.3	3.2	$\begin{cases} -44 \pm 10^a \\ -47 \pm 13^b \\ -56 \pm 5^c \\ -40 \pm 1^d \end{cases}$	-46 cf. -28 (band)
Ag, Au	Ag exp.	Au exp.		Ag exp.	Au exp.
B⟨100⟩	0.93	0.97		2.2 ± 0.3^a	8 ± 1^a
B⟨111⟩	0.86	0.83		0.8 ± 0.3^a	2.6 ± 0.4^a
R⟨100⟩	0.87	0.97		-2.4 ± 0.5^a	4 ± 1^a
D⟨110⟩	0.73	0.73			
N⟨111⟩	10.6	5.6		$\begin{cases} -73 \pm 5^a \\ -127 \pm 5^b \\ -108 \pm 25^c \end{cases}$	$\left.\begin{array}{l} -60 \pm 8^a \\ -73 \pm 8^b \\ -97 \pm 9^c \\ -100 \pm 10^e \end{array}\right\}$ cf. -100 (band)

Notes: The hydrostatic 'fitted' values (shown only for Cu) are those of Fawcett *et al.* (1980) based on suitable choice of phase shift derivatives; fits of similar quality were obtained by Templeton (1974) for Ag and Au as well as Cu. The uniaxial calculated values (for Cu) are essentially those of Fawcett *et al.*, but a computational error in the D⟨110⟩ entry has been corrected (J. M. Perz, private communication); they are the sums of the hydrostatic experimental values and the calculated effects of shear assuming the phase shifts are as in the unsheared metal. The band structure entries for hydrostatic strain are those of Davis *et al.* (1968) and for uniaxial strain those of Davis (1970). The hydrostatic experimental entries are those of Templeton (1974) and the errors are roughly ±1 in the last figure shown. The uniaxial experimental data are those of (*a*) Shoenberg and Watts (1967), (*b*) Aron (1972), (*c*) Slavin (1973) (*d*) Ruesink and Perz (1983), (*e*) Gamble and Watts (1972).

can be used to calculate the FS areas of the sheared lattice and hence the derivatives of the areas with respect to shear. It can be seen from table 5.6 that the shear contribution to $d \ln A / d \ln A_s$ is usually important and for some orbits (e.g. the neck) quite dominant. Ironically, the calculated values of $d \ln A / d \ln A_s$ for uniaxial stress agree best with the Shoenberg and Watts results, which, until they were supported by the recent magnetostriction results of Ruesink and Perz, seemed the least reliable!

A number of other metals have been studied whose FS can be successfully described by band structure calculations with phase shifts and Fermi energy as adjustable parameters. The most comprehensively studied so far has been tungsten (for details, see Fawcett *et al.* 1980). The pressure dependences of various dHvA frequencies have been determined both directly and indirectly (by combining the appropriate strain derivatives, as explained in §4.3.3) and several combinations of the indirect techniques have been used to determine the shear dependence. The various determinations of pressure dependence agree reasonably well with each other and with theoretical estimates (typically to within 5% or so). The shear derivative determinations by different methods agree quite well among themselves but the agreement with theory is much poorer than for pressure dependence, particularly for the smaller orbits. The calculations, however, involve simplifying assumptions such as the 'rigid muffin-tin-sphere approximation' and perhaps the discrepancies may point the way to developing more realistic and more accurate calculations.

5.4.4. Ferromagnetics (for a detailed discussion see Lonzarich 1980)

The presence of a large saturation magnetization in ferromagnetics introduces interesting new features into the stress dependence of the dHvA effect, as it does into the effect without stress (see §5.3.4.2). First there is the relatively obvious complication that the effective field is B (see (5.9)), and even though the applied field H_e stays constant, B changes when pressure (or any other form of stress) is applied, because the saturation moment M_s is pressure dependent. This causes a slight but significant shift of the dHvA phase with pressure which must be allowed for if an accurate estimate is to be made of the pressure dependence of the FS area A. A more fundamental feature is that there is a specifically 'magnetic' contribution to the pressure dependence of A in addition to the pressure dependence characteristic of a non-magnetic metal of similar band structure. The non-magnetic contribution can be estimated by suitably scaling the known pressure dependence of A in an appropriate non-magnetic transition metal (Mo or W can be compared with Fe and Cu with Ni) and the magnetic contribution can thus be separated out from the experimentally observed pressure dependence. It turns out that in Fe the magnetic contribution is the dominant one, but that in Ni the two contributions are comparable and of opposite sign, so that the resulting pressure dependence of A for the neck is much smaller than in Cu. As discussed in detail by Lonzarich, the size of the magnetic contribution to the pressure dependence provides valuable evidence on the basic nature of itinerant electron ferromagnetism. No study of the effect of stresses other than hydrostatic pressure has yet been made on any ferromagnetic metal.

5.5. Temperature dependence of the Fermi surface

As has already been mentioned in §5.2.1, there are various effects which make the Fermi surface slightly temperature dependent, so that the dHvA frequency F can change slightly over the temperature range in which oscillations are observable. Such an effect was first observed in the Zn 'needle' by Berlincourt and Steele (1954) who found that F fell by a factor of about two as the temperature rose from 4 to 60 K. A rather slighter temperature dependence probably occurs also for Bi (Dhillon and Shoenberg 1955) and Hg (Palin 1972) and probably quite large effects could be found in other metals which have unusually small pieces of Fermi surface. In such cases even very slight changes of lattice dimensions, such as due to thermal expansion, can cause disproportionately large changes in the dimensions of small FS pockets close to zone boundaries. Moreover, for such small pockets, the cyclotron mass tends to be small so that the dHvA oscillations can be followed to much higher temperatures than usual, with a correspondingly larger change of F. Another possible cause of a temperature dependent F is that the Fermi energy ζ is itself slightly temperature dependent. For a parabolic band $\zeta(T)$ is given approximately by

$$\zeta(T) = \zeta_0 \left(1 - \frac{\pi^2}{12} \left(\frac{kT}{\zeta_0} \right)^2 \right) \tag{5.25}$$

and this can become an appreciable effect if the dHvA effect can still be observed at a temperature T comparable with ζ_0/k.

For Zn, it turns out that thermal expansion is the major cause of the temperature dependence of F, since ζ_0/k is of order 300 K, so that at 60 K the temperature dependence of ζ accounts for only a few % of the observed large reduction in F. Indeed, within experimental uncertainties, it proves that the significant factor is the change of the axial ratio c/a due to thermal expansion. This was strikingly shown by O'Sullivan and Schirber (1966) who plotted the changes in F due to hydrostatic pressure and to temperature together on a single graph as a function of the corresponding changes of c/a. As can be seen from fig. 5.31, there is a strong implication that the effect of temperature is mainly due to the same cause as the effect of pressure which, as was discussed earlier, is mainly the change of the lattice dimensions.

For larger pieces of FS the effect of temperature is much smaller than in Zn, but with the precision of measurement now available, this effect is nevertheless amenable to observation. The necessary techniques were first developed by Lonzarich and Gold (1974) in connection with an investigation of ferromagnetics where, as we shall see below, the temperature dependence of F proves to be less than 1 part in 10^6 over the 10 K or so in which dHvA oscillations can be observed. They showed that to achieve significant results at this sort of level of accuracy it is essential to take account of a number of small effects, partly instrumental and partly due to approximations in the conventional theory, which give rise to slight apparent changes of F as T is raised. Fortunately most of these effects lead to linear changes of the observed F with T and so can be distinguished from any more fundamental changes associated with lattice vibrations (which should vary as T^4) or, in ferromagnetics, with temperature dependence of the spin magnetization (which varies roughly as T^2).

Recently Cooper (1979 and unpublished data of Lonzarich) has used these techniques to look at $F(T)$ in the Au neck and found a small but quite appreciable effect (see fig. 5.32). The interesting feature here is that although the effect is of the same order of magnitude as to be expected on the basis of

Fig. 5.32. Variation of dHvA frequency of the Au necks with increasing temperature compared with variation predicted from thermal expansion shown by broken line (Cooper 1979 and Lonzarich private communication). The points shown were taken in various runs on two samples at 71.5 kG.

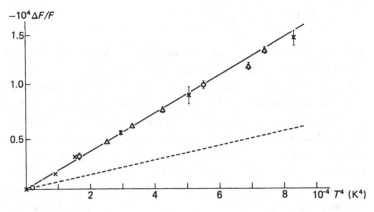

the thermal expansion and the known pressure dependence, the observed effect is appreciably (about 3 times) larger. Probably this indicates that with increase of T the band structure changes in a more fundamental way than merely due to the change in lattice dimensions. In Zn such effects are presumably slight, because it is so nearly a free electron metal, but in Au which is much less free electron-like, electron–phonon effects may well be more important. This may indeed be a profitable subject for further study in providing direct evidence of how the electron–phonon interaction modifies band structure.

As we mentioned above, this kind of experiment was originally developed as a means of studying ferromagnetism. For a ferromagnetic the main effect of temperature is expected to be in respect of the change of the saturation magnetization M_s it causes, and the importance of the experiment is that it provides direct evidence as to the basic mechanism of ferromagnetism. The negative results of the experiments (a change of F less than 1 in 10^6 over 10 K or so in both Fe (Lonzarich and Gold 1974) and Ni (Cooper 1979, Lonzarich 1980) indicate that Stoner 'single particle' excitations do not provide a significant contribution to the temperature dependence of the saturation magnetization. For a more detailed discussion of this topic see Lonzarich (1980).

5.6. Modification of Fermi surface by alloying

As mentioned in §1.1, one of the earliest experiments on the dHvA effect was a study of how the oscillations in Bi were modified by alloying (Shoenberg and Uddin 1936). These experiments demonstrated two striking features. The first, which proves to be characteristic for all metals, is that the amplitude is greatly reduced by even very slight alloying. This was later interpreted as a consequence of the extra scattering of electrons by the impurity atoms which increases the Dingle temperature. We postpone discussion of this aspect to §8.3 and only point out here that this rapid increase of Dingle temperature severely limits the possibility of observing dHvA oscillations in alloys (other, of course, than stoichiometric compounds, already briefly discussed in §5.3.6); typically the oscillations are completely damped out for concentrations of order 1% (atomic) of an impurity component in a pure host.

The second striking feature, which is specific to oscillations of very low frequency, such as in Bi, is that quite minute concentrations of a foreign atom are sufficient to cause considerable frequency changes. Thus only 0.005% Te was found to double the frequency of the electron oscillations of Bi while 0.07% Pb halved it. Qualitatively this great sensitivity reflects the fact that the Bi FS contains very few carriers ($\sim 10^{-5}$ electrons or holes per Bi atom). Thus the replacement of only 1 in 10^5 Bi atoms by atoms with one more (Te) or one less (Pb) valence electron per atom can drastically alter the FS, rather as a few drops of liquid can alter the level in the neck of a bottle considerably, while hardly affecting the volume of liquid in the bottle. We shall examine this idea more critically below.

One interest of such alloying experiments is in providing additional tests of any proposed band structure. Sometimes there may be doubt in the attribution of a particular branch of the F spectrum to a particular pocket of FS predicted by the band structure theory. The effect of alloying can help check the attribution by giving a definite indication of whether the branch is electron-like or hole-like according as the F value is increased (decreased) or decreased (increased) by adding atoms of valence greater (less) than that of the host. For instance the fact that the addition of Te (valence six) increases the frequencies associated with the three Bi ellipsoids while Pb (valence four) has the opposite effect confirms that these ellipsoids do indeed contain electrons; similarly what we have called the 'hole ellipsoid' is indeed affected by alloying in the opposite sense to the electron ellipsoids.

Since the early experiments on dilute alloys of Bi there have been many others on a variety of host metals and added foreign atoms. A summary of the literature is given by Coleridge (1980) who reviews the basic theory. Here we shall discuss mainly one important generalization, the 'rigid band model' (RBM), which provides at least a rough guide to the effect of alloying on the FS of the host metal and sometimes – particularly for small Fermi surfaces, as with Bi or Sb – can lead to a fair quantitative interpretation. This RBM supposes that the only effect of alloying is to add or remove electrons, (according to the valence difference between host and additive) without otherwise changing the bands. In fact, of course, alloying also modifies the lattice to some extent and so the RBM can be at best only a good approximation. A useful diagnostic test of its applicability is to add a constituent of the same valency as the host, e.g. Sb to Bi, in which case the RBM would predict no change of FS. Shoenberg and Uddin found indeed that Sb had about 100 times less effect on dHvA frequencies in Bi than did heterovalent Pb, Sn or Te, so that presumably the RBM is likely to be applicable for the much smaller concentrations of heterovalent additions which appreciably change the frequency.

For a metal such as Bi or Sb with equal numbers of carriers in the electron and hole bands, the RBM requires that

$$N_e - N_h = c\Delta Z \tag{5.26}$$

where N_e is the number per atom in the electron band, N_h in the hole band, c

5.6. Modification of FS by alloying

the atomic concentration of the added component and ΔZ the valency difference (e.g. 1 for Te but -1 for Pb or Sn). If we suppose that the shape of each part of the FS is unchanged, i.e. that alloying scales each branch of the F spectrum by the same factor at all orientations (and this has been demonstrated to be a good approximation for dilute alloys of both Bi (Bhargava 1966) and Sb (Harte, Priestley and Vuillemin 1978), we should have

$$\frac{N_e}{N} = \left(\frac{F_e}{F_{e0}}\right)^{3/2} \quad \text{and} \quad \frac{N_h}{N} = \left(\frac{F_h}{F_{e0}}\right)^{3/2} \tag{5.27}$$

when F_e and F_h refer to electrons and holes for an arbitrary orientation and N is the number (per atom) of electrons or holes in the pure host. This then offers an immediate quantitative test of the validity of the RBM: if N is known, (5.27) determines N_e and N_h from frequency measurements, and a test of (5.26) can then be made.

For dilute alloys of Bi, the tests so far made are not altogether conclusive, largely because of uncertainties in concentration estimates. The first experiment in which hole as well as electron frequencies were measured for Pb in Bi was that of Bhargava (1966), who found that for alloys in the range of atomic concentration up to 6×10^{-5}, $N_e - N_h$ was typically only about $-\frac{1}{2}c$, rather than $-c$ as (5.26) could suggest. It is, however, possible that the mean concentrations of his samples (measured by emission spectroscopy) were appreciably greater than the 'effective' concentrations which determine F, because if the sample is inhomogeneous, the Dingle factor will give much greater weight to the regions of lower concentration. The phase diagram of the Bi–Pb system suggests that very considerable inhomogeneity of composition may result in solidifying the alloy from the melt, and in the early experiments of Shoenberg and Uddin, (who deduced the concentrations simply from the amount of Pb put in at the start), the nominal concentrations to produce the same change of F were about ten times greater than the measured concentrations of Bhargava. The results of Brandt and Razumenko (1960), which agreed fairly well with those of Shoenberg and Uddin, may also have suffered from unreliable concentration determination, and possibly, in spite of the care taken by Bhargava to ensure homogeneity, his samples may not have been quite homogeneous enough to rely on his estimated concentration to better than a factor of two or so. The concentration difficulty can be avoided if $N_e - N_h$ is determined by the independent method of measuring the Hall effect at high fields though only partially if the sample is inhomogeneous. This approach was used by Bate and Einspruch (1967), who studied Bi–Sn alloys and obtained reasonably good agreement between the Hall effect determination of $N_e - N_h$ and the value derived from (5.27) (they used SdH rather than dHvA oscillations).

In spite of the various 'loose ends' there is no reason to doubt that the RBM applies reasonably well for dilute alloys of Bi, but it would be a good thing if a new and more thorough study were made with really reliably known concentrations and with measurements of both hole and electron oscillations for a range of concentrations of several additions of both greater and smaller valency than the host. More recently, detailed studies have been made by Dunsworth and Datars (1973) and Harte, Priestley and Vuillemin (1978) of dilute alloys of Sb and convincing evidence obtained for the validity of the RBM within experimental accuracy. In both studies it was first established that in a given alloy the frequencies associated with each piece of FS changed by the same factor, independent of orientation. Thus, as for Bi, the shapes of the various pieces of FS (in fact all nearly ellipsoidal) are unchanged by alloying. From the frequency factors F_e/F_{e0} and F_h/F_{h0} the values of N_e and N_h were deduced, using (5.27) and the value of N ($5.51 \times 10^{19}\,\mathrm{cm}^{-3}$ or 1.66×10^{-3} per atom) known from dHvA studies of pure Sb (e.g. Windmiller 1966), and it was possible to confirm the validity of (5.26) to within the experimental accuracy (of order 10%) for a range of concentrations of added Sn and Te (see Table 5.7, based on Harte *et al.*).

The next test of the RBM by Harte *et al.* was rather more subtle. This was to determine the changes $\Delta\zeta_e$ and $\Delta\zeta_h$ of the electron and hole Fermi energies and to show that they were indeed equal and opposite. This was achieved by measuring cyclotron masses m as well as frequencies F for each of a series of concentrations, and obtaining the change of Fermi energy at any concentration by the relation

$$\Delta\zeta \equiv \zeta(\text{alloy}) - \zeta(\text{Sb}) = \frac{\hbar^2}{2\pi} \int_{\text{Sb}}^{\text{Alloy}} \frac{\mathrm{d}A}{m(A)} \tag{5.28}$$

where A is the extremal area, given by $(2\pi e/c\hbar)F$. Once again, as can be seen from table 5.7, the results confirmed that $\Delta\zeta_e = -\Delta\zeta_h$ to within the experimental error of about 10%, though the systematic discrepancies of order 10% may point to incomplete validity of the RBM. Bhargava had made a similar but rather less convincing check in his study of Bi alloys, but had deduced the changes of Fermi energy not from measured cyclotron masses but by assuming the validity of the Lax model (see Appendix 1 and §5.3.5) with particular values of its parameters ζ_e, ζ_h and ε_g. For Sb it was possible to go the other way round, i.e. to demonstrate the approximate validity of the Lax model and determine its parameters from the measured cyclotron masses. Thus once the RBM has been established, alloying provides a kind of probe of the band structure by varying the level to which the bands are filled in a controlled fashion.

The FS of Sb is not only larger than that of Bi but rather more complicated in as much as there are three hole ellipsoids with non-parabolic k-dependence of energy rather than the single hole ellipsoid with parabolic dependence in Bi. Correspondingly the Lax model is more complicated and involves a second small gap ε_{gh} between the hole band and a close empty band above it as well as the gap ε_{ge} between the electron band and the filled band below it. If the Lax model is valid we should have from (A1.33) and (A1.34) that

$$\left(\frac{m}{m_0}\right)^2 = \frac{2m_b}{\pi m_0^2}\frac{\hbar^2}{\varepsilon_g} A + \left(\frac{m_b}{m_0}\right)^2 \tag{5.29}$$

and in fact a plot of $(m/m_0)^2$ against A gave quite a reasonably linear plot for both holes and electrons. From the intercepts and slopes m_b and ε_g could be determined and hence (by way of (A1.33) or (A1.34)) also the Fermi energies ζ_e, ζ_h of electrons and holes for pure Sb. The results ($\varepsilon_{ge} = 110 \pm 25$ meV, $\zeta_e = 150 \pm 10$ meV, $m_{be}/m_0 = 0.022 \pm 0.006$ and $\varepsilon_{gh} = 140 \pm$

Table 5.7. *Check of RBM and Lax model for dilute alloys of Sb with Sn and Te (Harte et al. 1978)*

10^3c	N_e/N	N_h/N	$(N_e - N_h)/c\Delta Z$	$\Delta\zeta_e$(meV)	$\Delta\zeta_h$(meV)
Sn ($\Delta Z = -1$)					
1.8	0.51(0.50)	1.69(1.67)	1.09	-39	43
2.6	0.35(0.34)	2.09(2.05)	1.11	-56	63
3.2	0.32(0.32)	2.18(2.16)	0.97	-61	68
3.7	0.16(0.15)	2.92(2.76)	1.24	-86	98
5.2	0.14(0.12)	3.02(3.00)	0.94	-92	103
5.3	0.10(0.09)	3.28(3.21)	1.00	-100	111
Te ($\Delta Z = 1$)					
1.2	1.40(1.41)	0.74(0.75)	0.92	23	-21

Notes: The results are presented here slightly differently from the original published form; a few of the original entries, for which data were incomplete or c was in doubt have been omitted. Here N is per atom and taken as 1.66×10^{-3} and the $\Delta\zeta$s are defined with the sign convention of (5.28). The values of N_e/N and N_h/N in brackets are calculated from (5.30) and agree quite reasonably with the values calculated from (5.27); the agreement is rather poorer if (as would seem more logical) a mean value of $\Delta\zeta$ is used in (5.30) (i.e. $\frac{1}{2}(\Delta\zeta_e - \Delta\zeta_h)$ instead of $\Delta\zeta_e$ and $\frac{1}{2}(\Delta\zeta_h - \Delta\zeta_e)$ instead of $\Delta\zeta_h$).

140 me V, $\zeta_h = 180 \pm 40$ me V, $m_{bh}/m_0 = 0.018 \pm 0.018$), though somewhat rough, were found to be reasonably consistent with other evidence (e.g. the electronic specific heat). An auxiliary check of the parameters is to calculate from them for each alloy the values of N_e and N_h by means of the Lax model relations

$$N_e/N = \left[\frac{(\zeta_e + \Delta\zeta_e)(\zeta_e + \Delta\zeta_e + \varepsilon_{ge})}{\zeta_e(\zeta_e + \varepsilon_{ge})}\right]^{3/2}$$

$$N_h/N = \left[\frac{(\zeta_h + \Delta\zeta_h)(\zeta_h + \Delta\zeta_h + \varepsilon_{gh})}{\zeta_h(\zeta_h + \varepsilon_{gh})}\right]^{3/2} \tag{5.30}$$

using the empirical values of $\Delta\zeta$ given in table 5.7. These estimates, shown in brackets, agree quite reasonably with the estimates based on (5.27).

As an example of a situation where the RBM provides only a rough description of the effect of alloying in a 'simple metal' we shall describe briefly the results of Holtham and Parsons (1976) on dilute alloys of In with Pb and Tl. Although In is tetragonal, its FS (van Weeren and Anderson 1973) is very similar to that of Al, which is cubic (see §5.3.3.1) and in particular there are four-sided rings in the third zone, which contribute 'medium' dHvA frequencies ($\sim 5 \times 10^6$ G from the maximum cross sections of the sides of the ring and low frequencies ($\sim 2 \times 10^5$ G) from the necks at the corners. Although calculation of the effect of alloying with Pb taking account only of the increase in valency (from 3 to $3 + c$ for atomic concentration c) gives quite a reasonable account of the increase of F (by about 2% for $c = 0.005$) for the medium frequencies, it predicts about double the observed increase (about 25% for $c = 0.005$) for the more sensitive low frequencies. However a more elaborate calculation allowing for the changes of lattice parameters, pseudopotential coefficients and spin-orbit coupling gives good agreement for both frequencies. The inadequacy of the simple RBM is demonstrated by the results for dilute alloys with Tl. Since Tl has the same valency as In, the RBM would predict no change in dHvA frequency, but in fact a decrease is observed nearly as big as the increase for adding Pb. The detailed calculations show that this decrease can be reasonably accounted for by the changes of lattice etc., though comparison with theory is somewhat complicated by uncertainty in the magnitudes of the lattice changes.

The first measurement of a frequency shift in a noble metal due to alloying was made with the pulsed field method by King-Smith (1965) who detected a decrease of about 0.2% in the neck frequency of Au for addition of 0.06% Pt (which has valence one less than Au) and 0.3% for addition of 0.3% of Ag which is homovalent. The Pt decrease is about two-thirds of

5.6. Modification of FS by alloying

what would be expected on the RBM and the small Ag decrease (no change would be expected on the strict RBM) is consistent with a linear interpolation between the neck frequencies of pure Au and pure Ag. Later precision data (e.g. Templeton and Coleridge 1975a; for full references see Coleridge 1980) on various dilute alloys of Cu, Ag and Au showed that in favourable circumstances the RBM gave a qualitative, and sometimes even a nearly quantitative, account of the frequency changes if lattice changes were taken into account. However the RBM begins to go badly wrong if lattice distortion is large or if ΔZ, the valence difference between impurity and host, exceeds 2 or is negative. There are also additional complications if the impurity is magnetic.

A more meaningful interpretation of the changes in the FS is provided by a parametrized band structure approach similar to that of §5.3.2, in which phase shifts characterizing the alloy are introduced as fitting parameters. In favourable systems the parameters required to fit the observed frequency changes at a number of orientations are found to be consistent, not only with the frequency changes, but also with the anisotropic scattering by the added atoms, as determined by Dingle temperature studies. However it will be convenient to defer discussion of this topic to §8.3 where the Dingle temperature results are discussed; table 8.2 illustrates the quality of the fits obtained and also indicates the predictions of the RBM. So far it is only for a few noble metal dilute alloy systems that there has been any systematic attempt to bring together frequency shifts and scattering probabilities (i.e. effectively real and imaginary parts of the self energy associated with alloying). In view of the promising results so far obtained, much more deserves to be done.

6

Magnetic interaction

6.1. Introduction

In the theoretical treatment of the de Haas–van Alphen effect as presented so far, it has been taken for granted that the magnetic field acting on the electrons is the same as the field of the magnet and that the field due to the oscillating magnetization $\tilde{M}$ of the metal can be neglected. In fact, however, even though the amplitude $|4\pi\tilde{M}|$ is never appreciable compared to H, it can, in conditions of high field and low temperature, become quite comparable to the period H^2/F (see table A7.1). In such conditions, which imply that $4\pi|\mathrm{d}\tilde{M}/\mathrm{d}H|$ is comparable to 2π, the field arising from the oscillating magnetization of the electrons might be expected to modify the form of the oscillations appreciably from that given by the LK formula, which was deduced ignoring any such contribution.

This was first discovered in the course of experiments on the dHvA effect in the noble metals (Shoenberg 1962) when it was noticed (see fig. 3.5) that the harmonic content of the oscillations was much stronger than predicted by the LK formula and also that the field and temperature dependence of the oscillation amplitude showed anomalies in conditions which made $4\pi|\mathrm{d}\tilde{M}/\mathrm{d}H|$ appreciable. It was eventually* realised that if the field acting on the electrons were B rather than H, the contribution from $4\pi\tilde{M}$ would provide a kind of 'feedback' effect which could indeed explain the observed anomalies, and this became known as the magnetic interaction (MI) effect.

We shall start by outlining the arguments which justify the replacement of H by B in the LK formula for $\tilde{M}$ and then explore the consequences of this modification. First we shall show how MI modifies the line shape of an originally single frequency dHvA oscillation, thus producing a rich harmonic content in appropriate conditions. The complications of sample shape and crystal anisotropy will then be considered and it will be shown that for sufficiently strong MI the sample may break up into domains

---

* Originally I thought only of the demagnetizing field $-4\pi n\tilde{M}$ as the 'feedback' contribution, but very soon afterwards A. B. Pippard reminded me of the well-known fact that the average field 'seen' by a charge in a material body is B, which provided the more significant feedback contribution $4\pi\tilde{M}$ to the effective field.

254

6.2. Justification of replacing H by B

(Condon 1966). When the dHvA oscillations contain more than a single frequency a rich variety of new effects can occur, such as frequency and amplitude modulation of one frequency by another and, more generally, the creation of combination frequencies and their harmonics. From this variety we shall select for discussion only relatively simple limiting situations which correspond, at least qualitatively, to experimentally observable effects.

It turns out that the idealized theory fails to account quantitatively for some of the observed oscillation line shapes and for the frequency and amplitude modulation actually observed, and that the basic cause of these discrepancies is phase inhomogeneity in the sample. The idealized theory for a homogeneous sample has to be modified to allow for the inhomogeneity and it will be shown that agreement with experiment is much improved if the necessary phase smearing (which usually provides the major contribution to the Dingle amplitude reduction factor in the LK formula) is applied *after* rather than *before* the magnetic interaction has been taken into account. The effect of MI on other oscillatory effects will then be discussed and finally it will be shown that in conditions of extremely high density, such as are relevant in astrophysics, MI may be capable of producing large magnetization without any applied field, i.e. a kind of ferromagnetism.

6.2. Justification of replacing H by B in the formula for $\tilde{M}$

Although it is well known that the spatial average magnetic field in bulk matter is B, it is not immediately clear how this should be taken into account in the theory. In fact the correct procedure is to replace H by B in the LK formula for the oscillating magnetization. Thus if we call the LK result $\tilde{M}_0(H)$, then the oscillatory magnetization $\tilde{M}$ allowing for magnetic interaction will be

$$\tilde{M} = \tilde{M}_0(B) \tag{6.1}$$

This has been justified by a microscopic analysis (Holstein, Norton and Pincus 1973) in which it is shown that this is the correct result if the magnetic interaction is treated by the Hartree approximation. A simpler argument has been given by Pippard (1980) who, following Van Vleck (1932), points out that electron diamagnetism results from a slight imbalance between the large diamagnetic moment of the electron orbits in the bulk of the metal and the large paramagnetic moment of electrons bouncing round the periphery of the metal. The dHvA effect arises from oscillations in the diamagnetic contribution from the interior and, since the fluctuations of magnetic field are on a much finer scale than a typical orbit size, this contribution is governed by the mean field B in the interior which determines the quantization of the energy levels of the internal electrons. The field which governs the motion of the bouncing electrons at the surface is intermediate between B and H, but this is irrelevant, since the paramagnetic contribution of these electrons does not contribute to the dHvA effect. Pippard (1963) had earlier justified (6.1) by a thermodynamic argument based on the idea that the entropy S should surely be determined by the average local field, i.e. B, so that

$$S = S_0(B) \tag{6.2}$$

where $S_0(H)$ is the entropy as given by the LK theory. It follows by thinking of the metal in a field B that

$$(\partial S/\partial B)_T = (\partial S_0(B)/\partial B)_T = (\partial M_0(B)/\partial T)_B$$

and since we also have

$$(\partial S/\partial H)_T = (\partial M/\partial T)_H = (\partial M/\partial T)_B(\partial B/\partial H)_T$$

we find by division

$$(\partial M/\partial T)_B = (\partial M_0(B)/\partial T)_B$$

which integrates to (6.1) as far as the oscillating part of M is concerned, since this part must vanish at sufficiently high T, so that any integration constant (which could be a function of B) must vanish too.

If (6.1) is accepted as the correct way of replacing H by B we can immediately determine how the thermodynamic potential $\tilde{\Omega}$ must be modified to take account of MI. Since

$$\tilde{M} = -(\partial \tilde{\Omega}/\partial H)_T \tag{6.3}$$

we have

$$\tilde{\Omega} = -\int \tilde{M}_0(B)\,\mathrm{d}H = -\int \tilde{M}_0(B)\,\mathrm{d}B + 4\pi \int \tilde{M}\,\mathrm{d}\tilde{M}$$

so that

$$\tilde{\Omega} = \tilde{\Omega}_0(B) + 2\pi \tilde{M}^2 \tag{6.4}$$

where $\tilde{\Omega}_0(B)$ is the expression obtained if H is replaced by B in the LK formula (2.15) for the thermodynamic potential.* Thus the oscillatory

---

* Note that the contribution of the steady part of M to B has been ignored here; this is usually amply justified.

thermodynamic potential is *not* correctly given if we merely replace H by B in the LK formula. Indeed the extra term $2\pi\tilde{M}^2$ is essential in conditions where MI is relevant; its contribution to $\tilde{M}$ is in fact $4\pi M \, d\tilde{M}/dH$ and, as we shall see below, it is just where $4\pi d\tilde{M}/dH$ becomes comparable to unity that MI becomes significant. The reason that $\tilde{\Omega}$ cannot be equated to $\tilde{\Omega}_0(B)$ is, of course, that the LK formula for the thermodynamic potential is based on the assumption of independent particles, which is no longer valid when magnetic interaction is taken into account. An almost obvious concomitant of (6.4) is that

$$\tilde{M} = -\left(\partial\tilde{\Omega}_0(B)/\partial B\right)_T \tag{6.5}$$

6.3. Theory of MI for a single frequency in absence of anisotropy and shape effects

Since we shall be concerned throughout only with the oscillatory part of M and Ω we shall for simplicity in writing omit the wavy superscript in what follows. We shall usually be concerned with only a few periods of oscillation, so it is convenient to write

$$H = H_0 + h \quad \text{and} \quad B = H_0 + b \tag{6.6}$$

and we shall consider dHvA oscillations of a single frequency F, so that without MI we should have

$$4\pi M = A\sin\left(\frac{2\pi F}{H} + \phi\right) = A\sin\left(\frac{2\pi Fh}{H_0^2}\right) \tag{6.7}$$

Here h and b are small increments to H_0, which is the major part of H or B, and which is chosen so that at H_0 the total dHvA phase $(2\pi F/H_0 + \phi)$ is an odd multiple of π (i.e so that for $H = H_0$, $M = 0$, $B = H_0$ and dM/dH is positive). If we now replace H by B, a sinusoidal term such as (6.7) in the LK formula becomes to a sufficient approximation

$$4\pi M = A\sin kb \tag{6.8}$$

where

$$k = 2\pi F/H_0^2 \tag{6.9}$$

Since for the present we shall ignore complications of shape and suppose the sample is a long rod oriented along the field, we may put

$$B = H + 4\pi M \quad \text{or} \quad b = h + 4\pi M \tag{6.10}$$

and we need not distinguish between H and the field H_e of the magnet. We also for the present ignore anisotropy and assume that M is parallel to H. Combining (6.9) and (6.10), we see that M is determined by the implicit equation

$$4\pi M = A \sin k(h + 4\pi M) \tag{6.11}$$

The criterion for MI to be significant (i.e. for the feedback provided by the extra term in the argument of the sin to be appreciable), is simply that the amplitude of the extra term $4\pi k M$ should become appreciable compared to unity i.e. that

$$4\pi |dM/dB| \gtrsim 1 \tag{6.12}$$

Reference to table A7.1 shows that this can indeed occur in practical situations, and we shall now investigate what happens when it does.

To solve (6.11) it is convenient to introduce 'reduced' notation:

$$y = 4\pi k M, \quad x = kh, \quad \theta = kb \quad \text{and} \quad a \equiv 4\pi |dM/dB| = kA \tag{6.13}$$

so that (6.10) and (6.11) become*

$$\theta = x + y \tag{6.14}$$

and

$$y = a \sin \theta = a \sin(x + y) \tag{6.15}$$

The solution of the implicit equation (6.15) is most easily found by an obvious graphical construction in which each point of the graph $y = a \sin x$ is shifted in the negative x direction by a distance equal to y. As shown in fig. 6.1, the distortion of the originally sinusoidal line shape becomes more and more severe as a approaches 1, and for $a > 1$, y (i.e. M) becomes multivalued. The corresponding variation of θ (i.e. B) is shown in fig. 6.2. The solution can also be easily computed by inverting (6.15) into the form

$$x = \sin^{-1}(y/a) - y \tag{6.16}$$

For $a < 1$ the solution for y can also be expressed explicitly as an

---

* In some of the original papers a different sign convention was used, in which a minus sign was attached to the l.h.s. of (6.11) and y was defined as $-4\pi k M$, so that $x + y$ in (6.14) and (6.15) became $x - y$. This was thought at the time to have some advantages, but probably the present notation is more straightforward and less likely to lead to confusion. However it must be remembered that at many places the present treatment leads to signs opposite to those of the original papers.

6.3. Theory of MI for a single frequency

expansion in harmonics with Bessel function coefficients.*

$$y = \sum_{r=1}^{\infty} (2/r) J_r(ra) \sin rx$$

and

$$dy/dx = 4\pi \, dM/dH = \sum_{r=1}^{\infty} 2J_r(ra) \cos rx$$

(6.17)

For $a \ll 1$ the Bessel functions can be usefully expanded in powers of a and

Fig. 6.1. Shape of dHvA oscillations with magnetic interaction for $a = 0.5$, 1, 2 and 10 plotted in reduced form; y is the reduced magnetization, x is the reduced field, and a is the reduced amplitude ($= 4\pi |dM/dB|$). For $a > 1$ the solution is multivalued but the broken line portions of the curve are never realized, so that y changes discontinuously at each multiple of 2π. Note that the scales for both y and x are different for $a \leqslant 1$ and for $a > 1$. The oscillations for $a = 10$ are shown over a wider range of x in order to make the periodicity more apparent. The notation P, Q, p, q is explained in the text.

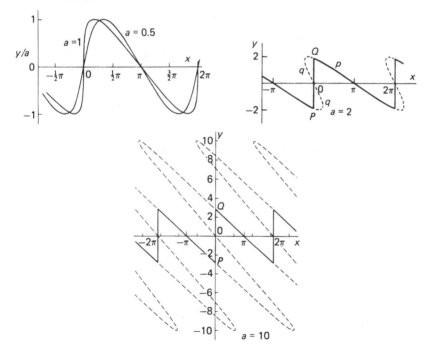

* It is interesting that (6.15) was an equation well known in an astronomical context as far back as the 18th century and that later Bessel (in 1824) introduced the functions J_r specifically to obtain y as an expansion in harmonics of x (see for instance Gray, Matthews and MacRobert 1952). I am grateful to Professor A. Seeger for drawing my attention to the result (6.17).

259

to terms in a^3 we find

and
$$
\left.
\begin{array}{l}
y = a\{(1 - \tfrac{1}{8}a^2)\sin x + \tfrac{1}{2}a\sin 2x + \tfrac{3}{8}a^2\sin 3x\} \\
\mathrm{d}y/\mathrm{d}x = a\{(1 - \tfrac{1}{8}a^2)\cos x + a\cos 2x + \tfrac{9}{8}a^2\cos 3x\}
\end{array}
\right\}
\qquad (6.18)
$$

This result can of course also be obtained directly from (6.15) by an iterative process of successive approximations in which y on the r.h.s. of (6.15) is treated as small.

We must now return to the question of what happens when $a > 1$ and the solution of (6.15) is multivalued. The clue to this problem is provided by considering the thermodynamic potential Ω as given by (6.4), and bearing in mind that the system will try to assume the state of lowest Ω (Pippard 1963, 1980). If we define a reduced thermodynamic potential Z as

$$
Z = 4\pi k^2 \Omega \qquad (6.19)
$$

and remember that $y = a\sin\theta$, (6.4) becomes

$$
Z = a\cos\theta + \tfrac{1}{2}a^2\sin^2\theta \qquad (6.20)
$$

Curves of Z against $x (= \theta - y)$ for various values of a can be easily constructed once y is known for given x and some are shown in fig. 6.3. For $a > 1$, when a point such as P in the oscillation of M is passed, the metal can lower its thermodynamical potential by undergoing a discontinuous transition from P to Q which have equal values of Z and then following the

Fig. 6.2. Variation of B with H plotted in reduced form (i.e. θ versus x) for $a = 0.5$, 1, 2 and 10; for simplicity the inaccessible parts of the curve for $a = 10$ are not shown, and the $a = 0.5$ curve is shown dotted to avoid confusion with the superimposed $a = 1$ curve.

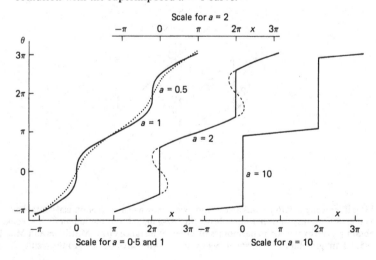

branch p rather than q of the thermodynamic potential. This is somewhat analogous to the behaviour of the p–v diagram for the gas–liquid phase transition at the point where the free energies of the two phases become equal. The question of how the transition at PQ actually occurs will be discussed later, but for the present we shall simply assume that it does occur and that oscillations of y (i.e. M), θ (i.e. B) and Z (i.e. Ω) have the forms indicated by the full curves in figs. 6.1, 6.2 and 6.3. For $a \gg 1$ it is easy to show from (6.16) (since y/a is small) that y, θ, and Z are given to a good approximation by

$$
\left.\begin{aligned}
y &= (a/a + 1)[(2r + 1)\pi - x] \\
\theta &= (2r + 1)\pi + [x - (2r + 1)\pi]/(a + 1) \\
Z &= -a\{1 - [(2r + 1)\pi - x]^2/2(a + 1)\}
\end{aligned}\right\} \quad \text{for} \quad 2r\pi \leqslant x \leqslant (2r + 2)\pi
$$

$$(6.21)$$

Fig. 6.3. Variation of the thermodynamic potential with H plotted in reduced form (i.e. z/a versus x) for $a = 0.5, 2$ and 10. For $a > 1$ only the full portions of the curves are realized. The notation P, Q, p, q is explained in the text. Note that the minimum value of Z/a is always -1.

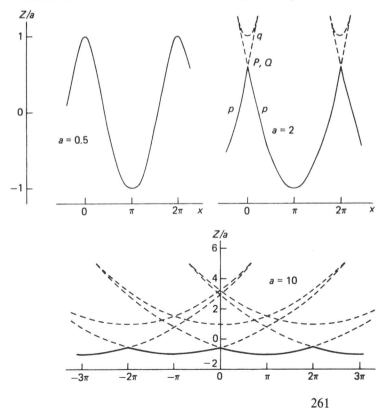

Note that for $a \gg 1$ the θ versus x (i.e. B versus H) curve assumes the form of a staircase. The triangular magnetization curve in this limit (cf. fig. 6.1 for $a = 10$) is very reminiscent of the magnetization curve of a superconductor, though of course the magnetization of a superconductor disappears at the critical field, so that there is no periodic continuation outside the range indicated. This similarity will be helpful when we come to consider sample shape effects.

The oscillations of dy/dx (i.e of $4\pi dM/dH$, which is the directly observed quantity in the field modulation technique for low modulation) are shown in fig. 6.4. We may note that for $a < 1$ the half peak to peak swing ('amp' for short) of the y oscillations is the same as without MI, i.e. just a, but for dy/dx the amp increases indefinitely as a approaches 1. Indeed it can be seen from (6.15) that

$$\frac{dy}{dx} = \frac{a\cos\theta}{1 - a\cos\theta} \tag{6.22}$$

so that the peak to peak swing is from $-a/(1 + a)$ to $a/(1 - a)$ giving a half peak to peak swing of

$$\text{amp}\left(\frac{dy}{dx}\right) = \frac{a}{1 - a^2} \tag{6.23}$$

which becomes infinite for $a = 1$. For $a > 1$, (6.22) applies only for the realizable parts of the oscillations, i.e. for a small range of θ around

Fig. 6.4. Variation of dy/dx ($= 4\pi dM/dH$) with x (i.e. H) for $a = 0.5, 2$ and 10. For $a > 1$ only the accessible parts of the curve are shown and dy/dx becomes infinite at $x = 0, 2\pi$ etc. as indicated by the broken vertical lines.

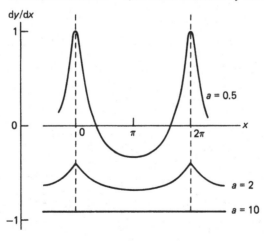

$(2r + 1)\pi$ for which $dy/dx = -a/(a + 1)$. At the discontinuities of the y versus x curve, dy/dx rises to infinity, like a δ-function.

It should be noticed that in all conditions of appreciable MI there is a left–right asymmetry of the M oscillation and an up–down asymmetry of dM/dH. These asymmetries can be of practical use in establishing the true sign of M and dM/dH as they appear in the output of an experimental detection system, the polarity of which is not otherwise obvious. In principle the absolute value of a can also be determined from the harmonic content of the oscillations, but as we shall see later this possibility is in practice complicated by phase smearing effects.

6.4. Crystal and shape anisotropy effects

Both crystal anisotropy and demagnetization (i.e. shape) effects modify the relations between B, H_e (the field of the magnet), H (the field in the material) and M, from the simple form (6.10). It is helpful first to consider the general situation in order to see in what conditions crystal and shape effects can be discussed separately. We shall assume throughout that the sample is ellipsoidal in shape, so that if the external field H_e is uniform over all space, H and M will be uniform within the sample. If, as in practice is inevitable, the sample is *not* a perfect ellipsoid, H and M will not be exactly uniform, but the predicted results for an exact ellipsoid approximating to the actual sample will still be qualitatively valid, apart from some phase smearing of the dHvA oscillations (if MI is appreciable).

Since usually it is the field H_e rather than H which is directly accessible to measurement, we need to express H in (6.10) in terms of H_e, and the general relation for the vector components is

$$H_a = H_{ea} - 4\pi n_{ab} M_b \tag{6.24}$$

where $4\pi n_{ab}$ is the demagnetization tensor (a, b go from 1 to 3) and the summation convention is implied. Thus

$$B_a = H_{ea} + 4\pi(\delta_{ab} - n_{ab})M_b \tag{6.25}$$

where $\delta_{ab} = 1$ for $a = b$ and 0 for $a \neq b$. Note that if we choose the principal axes of the ellipsoid as reference axes, n_{ab} is purely diagonal. We shall for simplicity discuss only two extreme situations in any detail (for a more general but not very transparent treatment, see Crabtree 1977). If the sample is a disc with H_e oriented in the plane of the disc, which is also a symmetry plane of the crystal, then M will also lie in the plane of the disc and demagnetization plays no role ($n_{11} = n_{22} = 0$, $n_{33} = 1$ but irrelevant); the direction of M relative to H_e is then determined entirely by crystal anisotropy. If on the other hand crystal anisotropy can be ignored (for instance by choosing the direction of H_e such that F is an extremum with respect to orientation) then only shape effects remain.

Before considering these opposite extremes it is convenient to write down here a formula for the thermodynamic potential Ω valid in a general anisotropic situation. This formula, first given by Condon (1966), is easily derived from the generalized vector form of (6.3), which is

$$M_a = -\partial\Omega/\partial H_{ea} \tag{6.26}$$

so that using (6.25) we have

$$\Omega = -\int M_a\, dH_{ea} = -\int M_a\, dB_a + 4\pi \int M_a(\delta_{ab} - n_{ab})\, dM_b$$

Since in general M_a is proportional to M_b when H_e varies, this becomes

$$\Omega = \Omega_0(B) + 2\pi(\delta_{ab} - n_{ab})M_a M_b \tag{6.27}$$

where we have used the vector form of (6.5) to get the first term on the r.h.s.

6.4.1. Crystal anisotropy

Since F varies with orientation there will be a component M_2 of M perpendicular to H_e as well as the parallel component M_1 and we can derive M_2 from the vector form of (6.5) i.e.

$$M_a = -\partial\Omega_0(B)/\partial B_a \tag{6.28}$$

i.e.

$$M_1 = -\partial\Omega_0(B)/\partial B_1 \quad \text{and} \quad M_2 = -\partial\Omega_0(B)/\partial B_2 \tag{6.29}$$

The function $\Omega_0(B)$ is really a function of F/B and F varies with B_2 because the resultant B (which to an adequate approximation has magnitude B_1) is tilted by a small angle B_2/B_1 from the field direction, so we find, exactly as in the absence of MI

$$M_2 = -\mu M_1 \tag{6.30}$$

where

$$\mu = \frac{1}{F}\frac{\partial F}{\partial \theta} \tag{6.31}$$

(θ being an angular variable in the $M_1 M_2$ plane)

The important novel feature, however, is that the tilt of B away from H_e produces a slight change in F, proportional to M, and this change gives an additional feedback term. For the special case of all the vectors lying in the

6.4. Crystal and shape anisotropy effects

plane of a disc sample, we have

$$B_2 = 4\pi M_2$$

so the small angle of tilt is given by

$$\Delta\theta = 4\pi M_2/H = -4\pi\mu M_1/H \tag{6.32}$$

and F becomes $F(1 - 4\pi\mu^2 M_1/H)$. If, as before, we measure fields from a suitably chosen origin H_0 and put in the modified value of F, the implicit equation determining M_1 reduces to

$$4\pi M_1 = A \sin k(h + 4\pi(1 + \mu^2)M_1) \tag{6.33}$$

where k is defined as before by (6.9). This can be expressed in the same reduced form (6.15) as before, i.e.

$$y = a \sin \theta = a \sin(x + y),$$

but now the reduced parameters, other than x, have different meanings:

$$y = 4\pi M_1 k(1 + \mu^2), \quad x = kh, \quad \theta = k(b + 4\pi\mu^2 M_1),$$
$$a = kA(1 + \mu^2) \tag{6.34}$$

The thermodynamic discussion is also much as before. From (6.27) and (6.28) we obtain

$$\Omega = -\int M_1(B)\,dB_1 - \int M_2(B)\,dB_2 + 2\pi(M_1^2 + M_2^2) \tag{6.35}$$

and since $B_2 = 4\pi M_2$, the second and fourth terms on the r.h.s. cancel. After a little manipulation and use of (6.33) and (6.34) we find the same result (6.20) for the reduced thermodynamic potential Z as before, i.e. $Z = a\cos\theta + \frac{1}{2}a^2 \sin^2\theta$, but Z is now defined differently as

$$Z = 4\pi k^2(1 + \mu^2)\Omega \tag{6.36}$$

The solution of the implicit equation (6.15) is exactly the same as that discussed in §6.3 and, just as before, only parts of the y–x and θ–x curves are stable for $a > 1$, with discontinuities occurring when x is a multiple of 2π. However there are some differences in the physical meaning of the results. Thus for given A the interaction parameter a is greater by the factor $(1 + \mu^2)$ as compared with the isotropic case. This factor can be as high as 5 for the Be cigar oscillations, over a range of orientations, and in such a case MI sets in for much weaker amplitudes of M_1 than would be the case for an isotropic metal. However, often μ^2 is quite small: in the noble metals it is at most about 0.2 and of course it always vanishes at turning points in the

265

angular dependence of F. It should be noticed too that the perpendicular component M_2, which determines the torque on the sample, is given by (6.30) and is in a sense a 'slave' to M_1, which is the 'master'. Thus the line shape of the M_2 oscillations is simply a scaled version of that of M_1, which is determined by the solution of the MI equation (6.15). This is true however small the amplitude of M_2 may be (if μ is small), so that strong MI can be observed in the torque oscillations even though their amplitude may be arbitrarily small. In the limit $a \gg 1$ where the oscillations of M_1 have triangular shape, their amplitude is smaller by a factor $(1 + \mu^2)$ as compared with the isotropic case ($\mu = 0$). A consequence of this is that the staircase-like variation of b with h (fig. 6.2) is modified, the nearly horizontal portions becoming steeper with smaller vertical rises in between. This can also be seen by noticing that θ as defined in (6.34) is no longer proportional to b; θ itself still follows the rectangular staircase of fig. 6.3 but kb, which is given by

$$kb = \theta - \frac{\mu^2 y}{1 + \mu^2} \tag{6.37}$$

has a variation with kh which approaches more and more closely to the 45° linear relation of a non-magnetic material as μ increases. These effects of crystal anisotropy on MI have not so far been experimentally observed. It should be noted that for geometries other than the simple one considered here, the factor $(1 + \mu^2)$ is reduced by shape effects and we shall give some examples later.

6.4.2. Effect of shape

We shall now consider the opposite extreme of ignoring crystal anisotropy and discuss the effects of sample shape *per se*; for simplicity we shall consider only ellipsoidal shapes. In the absence of crystal anisotropy we need not worry about any perpendicular component of the demagnetizing field, since the rotation of the B direction it causes does not change F and is therefore irrelevant. It is not difficult to show that if H_e is in a direction other than that of a principal axis of the ellipsoid, M and B (which are parallel since M is determined by B and there is no crystal anisotropy) are slightly tilted from H_e. The angle of tilt is $\sin \theta \cos \theta (n_1 - n_2) 4\pi M/B$ if H_e lies in the 1, 2 plane at angle θ to axis 1, where 1 and 2 are two of the principal axes of the ellipsoid. Thus the component of M perpendicular to H_e is at most of order M^2/B and can be safely ignored (e.g. the couple due to it is negligible).

If then we ignore any perpendicular component, (6.25) reduces to

$$B = H + 4\pi M = H_e + 4\pi(1 - n)M \tag{6.38}$$

6.4. Crystal and shape anisotropy effects

where $4\pi n$ is the demagnetizing coefficient with n given by

$$n = n_{11}\lambda_1^2 + n_{22}\lambda_2^2 + n_{33}\lambda_3^2 \tag{6.39}$$

if M has direction cosines $(\lambda_1, \lambda_2, \lambda_3)$ and the axes of reference are the principal axes of the ellipsoid. For the limiting case of a long rod at angle θ to the field, $n = \frac{1}{2}\sin^2\theta$ and for a thin plate with the field at angle θ to the plane of the plate $n = \sin^2\theta$.

Thus (6.11) is modified to

$$4\pi M = A \sin k(h_e + 4\pi(1-n)M) \tag{6.40}$$

where

$$H_e = H_0 + h_e$$

and H_0 is chosen as before.

At first sight the effect of the new factor $(1 - n)$ would seem to be merely to weaken the magnetic interaction and indeed to eliminate it altogether in the limiting case of a disc-shaped sample with H_e normal to its plane, for which $n = 1$. However, as was first pointed out by Condon (1966), this is only true if the homogeneously magnetized state (with M given by the solution of (6.40)) is indeed that of lowest free energy. It turns out that if MI is strong enough, a state of lower free energy can be achieved over part of an oscillation cycle by the sample breaking up into domains for which the local value of M alternates in sign from one domain to the next (fig. 6.8). As the external field is increased, the proportion of $+$ to $-$ domains increases steadily from 0 to 100% and the mean M of the sample varies linearly with H_e (fig. 6.5). The questions of how the domains are formed and of their size and arrangement will be discussed later and for the present we shall *assume* that domains do form if the free energy is thereby lowered and investigate the conditions for this to happen.

Before discussing this domain problem, let us briefly consider the solution of (6.40) for *weak* MI. In reduced form the equation is

$$y = a \sin\theta = a \sin(x_e + (1-n)y) \tag{6.41}$$

where $x_e = kh_e$ and y, a and θ are defined as before. This can be rewritten in the same form as (6.15) i.e.

$$y' = a' \sin\theta = a' \sin(x_e + y') \tag{6.42}$$

if we define

$$y' = (1-n)y, \qquad a' = (1-n)a \tag{6.43}$$

and the weak MI solutions are exactly as before, i.e. as in fig. 6.1 and in (6.17), (6.18), (6.22) and (6.23) with y replaced by y' and a by a'.

As a or a' increase, the solution of (6.42) remains single valued until $a(1 - n)$, i.e. $a' > 1$. However if we think of M as a function of h rather than h_e, i.e. of y as a function of x rather than x_e, (6.41) returns to the old form

$$y = a \sin(x + y)$$

which becomes multivalued as soon as $a > 1$, whatever the value of n. Two examples with $a(1 - n) < 1$, though $a > 1$, are illustrated in fig. 6.5. The physical significance of this can be understood if, following Condon, we think of the sample as made up of many parallel thin rods (or thin plates)

Fig. 6.5. Shape effect on magnetic oscillation form for $a > 1$. Curves of y v. x_e (i.e. M v. H_e) are shown for (a) $n = 0.6$, $a = 2$; (b) $n = 0.6$, $a = \infty$; (c) $n = 1$, $a = 2$; (d) $n = 1$, $a = \infty$. Only half cycles are shown for (c) and (d). The sample breaks up into domains between P and Q, and the average magnetization follows the straight line PQ rather than the broken curve derived from the MI relation.

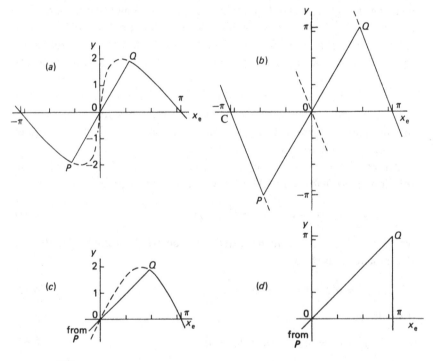

6.4. Crystal and shape anisotropy effects

along the field direction. Each of these rods (or plates) experiences the field H rather than H_e and so, if $a > 1$, can for $h = 0$ (or $2r\pi/k$) exist in either of two states of equal free energy. For a finite demagnetizing coefficient, the field h can stay constant at the critical value as h_e increases, provided M varies i.e. provided

$$h_e - 4\pi nM = 0 \quad (\text{or } 2r\pi/k)$$

or

$$x_e - ny = 0 \quad (\text{or } 2r\pi) \tag{6.44}$$

where r is any integer. In what follows we shall, for simplicity, consider only the first cycle, i.e. take $r = 0$. Thus from the point on the magnetization cycle where x first vanishes, i.e. the point P in fig. 6.5 such that

$$y = a \sin y \tag{6.44a}$$

(which has non-zero solutions *only* for $a > 1$), y can vary linearly with x_e according to (6.44) until the next non-zero solution of (6.44a), i.e. the point Q, is reached. Along the line PQ the sample will be broken up into domains, with an increasing proportion of $+$ rather than $-$ domains as we go along PQ until 100% $+$ is reached at Q.

It is important to note that over this domain part of the magnetization curve the observed y (i.e. M) and θ (i.e. B) are *average* values; the values in the individual domains, as indicated in fig. 6.6, are those at P and Q. Thus as the external field is varied there are periodic ranges of the local B which are missing, in the sense that no individual part of the sample experiences them. This idea, emphasized by Pippard (1980), will prove significant later in the discussion of certain phenomena in MI conditions.

The fact that the free energy is lowered by breaking up into domains, i.e. by the mean magnetization following the straight line PQ (equation (6.44)) rather than the solution of (6.41), is almost obvious if it is recalled that when demagnetization effects matter (6.3) must be replaced by (6.26) so that the change in Ω for a change of H_e is given by

$$\Delta\Omega = -\int M \, dH_e$$

or in reduced form

$$\Delta Z = -\int y \, dx_e \tag{6.45}$$

Thus $\Delta\Omega$ is proportional to the area under the y–x_e curve and the increase of Ω or Z from P is evidently less if the straight line PQ is followed, rather

Magnetic interaction

than the curved solution of (6.41). At Q, of course, $\Delta\Omega$ again vanishes corresponding to the sample being again in a uniform state.

The value of Ω is given by (6.27) which in the absence of domains becomes

$$\Omega = \Omega_0(B) + 2\pi(1 - n)M^2 \tag{6.46}$$

or in reduced form

$$Z = a\cos\theta + \tfrac{1}{2}(1 - n)a^2 \sin^2\theta \tag{6.47}$$

Fig. 6.6. In the domain region M and B (i.e. y and θ) stay constant within each individual domain as the external field (i.e. x_e) varies. This is indicated by the broken horizontal lines in (a) the y, x_e curve and (b) the θ, x_e curve (note that (a) is essentially the same as fig. 6.5a). It should be emphasized that the values of θ (i.e. of B) between these lines do not occur at any individual point of the sample. However the proportion of Q to P domains increases from 0 at P to 100% at Q as x_e increases and the average values of y and θ follow the dotted lines. Graph (c) shows y as a function of θ; the range between P and Q is never realized at any individual point of the sample, though the *average* values of y and θ over the sample follow the dotted line. All the graphs are for $a = 2$ and $n = 0.6$, though the form of (c) is independent of n.

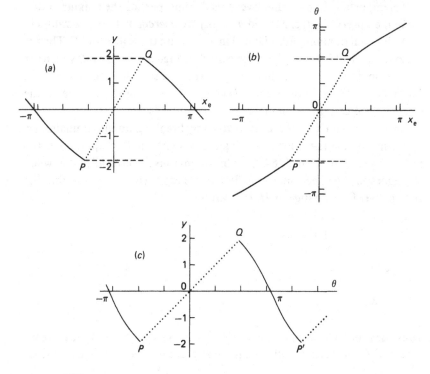

270

6.4. Crystal and shape anisotropy effects

The solution of (6.41) gives y and θ in terms of x_e and a curve of Z against x_e can then be constructed (fig. 6.7). Between P and Q the linear y–x_e relation implies a parabolic increase ΔZ (domain) of Z given by (6.45), and using (6.44) we find

$$\Delta Z(\text{domain}) = \frac{1}{2n}(x_{e0}^2 - x_e^2) \qquad (6.48)$$

where x_{e0} is the value of x_e at P or Q. As can be seen from fig. 6.7, the Z curve for the domain state lies, as it must, below that for the uniformly magnetized state. The two curves of course touch at P and Q since the slopes of the Z curves give M, which is the same for the two alternatives at both P and Q.

Returning to the form of the magnetization curve in the domain regime it is interesting to consider the case $a \gg 1$. The appropriate solutions of $y = a \sin y$ which determine the points P and Q are then approximately

$$y = \pm a\pi/(a + 1) \simeq \pm\pi \qquad (6.49)$$

and the corresponding values of x_e are given by ny. Between P and Q, the magnetization varies linearly in accordance with (6.44) and the remainder of the cycle consists of two nearly straight lines to meet the x_e axis at $\pm\pi$ (see fig. 6.5(b)). For the limiting case of a disc ($n = 1$) for $a \gg 1$, the domain region occupies the whole cycle and the magnetization curve becomes a mirror image of that for $n = 0$ (reflected about the line $x_e = \pi/2$). Once

Fig. 6.7. Variation of the reduced thermodynamic potential Z with x_e, the reduced external field for $a = 2, n = 0.6$. Between P and Q the value of Z is lowered from the broken to the dotted curve if the sample breaks up into domains. The broken and dotted curves touch at P and Q ($Z = 0.08$).

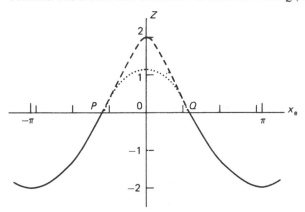

Magnetic interaction

again there is a striking similarity between this $a \gg 1$ situation and a superconductor, if we consider only the negative half of the dHvA cycle and imagine it scaled so that the half period corresponds to the critical field H_c of the superconductor; we must also suppose that the origin of the superconducting magnetization curve lies at C (i.e. at $x_e = -\pi$). For demagnetizing coefficient n, the superconducting sample breaks up into domains at the same point P as for the dHvA effect, with the field H staying at just H_c over the domain region (the 'intermediate state'). The values of $4\pi dM/dH_e$ over the sections CP and PO are exactly the same for the superconducting and dHvA situations, i.e.

$$4\pi dM/dH_e = -1/(1-n) \quad \text{for} \quad CP \quad \text{and} = 1/n \text{ for } PO \qquad (6.50)$$

But of course the superconducting magnetization disappears at O (i.e. for $H_e = H_c$) and stays zero for further increase of H, in contrast to the dHvA effect which continues periodically.

Let us now return briefly to the problem of crystal anisotropy when shape is also relevant. For simplicity we consider only the two examples of a sphere ($n_{11} = n_{22} = n_{33} = \frac{1}{3}$) and a long rod with H_e along its length ($n_{11} = 0, n_{22} = n_{33} = \frac{1}{2}$). It is not difficult to consider more general situations, but the resulting expressions are rather unwieldy and do not illustrate any new principle. If we choose the 1,2 plane to contain M_1 and M_2 we can consider both the sphere and the rod together, using the abbreviations $n_1 = n_{11}$ ($=\frac{1}{3}$ for sphere, 0 for rod) and $n_2 = n_{22}$ ($=\frac{1}{3}$ for sphere, $\frac{1}{2}$ for rod). Because of the demagnetizing field in the 2 direction we now have

$$B_2 = 4\pi(1 - n_2)M_2$$

so that the tilt angle of (6.32) must be multiplied by $(1 - n_2)$. Also if we wish to use h_e rather than h in the analogue of (6.33) we must replace h by $h_e - 4\pi n_1 M_1$. Thus (6.33) is modified to

$$4\pi M_1 = A \sin k[h_e + 4\pi M_1((1 - n_1) + (1 - n_2)\mu^2)] \qquad (6.51)$$

The factor of $4\pi M_1$ on the r.h.s. is $\frac{2}{3}(1 + \mu^2)$ for a sphere and $(1 + \frac{1}{2}\mu^2)$ for a rod, as compared with $(1 + \mu^2)$ for the disc ($n_1 = n_2 = 0$). Without crystal anisotropy the factor would have been $\frac{2}{3}$, 1 and 1 respectively. However, following Condon's argument, instability and break-up into domains is governed by the relation between M_1 and h rather then h_e, which is

$$4\pi M_1 = A \sin k[h + 4\pi M_1(1 + (1 - n_2)\mu^2)] \qquad (6.52)$$

and this sets in when $Ak(1 + (1 - n_2)\mu^2) \geqslant 1$. Thus the effect of crystal anisotropy in making MI instability set in for weaker amplitude A is more marked for the disc $\left(Ak(1 + \mu^2) \geqslant 1\right)$ than for the sphere $\left(Ak(1 + \tfrac{2}{3}\mu^2) \geqslant 1\right)$ or the rod $\left(Ak(1 + \tfrac{1}{2}\mu^2) \geqslant 1\right)$.

6.4.2.1. Domain structure: theory

The size and shape of the domains in the dHvA oscillations are governed by considerations very similar to those in other domain situations (e.g. for ferromagnetics and superconductors). In order to avoid unfavourable magnetostatic energy at the boundaries between domains, the boundaries must run nearly parallel to the applied field, as in fig. 6.8, though the arrangement will be distorted near the sample surfaces and 'branching' may occur to reduce the free energy slightly below that of the simple pattern of fig. 6.8. Since such distortions will not affect the orders of magnitude of the domain sizes, we shall ignore them in our essentially dimensional treatment of the problem. This treatment is similar to one given by Pippard (1980) and agrees qualitatively with a more sophisticated treatment by Condon (1966). The argument starts by assuming that the 'wall' thickness in the boundary layer is of order D, the orbit diameter (the orbit is assumed circular for simplicity), and, since the magnetization is neither parallel nor antiparallel to the field in this thickness there is an unfavourable free energy per unit area of the wall. This would argue for having thick domains with correspondingly few walls, but against this must be set the unfavourable magnetostatic energy of the free poles where the domains meet the surface. The magnetostatic energy is reduced by having thin domains (and correspondingly many of them) so that the field of the poles falls off as rapidly as possible away from the surface. Minimization of the sum of these two unfavourable contributions leads to an optimum domain size which proves to be of order of the geometric mean of the sample thickness Z and the orbit diameter D.

To simplify the argument we suppose the sample is a thin plate normal to

Fig. 6.8. Schematic sketch of Condon domains in a plate-like sample normal to H. The domain period X is greatly exaggerated compared with the thickness Z, and Z is exaggerated compared with Y (in fact the theory assumes $X \ll Z \ll Y$).

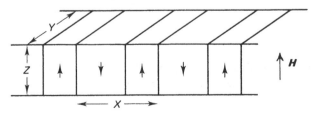

the field, and that we are at the 'half way' point O in the cycle so that the parallel and antiparallel domains are equally wide. We also assume that the interaction parameter a is so large that the oscillation has the limiting form shown in fig. 6.5d. The extra free energy in the wall between antiparallel domains can be roughly estimated by supposing that over a distance D the extra free energy is of the same order of magnitude as that corresponding to $Z/a = 1$ in fig. 6.3, i.e. $a/4\pi k^2$ per unit volume, or $Da/4\pi k^2$ per unit area of wall, where k is as defined in (6.9). Since there are $2/X$ walls per unit width of sample, the extra contribution of the wall energy to the total free energy per unit volume is of order

$$\Delta_1 = aD/2\pi k^2 X \tag{6.53}$$

The magnetostatic energy of the system of alternating strips of $+$ and $-$ poles has been calculated before in connection with the ferromagnetic problem (Kittel 1949, p. 558; see also Appendix 12). For a pole density per unit area of $\pm M_0$ or $\pm 1/(4k)$, this contribution per unit volume of the sample (bearing in mind that both the top and bottom surface of the sample contribute) is given by

$$\Delta_2 = 0.8525X/16Zk^2 \tag{6.54}$$

The minimum of $\Delta_1 + \Delta_2$ with respect to X occurs when $\Delta_1 = \Delta_2$ and this gives

$$X = \alpha(ZD)^{1/2} \tag{6.55}$$

where α is given as $1.73a^{1/2}$ by our calculation, but in view of the crudeness of the calculation, α is better regarded as a dimensionless constant which increases slowly with a. Condon's more detailed calculation gives α as a function of a, increasing from $\alpha = 0$ for $a = 1$, to $\alpha = 2$ for $a = 4$ and $\alpha = 3.5$ for $a = 10$. In any case the concept of an orbit diameter is somewhat vague except for circular orbits and evidently the direction in which the domain walls run will depend on the orbit shape. In various experiments where domain structure has been relevant (see below) the orbit diameter D was of order 1 or 2×10^{-4} cm (e.g. for Be at 23 kG or Ag at 90 kG), Z was typically 0.1 cm and $a \sim 3$, so we find, using Condon's estimate of $\alpha \sim 2$

$$X \sim 7 \times 10^{-3} \, \text{cm} \tag{6.56}$$

It is reassuring that this domain repeat period is much larger than the assumed wall thickness D, so that the concept of clearly distinguishable domains makes sense.

274

6.4.2.2. Domain structure: experiment

The scale and pattern of the domain structure in ferromagnets and the intermediate state of superconductors has been revealed by various techniques, most strikingly by the Bitter patterns displayed when magnetic powder is spread on the sample surface. Such techniques have not yet been applied to demonstrate the domain patterns in the MI situation though in fact the scale of the patterns, according to (6.44), should be quite comparable to those observed in ferromagnetics and superconductors. Hopefully the additional complications of high magnetic field and rather smaller sample dimensions will be overcome before too long. Direct evidence that the sample breaks up into macroscopic regions with different values of B has however been obtained in an ingenious n.m.r. experiment by Condon and Walstedt (1968), and of course there is plenty of indirect evidence supporting the idea of a domain structure from the line shapes of the magnetic oscillations in appropriate conditions (see §6.5).

In the n.m.r. experiment it was shown that over the part of a dHvA belly oscillation in Ag for which domains are predicted, two distinct n.m.r. resonances could be observed, corresponding to the different B values in the + and − domains. These values of B are

$$B = H \pm 4\pi M_0 \tag{6.57}$$

where M_0 is the magnetization at P in fig. 6.6 (i.e. such that $y_0 \, (= 4\pi k M_0)$ is the solution of $y = a \sin y$, and the field H, which stays constant over the domain part of the cycle as H_e varies, is H_0 or H_0 increased by any whole number of periods.

Condon and Walstedt's results are illustrated in fig. 6.9. The sample was a thin silver crystal (8 mm × 8 mm × 0.8 mm) and a field of about 90 kG was applied normal to the plate, in a $\langle 100 \rangle$ direction. Two reference oscillators were used whose frequencies corresponded approximately to the resonances expected at the two fields specified by (6.57). The exact resonance frequency was indicated by the beating of the free precession signal from the Ag nucleus with the reference oscillation. The existence of domains is clearly demonstrated by the *simultaneous* occurrence of both resonance frequencies over just the appropriate part of the dHvA cycle, while only a single resonance frequency is observed over the part of the cycle without domains. From the field difference ΔB, corresponding to the resonant frequency difference Δv over the domain region, the value of a can be deduced. Thus from the definition of y_0 we have

$$a = y_0/\sin y_0 \tag{6.58}$$

and y_0 is given by $\pi \Delta B/\Delta H$, where ΔB is the field difference corresponding

275

to the observed constant frequency difference over the domain region and ΔH is the dHvA period (i.e. $2\pi/k$). The value of a was found to be about 2.6, which is reasonable for a good sample in the conditions of the experiment (90 kG and 1.4 K). It can be seen that the experimental points in fig. 6.9 lie close to the predicted curve for this value of a, taking into account that the value of n for the plate is appreciably different from 1 (the inscribed oblate ellipsoid has $n \sim 0.84$). Some slight discrepancies may be due to complications from the simultaneous dHvA oscillations coming from the rosette orbit. Further confirmation of the existence of the domains is provided by the variation of the amplitudes of the two resonances, which do indeed follow the varying proportions of the $+$ and $-$ domains just about as they should. Outside the domain region the amplitudes fall off because B varies with H_e so that the fixed driving frequency gets further and further off resonance.

Fig. 6.9. An n.m.r. experiment demonstrating the reality of domains in Ag (after Condon and Walstedt 1968). The experimental points ($\bigcirc$, $\times$: different reference frequencies, $+$: very weak signals) indicate the local field B as determined by n.m.r. for varying values of the applied field H_e. Both fields are measured from a zero near the centre of a dHvA period ΔH. The significant feature is that two values of B differing by ΔB are obtained over the central region of the period where domains are predicted, but only one value elsewhere. The broken line indicates the predicted variation, which is qualitatively similar to that of fig. 6.6b except that here $a = 2.6$ and $n = 0.84$, for which $\Delta B/\Delta H_e = 0.69$ and dB/dH_e over the non-domain sections is about 0.73.

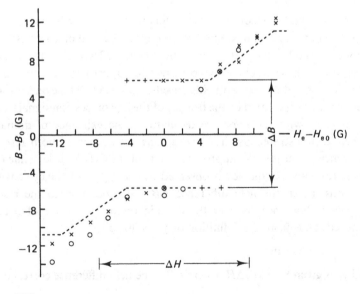

6.5. Qualitative verification of theory for a single frequency

As already mentioned, it was the pulsed field experiments on the noble metals (Shoenberg 1962) which first led to the idea of magnetic interaction and similar effects were observed soon afterwards in Be (Plummer and Gordon 1964). Although the strong harmonic content and anomalous temperature and field dependence of amplitude agreed roughly with the theoretical predictions of §6.3, the early measurements were insufficient to test the theory at all critically. It was only somewhat later that more systematic experiments were undertaken (Shoenberg and Vuillemin 1966, Broshar, McCombe and Seidel 1966, Condon 1966) and these, while confirming the general features of the theory, still left some puzzling discrepancies to be explained. It was fairly clear that the origin of these discrepancies lay in the lack of the perfect homogeneity of field and of sample tacitly assumed in the theory, but it was only a good deal later (Shoenberg 1976) that the phase smearing effects of inhomogeneity were better understood (see §6.6).

We shall first discuss the experiments of Shoenberg and Vuillemin (1966) on Au and Ag by the modulation method, in which rod-shaped samples were used, so that demagnetization effects were not very important. For the field along $\langle 111 \rangle$, the belly provides the dominant oscillation in dM/dH_e and at the working field of about $50\,\text{kG}$ the MI parameter a passes through 1 as T is lowered a little below $2\,\text{K}$. The simultaneous presence of the much lower frequency neck oscillations modulates both the amplitude and frequency of the belly oscillations, but we shall postpone discussion of these modulation effects (see §6.8) and concentrate for the present on the region of the neck cycle where the largest belly amplitude is observed. As can be seen from fig. 6.10, lowering of T changes the oscillation line shape of dM/dH_e quite dramatically, from a slightly (but appreciably) distorted sinusoid at $2.18\,\text{K}$ to an extremely distorted one at $1.39\,\text{K}$, with narrow pointed positive peaks and almost flat bottoms.

As can be seen from fig. 6.11, the line shapes at both temperatures are not too unlike those predicted for a sample of the appropriate small demagnetizing coefficient ($n = 0.075$) with an MI parameter of $a = 3$ at $1.39\,\text{K}$ and 0.35 at $2.18\,\text{K}$, i.e. varying with T in accordance with the known cyclotron mass ($m/m_0 = 1.09$). It is plausible that the predicted narrow peaks for $a = 3$ should be much reduced in height and broadened by phase smearing, but the flat bottoms much less seriously affected than the peaks. Thus the theory can be tested to some extent by ignoring phase smearing and studying the depth Y_b of the flat bottom of the $4\pi\,dM/dH_e$ cycle and its temperature variation. The analogue of (6.22) for non-zero n, valid over

Fig. 6.10. Belly oscillations in Au $\langle 111 \rangle$ at (a) 1.39 K and (b) 2.18 K for $H = 48.5$ kG (Shoenberg and Vuillemin 1966 (SV)). The oscillation period is about 5.1 G and the period of the amplitude modulation is just that of the neck oscillations (whose amplitude is too weak to show on the scale of the diagram). In both (a) and (b) a few oscillations around the maximum amplitude are shown below on an expanded scale. The scale of $Y = 4\pi dM/dH_e$ and the position of the zero line of the oscillations are only approximate; the field scales are different in (a) and (b).

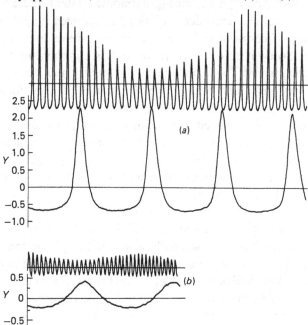

Fig. 6.11. Graphs of dy/dx_e (i.e. $4\pi dM/dH_e$) versus x_e (i.e. H_e); $n = 0.075$, $a = 3$ and 0.35 (roughly corresponding to the conditions of fig. 6.10). The graphs are calculated from (6.59) except in the domain region for $a = 3$, where (6.44) is used.

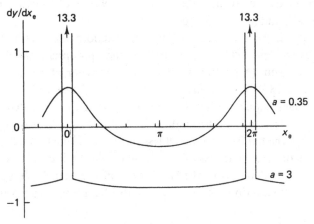

the accessible range of θ (i.e. no domains) is

$$4\pi \, \mathrm{d}M/\mathrm{d}H_e = \mathrm{d}y/\mathrm{d}x = a\cos\theta/(1 - a(1 - n)\cos\theta) \qquad (6.59)$$

Thus the depth of the flat bottom Y_b for $\theta = \pi$ is given by

$$Y_b = a/(1 + (1 - n)a) \qquad (6.60)$$

from which a can be deduced if the calibration of the equipment is accurately known, so that the absolute value of Y_b is determined. In fact the calibration was known only rather roughly so that an alternative procedure was devised based on the temperature dependence of Y_b. If the cyclotron mass is reliably known, the ratio $a(T_0)/a(T)$, where T_0 is any fixed temperature (we shall take it as the lowest temperature used, which was 1.39 K) can be regarded as a reliably known parameter. We can then rewrite (6.60) as

$$\frac{1}{Y_b'(T)} = \frac{c}{a(T_0)}\left(\frac{a(T_0)}{a(T)}\right) + c(1 - n) \qquad (6.61)$$

where $Y_b'(T)$ is the *observed* signal and c the calibration constant such that

$$Y_b = cY_b'$$

Thus a plot of $1/Y_b'(T)$ against $a(T_0)/a(T)$ should give a linear plot, from whose slope and intercept both c and $a(T_0)$ can be determined.* Such a plot is shown in fig. 6.12 and is indeed reasonably linear, giving $a = 3.3$ at $T = 1.39$ K and a value of c which agrees within a few per cent with the rough value based on the coil dimensions and the circuit constants. Similar results were obtained with Ag, indicating even higher values of a (~ 11 at 1.33 K), though phase smearing of the positive peaks was even worse than for Au. The fact that the a values for both Au and Ag were roughly consistent with those predicted by the LK formula assuming a plausible Dingle temperature† supported the general validity of the theory, but the confirmation was not very decisive. Much of the difficulty of these 1966 experiments arose from inadequate homogeneity of the magnet and from difficulties in establishing a reliable base line from which Y_b was measured. As we shall see later when we consider the effect of phase smearing, the fair

---

* In the published paper the analysis was slightly modified to correct for the presence of the neck oscillations, but for the sake of simplicity we ignore this complication here. This leads to some slight differences between numerical values quoted here and those in the published paper.
† The Dingle temperature was not actually measured; it was only later (Bibby and Shoenberg 1979) that it was realized that this could be done from field dependence measurements at high temperatures in the absence of MI.

linearity of such plots as shown in fig. 6.12 may be somewhat fortuitous and a consequence of incorrect choice of baseline.

One interesting feature of the experiments was the behaviour of the positive peak (the 'spike') as a function of the modulating field. This was found to be appreciably non-linear, as shown in fig. 6.13, and this can perhaps be interpreted as a consequence of the fact that over the region of the spike the sample has a domain structure. If the sample is not perfectly homogeneous there may be local obstacles to the free movement of the domain walls when the field varies and perhaps at sufficiently low modulation some of these obstacles cannot be overcome, resulting in a lowering of the apparent slope susceptibility. Some supporting evidence for such a domain 'pinning' interpretation comes from the area under the dM/dH curve, which should of course be exactly zero over a cycle if measured from the correct baseline. In fact the area under the spike often appeared to be rather less than required to compensate the area under the flat bottom and the discrepancy was worse at lower modulations. Later experiments on Ag disc normal to the field by Smith (1974) showed similar effects, and there was qualitative agreement with a calculation based on an *ad hoc* model of the pinning. Here again further experiments using better

Fig. 6.12. Plot of $1/Y'_b$ (i.e. $1/4\pi \mathrm{d}M/\mathrm{d}H_e$) in arbitrary units) as a function of $a(1.39\,\mathrm{K})/a(T)$ (after SV 1966, but with some reinterpretation of the original data). The temperatures of the various points are indicated on the upper scale. The broken lines are calculated from the limiting formulae (6.61) (high T) and (6.81) (low T) assuming $c = 0.29$, $a_0(1.39\,\mathrm{K}) = 5.34$, $\gamma = 0.57$ as in fig. 6.17a. The full line is a rough linear fit to all the points and would imply $c = 0.32$ and $a(1.39\,\mathrm{K}) = 3.3$

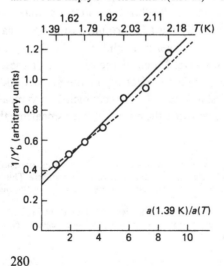

techniques and more varied sample conditions might provide valuable evidence on the details of the domain structure; the main difficulty of such experiments of course is that extremely high sensitivity (i.e. signal to noise ratio) is required to work at the very low modulations involved.

The experiments of Plummer and Gordon (1964) demonstrated strong MI in the dHvA oscillations arising from the Be cigar, and these oscillations have the advantage of showing a long beat, so that at low T, as the field varies, the effective a changes from a value well below unity at the beat minimum to one well into the MI range at the beat maximum. This was very prettily shown in the later experiments of Broshar, McCombe and Seidel (1966) using the modulation method and of Condon (1966) using the torque method. As can be seen from fig. 6.14, the beat pattern has more or less the usual sinusoidal form at 4 K, where a remains small even at the beat maximum, but at 0.35 K the amplitude around the beat maximum becomes roughly independent of a because of strong MI.

The particular interest of Condon's experiment was in demonstrating the effect of demagnetization coefficient in conditions of fairly strong MI. As can be seen from fig. 6.15, the oscillations of torque near the beat maximum do approach the ideal limiting forms for a long rod and a disc discussed

Fig. 6.13. Variation of output signal Y' (arbitrary units) with modulation amplitude h for 'spike' and flat bottom of the Au sample of figs 6.10 and 6.12 (SV 1966). At the highest value shown of $2\pi Fh/H^2 (= z)$, $J_1(z)$ differs from $\frac{1}{2}z$ by only about 1%. The ratio of spike height to flat bottom was not quite as high in this experiment as in that of fig. 6.10 (when the modulation amplitude was about the same as the greatest shown here); this was either because the sample had deteriorated or because of slightly imperfect setting of the sample.

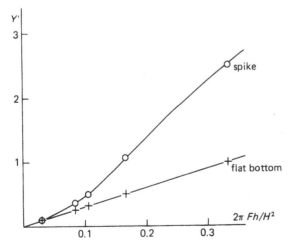

Fig. 6.14. Beats in oscillations of dM/dH_e in Be (Broshar *et al.* 1966). At 4 K, MI is not appreciable even at beat maximum, but at 0.35 K, MI is so strong that the amplitude hardly varies over a considerable range around the beat maximum. The sample was an ellipsoid of revolution (axial ratio ~ 1.6) with its long axis along the hexagonal axis and along **H**.

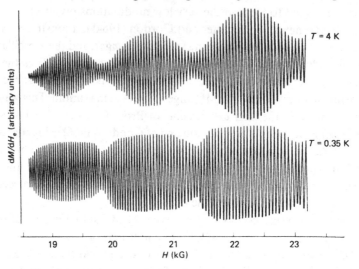

Fig. 6.15. Beats in torque oscillations in Be at 1.4 K for two samples: (*a*) a rod with its length along the hexad axis and (*b*) a disc with its plane normal to the hexad axis (Condon 1966). The field was at 5° to the hexad in both measurements and varied from 24.9 to 30.0 kG across the picture. In each case the lower diagram shows a few oscillations on an expanded scale near the beat maximum.

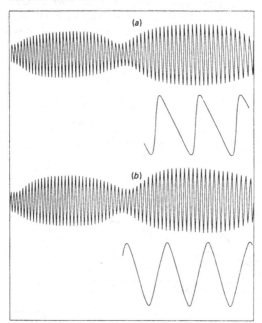

earlier, and it was these observations which led Condon to propose a domain structure over part of each cycle, indeed over most of each cycle for the disc sample. In torque experiments it is of course the perpendicular component M_2 that is measured, but as explained in §6.4.1 the MI effects in M_2 should be proportional to those in the parallel component M_1, so measurements of M_2 test the theory just as well as those on M_1. It should also have been possible to check the theory of the effect of crystalline anisotropy as regards the modification of the interaction parameter by the factor $(1 + \mu^2)$, but in the conditions of Condon's experiments μ^2 was probably too small to have much effect. Just as for the noble metals, the detailed interpretation of the Be results was bedevilled by phase smearing effects and we shall now consider how the theory needs to be modified to take account of inhomogeneity.

6.6. Phase smearing in MI conditions

We have already dealt with the theory of phase smearing in §2.3.7.3 in conditions where MI is negligible and there it was pointed out that often the dominant contribution to the Dingle damping factor is indeed phase smearing associated with sample inhomogeneity, rather than broadening of the Landau levels due to finite scattering time. However if MI is important the problem of sample inhomogeneity (or of field inhomogeneity) becomes much more complicated. It is no longer correct to calculate the effect of MI as if the sample was homogeneous with a magnetization corresponding to the average of M. The difficulty is that the local value of B is determined not by the average value of M and (for a finite demagnetizing coefficient) by the pole density on the sample surface, but also by the local environment of pole density div M throughout the volume of the sample. In detail, the determination of M as a function of the variable dHvA frequency F through the sample leads to a complicated implicit equation for which an explicit solution has yet to be found, if indeed such a solution exists. However a 'mean field' approach (Shoenberg 1976) leads to a solution which appears to be a fair approximation to the truth and is certainly a better approximation than ignoring the inhomogeneity. In what follows we shall concern ourselves only with sample inhomogeneity; a proper discussion of field inhomogeneity is if anything more difficult.

We suppose that locally, over a region R (see fig. 6.16), large compared with the orbit size, the magnetization has the 'ideal' value $M_0(B)$ predicted by the LK formula with H replaced by B, and including a Dingle damping factor which takes account only of electron scattering by impurities. The actual Dingle factor (which could be measured at temperatures high enough to make MI negligible) will be supposed to be predominantly due to

Magnetic interaction

amplitude reduction by smearing over the variable phase of $M_0(B)$ through the sample. We next assume that outside the region R the magnetization can be treated as having a uniform value M, the average magnetization of the whole sample. If the region R is taken as a spheroid of revolution with its axis parallel to H_e (and also to M and M_0 if we ignore anisotropic effects), with demagnetizing coefficient $4\pi n'$, we then have that inside R, B is given by

$$B = H_e - 4\pi(n - n')M + 4\pi(1 - n')M_0$$

or

$$B = H_e + 4\pi(1 - n)M_0 + 4\pi(n - n')(M_0 - M) \qquad (6.62)$$

It seems plausible to guess that on average n' might be something like $\frac{1}{3}$, i.e. that on average the regions of homogeneous phase are something like spherical, so if the sample is spherical or elongated in the field direct, $(n - n')$ should be appreciably smaller than $(1 - n)$. Moreover if we define a reduction factor γ to take account of phase smearing, i.e.

$$\gamma = |M|/|M_0| \qquad (6.63)$$

and suppose that the phase smearing is symmetrical, so that M has the same phase as M_0, then

$$M_0 - M = (1 - \gamma)M_0$$

Thus except for rather severe phase smearing (say $\gamma < \frac{1}{2}$), and rather flat-shaped samples, the last term of (6.62) should be a good deal smaller than

Fig. 6.16. Illustrating ellipsoidal sample of mean magnetization M in a field H_e. Because of sample inhomogeneity different regions have magnetizations which differ in phase, though not in amplitude, and one such region R of magnetization M_0 is indicated schematically (Shoenberg 1976).

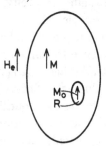

6.6. Phase smearing in MI conditions

the second term and we shall simply ignore it. Although this is rather a drastic approximation it has the merit that it enables an explicit calculation to be made which takes at least some account of the inhomogeneity and, as we shall see below, does seem to agree quite reasonably with observation.

We shall therefore assume that within R we have simply

$$B = H_e + 4\pi(1 - n)M_0, \tag{6.64}$$

then solve the MI problem for M_0 and finally phase smear M_0 to give the average value of M which is of course the observable quantity. The implicit equation to be solved is now

$$4\pi M_0 = A_0 \sin k\big(h_e + 4\pi(1 - n)M_0\big) \tag{6.65}$$

instead of (6.40) and if (analogously to the definitions of y' and a' in (6.42) and (6.43)) we define

$$y_0 = 4\pi M_0 k(1 - n), \quad x_e = kh_e$$

and

$$\alpha_0 = A_0 k(1 - n) = a_0(1 - n) \tag{6.66}$$

this assumes the standard form

$$y_0 = \alpha_0 \sin(x_e + y_0) \tag{6.67}$$

For weak or moderate MI (a_0 must be < 1 to avoid domain formation), the solution as before (see (6.17)) is

$$y_0 = \sum_{r=1}^{\infty} (2/r) J_r(r\alpha_0) \sin r x_e \tag{6.68}$$

We must now phase smear this result in some appropriate way. The simplest is to assume that the distribution of phase is Lorentzian, which is consistent with the fact that Dingle plots are usually linear (see §2.3.7.3). With this assumption, if the amplitude reduction factor for the fundamental is γ, then it will be γ^r for the rth harmonic and so the phase smeared result is simply

$$y = \sum_{r=1}^{\infty} (2/r)\gamma^r J_r(r\alpha_0) \sin r x_e \tag{6.69}$$

where y is defined as $4\pi k(1 - n)M$ and M, of course, is the average magnetization of the sample. If we had applied the phase smearing before rather than *after* applying MI, as was done originally, the suffixes zero would not have appeared in (6.65) and (6.68) and our final result would have

285

been

$$y = \sum_{r=1}^{\infty} (2/r) J_r(r\alpha_0\gamma) \sin rx_e \qquad (6.70)$$

since the amplitude A_0 would have been replaced by $A = \gamma A_0$ and α_0 by $\alpha = \gamma \alpha_0$.

Evidently, if the MI is very weak, i.e. $\alpha_0 \ll 1$, the two formulations become identical, since the leading term in the expansion of $J_r(z)$ is proportional to z^r, but when a_0 and α are comparable with unity ('moderate MI) the amplitudes predicted by the new treatment (i.e. by (6.69)) may be appreciably lower than those predicted by the old (i.e. by (6.70)). For example if $\alpha_0 = 0.9$ and $\gamma = 0.5$, (6.69) predicts a fundamental amplitude (i.e. for $r = 1$) about 8% lower than (6.70); thus if a is deduced from the measured fundamental amplitude, the use of the old treatment, i.e. (6.70) will give an underestimate of a.

Another procedure for estimating the value of a in not too strong MI conditions is to measure the half peak to peak absolute amplitude, z, of the oscillations in Y, defined as $4\pi \, \mathrm{d}M/\mathrm{d}H_e$ or $\mathrm{d}y/\mathrm{d}x_e$ (note that Y is differently defined in Shoenberg 1976); this is rather simpler than Fourier analyzing the oscillations to get the separate harmonic amplitudes. A simple calculation of Y and hence z as a function of a, based on (6.69), can be inverted by an iterative procedure to show that

$$a = z\left\{1 - \left(\frac{9}{8} - \frac{1}{8\gamma^2}\right)(1-n)^2 z^2 + \left(\frac{833}{384} - \frac{27}{128\gamma^2} + \frac{1}{24\gamma^4}\right)\right.$$

$$\left. \times (1-n)^4 z^4 + \cdots \right\} \qquad (6.71)$$

while if (6.70) is used

$$a = z\{1 - (1-n)^2 z^2 + 2(1-n)^4 z^4 + \cdots\} \qquad (6.72)$$

In some experiments on gold spheres, Bibby and Shoenberg (1979) showed that the use of (6.71) rather than (6.72) gave a straighter Dingle plot for the a values at different fields, thus suggesting that the new procedure was the more correct one.

Yet another method of estimating a from experiment is to measure the ratio R of the second harmonic amplitude in Y to that of the fundamental. On the old procedure, it is easy to deduce from (6.70) that

$$a(1-n) = R\left(1 + \frac{5}{24}R^2 + \frac{69}{576}R^4 + \cdots\right) \qquad (6.73)$$

286

6.6. Phase smearing in MI conditions

while if the new procedure based on (6.69) is used, this is modified to

$$a(1 - n) = R\left(1 + \frac{5}{24}\frac{R^2}{\gamma^2} + \frac{69}{576}\frac{R^4}{\gamma^4} + \cdots\right) \tag{6.74}$$

and, again, appreciably different estimates of a are obtained if γ is small and R is greater than say 0.3. However there is no systematic experimental evidence to demonstrate that (6.74) gives better estimates than (6.73).

When we come to 'strong' MI i.e. $a_0 > 1$, the differences between the old and new procedures of phase smearing become more striking. We shall discuss only the limiting case $a_0 \gg 1$, for which an explicit formula can be given describing the line shape after phase smearing. For intermediate values of a_0 the MI problem has first to be solved on the lines of §6.4 and the resulting oscillations must then be convoluted with the phase distribution function; even for a Lorentzian distribution no explicit formula can be given and the form of the oscillation must be computed for each particular case.

For $a_0 \gg 1$, the curve of $Y_0 (= 4\pi\, dM_0/dH_e)$ as a function of H_e has approximately a rectangular form such as shown in fig. 6.11 and given by

$$Y_0 = 1/n \quad \text{for} \quad -\mu < x_e < \mu. \tag{6.75}$$

in the 'spike' or domain part of the cycle, where

$$\mu = na_0\pi/(1 + a_0) \tag{6.76}$$

and

$$Y_0 = -a_0/(1 + a_0(1 - n)) \quad \text{for} \quad \mu < x_e < 2\pi - \mu \tag{6.77}$$

in the diamagnetic part of the cycle.

The phase smearing can be done either by direct convolution of (6.75) and (6.77) (indefinitely repeated periodically in both the $+$ and $-$ directions of x_e) with the appropriate Lorentzian or, rather more simply, by Fourier analyzing (6.75) and (6.77), reducing the amplitude of the rth harmonic by the factor γ^r and finally synthesising the reduced harmonics. The result is

$$Y = \frac{a_0}{\mu(1 + a_0(1 - n))}\left\{\tan^{-1}\frac{\gamma \sin(x_e + \mu)}{1 - \gamma \cos(x_e + \mu)}\right.$$

$$\left. - \tan^{-1}\frac{\gamma \sin(x_e - \mu)}{1 - \gamma \cos(x_e - \mu)}\right\} \tag{6.78}$$

287

As can be seen from fig. 6.17, this formula can be fitted very satisfactorily to the experimental points by suitable adjustment of γ alone, with a_0 (which determines μ, since n is known) calculated from the LK formula. For the Au rod of the Shoenberg and Vuillemin experiments γ was not determined independently, so that the good fit might be somewhat fortuitous, but the good fit to Bibby's (1976) Au sphere is more significant since γ was determined independently as 0.50 from a Dingle plot at 3 K, where MI was unimportant. The value of γ used to give the fit of fig. 6.17(b) was 0.58 and the slight difference from the directly observed 0.50 may well be due to a small electron scattering contribution to the overall Dingle reduction factor. Indeed if it is supposed that electron scattering reduces the fundamental amplitude by a factor $\gamma' = 0.85$, then we should have included this in a_0, giving $a_0 = 9.6$ (instead of the 11.3 that was used in the fit) and it turns out that this value with a phase smearing γ of 0.59 gives a fit practically indistinguishable from that shown. This is very satisfactory since $0.85 \times 0.59 = 0.50$, which is just the independently determined overall reduction factor; moreover the assumption $\gamma' = 0.85$ is reasonably consistent with what might be expected from the electrical resistivity.

We now see that the provisional analysis by which the value of a was estimated from the values of Y_b of the flat bottoms (see (6.60)) was only approximately valid. From (6.78) the value of Y_b (which is the value of Y for

Fig. 6.17. Belly oscillation line shape for rather strong MI in Au with H along $\langle 111 \rangle$ (Shoenberg 1976). (a) For a rod sample with $n = 0.075$; the curve is experimental, taken from fig. 6.10a ($T = 1.39$ K, $H = 48.5$ kG) and the points are calculated from (6.78) assuming $a_0 = 5.34$ (as given by the LK formula) and $\gamma = 0.57$. (b) For a spherical sample at $T = 1.21$ K and $H = 84.5$ kG; the curve is again experimental (Bibby 1976) and the points calculated from (6.78) assuming $a_0 = 11.3$ (as given by the LK formula) and $\gamma = 0.58$.

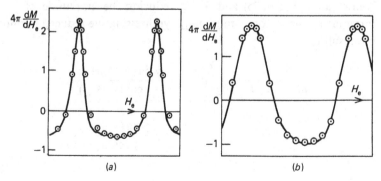

6.6. Phase smearing in MI conditions

$x_e = \pi$) for strong MI should be

$$Y_b = \frac{2a_0}{\mu(1 + a_0(1 - n))} \tan^{-1} \frac{\gamma \sin \mu}{1 + \gamma \cos \mu} \tag{6.79}$$

and if μ is small enough (for the Au rod with $n = 0.075$ and $a_0 = 5.34$ we have $\mu = 0.23$), this reduces to

$$Y_b = \frac{2a_0\gamma}{(1 + a_0(1 - n))(1 + \gamma)} \tag{6.80}$$

which is smaller than the prediction (6.60) by the factor $(2/(1 + \gamma)) \times (1 + a(1 - n))/(1 + a_0(1 - n))$. This becomes 1 for $\gamma = 1$, as it obviously should, and becomes smaller as γ decreases; for the conditions of the Au rod it is about 0.8, but (for the same a_0) it would be only about 0.5 for $\gamma = 0.1$. The form of the linear plot predicted by (6.61) should be valid only in the limit of weak MI, i.e. at high temperatures, while for the opposite limit of strong MI it should be replaced by

$$\frac{1}{Y_b'} = \frac{c}{a(T_0)} \frac{(1 + \gamma)}{2} \frac{a(T_0)}{a(T)} + \frac{c(1 + \gamma)}{2\gamma}(1 - n), \tag{6.81}$$

which has a lower slope and a higher intercept. The experimental uncertainty in the position of the baseline from which Y_b' is measured makes it difficult to confirm that there is really a change of slope and intercept between the low and high temperature regions of the plot, especially since, for the data of fig. 6.12, (6.81) should be valid only for the lowest two or three points. However within the limits of the baseline uncertainty it is quite possible to reinterpret the data consistently with the values of a and γ used in fitting the line shape of fig. 6.17(a); this is indicated schematically by the broken lines of fig. 6.12 which represent the two limiting formulae (6.61) and (6.81). The data of Shoenberg and Templeton (1968) on a Ag rod for which a_0 may have been as high as 20, can also be reinterpreted in this way if a somewhat different criterion is used for determining the base line.

As already mentioned, phase smearing due to inhomogeneity of the field rather than of the sample is a difficult problem in strong MI conditions, and we shall not attempt to discuss it beyond the qualitative remark that probably the effect of slight field inhomogeneity is roughly similar to that of sample inhomogeneity. One other cause of phase smearing is non-ellipsoidal sample shape. This however causes phase smearing only when MI is appreciable (since in the absence of MI sample shape has no effect at all) so that the reduction factor γ becomes different at high and low temperatures (more reduction at low temperatures when B becomes inhomogeneous because of shape).

6.7. Magnetic interaction for several de Haas–van Alphen frequencies

If, as often happens, the oscillations contain a number of frequencies, each oscillatory term in M of a particular frequency is periodic in $1/B$ and so involves all the other frequencies. Thus for a homogeneous sample

$$4\pi M = 4\pi \sum M_r = \sum A_r \sin\left(\frac{2\pi F_r}{B} + \phi_r\right) \tag{6.82}$$

and

$$B = H_e + 4\pi(1 - n)\sum M_r \tag{6.83}$$

The solution of these equations to give M as a function of H_e can always be computed by the 'graphical' method mentioned in §6.3. Thus if the A_rs and ϕ_rs are known, $4\pi M$ can be computed as a function of B and the corresponding value of H_e obtained by subtracting $4\pi M$ from B. Just as for a single frequency, if the MI becomes strong enough, only parts of the oscillation cycles are stable and criteria for the regions of domain formation can also be established. However the general case of many frequencies cannot be usefully pursued very far analytically and we shall consider only a few relatively simple special cases which relate to practical situations and serve to illustrate some of the interesting effects that can occur.

Most of these special cases are for the interaction of just two frequencies (Shoenberg 1968) and we shall discuss the amplitude and frequency modulation of a high frequency by a low one, the generation of a difference frequency by the interaction of two nearly equal frequencies and the apparent suppression of a relatively weak oscillation by a strong one of lower frequency. Finally we shall consider the mutual MI of the fundamental of a de Haas–van Alphen frequency with all its harmonics as given by the LK formula (Phillips and Gold 1969). The complicating effects of magnetic anisotropy, shape and phase smearing will be considered at appropriate places.

6.7.1. General formulation for MI with two frequencies

Instead of (6.8) we now have

$$\left.\begin{array}{l} 4\pi M = A \sin kb + A' \sin k'b \\ b = h + 4\pi M, \end{array}\right\} \tag{6.84}$$

but if we wish to treat the two frequencies on an equal footing it is no longer convenient to use a reduced notation. The solution of this implicit equation for the case of weak or moderate MI (i.e. $(kA + k'A') < 1$) is most compactly expressed* in terms of the differential susceptibility $4\pi\,\mathrm{d}M/\mathrm{d}H$ as

$$4\pi\,\mathrm{d}M/\mathrm{d}H = -1 + \sum_{i=-\infty}^{\infty}\sum_{j=-\infty}^{\infty} J_i\big(A(ik + jk')\big)$$

$$\times\; J_j\big(A'(ik + jk')\big)\cos\big((ik + jk')h\big)$$

(6.85)

In (6.85) the -1 on the r.h.s. cancels the term $i = j = 0$ and it should also be noted that each term of the form $\cos((ik + jk')h)$ is repeated as $\cos((-ik - jk')h)$. The result is easily generalized to take account of sample shape by adding a factor $(1 - n)$ to $\mathrm{d}M/\mathrm{d}H$, M, A and A' and replacing H by H_e and h by h_e. Another generalization which will be useful in discussing phase smearing is to add phase constants ϕ and ϕ' to kb and $k'b$ respectively in (6.84); (6.85) is then modified by the addition of a phase constant $(i\phi + j\phi')$.

If we examine the form of (6.85) we see that it contains not only the original frequencies k and k' and all their harmonics, but also all the combination and difference frequencies of k and k' and their harmonics. The general formula is useful as a starting point for a detailed investigation of any particular situation but often, and this will be so for the special situations we shall consider, only a few of the terms have any appreciable amplitude. The physics is then made more transparent by introducing the special features of the situation at the start, rather than using the general solution, thus simplifying the analysis.

The general formulation for strong MI is even more difficult since it involves the question of stability against domain formation and we shall not attempt it, but consider only the simpler special situations. For both weak and strong MI in the region where there is no domain formation one useful result is obtained immediately by differentiating (6.84). This is

$$4\pi\,\mathrm{d}M/\mathrm{d}H = \frac{4\pi\,\mathrm{d}M/\mathrm{d}B}{1 - 4\pi\,\mathrm{d}M/\mathrm{d}B} = \frac{a\cos kb + a'\cos k'b}{1 - (a\cos kb + a'\cos k'b)} \quad (6.86)$$

where

$$a = Ak \quad \text{and} \quad a' = A'k'.$$

---

* I am grateful to Professor S. Nakajima for showing me this result in 1965. As quoted in equation (15) of Shoenberg (1968) the symmetry of k and k', evident in (6.85), is somewhat masked by the unsymmetrical reduced notation which was used.

This is the analogue of (6.22) and is useful in calculating peak to peak excursions for weak MI, while for both weak and strong MI it gives the depth of the flat bottom of the dominant oscillation.

6.7.1.1. MI of a high and a low frequency

Where one of the two frequencies is much lower than the other and has sufficiently low amplitude–as in the illustration of fig. 6.10–the treatment can be much simplified and in order to make its significance more transparent we shall somewhat oversimplify it by omitting higher order terms which do not change the qualitative character of the results.

Let us suppose that in (6.84) $k' \ll k$ and that *both Ak and $A'k$* are small ($\ll 1$); we can then expand (6.84) and retaining only the lowest order terms of interest, we find

$$4\pi M = A\{\sin kh + \tfrac{1}{2}A'(k + k')\sin(k + k')h$$
$$- \tfrac{1}{2}A'(k - k')\sin(k - k')h\} \qquad (6.87)$$

This expression, in which all LF terms, all harmonics of k and also all higher side-band frequencies $k \pm rk'$ have been omitted, can of course also be derived from the general formula (6.85). The terms $\tfrac{1}{2}A'k(\sin(k + k')h - \sin(k - k')h)$ are just the first order side bands of a weak* frequency modulation of the form $\sin(kh + A'k\sin k'h)$, i.e. frequency modulation of the HF by the LF, while the small extra terms $\tfrac{1}{2}A'k'(\sin(k + k')h + \sin(k - k')h)$, which come from the HF contribution to the phase of the LF term in (6.84), represent a weak amplitude modulation.

The strength of modulation of frequency or amplitude is conveniently measured by the parameters

$$m_f = (f_{max} - f_{min})/(f_{max} + f_{min}) \qquad (6.88)$$

$$\mu_a = (|M|_{max} - |M|_{min})/(|M|_{max} + |M|_{min}) \qquad (6.89)$$

$$m_a = (|dM/dH|_{max} - |dM/dH|_{min})/(|dM/dH|_{max} + |dM/dH|_{min}), \qquad (6.90)$$

where the frequency f is defined as the reciprocal of an individual HF period, and $|M|$ and $|dM/dH|$ refer to just the HF part and are defined as half the peak-to-peak excursion in an oscillation. It can easily be seen from (6.87) that we should have

$$m_f = \mu_a = \tfrac{1}{2}m_a = A'k' \equiv a' \qquad \textbf{(6.91)}$$

---

* For stronger modulation, $A'k$ in (6.87) must be replaced by $2J_1(A'k)$ and of couse the higher side bands of (6.85) may become significant.

292

6.7.   *MI for several dHvA frequencies*

The amplitude modulation of dM/dH (i.e. m_a) is of course stronger than that of M (i.e. μ_a) because the HF frequency varies together with the HF amplitude as we go through an LF cycle.

These predictions proved to be only partially true. In the experiments of Shoenberg and Vuillemin (1966) a frequency modulation was indeed observed, but it was about 30% stronger than the directly measured amplitude of $4\pi\,dM'/dH$ (i.e. $A'k'$) and, even more strikingly, the observed amplitude modulation was much stronger than the prediction of (6.91). The theory might be expected to apply, at least roughly, to the higher temperature dM/dH oscillations in fig. 6.10, for which the directly measured value of $A'k'$ was about 0.02, yet the modulation m_a was something like 0.25, or six times stronger than predicted by (6.91).

The explanation of these discrepancies almost certainly lies in the inhomogeneity of the sample (and perhaps also of the field). The observed stronger frequency modulation suggested that the measured frequency modulation characterizes some kind of 'internal' value of $4\pi\,dM'/dH$ within each homogeneous sub-unit of the sample, while the measured amplitude of $4\pi\,dM'/dH$ is a phase smeared average. As we shall show, development of this idea does lead to a plausible explanation of the amplitude modulation, if it is supposed that the phases of the HF and LF are to some extent correlated as we go from one part of the sample to another.

Let us then modify (6.84) in the spirit of the new procedure of §6.6, i.e. by phase smearing *after*, rather than *before*, MI. To do this we introduce phase constants ϕ and ϕ', which we shall suppose are constants within any small homogeneous region of the sample, but vary from region to region. Within each region the amplitudes of the two frequencies will have the ideal LK values A_0 and A'_0. Thus the magnetization M_0 within such a region will be

$$4\pi M_0 = A_0 \sin(kb + \phi) + A'_0 \sin(k'b + \phi') \tag{6.92}$$

Making the same approximations as before we find, instead of (6.87)

$$4\pi M_0 = A_0\{\sin(kh + \phi) + \tfrac{1}{2}A'_0(k + k')\sin[(k + k')h + \phi + \phi']$$
$$- \tfrac{1}{2}A'_0(k - k')\sin[(k - k')h + \phi - \phi']\} \tag{6.93}$$

and we are now ready to consider the effects of phase smearing, i.e. averaging over the phases ϕ and ϕ'. Everything now depends on how these phases are correlated and we start by considering two extreme situations, the one of complete correlation and the other of complete *lack* of correlation. For complete correlation we would have simply

$$\phi' = \lambda\phi \tag{6.94}$$

293

Magnetic interaction

where λ is an appropriate constant which may be positive or negative. If phase smearing reduces the HF and LF amplitudes in the absence of MI by factors u and v respectively (these factors can be estimated from Dingle plots at relatively high tempertures), then for a Lorentzian distribution of ϕ or ϕ' we have

$$v = u^{|\lambda|} \qquad (6.95)$$

The effect of phase smearing on (6.93) is to reduce the first term by a factor u, the second by a factor $u^{|1+\lambda|}$ and the third by a factor $u^{|1-\lambda|}$. The values of these last two factors in terms of u and v depend both on the sign of λ and on whether λ is smaller or greater than 1, as indicated in table 6.1. If we introduce the notation

$$p = u^{|1+\lambda|}/uv, \qquad q = u^{|1-\lambda|}/uv \qquad (6.96)$$

then p and q have the values* shown in table 6.1, and it is easily shown that phase smearing reduces (6.93) to

$$4\pi M = A\{\sin kh + \tfrac{1}{2}pA'(k + k')\sin(k + k')h$$
$$- \tfrac{1}{2}qA'(k - k')\sin(k - k')h\} \qquad (6.97)$$

where

$$A = A_0 u, \qquad A' = A'_0 v \qquad (6.98)$$

The presence of p and q in (6.97) as compared with (6.87) can drastically change the amplitude modulation m_a and also appreciably change the frequency modulation m_f from the predictions of (6.91). Thus it is easily

Table 6.1. *Values of side-band reduction factors*

	$u^{|1+\lambda|}$	p	$u^{|1-\lambda|}$	q
$1 > \lambda > 0$	uv	1	u/v	$1/v^2$
$\lambda > 1$	uv	1	v/u	$1/u^2$
$-1 < \lambda < 0$	u/v	$1/v^2$	uv	1
$\lambda < -1$	v/u	$1/u^2$	uv	1

* The quantities p and q in Shoenberg (1976) are somewhat differently defined, being just v times those of (6.96).

294

shown that (6.97) gives

$$m_f = \tfrac{1}{2} A' k' \left(p \left(1 + \frac{k'}{k} \right) + q \left(1 - \frac{k'}{k} \right) \right) \tag{6.99}$$

and

$$m_a = \tfrac{1}{2} A' k \left(p \left(1 + \frac{2k'}{k} \right) - q \left(1 - \frac{2k'}{k} \right) \right) \tag{6.100}*$$

These reduce to (6.91) if $p = q = 1$, in which case m_a is the difference of two nearly equal quantities, but as soon as p differs appreciably from q the magnitude of m_a is greatly increased, and it may become negative if $p < q$.

Let us now consider the opposite extreme case in which there is no correlation at all between ϕ and ϕ'. The phase smearings over ϕ and ϕ' are then quite independent and their effects are multiplicative, so that both the second and third terms of (6.93) are multiplied by the same reduction factor uv and (6.93) reduces to just the same result, (6.87), as if we had introduced MI after rather than before phase smearing† or as if we had put $p = q = 1$. Thus completely uncorrelated phase smearing predicts just the modulations (6.91), which as we have already mentioned, do not agree with experiment.

We have already seen that in the Shoenberg and Vuillemin experiments m_f was appreciably bigger and m_a considerably bigger than predicted by (6.91), i.e. by the conventional treatment, or within our approximations, by the assumption of uncorrelated phase smearing, but since no measurements were made at high temperatures to determine u and v and, moreover, the field homogeneity was probably not adequate, these experiments are not very suitable for testing the theory. More direct checks were possible in some experiments by Bibby (1976) on Au spheres and these are summarized in table 6.2.

In these experiments the temperature was deliberately made high enough to ensure that our neglect of higher order terms was a reasonable approximation. The price to be paid for this simplification was that the frequency modulation became very weak and a special trick was necessary

---

* Note that as defined in (6.90), m_a is always positive, but when there is some phase correlation it is convenient to attach a sign to m_a according as the maximum of $|dM/dH|$ occurs at the maximum of the LF oscillation of dM'/dH (positive sign) or at the minimum (negative sign). This convention has been implicitly assumed in the derivation of (6.100).

† Note however that this is only exactly true with our crude approximation of ignoring the higher order side bands. The more general case is briefly discussed in the Appendix to Shoenberg (1976).

to measure m_f at all accurately. This was to measure carefully the field modulation amplitude required to make the belly signal vanish (at a 'Bessel zero', see §3.4.3.1); since this modulation amplitude is directly proportional to a dHvA period and can be rather precisely measured, it provides a sensitive means of following the small ($\sim 1\%$) changes of the dHvA frequency as we go through the neck cycle (see Appendix 13 for details of this method).

It can be seen from table 6.2 that the observed values of m_f are always larger than predicted by the uncorrelated assumption and sometimes are quite close to the correlated prediction. The observed values of m_a, however, though always much larger than those predicted by the uncorrelated assumption all fall a good deal short of the fully correlated prediction. Probably this implies that the correlation between ϕ and ϕ' is only partial in the sense that the correlation factor λ varies appreciably from point to point in the sample. However, it must be remembered that our theoretical treatment is based on many simplifying assumptions, some of which may be too crude, as for instance the assumption of an exactly Lorentzian

Table 6.2. *Frequency and amplitude modulation*

		Sample A $\langle 111 \rangle$ 3 K, 68 kG	Sample B $\langle 111 \rangle$ 3 K, 68 kG	Sample B 6° off $\langle 111 \rangle$ 2.17 K, 67.3 kG
$Ak(1-n) \times 10^3$		48	78	6.4
$A'k'(1-n) \times 10^3$		5.2	9.0	9.3
u		0.34	0.56	0.064
v		0.39	0.68	0.60
λ		-0.87	-0.67	0.19
p, q correlated		6.6, 1	2.2, 1	1, 2.8
p, q uncorrelated		1, 1	1, 1	1, 1
m_f	correlated	0.02	0.014	0.017
	uncorrelated	0.0052	0.0090	0.0093
	observed	0.0079	0.0147	0.0156
m_a	correlated	0.46	0.18	-0.21
	uncorrelated	0.010	0.018	0.019
	observed	0.083	0.07	-0.040

Notes: k'/k is 0.0341 for the first two columns, but 0.0346 for the third. The sign of λ is inferred according as the maximum belly amplitude $|\mathrm{d}M/\mathrm{d}H|_{max}$ coincides with the maximum or minimum of the neck oscillation of $\mathrm{d}M/\mathrm{d}H_e$ (λ negative or positive respectively). The values of λ are calculated from (6.95). The predicted values of m_f and m_a are calculated from (6.99) and (6.100) using the appropriate values of p and q (taken from table 6.1) and putting in the factor $(1-n)$, i.e. $\frac{2}{3}$, to allow for the spherical shape. Note that p and q are differently defined in Shoenberg (1976).

distribution of phases, so it is difficult to assess the quantitative reliability of the formulae.

Some clues to the mechanism of the phase smearing are provided by the magnitude and sign of λ as determined from the reduction factors u and v using (6.95) and from the phase of the amplitude modulation in relation to the neck (LF) cycle. If the phase smearing is due to variable orientation, i.e. mosaic structure around $\langle 111 \rangle$, we should have $\lambda \sim 0.1$, if due to field inhomogeneity (though of course in that case a Lorentz distribution is not appropriate), $\lambda \sim 0.03$, if due to variable tensions (along $\langle 111 \rangle$), $\lambda \sim -0.3$ and if due to pure shear (i.e. the effect of a tension along $\langle 111 \rangle$ less the effect of the corresponding change or volume) $\lambda \sim -1$. Thus the high and negative values of λ for the two $\langle 111 \rangle$ samples suggest that here the main cause of the phase smearing is a variable strain, though the evidence is too slight to point to any particular kind of strain. For the sample which is 6° off $\langle 111 \rangle$ the positive sign of λ suggests that the dominant phase smearing mechanism is that of variable orientation, though probably variable strain also contributes, and spoils the close correlation between ϕ and ϕ' that would be expected from orientation variation alone. The strain system is probably one associated with dislocations (see chapter 8) and it is plausible that the ϕ and ϕ' associated with strain alone should not be completely correlated, because the local value of λ depends on the precise nature of the local strain in relation to the crystal axes and this is unlikely to be the same everywhere.

Other aspects of the MI effects of an LF oscillation on an HF have been studied in Pb. In this system the LF γ oscillations are relatively much stronger than the neck oscillations in Au and moreover show a long beat (they consist of two slightly differing frequencies) so that the effect of LF amplitude is conveniently studied by merely varying the field through a beat cycle. On the other hand the ratio of frequencies of the HF α oscillations and the LF γs is only about 8.8 as compared with 30 for the belly neck ratio in Au, so some of our approximations based on assuming $k/k' \gg 1$ are less valid for Pb than for Au. Some ingenious experiments by Aoki and Ogawa (1978, 1979) on the α and γ oscillations of Pb have not only demonstrated the validity of the general formula (6.85) but also shown that their results require the MI to be introduced before rather than after phase smearing. In their experiments the oscillations were Fourier analysed so that the amplitudes of individual Fourier components in (6.85) and their dependence on the LF amplitude could be studied.

The most striking manifestation of MI was the demonstration that a sufficiently strong LF oscillation can practically annihilate the central HF frequency (though not of course the side bands). Thus if T_α and T_γ are the

Magnetic interaction

amplitudes of the pure k and k' components in the Fourier analysis of the $4\pi\,dM/dH$ oscillations, then (6.85) as it stands predicts*

$$T_\alpha(\text{old}) = |2J_1(Ak)\,J_0(A'k)| \tag{6.101}$$

$$T_\gamma(\text{old}) = |2J_0(Ak')\,J_1(A'k')| \tag{6.102}$$

If, however, the new treatment is applied, in which phases $i\phi + j\phi'$ are introduced into the combination tone of frequency $ik + jk'$ in (6.85), with A and A' replaced by A_0 and A'_0 and phase smearing only then applied, we find

$$T_\alpha(\text{new}) = |2uJ_1(A_0k)\,J_0(A'_0k)| \tag{6.103}$$

$$T_\gamma(\text{new}) = |2vJ_0(A_0k')\,J_1(A'_0k')| \tag{6.104}$$

where u and v are the reduction factors caused by phase smearing over ϕ and ϕ' respectively (i.e. such that $A = uA_0$ and $A' = vA'_0$). Note that the question of phase correlation does not arise here because the phase smearing in each case is over the single phase ϕ or ϕ', rather than a combination of them. In these experiments A_0k, A_0k' and A'_0k' (and *a fortiori* Ak, Ak', $A'k'$) but *not* A'_0k or $A'k$ were sufficiently small to permit retaining only the leading term in the relevant Bessel functions so the above equations reduce to

$$T_\alpha(\text{old}) = Ak|\,J_0(A'k)| \tag{6.105}$$

$$T_\alpha(\text{new}) = Ak|\,J_0(A'_0k)| \tag{6.106}$$

and

$$T_\gamma(\text{old}) = T_\gamma(\text{new}) = A'k' \tag{6.107}$$

Thus both the old and the new treatments predict that as A' or A'_0 varies through a beat cycle of the LF γ oscillations, T_α should decrease and pass through zero when the argument of J_0 becomes equal to 2.4 (the corresponding effect in Au is much less striking, since A'_0k for Au barely exceeds 1.5 at best). We notice that because $A'k'$ is small, MI does not affect the Fourier component of frequency k', so that T_γ should be just $A'k'$ on either treatment and hence directly measures A', since k' is of course known. As can be seen from fig. 6.18 something like the prediction of (6.105) or (6.106) does really happen, in as far as T_α passes through a deep minimum when plotted against T_γ over a beat cycle. Aoki and Ogawa also measured

* For simplicity, the factor $(1 - n)$ is omitted in all the analysis which follows, but it is taken into account in the comparison with experiment shown in fig. 6.18.

the absolute values of T_γ, i.e. of A', so it is possible to distinguish between the predictions of the old and the new treatments as regards the position of the minimum. As shown in fig. 6.18, the experimental results are much closer to the prediction of the new treatment. The fact that only a deep minimum rather than a zero occurs for T_α as a function of T_γ is probably mainly because at least five γ cycles had to be used in the Fourier analysis from which T_α and T_γ were extracted. This is an appreciable proportion of a beat cycle so that effectively the Fourier analysis gives a value of T_α corresponding to smearing $J_0(A'_0 k)$ over an appreciable range of A'_0. In any case it should not be forgotten that the new treatment is based on highly idealized assumptions and is unlikely to provide more than a good approximation to the truth.

A rather different kind of experiment illustrating the importance of shape effects in MI is that of Everett and Grenier (1977). They used a disc-shaped Pb sample prepared so that one of the $\langle 110 \rangle$ axes lay in a direction

Fig. 6.18. MI effect of low frequency (γ) on high frequency (α) oscillations in Pb $\langle 110 \rangle$ (after Aoki and Ogawa 1978 but with modified notation) at $T = 1.04\,\mathrm{K}$. The theoretical curves are calculated from (6.105) (broken curve) and (6.106) (full curve) assuming an absolute calibration of the T_γ scale which may not be very precise. The definition of T_γ is really $(1 - n)A'k'$ rather than simply $A'k'$ and the calibration assumes $n = 0.24$. The assumed value 0.54 of the Dingle reduction factor γ (needed to get A'_0 from the known A') may also be somewhat rough. For simplicity the calculated curves ignore the slight variation of field through the beat cycle of the γ oscillations, which goes from 38.8 kG at the node ($T_\gamma \sim 0.002$) to 40.7 kG at the antinode ($T_\gamma \sim 0.16$).

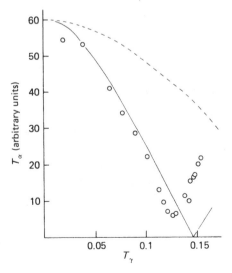

299

of small demagnetizing coefficient ($n = n_1$, parallel to the disc plane) and another $\langle 110 \rangle$ in a direction of much larger demagnetizing coefficient ($n = n_2$, normal to the disc plane). Because of the factor $(1 - n)$ which should multiply A and A' in all the analysis if n is not negligible, the frequency modulation m_f should be stronger in the small than in the large n direction and this they did indeed find. Unfortunately, however, their detailed results on m_f cannot be adequately interpreted with the simplifying assumptions which lead to (6.99) because the omitted higher order terms are not negligible. If the higher order terms are retained (as outlined in the appendix to Shoenberg 1976) a plausible interpretation in terms of the new treatment becomes possible, but rather *ad hoc* assumptions about the degree of phase correlation have to be made. It is in any case clear that their results cannot be explained by the old treatment, but it is difficult to evaluate how well the new treatment fits, in the absence of absolute amplitude measurements and independent information on the phase smearing reduction factors. Other systems in which modulation of an HF by an LF oscillation has been observed are the HF second zone oscillations and the LF third zone oscillations in Al (Anderson and Lane 1970) and In (van Weeren and Anderson 1973) but these have not yet been systematically studied and it is not certain that MI is the only relevant mechanism responsible for the modulation.

Field inhomogeneity has been mentioned several times as a cause of phase smearing, but we shall consider it only briefly since it is an instrumental rather than a fundamental effect. It differs from the other kinds of phase smearing mainly in that the smearing distribution function over the sample is not anything like Lorentzian. A simple situation which can be analyzed precisely is when the field gradient and the sample cross-section are both constant, with a field difference ΔH over the sample. The phase distribution is then of the 'top-hat' kind considered in §2.3.7.3 and the amplitude reduction factor for a single frequency F is that of (2.143), i.e.

$$R_{\Delta H} = \sin \lambda / \lambda \quad \text{with} \quad \lambda = \pi F \, \Delta H / H^2 \qquad (6.108)$$

Evidently, in the neighbourhood of $\lambda = \pi$ i.e. of $\Delta H = H^2/F$, where the amplitude becomes small, it will be very sensitive to small changes of the effective F. From (6.108) it follows that if F is modulated slightly with modulation parameter m_f (which is just $\delta F/F$ if $|\delta F|$ is the maximum departure of F from its mean), then the amplitude modulation m_a (from this cause by itself) will be approximately

$$m_a = \left| \frac{\lambda}{\tan \lambda} - 1 \right| m_f \qquad (6.109)$$

and we see that the amplitude modulation can be made to 'blow up' if the field inhomogeneity is such as to make λ approach π. Another instrumental effect which can magnify the apparent amplitude modulation is the use of large modulating fields in the field modulation technique, as discussed in Appendix 13. Both these instrumental effects have been studied experimentally by Hörnfeldt, Ketterson and Windmiller (1973) and Alles and Lowndes (1973) and found to be broadly in agreement with theoretical prediction. Such studies are of course useful in defining the experimental conditions required to avoid confusing more fundamental studies by such instrumental effects.

Finally we shall consider briefly what happens when the amplitude of the HF is *not* low, so that it shows strong 'self interaction' (as at low temperatures in Au, see fig. 6.10(a)). Qualitatively we would expect, as for the weak MI case, that amplitude modulation would be negligible without phase correlation, but a rigorous calculation leads to complicated expressions with no very transparent interpretation. It should be possible to obtain a computer solution of the basic implicit equation (6.84) which would show how the oscillations varied over an LF cycle taking account of the thermodynamic instability within each HF cycle. If however arbitrary phases ϕ and ϕ' are included in (6.84) and correlated smearing over these phases is to be carried out, the problem becomes much more difficult.

A rather simple-minded approach is to consider the LF as a weak perturbation of the solution of the strong MI problem for the HF alone, i.e. of the Fourier components of the triangular wave shape of $\tilde{M}$ given by (6.21). In this approach we modify

$$4\pi M_0 = \sum_p (2/pk)(-1)^p \sin pkh \tag{6.110}$$

which is the Fourier expansion of the HF magnetization (before phase smearing) in the limit $a \to \infty$, by adding the contribution to B of the LF magnetization. This gives

$$4\pi M_0 = \sum (2/pk)(-1)^p \sin\bigl(pk(h + A'\sin k'h)\bigr) \tag{6.111}$$

and if the LF is weak enough in the sense $pkA' \ll 1$ this approach, with correlated phase smearing, can be worked out exactly and does give strong amplitude modulation (in contrast to uncorrelated phase smearing). This calculation fails, however, to account adequately for a striking feature of the experimental observation in Au (see fig. 6.10(a)), namely that it is only the sharp peaks that are appreciably modulated by the LF while the flat bottoms stay almost constant. The calculation predicts that the flat bottoms should be modulated more weakly than the sharp peaks by a factor $(1 - u)/(1 + u)$ where u is the phase smearing reduction ratio for the HF. Since u was about 0.6, this factor was something like $\frac{1}{4}$ for the belly oscillations, while as can be seen from fig. 6.10(a) the relative variation in flat bottom level is at most 1/25 of that in the sharp peak. In fact, however, the perturbation calculation is unlikely to be quantitatively valid for the actual experimental conditions. Thus the criterion for 'weak' perturbation is $A'k \ll 1$, while in fact $A'k$ was about 0.9. Moreover we know from the high temperature (weak MI) results that only partial, rather than complete phase correlation actually occurs.

6.7.1.2. MI of two nearly equal frequencies

This problem aroused a good deal of interest soon after MI was first considered, because the experimental observations seemed completely at variance with theoretical prediction. The most puzzling feature of the experiments was the observation of a *difference* frequency of amplitude much stronger than predicted by a straightforward application of MI theory. Thus in the noble metals Joseph and Thorsen (1965) could sometimes observe *only* the low difference frequency coming from the nearly equal central and non-central bellies rather than the belly frequencies themselves. For Be, Plummer and Gordon (1964, 1966) observed a strong difference frequency between the central and non-central frequencies of the cigar, while Condon (1966) reported a particularly strong difference frequency when magnetothermal oscillations were observed. It turns out that some of these results are basically instrumental and the apparent disagreements with theory evaporate when the detailed experimental conditions are taken into account. We shall now outline the theoretical predictions, but limit ourselves for the moment to oscillations of M; magnetothermal oscillations will be discussed separately in §6.8.

Since k' is nearly equal to k it is convenient to replace kb and $k'b$ by $\bar{k}b + \frac{1}{2}\psi$ and $\bar{k}b - \frac{1}{2}\psi$ in (6.84), where

$$\psi = (k - k')b \simeq (k - k')h \quad \text{and} \quad \bar{k} = \frac{1}{2}(k + k') \tag{6.112}$$

i.e. ψ is the slowly varying phase of the beat cycle between k and k'. We can then write (6.84) as if there were only a single frequency $\bar{k}$, but a slowly varying amplitude and phase

$$4\pi M = C \sin(\bar{k}b + u) \tag{6.113}$$

where

$$C = (A^2 + A'^2 + 2AA' \cos \psi)^{1/2} \tag{6.114}$$

and

$$\tan u = \frac{A - A'}{A + A'} \tan \frac{\psi}{2}.$$ (6.115)

(note that for $A = A'$, $C = 2A \cos \frac{1}{2}\psi$, $u = 0$).

If we now make the very reasonable approximation that during any single cycle of oscillation u and C can be treated as constant, the problem becomes identical with the one of a single frequency which we have already discussed in §6.3. If Ck is comparable to 1, the line shape of the oscillation of M will become more and more distorted as C increases through the best cycle, but the $+$ and $-$ excursions of M will always remain symmetrical so that there should be no component of the difference frequency $k - k'$. This is still true when $Ck > 1$, even though parts of the oscillation become unstable.

For $Ck < 1$ there is no need for the approximation that C and u remain constant during a cycle and, as we saw earlier, (6.84) can be solved exactly in the form (6.85). The Fourier analysed oscillations now do show a small component of frequency $(k - k')$; if we integrate the relevant term in (6.85) we find $\bigl($treating $(k - k')A$ as small$\bigr)$

$$(4\pi M)_{k-k'} = -\tfrac{1}{2}(k - k')AA' \sin(k - k')h$$ (6.116)

The amplitude of this difference frequency is indeed small compared with the maximum amplitude $C_{\max}$; the ratio is

$$\frac{|M|_{k-k'}}{C_{\max}} = \tfrac{1}{2}(k - k')\frac{AA'}{A + A'}$$ (6.117)

This is approximately $\tfrac{1}{2}(k - k')A'$ if $A'/A \ll 1$, as in the noble metal situation.

A more detailed examination of the strong MI situation (Shoenberg 1968) shows that there too there should be a weak difference frequency component. The ratio of the amplitude of this component to the amplitude of the main component (defined as half the peak-to-peak of the triangular oscillation) is for $A'/A \ll 1$ approximately

$$\frac{|M|_{k-k'}}{|M|_{\text{strong MI}}} \simeq (k - k')A'/\pi$$ (6.118)

The ratios in (6.117) and (6.118) are of course reduced by a further factor $(k - k')/k$ if it is dM/dH rather than M which is considered.

It was at first puzzling, then, why Plummer and Gordon (1964) should have observed such a considerable amplitude of difference frequency superimposed on the beats of the two nearly equal frequencies of the Be cigar, especially since it was dM/dH that they measured. The probable explanation (Plummer and Gordon 1966, Knecht, Lonzarich, Perz and Shoenberg 1977) is that this was an instrumental effect, a subtle consequence of the role of eddy currents in the field modulation method. As long as the field modulating frequency ω is low enough to make the skin depth large compared to the sample dimensions, the output signal should be accurately proportional to dM/dH, but once the skin depth is comparable to or smaller than the sample dimensions, the relation between the output and dM/dH is no longer linear if $4\pi\,dM/dH \gtrsim 1$, i.e. just for the conditions of MI. For strong MI in this regime, the sharp peaks of the $4\pi\,dM/dH$ cycles show up relatively more strongly in the output signal than do the flat bottoms, and so the apparent base line of the oscillations is lifted by an amount which varies through the beat cycle and gives the appearance of a relatively strong difference frequency component.

The measurements of Joseph and Thorsen (1965) which showed a strong difference frequency coming from the central and non-central bellies of the noble metals, were made by the torque method and are explainable in rather a different way (for a discussion see Shoenberg 1968, and Shoenberg and Templeton 1968). Thus the torque method measures the perpendicular component M_2 rather than the parallel component M_1 and we shall now show that in strong MI conditions the oscillations of M_2 do contain a much more appreciable difference frequency component than those of M_1.

As explained earlier, the MI problem for M_1 is not much affected by the existence of anisotropy, apart from the need to include an additional factor $(1 + \mu^2)$ in the basic equation, where $\mu = (1/F)(\partial F/\partial\theta)$ (see 6.33). If μ is small (as for the noble metal bellies) this factor can, however, be ignored and we shall do so for the sake of simplicity. We then have (6.113) for M_1 i.e.

$$4\pi M_1 = C_1 \sin(\bar{k}b + u_1) \tag{6.119}$$

with

$$C_1 = (A^2 + A'^2 + 2AA'\cos\psi)^{1/2}$$

and

$$\tan u_1 = \left(\frac{A - A'}{A + A'}\right)\tan\frac{\psi}{2} \tag{6.120}$$

For the perpendicular component M_2, however, the essential novelty here, as compared with (6.30), is that μ may be different for the two oscillations involved. For the central and non-central belly, the values μ and μ' can be considerably different, a typical value of μ'/μ being 2. It is this feature which

6.7. MI for several dHvA frequencies

leads to generation of a difference frequency in strong MI conditions. It is convenient to use the notation

$$\rho = \mu'/\mu \tag{6.121}$$

and we then have (dropping the bar over the k from now on)

$$4\pi M_2 = C_2 \sin(kb + u_2) \tag{6.122}$$

with

$$C_2 = \mu(A^2 + \rho^2 A'^2 + 2\rho AA' \cos\psi)^{1/2}$$

$$\tan u_2 = \left(\frac{A - \rho A'}{A + \rho A'}\right)\tan\frac{\psi}{2} \tag{6.123}$$

Just as before, the parallel component is the 'master' and the perpendicular component is the 'slave' i.e. the solution of (6.119) for b in terms of h must be substituted in (6.122) to give the behaviour of M_2 in terms of h.

Provided MI is not too strong (i.e. $C_1 k < 1$) it is evident that within the approximation of treating the Cs and us as constants during a single cycle, there can be no appreciable amplitude of difference frequency; thus, just as before, the $+$ and $-$ excursions of M_2 are almost symmetrical. The situation changes however when $C_1 k > 1$, since then only part of the cycle of (6.119) (or (6.122)) is stable. For simplicity we consider only the case of very strong MI ($C_1 k \gg 1$), when the solutions of (6.119) for M_1 and for kb are as shown in figs. 6.1 and 6.2. Explicitly we have

$$4\pi M_1 = \frac{C_1 k}{1 + C_1 k}[\pi - (kh + u_1)] \tag{6.124}$$

and

$$kb + u_1 = \pi - [\pi - (kh + u_1)]/(1 + C_1 k) \tag{6.125}$$

between $kh + u_1 = 0$ and 2π (our formulation is for only this one cycle and ignores demagnetizing effects), followed by a jump in $4\pi k M_1$ of $2\pi C_1 k/(1 + C_1 k)$ to start the next cycle. If we substitute (6.125) into (6.122) and bear in mind that $kb + u_1$ is close to π over the whole range of $kh + u_1$, we find

$$4\pi M_2 = C_2 \left\{ \frac{\pi - (kh + u_1)}{1 + C_1 k} \cos(u_2 - u_1) - \sin(u_2 - u_1) \right\} \tag{6.126}$$

If, moreover, $u_2 - u_1 \ll 1$, which is true if $A'/A \ll 1$, as in the noble metal

305

Magnetic interaction

situation, this is approximately equivalent to

$$4\pi M_2 = C_2\left\{\frac{\pi-(kh+u_1)}{1+C_1k}+(\rho-1)\frac{A'}{A}\sin\psi\right\} \tag{6.127}$$

i.e. a scaled version of the M_1 oscillations (but with beats given approximately by C_2/C_1) riding on top of the $\sin\psi$ difference frequency term. It should be noticed that the difference frequency oscillation is in quadrature to the beat pattern, i.e. it goes through its zero when the beat has a node or an antinode; this is in contrast to the difference frequency generated by the eddy current effect, where the difference frequency is 'in phase' with the beat pattern. The ratio of the difference frequency amplitude to that of the triangular oscillations (we define amplitude here as half the peak-to-peak) is approximately

$$|M_2|_{k-k'}/|M_2|_k \simeq (\rho-1)A'k/\pi \tag{6.128}$$

(since $C_1\simeq A$ if $A'/A\ll 1$). This ratio, though not very large in itself (for Ag at 50 kG and 1 K, $\rho-1\sim 1$, $A'k\sim 1$ so the ratio ~ 0.3) is much larger than the predicted ratio for M_1, which is roughly $A'(k-k')/\pi$ (see (6.118)). The results of Joseph and Thorsen's torque measurement on the difference frequency in Ag were not presented in sufficient detail to compare with the above estimates but various clues suggest that the amplitudes they observed were sometimes considerably larger than could be expected from (6.128).

As suggested by Vanderkooy and Datars (1968) this may well have been due to the 'torque interaction' (TI) effect, essentially a purely instrumental effect, which was mentioned in §3.3.2 and which we shall now briefly discuss. Because of the finite twist of the measuring system when a torque is applied to it and because of the variation of dHvA frequency F with twist, there is a 'feedback' effect and the torque T is given by the implicit equation

$$T = \gamma\sin\left(\frac{2\pi F}{H}+\phi+\frac{2\pi}{H}\frac{dF}{d\theta}\frac{T}{c}\right), \tag{6.129}$$

where c is the torque per unit angle of twist. This is in fact identical in form with the MI equation, as is evident if we introduce an interaction parameter p defined as in (3.6), by

$$p = \frac{2\pi\gamma}{c}\frac{F}{H}\left(\frac{1}{F}\frac{dF}{d\theta}\right) \tag{6.130}$$

and also the dimensionless parameters

$$y = -\frac{2\pi}{c}\frac{F}{H}\left(\frac{1}{F}\frac{dF}{d\theta}\right)T, \qquad x = kh \tag{6.131}$$

306

6.7. MI for several dHvA frequencies

The reduced form of (6.129) is then

$$y = p\sin(x + y) \tag{6.132}$$

just as for MI.

Just as with MI, the resulting line shape of the oscillations of y (i.e of torque) become more and more distorted as p increases, but for $p \geqslant 1$, TI behaves quite differently from MI. In both TI and MI the curve of y versus x becomes multivalued for p or $a \geqslant 1$, but for TI there is no way of breaking through to a configuration of lower free energy until the top (or bottom) of the cycle has been passed. As shown in fig. 6.19(a) this leads to considerable offset of the torque oscillations and considerable hysteresis effects. The significant aspect in our present context is that the positive and negative excursions of T are unequal and the mean level depends strongly on the magnitude of p, i.e. on the torque amplitude. Thus as we go through a beat cycle and p varies, the mean level will oscillate at the difference frequency provided $p > 1$ over much of the cycle (see fig. 6.19(b)). The detailed theory of TI is, however, more complicated (Vanderkooy and Datars 1968) and cannot be described simply by variation of a simple parameter p through the beat cycle; this is because $(1/F)\,dF/d\theta$ may be different for the two beating frequencies and an illustration of how this affects the oscillation is shown in fig. 6.19(c). A further complication occurs if the field is swept too rapidly and there is an appreciable time lag between torque and angular response. The exact conditions of Joseph and Thorsen's torque measurements are not known but from internal evidence it seems likely that the difference frequency they observed was generated predominantly by TI rather than MI.

An attempt was made by Shoenberg and Templeton (1968) to check the theory of how a genuine difference frequency is generated in M_2, i.e. to check (6.127) by a method other than that of torque measurement, and rough confirmation was indeed obtained by the field modulation method with a pick-up coil perpendicular to H. The confirmation was only rough, however, because the output signal is proportional to dM_2/dH rather than M_2 and the signal corresponding to the low difference frequency was very weak. In this experiment it was possible also to observe the difference frequency component in M_1 and to test the validity of (6.118). Although once again the measurements were rather rough, the feeble observed amplitudes were usually several times larger than those predicted by (6.118). This discrepancy might be partly a consequence of the eddy current effect mentioned above, if the modulation frequency was not quite low enough. It might also be partly due to strongly correlated phase smearing, which could reduce the damping of the difference frequency as compared

Magnetic interaction

Fig. 6.19. Torque interaction. (*a*) A single frequency with $p = 5$; the computer-plotted points are solutions of $y = 5 \sin(x + y)$, where x is the reduced field and y the reduced torque. For increasing x the system becomes unstable at points such as P and 'snaps' to the next higher branch, while for decreasing x the snaps occur at points such as Q. For $p \gg 1$ the mean position of the sawtooth oscillations is $p(2p - \pi)/(2p + \pi)$ below or above the line of zero torque, according as x is increasing or decreasing, (*b*) Two beating frequencies with the same sensitivity to angular change; the points are solutions of $y = 5 \sin(x + y) + \sin[1.1(x + y)]$. The mean points of the snaps now generate a difference frequency oscillation (broken lines) with a period of 20π. (*c*) As (*b*) but with an enhanced angular sensitivity of the subsidiary (higher) frequency; the points are solutions of $y = 5 \sin(x + y) + \sin(1.1x + 2y)$ and it can be seen that the difference frequency oscillations in this more realistic case have a more complicated form. The computing was kindly carried out by Dr. G. G. Lonzarich.

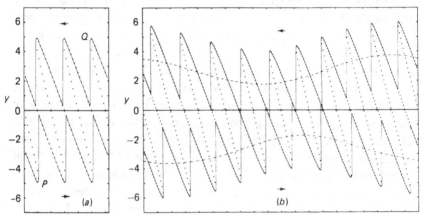

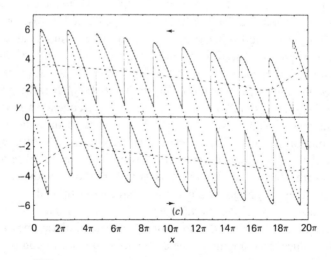

308

with that of the mean frequency. However neither of these explanations can account very convincingly for the magnitude of the discrepancy and further experiments would be desirable.

6.7.1.3. Modification of strong MI in one frequency by a weak second higher frequency

For the $\langle 111 \rangle$ direction in the noble metals the large amplitude belly oscillations should be accompanied by weaker six-cornered rosette oscillations of about 1.9 times the belly frequency, but the existence of these higher frequency oscillations has never been very convincingly demonstrated, except perhaps in Ag (see below). The difficulty in observing these oscillations is that they have appreciable amplitude only in conditions such that the dominant belly oscillations show strong MI and they might then be overshadowed by the strong belly harmonic (i.e. the $2k$ frequency). We shall now show that with strong MI in the dominant frequency a weaker additional high frequency is in fact very much obscured and should show up only in a careful Fourier analysis of the oscillations.

Starting as before from (6.84), i.e.

$$4\pi M = A \sin kb + A' \sin k'b \tag{6.133}$$

where now $k' > k$, and assuming that $Ak \gg 1$ but $A'k' \ll 1$, we can treat the A' term as a perturbation on the strong MI solution of the single frequency problem, which is that kb differs little from π over the range $0 < kh < 2\pi$, so that

$$\sin kb \doteqdot \pi - kb$$

Thus (6.133) becomes

$$4\pi M = A(\pi - kb) + A' \sin k'b$$

and making use of

$$b = h + 4\pi M$$

we obtain

$$4\pi M = \frac{A}{1 + kA}(\pi - kh) + \frac{A'}{1 + kA} \sin k'b \tag{6.134}$$

and

$$b = \frac{h + A\pi}{1 + kA} + \frac{A' \sin k'b}{1 + kA} \tag{6.135}$$

Since $k'A' \ll 1$, we can to a first approximation neglect the second term of

(6.135) compared with the first when we substitute into the second term of (6.134) and we find for the range $0 < kh < 2\pi$

$$4\pi M = \frac{A}{(1 + kA)}(\pi - kh) + \frac{A'}{(1 + kA)}\sin\left(\frac{k'h}{1 + kA} + \psi\right) \quad (6.136)$$

where

$$\psi = k'A\pi/(1 + kA)$$

We see then that the strong MI of the dominant frequency has drastically reduced both the apparent amplitude and the apparent frequency of the subsidiary higher frequency oscillation (since $kA \gg 1$). Thus the subsidiary frequency would not be noticeable by casual inspection of the dominant triangular oscillations on which it rides in much diminished style. Indeed if $k'/k < 1 + kA$ (as would usually be the case in attempts to observe the six-cornered rosette), there would not be a complete subsidiary cycle in the course of a single dominant oscillation. This does not however mean that the subsidiary frequency is undetectable. A more detailed analysis (Shoenberg 1968) shows that the 'jump point' of the triangular cycle is slightly shifted by the presence of the subsidiary oscillation; instead of occurring whenever $kh = 2r\pi$, the jump occurs approximately for

$$kh = 2r\pi - k'A \sin k'h \quad (6.137)$$

and we have essentially a frequency modulation of the triangular dominant oscillations by the subsidiary ones. If we then Fourier analyse the triangular oscillations, the subsidiary oscillations should show up as combination frequency components. For the situations of interest we have in fact $k' = (2 - \varepsilon)k$ where $\varepsilon \ll 1$ and it can be shown that, in the Fourier analysis, each harmonic of frequency pk should have side bands of frequencies $(p \pm \varepsilon)k$ with amplitudes roughly $pA'/2A$ times that of the pk frequency; this gives amplitudes of approximately $A'/2Ak$ for all p. In particular, the side band $(2 - \varepsilon)k$, which is just the subsidiary frequency k', appears in the Fourier analysis with its original amplitude reduced by the large factor $2Ak$. Some evidence of such side bands was obtained in unpublished preliminary experiments by Lee, McEwen and Vanderkooy (quoted by Halse 1969, see p. 520) and the frequency of the six-cornered rosette obtained in this way agreed well with the value calculated from the known Fermi surface; however, various confirmatory checks were never carried out, and moreover no results were obtained for Cu and Au. Quite recently somewhat more detailed observations of a similar kind, but again only for Ag, have been made in the high field installation of the University of Amsterdam (L. W. Roeland, private communication).

6.7. MI for several dHvA frequencies

6.7.2. MI of a single frequency and its harmonics

In §6.3 we discussed the consequences of MI for an oscillation of a single frequency F, but ignored the fact that even without MI, F is accompanied by all its harmonics rF (it will be convenient sometimes to refer to these as 'Lifshitz–Kosevich' or 'LK' harmonics to distinguish them from the harmonics generated by MI). The neglect of the LK harmonics was quite justified in discussing the noble metal belly oscillations, since in typical practical conditions the LK harmonics are indeed very feeble. Often, however, this is not so and we shall now discuss the effect of MI on the combination of a single frequency F with all its LK harmonics. The problem then is to solve the implicit equation obtained if H in the periodic terms of (2.152) is replaced by B. If we write $B = B_0 + b$ and choose B_0 to make F/B_0 an integer, the modified LK formula (2.152) for the parallel component of oscillatory magnetization can be written as

$$4\pi M = -\sum_{r=1}^{\infty} A_r \sin\left(rkb + \frac{\pi}{4}\right) \tag{6.138}$$

where A_r is the coefficient of $\sin((2\pi rF/H) - (\pi/4))$ in (2.152)*.

For moderate and weak MI a solution of (6.138), i.e. a formula for M in terms of $h(=H - B_0)$, can be obtained by an iterative method of successive approximation due to Phillips and Gold (1969). The idea assumes a hierarchy of amplitudes based on the Dingle reduction factor γ. Thus A_r is proportional to γ^r and so is considered to be of the same order of magnitude as A_1^r; in this sense any term of the form $(A_i)^\alpha(A_j)^\beta$ is considered to be of order $i\alpha + j\beta$. In each iteration only terms up to a certain order r are included and these are then substituted into b to give the next approximation. Thus, using an obvious notation

$$4\pi M(1) = -A_1 \sin\left(kh + \frac{\pi}{4}\right), \tag{6.139}$$

$$4\pi M(2) = -A_1 \sin\left(k(h + 4\pi M(1)) + \frac{\pi}{4}\right)$$
$$-A_2 \sin\left(2kh + \frac{\pi}{4}\right)$$

* In our earlier discussion of a single frequency the phase constant could be omitted simply by choosing B_0 slightly differently but here, because of the harmonics, it is essential to include phase constants in one way or another. For simplicity we have omitted a $\pm$ in our formulation, thus implying that the Fermi surface is convex; if in fact it is concave the sign of $\frac{1}{4}\pi$ must be reversed throughout.

311

which reduces to

$$4\pi M(2) = -A_1 \sin\left(kh + \frac{\pi}{4}\right) + \left(\frac{A_1^2}{2\sqrt{2}} - A_2\right) \sin\left(2kh + \frac{\pi}{4}\right)$$

$$+ \frac{A_1^2}{2\sqrt{2}} \cos\left(2kh + \frac{\pi}{4}\right) \tag{6.140}$$

and so on. It can be seen that the coefficients of the harmonics rapidly become more complicated with successive iterations and Phillips and Gold went as far as $M(4)$, i.e. to fourth order terms in the sense explained above. A rather higher approximation was needed in the interpretation of some experiments on Na (Perz and Shoenberg 1976) and expressions for the coefficients of the first four harmonics up to the eighth order (in our special sense) were derived using a computer program. These complicated expressions will not be reproduced here, but we may mention that the formulation included allowance for a finite demagnetizing coefficient and also that the results are easily adapted to either the old or the new treatments of phase smearing. The main point of these rather cumbersome calculations was to permit interpretation of a Fourier analysis of the observed oscillations (i.e. of M or dM/dH regarded as a periodic function of h). As can be seen even from the relatively simple expression for $M(2)$, the phases of the harmonics are no longer simple if the LK harmonics are comparable with what we might call the MI harmonics. We shall return to this topic in chapter 9 when we come to consider how the spin-splitting factor g can be extracted from the Fourier components of the observed oscillations if the MI is not too strong, or from the detailed line shape of the rather exotic looking oscillations in strong MI conditions.

6.8. MI in oscillations of other thermodynamic properties

The relations derived in chapter 4 between oscillations in various thermodynamic properties (temperature, magnetostriction and elastic constants) and those of M still hold good in conditions of magnetic interaction. Thus, provided the MI problem has been solved to give M as a function of H (or of H_e if sample shape is relevant), the derivation of the oscillations of other thermodynamic properties is straightforward. The oscillations of ultrasonic velocity, however, require special consideration since, because of the high sound frequency, the oscillations of the relevant elastic constants with H prove to be significantly different from those of the static elastic constants.

6.8. Oscillations of other thermodynamic properties

6.8.1. MI in the magnetothermal effect

As we saw in §4.2.1, the oscillatory temperature variation $\widetilde{\Delta T}$ is given essentially by the temperature derivative of the integral of $\tilde{M}$ with respect to H (see (4.3)). For weak MI this implies that the harmonics of a single basic frequency are relatively about as strong in ΔT as in $\tilde{M}$ (since the weakening due to integration is compensated by the enhancement due to the temperature differentiation). However there are novel features in conditions of strong MI, such as a cusp-like line shape and a progressive weakening of the $\widetilde{\Delta T}$ oscillations as the interaction parameter increases beyond 1. If there are two nearly equal frequencies present, giving rise to beats, the difference frequency oscillations generated by MI, which as we saw in §6.7.1.2, are of only feeble amplitude in $\tilde{M}$, are much more significant in $\widetilde{\Delta T}$ and for strong MI eventually become dominant. In our analysis of these features we shall, for simplicity, ignore the complications of anisotropy and shape effects. Most of the theoretical results were first derived by Condon (1966).

6.8.1.1. Magnetothermal effect for a single frequency in MI conditions

For weak MI $(a < 1)$ we may use the explicit solution (6.17) and find from (4.3) that

$$\widetilde{\Delta T} = \left(-\frac{\partial \ln a}{\partial T} \right)\left(\frac{-T}{4\pi k^2 c} \sum_{r=1}^{\infty} \frac{2a}{r} J'_r(ra)\cos rkh \right) \tag{6.141}$$

Here the minus sign has been included in the first bracket to make it a positive quantity and the base line of $\widetilde{\Delta T}$ has been chosen to eliminate any steady contribution. If (6.141) is expanded in powers of a, and terms higher than a^3 can be neglected we find

$$\widetilde{\Delta T} = \left(-\frac{\partial \ln a}{\partial T} \right)\left(\frac{-Ta}{4\pi k^2 c} \right)[(1 - \tfrac{3}{8}a^2)\cos kh$$

$$+ \tfrac{1}{2}a\cos 2kh + \tfrac{3}{8}a^2 \cos 3kh] \tag{6.142}$$

i.e. to this approximation, nearly the same harmonic strengths occur in $\widetilde{\Delta T}$ as in $\tilde{M}$ (see (6.18)), but the phases are all shifted by 90°.

In the limit of strong MI, the oscillations of $\tilde{M}$ have triangular wave shape (see (6.21)) i.e.

$$4\pi kM = \frac{a}{a + 1}(\pi - kh) \tag{6.143}$$

313

Magnetic interaction

for $0 < kh < 2\pi$. Thus from (4.3), we find, if ΔT is measured from $h = 0$

$$\Delta T = \left(-\frac{\partial \ln a}{\partial T}\right)\left(\frac{Ta}{8\pi k^2 c(1+a)^2}\right)[\pi^2 - (kh - \pi)^2] \qquad \textbf{(6.144)}$$

and this variation is repeated periodically, as shown in fig. 6.20. It can be seen that as a increases, the amplitude of these cusp-shaped oscillations decreases because of the factor $a/(a+1)^2$. The half peak-to-peak amplitude is in fact $[\frac{1}{2}\pi/(1+a)]^2$ times what it would have been if MI had not intervened. Even without MI, the amplitude of $\tilde{\Delta T}$ falls as T is lowered because the factor $(-\partial \ln a/\partial T)$ falls off; the additional factor $[\frac{1}{2}\pi/(1+a)]^2$ makes it fall off even more rapidly.

The result (6.144) can also be derived directly from the reduced thermodynamic potential since

$$\Delta T = -\frac{T}{c}\Delta S = \frac{T}{c}\Delta\left(\frac{\partial\Omega}{\partial T}\right)_H = \frac{T}{4\pi k^2 c}\Delta\left(\frac{\partial Z}{\partial T}\right)_h$$

$$= \left(\frac{-T}{4\pi k^2 c}\frac{\partial \ln a}{\partial T}\right)\left(-a\Delta\frac{\partial Z}{\partial a}\right) \qquad \textbf{(6.145)}$$

where the Δs refer to changes with H or h from some reference field.

Fig. 6.20. Magnetothermal effect with MI; ΔT is the temperature difference measured from $h = 0$ and the constant of proportionality α is $-4\pi k^2 c/(\mathrm{d}\ln a/\mathrm{d}T)$. The curve for $a = 0.3$ is calculated from (6.142) (but with appropriately shifted base line); the curves for $a = 2$ and 10 are calculated from (6.144). Note that the zero of ΔT is different for each curve and that $x = kh$.

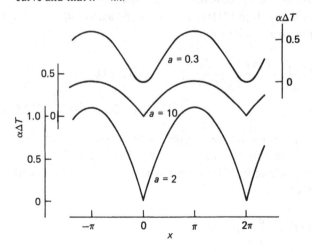

6.8. Oscillations of other thermodynamic properties

Substituting from (6.21) for Z in conditions of strong MI, we have

$$\Delta T = \left(\frac{-T}{4\pi k^2 c}\frac{\partial \ln a}{\partial T}\right)\left\{\frac{-a}{2(1+a)^2}(kh-(2r+1)\pi)^2\right.$$
$$\left. + a + a\left(\frac{\partial Z}{\partial a}\right)_{h=0}\cdot\right\} \qquad (6.146)$$

for each cycle such that $2r\pi \leqslant kh \leqslant (2r+2)\pi$ and ΔT is measured from $h = 0$. Since all the cycles are identical and $(\partial Z/\partial a)_{h=0} = \frac{1}{2}(\pi^2/(a+1)^2)-1$ (6.146) is identical with (6.144).

This alternative derivation emphasizes that the reduction of amplitude of $\widetilde{\Delta T}$ with increasing a for strong MI comes from the increasing gaps in the permitted ranges of b and the corresponding excision of parts of the cycle of Z. The purpose of the slightly awkward formulation of (6.146) is to permit discussion of a range of cycles such as occurs in the beats of two nearly equal frequencies.

6.8.1.2. Magnetothermal effect for two close frequencies in MI conditions

This is the situation already considered in §6.7.1.2 where k and k' in (6.84) are nearly equal and as in (6.113) it is convenient to write

$$4\pi M = A \sin kb + A' \sin k'b = C \sin(\bar{k}b + u) \qquad (6.147)$$

with C and u as specified in (6.114) and (6.115). As discussed earlier, if the MI is weak, there is only a feeble component of the $(k - k')$ difference frequency in $\tilde{M}$, given by (6.116). However because of the integration with respect to h, this $(k - k')$ frequency is much stronger in the oscillations of ΔT. Indeed it is easy to show that if $\partial \ln A/\partial T = \partial \ln A'/\partial T = \partial \ln a/\partial T$ (but this is only approximately valid for the beating frequencies of the Be cigar)

$$(\widetilde{\Delta T})_{k-k'} = \left(-\frac{\partial \ln a}{\partial T}\right)\left(\frac{TAA'}{4\pi c}\right)\cos(k-k')h \qquad (6.148)$$

where $a = Ck$ is the 'beating' amplitude. If we compare this with the leading term of (6.142), the weak MI result for a single frequency $\bar{k}$ of amplitude C, we see that the ratio

$$|\widetilde{\Delta T}|_{k-k'}/|\widetilde{\Delta T}|_k = kAA'/C \qquad (6.149)$$

This ratio varies from $kAA'/(A + A')$ at beat maximum to $kAA'/(A - A')$ and can become appreciable if MI is weak, but not too weak, i.e if kA and

315

kA' are comparable and not much less than 1. This situation is illustrated in Fig. 6.21(a).

If the MI is strong throughout the beat cycle, the $(k - k')$ difference frequency shows up even more strongly, as can be seen by adapting (6.146) to the situation of an amplitude a which varies gradually, i.e. for $a = Ck$ and $a \gg 1$ even for the beat minimum. The important differences are that the last term in (6.146) will now be a constant, depending on the field from which ΔT is reckoned, and that the phase constant u (which varies through the beat cycle) must be included. We thus obtain

$$\Delta T = \left(\frac{-T}{4\pi k^2 c} \frac{\partial \ln a}{\partial T} \right) \left\{ \frac{-a}{2(1 + a)^2} [kh + u - (2r + 1)\pi]^2 \right.$$

$$\left. + a - a_0 + \frac{1}{2} \frac{\pi^2 a_0}{(1 + a_0)^2} \right\} \tag{6.150}$$

for the range $2r\pi \leqslant kh + u \leqslant (2r + 2)\pi$, if a_0 is value of a at the reference field $h = 0$ (beat maximum) from which ΔT is measured.

The first term is essentially the same as the ΔT oscillation of a single frequency as in (6.144) (apart from the variation of phase and amplitude through the beat cycle), but the important novelty is the presence of the dominant second term, i.e. a, which, of course, varies with h through the beat cycle from $(A + A')k$ at the beat maximum to $(A - A')k$ at the beat minimum. The appearance of the oscillations if the MI is strong even at the beat minimum is shown in fig. 6.21(b).

If the value of a changes in the course of the beat from $a \gg 1$ (strong MI) to $a < 1$ (weak MI) the situation is rather more complicated, since the oscillations pass through the intermediate MI situation where explicit formulae are no longer available. A rough idea of what happens can, however, be obtained by extrapolating each formula a little beyond the range in which it is valid and joining the graphs smoothly together as is done in fig. 6.21(c). If necessary, of course, a more exact calculation could be carried out by computer.

A number of experiments have been carried out on the beating oscillations of the Be cigar and, as can be seen from fig. 6.22 due to Halloran and Hsu (1965) and quoted in more detail by Condon (1966), they do show most of the salient features that are predicted by the theory. These are the generation of a dominant difference frequency as MI gets stronger, the weakening of the basic frequency oscillations as the temperature is lowered beyond a certain point, the reversal of beat maximum and beat minimum when MI is strong over the whole beat cycle and the peculiar envelope shape when the beat carries the oscillation from weak to strong MI. It is

6.8. Oscillations of other thermodynamic properties

Fig. 6.21. Magnetothermal effect for two nearly equal frequencies with MI; the scale of ΔT is as in fig. 6.20. (a) Weak MI (high T); amplitudes $Ak = 0.5$, $A'k' = 0.3$. The curve is based on a simplified calculation in which only the k, k' and $k - k'$ frequencies are included. The dotted curve is the line of mid-points of the envelope (i.e. the difference frequency) and a few individual oscillations are indicated schematically. The beat period contains about 34 oscillations. (b) Strong MI (low T); $Ak = 5$, $A'k' = 3$. The individual oscillations are now cusp shaped (cf. fig. 6.20) and the amplitude is biggest at the beat minimum. (c) Fairly strong MI at beat maximum and fairly weak MI at beat minimum (medium T); $Ak = 1.25$, $A'k' = 0.75$. The full curves represent the appropriate limiting formulae, and the weak MI curve has been lowered to meet the strong MI curve in a reasonable way. In fact neither limiting formula is quite valid, so the diagram for these values of Ak and $A'k'$ is merely schematic. Note that the line shape of the oscillations changes from cusp-like at beat maximum to sinusoidal at beat minimum.

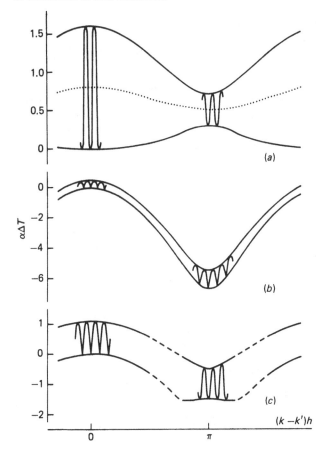

difficult on the evidence available to attempt any detailed quantitative comparison between experiment and theory, partly because the sample of fig. 6.22 was of non-ellipsoidal shape (it was a cube), partly because of phase smearing effects and partly because no magnetization data on the same samples in identical conditions are available. One feature of the theory which is not evident in the published data is the parabolic cusp-like line shape of the basic $\widetilde{\Delta T}$ oscillations predicted in conditions of strong MI. Further experiments to verify the theory in more detail would be desirable.

6.8.2. MI in oscillations of magnetostriction and elastic constants

The relation between magnetostriction and magnetization oscillations derived in chapter 4 is still valid in the presence of MI. This can be seen most simply from the general thermodynamic relation

$$(\partial \varepsilon_{ik}/\partial H_e) = (\partial M/\partial \sigma_{ik}) \qquad (6.151)$$

which is always valid, whatever the mechanism producing M. Thus if the

Fig. 6.22. Magnetothermal oscillations in Be at various temperatures (after Condon 1966). It can be seen that these experimental curves do show the main qualitative features of the calculated curves of fig. 6.21. The upward trend of the curves is due to magnetoresistance of the carbon thermometer and also to slow temperature drift; ΔT decreases upward.

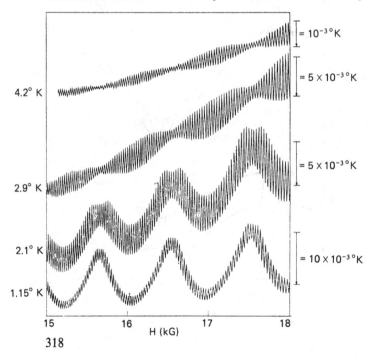

solution of the MI problem has the form (4.18), i.e.

$$\tilde{M} = \sum_r M_r \sin\left(\frac{2\pi F_r}{H_e} + \phi_r\right) = \sum \tilde{M}_r,$$

in which all the harmonics and combination frequencies of the basic frequencies are included, application of (6.151) with the usual assumption that $F_r/H_e \gg 1$ leads immediately to (4.20), i.e.

$$\tilde{\varepsilon}_{ik} = -\sum_r (\partial \ln F_r/\partial \sigma_{ik}) \tilde{M}_r H \tag{6.152}$$

(the distinction between H and H_e is of course unimportant in the final result). The stress derivative $(\partial \ln F_r/\partial \sigma_{ik})$ will of course be the same for all the harmonics of any particular 'basic' frequency, so the line shape of any basic frequency will be the same in the oscillations of magnetostriction and of M. Combination terms which involve two basic frequencies will, however, have different relative amplitudes because the stress derivatives of the F_rs involved will in general be different. These considerations were first discussed by Pudalov and Khaikin (1974) in connection with measurements of oscillatory magnetostriction in tin, where they observed moderately strong line shape distortion due to MI. Just as with the $\tilde{M}$ oscillations, the line shape of the magnetostriction oscillations of a single basic frequency can be used to estimate the absolute amplitude of the oscillations in $\tilde{M}$. If then the absolute amplitude of the magnetostriction oscillations is known, the stress derivative of F is determined from (6.152). In practice, as we saw earlier, this method of determining the absolute amplitude of $\tilde{M}$ is made somewhat uncertain by phase smearing due to sample inhomogeneity, unless the MI is very weak, when of course the accuracy is poorer.

Oscillations of the elastic constants should also obey the formulae of chapter 4 (i.e. (4.24) and (4.26)) in MI conditions, provided the stress does not vary too rapidly with time. This proviso is, however, *not* satisfied for the stress changes in an ultrasonic wave and so the oscillations in the velocity of sound in MI conditions require special consideration (Testardi and Condon 1970). The significant point is that if the stress or the strain changes too rapidly, eddy currents are induced in the sample and these tend to keep B constant as M changes. Thus, in the high frequency changes of an ultrasonic wave, the elastic constants should be calculated at constant B rather than, as was done in chapter 4, at constant H. If MI is negligible, the difference between B and H is also negligible and our previous results remain valid, but if MI is appreciable or strong, we need to think again.*

---

* To avoid undue complication we shall ignore the Alpher–Rubin effect mentioned in §4.3.3. This is in fact justified for propagation of longitudinal sound along a symmetry axis.

Magnetic interaction

Returning to the derivation of (4.26), the oscillatory dependence of the stress σ_{ik} for constant strain as the field is varied, is given by the analogue of the magnetostriction formula (6.152), i.e.

$$\tilde{\sigma}_{ik} = \sum_r \frac{\partial \ln F_r}{\partial \varepsilon_{ik}} \tilde{M}_r H \tag{6.153}$$

where, as before, the M_r are the various components of M as a function of H_e in the solution of the MI problem. The further analysis is straightforward only if there is a single basic frequency, so that the rs refer only to the harmonics of the basic frequency and $F_r = rF$. The derivatives $\partial \ln F_r / \partial \varepsilon_{ik}$ are then independent of r and we can write

$$\tilde{\sigma}_{ik} = \frac{\partial \ln F}{\partial \varepsilon_{ik}} \tilde{M} H \tag{6.154}$$

We must now differentiate (6.154) at constant B rather than constant H and find

$$\tilde{c}_{ikpq} = \left(\frac{\partial \tilde{\sigma}_{ik}}{\partial \varepsilon_{pq}}\right)_B = \left(\frac{\partial \ln F}{\partial \varepsilon_{ik}}\right)\left(\frac{\partial \tilde{M}}{\partial \varepsilon_{pq}}\right)_B H \tag{6.155}$$

and since $H = B - 4\pi M$ (we ignore shape effects for simplicity) it is easily shown that

$$(\partial \tilde{M}/\partial \varepsilon_{pq})_B = (\partial \tilde{M}/\varepsilon_{pq})_H/(dB/dH) \tag{6.156}$$

Moreover, on our assumption that $F_r = rF$, we have

$$(\partial \tilde{M}/\partial \varepsilon_{pq})_H = -H(\partial \ln F/\partial \varepsilon_{pq})(d\tilde{M}/dH)$$

so we finally obtain the modified form of (4.26)

$$\tilde{c}_{ikpq} = -\left(\frac{\partial \ln F}{\partial \varepsilon_{ik}}\right)\left(\frac{\partial \ln F}{\partial \varepsilon_{pq}}\right) H^2 \left(\frac{d\tilde{M}}{dB}\right) \tag{6.157}$$

In a series of elegant experiments, again on the Be cigar oscillations, Testardi and Condon (1970) arranged to measure $d\tilde{M}/dB$ directly (see §6.10) together with the oscillations of the 20 MHz ultrasonic velocity. If we make the plausible assumption once again that it is a good approximation to treat the beating oscillations as if they had only a single basic frequency, but a variable amplitude and phase, we should then expect from (6.157) that the velocity oscillations should be just a scaled version of the $d\tilde{M}/dB$ oscillations. As can be seen from fig. 6.23, this is true at 4.2 K where MI is appreciable but not strong enough to cause domain formation, even at the beat maximum. At 1.4 K, however, where domain formation should occur

320

6.8. Oscillations of other thermodynamic properties

Fig. 6.23. Oscillations of $d\tilde{M}/dB$ and $\tilde{v}/v$ at 4.2 K (weak MI) and 1.4 K (strong MI) in Be (after Testardi and Condon 1970). The horizontal lines in the $\tilde{v}/v$ curves indicate the origin of $\tilde{v}$.

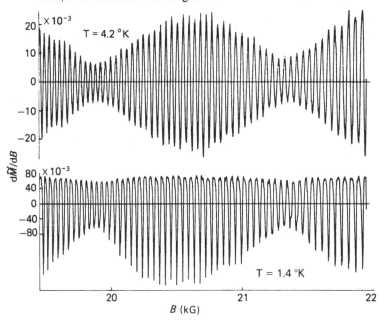

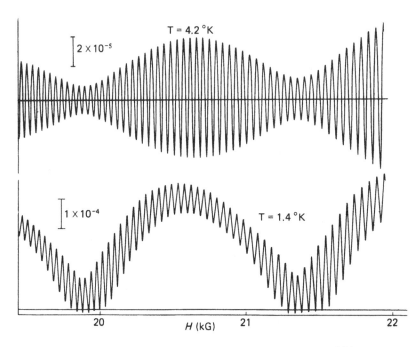

over nearly the whole cycle, the velocity oscillations do not any longer vary in the same way as those of $d\tilde{M}/dB$, and indeed show a strong difference frequency component.

Testardi and Condon suggest that the reason for this discrepancy is that the domain boundaries are unable to move at anything like the ultrasonic frequency* and thus over the domain part of the cycle the appropriate value of $d\tilde{M}/dB$ to put into (6.157) is the diamagnetic value on either side of the domain part. As can be seen from (6.15), $-4\pi\,d\tilde{M}/dB$ (i.e. $-dy/d\theta) = a$ in the middle of the non-domain range and for $a \gg 1$ falls only slightly below a at the limits of this range (cf. Q and P' in fig. 6.6c). This implies that the velocity oscillations should correspond to only the small changes of $-dy/d\theta$ from a and should have a strong positive bias from the 'true' base line (i.e. corresponding to zero velocity change) and this is indeed found to be so, as can be seen from fig. 6.23. As we go through the beat cycle this bias varies and so generates the observed difference frequency. Finally it may be mentioned that the $d\tilde{M}/dB$ oscillations shown in fig. 6.23 are themselves reasonably in accord with MI theory (i.e. $4\pi\,dM/dB$ is 1 in the domain region and is $a\cos\theta$ elsewhere, i.e. close to $-a$ for $a \gg 1$), though there are some discrepancies in detail, probably due to phase smearing and to the non-ellipsoidal shape of the sample.

6.9. Other oscillations in MI conditions (Pippard 1980)

So far we have discussed the effect of MI only on oscillations which are thermodynamically linked with those of free energy and magnetization. However, it is not difficult to extend the discussion to other oscillatory effects of the same periodicity, e.g. the SdH effect or GQO provided it can be assumed that the only effect of MI is to make the oscillations a function of B rather than of H (or H_e if sample shape is relevant). From the theory already developed for the variation of B with H_e in MI conditions the form of the oscillations as a function of H_e can then be obtained. The significant feature of strong MI is that most of the range of B is excised as H_e varies (see fig. 6.6b) and consequently (as with the dHvA effect) only small parts of the oscillations survive.

This is illustrated in fig. 6.24 for an oscillation which in our reduced notation would have been $U = u\sin(x + \phi)$ in the absence of MI, with a phase difference ϕ from the dHvA oscillations $y = a\sin x$. With MI this becomes $U = u\sin(\theta + \phi)$ and this can be expressed as a function of x_e, as explained in the caption of fig. 6.24 which assumes the strong MI limit. The

---

* An order of magnitude estimate shows that the eddy currents induced by domain wall motion would prevent motion at any driving frequency much higher than 100 kHz.

6.9. *Other oscillations in MI conditions*

very narrow permitted regions of θ (the non-domain ranges) are considerably spread out, e.g. between Q and P' when translated in terms of x_e, while over the excised regions of θ (contracted to the broken line parts of the oscillations, e.g. between P and Q) U is a mean of $U(P)$ and $U(Q)$ weighted according to the relative proportions of the P and Q domains. Fig. 6.24 shows that the reduction of amplitude due to the forbidden ranges of B depends strongly on the phase ϕ and the reduction is most severe for $\phi = \frac{1}{2}\pi$, as would be characteristic of SdH oscillations which have the phase of dM/dH (see §4.5 and (2.166)), or for $\phi = -\frac{1}{2}\pi$.

As mentioned in §4.5.1 there are few systematic results on magnetoresistance oscillations which are not due to magnetic breakdown. Indeed, the 'pure' SdH effect is strong only for small pieces of FS and for these the dHvA amplitudes are too weak to produce appreciable MI.

Fig. 6.24. Effect of strong MI on an oscillation which would be $U = u$ $\sin(x + \phi)$ in the absence of MI. With strong MI it is assumed that this becomes $U = u\sin(\theta + \phi)$, but with appropriate ranges of θ (i.e. of B) excised (see fig. 6.6) where the sample breaks up into domains. To calculate U as a function of x_e over the non-domain range (from Q to P') the solution of (6.41) for θ close to π is used; explicitly this gives $U = u\sin[(\pi - x_e)/(1 + \acute{a}(1 - n)) - \phi]$ for x_e from $na\pi/(1 + a)$ to $2\pi - na\pi/(1 + a)$. Over the domain region (from P to Q) the value of U is a weighted mean of the values $-u\sin[(\pi/1 + a) + \phi]$ at P or P' and $u\sin[(\pi/1 + a) - \phi]$ at Q. To a good approximation this gives $U = u[(x_e/na)\cos\phi - \sin\phi]$ between P and Q. The curves are calculated for $n = \frac{1}{2}$ (appropriate for transverse magnetoresistance) and $a = 10$, with $\phi = 0$, $-\pi/3$ and $\pi/2$. The oscillations for $\phi = 0$ are identical in form with those of $\tilde{M}$ (i.e. similar to fig. 6.5 with appropriately changed a and n); those for $\phi = \pi/2$ should correspond to $\tilde{\rho}/\rho$ in the SdH effect, with all but the bottoms of the oscillations cut out.

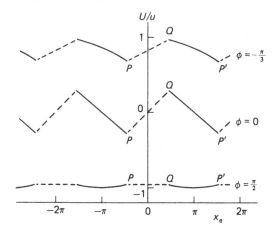

323

However the idea outlined above of excision of the 'forbidden' ranges of B in strong MI conditions applies just as well even if the magnetoresistance oscillations are due to magnetic breakdown, rather than to oscillations in the density of states. A striking illustration of this (fig. 6.25) was observed by Reed and Condon (1970) in the magnetoresistance oscillations associated with the Be cigar. These oscillations of resistance ρ come from magnetic

Fig. 6.25. Transverse magnetoresistance ($\tilde{\rho}/\rho_0$) and $\mathrm{d}\tilde{M}/\mathrm{d}H$ oscillations in Be for H along hexad axis (a) 4.2 K, (b) 1.35 K (after Reed and Condon 1970). Close examination of the oscillations shown that $\tilde{\rho}$ has a frequency which is 33/34 that of $\mathrm{d}\tilde{M}/\mathrm{d}H$; this is because $\tilde{\rho}$ comes from breakdown at the central section of the cigar, rather than the non-central one which dominates in $\mathrm{d}\tilde{M}/\mathrm{d}H$. The excursions of $\tilde{\rho}/\rho_0$, and the mean $\rho(H)/\rho_0$ in the field are both of order 10^3 (ρ_0 is the resistivity for $H = 0$).

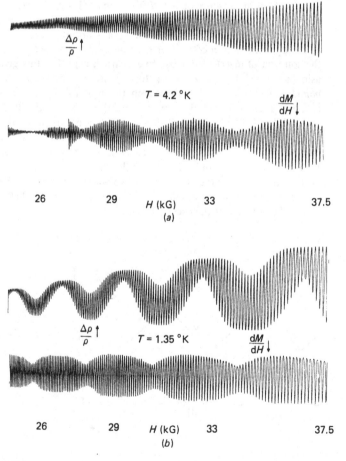

breakdown at the central section of the cigar (see fig. 5.22), which provides the weaker and lower frequency component of the beating oscillations in $\tilde{M}$. The magnetic breakdown mechanism of these oscillations need not concern us here (see chapter 7), but since they come from only one of the two extremal areas relevant to the dHvA effect, they show no appreciable beats at 4.2 K, where MI is insignificant (fig. 6.25a). However, when the temperature is lowered and MI becomes strong over the whole beat cycle, the amplitude of the resistance oscillations is cut down as discussed above. Moreover because of the slight frequency difference between the oscillations of ρ and of M, the phase of $\tilde{\rho}$ with respect to that of $\tilde{M}$ varies from beat minimum to beat maximum, so that the 'excision' moves from the top of the resistance oscillations at beat minimum ($\phi = -\frac{1}{2}\pi$) to the bottom at beat maximum ($\phi = \frac{1}{2}\pi$) (fig. 6.24). The detailed working out of this interpretation is complicated, since the approximations of fig. 6.24 are not really valid, particularly at beat minimum, but Reed and Condon were able with plausible assumptions to compute oscillations which resembled those observed fairly closely. One feature of the magnetic breakdown is that the contribution to $\tilde{M}$ of the central area of the cigar becomes progressively weaker as the field is increased, so that the beats in $\tilde{M}$ should gradually disappear. This is indeed reflected in a corresponding disappearance of the beats in the resistance oscillation beyond about 60 kG (see their fig. 3), but no details of the $\tilde{M}$ oscillations are shown at these high fields, so it is not possible to check whether the resistance oscillation amplitude is behaving as it should. It would be interesting too to check how far the oscillation line shape conforms to the predictions of fig. 6.24 in the beating range.

It would be well worth improving the technique of measuring weak oscillatory magnetoresistance so that the theory could be checked in a rather more straightforward situation, such as that of the belly oscillations in a noble metal, without the complications of magnetic breakdown and of beats. The sensitivity required would, however, be even higher than that suggested by the discussion of §4.5 because of the predicted amplitude reduction under strong MI conditions.

The effect of forbidden ranges of B in strong MI conditions should be even more severe in the giant quantum oscillations, though such an effect does not appear to have been observed as yet. As discussed in §4.6, the GQO in 'ideal' conditions should be a series of δ-function spikes in the ultrasonic attenuation, which occur whenever a Landau cylinder passes through the FS at very nearly the extremum of its cross-sectional area, i.e. whenever

$$F/B = n + \tfrac{1}{2}$$

if we suppose B rather than H is the operative field. However as illustrated in fig. 6.26, for $a \gg 1$ only the small range from

$$\frac{F}{B} = n \pm \frac{1}{8} - \frac{1}{2(1+a)} \quad \text{to} \quad n \pm \frac{1}{8} + \frac{1}{2(1+a)}$$

corresponds to stable values of B ($\pm$ according as the extremum is a maximum ($+$) or a minimum ($-$)). The *average* value of B in the sample can of course assume intermediate values (over a range of values of H_e, if there is a demagnetizing coefficient) but at any particular point, i.e. in any one domain, no intermediate values of B can occur. Thus in 'ideal' conditions the crucial values of B necessary for producing the GQO never occur at all and the oscillations should disappear altogether, rather than partially, as in magnetoresistance oscillations, which vary continuously with B. In fact, as we saw in §4.6.1, for a variety of reasons the giant peaks are usually smeared out into almost sinusoidal oscillations (see (4.64)), which like the oscillations of magnetoresistance are in quadrature to the oscillations of M, so that the effect of MI in these conditions should be similar to that of the magnetoresistance oscillations, i.e. a progressive reduction in amplitude as the MI parameter a is increased beyond 1.

Fig. 6.26. Illustrating how strong MI can practically kill giant quantum oscillations. The realizable narrow ranges of B are indicated for $a = 2$ (plain) and $a = 10$ (shaded): for a maximum cross-section these are between $n + \frac{1}{8} \mp 1/2(1 + a)$ (to avoid confusion the allowed ranges for a minimum are not shown). It can be seen that even for considerable broadening, the GQO peaks at $n + \frac{1}{2}$ fall almost completely in the forbidden range, even for $a = 2$. To get the true line shapes as a function of H_e of the small bits falling in the allowed ranges, the scale must be translated from B to H_e along the lines explained in the caption to fig. 6.24.

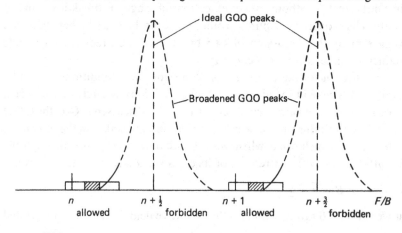

6.10. How to avoid MI (within limits)

It is sometimes advantageous to avoid the complications of MI in order to study more directly what the LK theory predicts for M as a function of B. This can be achieved by feeding back into an extra coil round the sample a current proportional to $\tilde{M}$, with the constant of proportionality and the sign chosen to produce a field $-4\pi(1-n)\tilde{M}$. In this way the B field in the sample can be maintained at exactly H_e and if H_e is modulated, the signal from the pick-up coil will be determined by M as given by the LK formula with H replaced by B. For instance, at low modulation and ω detection, the signal will be proportional to dM/dB and the distortion of MI should effectively be avoided.

All this is only true however as long as $|4\pi\,d\tilde{M}/dB| < 1$. Once $|4\pi\,d\tilde{M}/dB| > 1$, the sample inevitably breaks up into domains over part of each dHvA cycle and with feedback we obtain effectively M as a function of the average B of the sample (cf. fig. 6.6c). Thus it is only for $|4\pi\,d\tilde{M}/dB| < 1$ that MI can be altogether avoided.

As already mentioned (§6.8.2) this idea was used by Testardi and Condon (1970) in their study of Be and they did indeed demonstrate that it is only for 'weak' MI that MI is avoided (fig. 6.23). More recently Gold and Van Schyndel (1981), using a somewhat more flexible feedback arrangement, showed that the side bands $\alpha \pm n\gamma$ due to weak MI between the high frequency α and the low frequency γ oscillations in Pb could be practically eliminated by choosing the feedback constant appropriately. For this setting, the MI harmonics of the γ oscillations were almost negligible so that the pure LK harmonics could be studied directly without the complicated calculations that would otherwise have been needed.

Although this method of avoiding MI (or at least weak MI) is very simple in principle for a perfect sample, it does not work quite so simply in practice if the sample is inhomogeneous. This is because it is inevitably the *average* $\tilde{M}$ which is fed back and it is impossible to cancel the *local* $\tilde{M}$ perfectly at each point of the sample. It is probably some such slight residual MI effect which made it impossible to eliminate the side bands in Gold and van Schyndel's experiment quite completely, though they were considerably reduced.

6.11. The LOFER state; MI in the stars?

Since the magnetization M is a function of $B = H + 4\pi M$ the possibility exists, at least in principle, that a permanent magnetization might be produced by the action of the $4\pi M$ field alone, in zero applied field. This 'bootstrap' mechanism, analogous to that which causes electron spin ferromagnetism, was first proposed by Lee, Canuto, Chiu and Chiuderi (1969) who suggested the name LOFER ('Landau orbital ferromagnetism') for such a state if it occurred. Calculation (Lee 1970) shows that for metallic electrons the effect could occur only at extremely low temperatures (below 10^{-7} K in the most favourable conditions) and would produce only a very feeble permanent magnetization. If however the electron density is enormously greater, as in certain stars, the critical temperature for the LOFER state becomes much higher and the permanent magnetic moment produced much greater. It was in this astrophysical context that the LOFER idea was first proposed by Lee *et al.*, with the suggestion that it might provide an important mechanism for generating high magnetic fields in certain stars. In fact, as we shall mention below, considerations of domain formation and of relativistic degeneracy make it unlikely that the mechanism is of astrophysical significance, except possibly in rather special conditions (see also O'Connell & Roussel 1971).

The calculation of the conditions for the LOFER state is straightforward. If we write the LK formula (2.152) with H replaced by $4\pi M$ for a free electron gas we find

$$M^{1/2} = 6.3 \times 10^{-11} \frac{Fz}{\sinh z} \exp(-1.47 \times 10^5 x/4\pi M)$$

$$\times \sin\left(\frac{F}{2M} + \phi\right) \tag{6.158}$$

where

$$z = 1.47 \times 10^5 \, T/4\pi M \tag{6.159}$$

The most favourable (though, as we shall see, quite unrealistic) situation is for $x = 0$, i.e. an 'ideal' metal, and at $T = 0$, i.e. for $z = 0$ and $z/\sinh z = 1$. In these ideal conditions (6.158) can be written as

$$M = 3.9 \times 10^{-21} F^2 \sin^2[\tfrac{1}{2}(F/M) + \phi] \tag{6.160}$$

and clearly there will be an infinite set of solutions for M, all less than the value M_0 given by

$$M_0 = 3.9 \times 10^{-21} F^2 \tag{6.161}$$

(this can be seen by considering the intersections of the straight line $y = M$ with the curve $y = M_0 \sin^2\left(\tfrac{1}{2}(F/M) + \phi\right)$). For a free electron gas with electron density equal to that in Cu, $F = 6.1 \times 10^8$ and we find

$$M_0 = 1.5 \times 10^{-3} \, \text{G} \tag{6.162}$$

If, however, we include the Dingle factor, the LOFER state is no longer possible, even at $T = 0$, unless the Dingle temperature is extremely small. Thus instead of (6.160) we would have (putting $F = 6.1 \times 10^8$)

$$M = 1.5 \times 10^{-3} \exp(-2.34 \times 10^4 \, x/M) \sin^2\left(\tfrac{1}{2}(F/M) + \phi\right)$$

and a solution is now possible only if x is small enough. In fact it is easily shown that the condition is approximately

$$x < 1.6 \times 10^{-8} \, \text{K}$$

which implies a quite unattainable degree of perfection. If, nevertheless, it were possible to have x just smaller than this critical value, the highest value of M would be just below half the M_0 of (6.162) and only a few lower solutions would remain.

If we consider the effect of a finite temperature T, it is easy to see that, just as with a finite Dingle temperature, a solution is possible only if T is less than a critical temperature T_c. Thus if z is not vanishingly small, (6.160) is replaced by

$$M = M_0(z^2/\sinh^2 z)\sin^2\left(\tfrac{1}{2}(F/M) + \phi\right) \tag{6.163}$$

and bearing in mind the definition of z, a solution for M can be found only if

$$T < 8.5 \times 10^{-5} \, M_0(z^3/\sinh^2 z) \tag{6.164}$$

Since the maximum value of $z^3/\sinh^2 z$ is 0.76 (for $z = 1.29$), we see that a solution occurs only if $T < T_c$, where

$$T_c = 6.5 \times 10^{-5} \, M_0 \simeq 10^{-7} \, \text{K}, \tag{6.165}$$

and at T_c the highest solution for M is about $0.59 \, M_0$. As for a finite x, only a finite number of solutions for M exist if T is finite and only a few remain as T approaches T_c.

It is easy to show that for a metal with a different FS, different electron density, different cyclotron mass and different spin-splitting factor from those we have assumed above, the M_0 and T_c of (6.162) and (6.165) are modified to

$$M_0 = 1.5 \times 10^{-3} (F/F_0)^2(m/m_0)^{-2}(2\pi/A'')\cos^2(\tfrac{1}{2}\pi gm/m_0) \tag{6.166}$$

$$T_c = 10^{-7} (F/F_0)^2(m/m_0)^{-3}(2\pi/A'')\cos^2(\tfrac{1}{2}\pi gm/m_0) \tag{6.167}$$

where F, m, A'' and g are the parameters in (2.152) for the actual metal, while F_0 *and* m_0 refer to the free electron metal assumed so far. At best (for the $\langle 111 \rangle$ belly of Ag) these factors increase M_0 by about 5 and T_c by about 6, though for a favourable orientation of Bi they reduce M_0 and T_c by factors of about 3×10^{-3} and 0.4 respectively. Thus the free electron model does not greatly exaggerate the difficulties of achieving the LOFER state in a real metal.

Even if the extremely demanding conditions on temperature and sample perfection could somehow be met, it is still open to question whether a macroscopic permanent magnetization would actually occur. Lee (1970) shows that each LOFER solution for M should be stable in the sense that an energy barrier must be overcome to reach the next solution and he suggests that extremely long times would be required for the barrier to be crossed by a thermal fluctuation. We might then suppose that the highest possible value of M could be achieved by gradually reducing a superimposed applied field till the $4\pi M$ field took over. Even so it is quite possible that any permanent magnetization would be neutralized by domain formation once any 'guiding' applied field had been removed.

Although all the above discussion is rather academic in the context of real metals, it becomes more realistic in the astrophysical context because of the much higher electron densities which occur in stellar materials. As can be seen from (6.166) and (6.167), both M_0 and T_c are proportional to F^2, which in turn varies as $n^{4/3}$, where n is the electron density. Since relevant electron densities may be as much as 10^8 times higher than in Cu, we see that T_c could be as high as 10^3 or 10^4 K while $4\pi M_0$ could be as high as 10^8 or 10^9 G. Actually T_c and $4\pi M_0$ increase rather more slowly with n beyond about $n \sim 10^{29}$ cm^{-3} because of relativistic degeneracy (Kumar and Lee 1970, O'Connell & Roussel 1971). However, the predicted orders of magnitude of T_c and $4\pi M_0$ are still high enough to suggest that the LOFER state might perhaps play a part in explaining the very high magnetic fields in certain very dense stars as they cool. For a more detailed discussion of this very speculative idea the papers cited above should be consulted.

7

Magnetic breakdown

7.1. Introduction

As already mentioned (see pp. 17, 219), the idea of magnetic breakdown (MB) first arose out of Priestley's (1963) observation of a 'giant' orbit frequency in the de Haas–van Alphen spectrum of Mg. The existence of an orbit area larger than the hexagonal cross-section of the Brillouin zone seemed impossible and was reported in a progress review at the Cooperstown Conference as one of the Fermiological mysteries that needed explaining (Shoenberg 1960b). The explanation was given very soon afterwards by Cohen and Falicov (1961), who pointed out that in a sufficiently strong magnetic field electrons could tunnel from an orbit on one part of a Fermi surface to an orbit on another separated from the first by a small energy gap. This 'magnetic breakdown', as they called it, brings into existence new orbits, such as Priestley's giant orbit in Mg, by tunnelling across the band gaps at the appropriate points (see fig. 7.1).

The criterion for MB originally given by Cohen and Falicov was that the separation $\hbar\omega_c$ between Landau levels should become comparable to or greater than the energy gap ε_g, but Blount (1962) showed that this was not quite right and should be replaced by the much milder criterion

$$\hbar\omega_c \gtrsim \varepsilon_g^2/\zeta \tag{7.1}$$

where ζ is the Fermi energy. Since typically ε_g/ζ may be less than 0.01, this is much less demanding on the magnetic field required and in favourable situations appreciable MB can occur in fields as low as a few kG.

If MB is indeed appreciable, i.e. if (7.1) is satisfied, the concept of well defined semi-classical trajectories on the FS, which all our discussion has so far assumed, becomes untenable. The concept is still valid in the limit of such low fields that MB is inoperative and also in the limit of fields so high that the band gaps are completely broken down. But often we have to deal with an intermediate situation in which features of both limits appear, as well as new effects arising from the 'mixing'. In the dHvA effect this leads to a complicated frequency spectrum showing combination and difference frequencies of the frequencies characteristic of the low field regime. In this respect the consequences of MB are somewhat similar to those of MI, though in principle it is possible to distinguish between them by various criteria which will be briefly reviewed in §7.4.

As was pointed out by Cohen and Falicov when they first introduced the idea of MB, it should also produce dramatic changes in the galvanomagnetic behaviour and these have indeed been observed. Thus orbits which are open at low fields may give way to closed ones as MB sets in with increasing field, or, in other situations, 'hole' orbits give way to 'electron' ones, thus upsetting the compensation. In consequence a magnetoresistance which starts off as if it would increase without limit, corresponding to the low field topology of the FS, turns over as MB sets in and eventually saturates. There are also striking oscillatory galvanomagnetic effects in the MB regime if the metal is pure enough. These can arise both from MB involving small closed orbits (for a review see Stark and Falicov 1967a) or from MB between open orbits, where the effect is rather similar to that of an interferometer (Stark and Friedberg 1971).

The formal quantum mechanics involved in the theory of MB raises difficult questions of principle, essentially because of the incompatibility in detail between the quantization due to the lattice structure of the metal and that due to the magnetic field. Fortunately, however, intuitive ideas based on switching probabilities at junctions in a coupled network of orbits have

Fig. 7.1. Extremal section of the circular network for the free electron-like FS of a hexagonal metal normal to the hexad axis. At low fields the departures from free electron behaviour break up the network into the large hole orbits such as (a) and the small triangular orbits (the needle in Zn or the cigar in Mg) such as (b) (though the amplitude due to (a) is very small). At higher fields, as magnetic breakdown sets in, the 'giant' circular orbit (c) becomes viable with increasing probability.

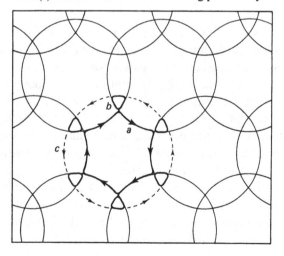

proved to give a very good description of many of the observed phenomena. These ideas, originally proposed by Pippard (1962, 1964, 1965*a*, 1969) still involve quite complicated analysis, particularly in the discussion of transport effects. We shall, however, not attempt to go much beyond describing the salient features of the phenomena and the qualitative nature of the theoretical explanations, or to review all the examples of MB that have been observed experimentally. For fuller details and more complete references, the review article of Stark and Falicov (1967*a*) should be consulted.

7.2. The probability of MB and the Blount criterion

The probability of MB and hence the criterion for an appreciable effect may be most simply understood by a simple diffraction argument given by Stark and Falicov (1967*a*). It is essentially because of the Bragg reflection of electrons of wave vector k_F travelling in appropriate directions that the FS of a polyvalent metal is broken into separate sheets in different zones. In the presence of a magnetic field, however, the electron paths are curved, so that an electron can travel only a finite distance $R \, \Delta\theta$ in the small range of directions $\Delta\theta$ over which Bragg reflection is possible, if R is the radius of the electron path (see fig. 7.2(*a*)). Thus if ξ is the extinction length in the

Fig. 7.2. (*a*) Illustrating Bragg reflection and transmission of electrons on a curved trajectory. After travelling a distance d through the lattice the direction of the trajectory has turned through an angle $\Delta\theta = d/R$, where R is the radius of the trajectory and only a fraction $e^{-d/\xi}$ of the electrons will continue to travel on the trajectory, the remainder $(1 - e^{-d/\xi})$ being reflected. (*b*) Illustrating the separation Δk between two orbits in k-space, which are distinct at low fields, but connected by MB (broken lines) at high fields.

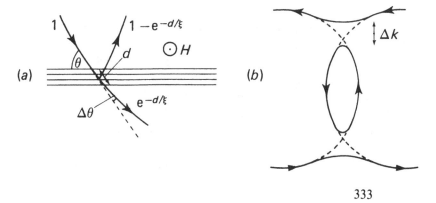

'forbidden' region, a fraction $\exp(-R\,\Delta\theta/\xi)$ of the electrons will be able to continue on the free electron path, without Bragg reflection, and this then is the probability of magnetic breakdown.

The orders of magnitude of both $\Delta\theta$ and ξ can be estimated in terms of the separation Δk between the two branches of the FS in the absence of MB (fig. 7.2(b)). The order of magnitude of this Δk is given by

$$\Delta k/k_F \sim \varepsilon_g/\zeta \tag{7.2}$$

where ε_g is the energy gap at the zone boundary, ζ the Fermi energy and k_F the radius of the free electron sphere.

Since the condition of Bragg reflection gives $\sin\theta \propto 1/k_F$, it is plausible to associate the range $\Delta\theta$ over which appreciable Bragg reflection occurs with the 'uncertainty' Δk specified by (7.2), and we have, apart from factors of order unity

$$\Delta\theta \sim \Delta k/k_F \tag{7.3}$$

Similarly the extinction distance ξ may plausibly be associated with Δk by

$$\xi \sim 1/\Delta k \tag{7.4}$$

We have also (for free electrons) that

$$R = v_F/\omega_c = c\hbar k_F/eH \tag{7.5}$$

and

$$\zeta = \hbar^2 k_F^2/2m \tag{7.6}$$

so that the probability P of MB can be expressed as

$$P = e^{-H_0/H} \tag{7.7}$$

with the breakdown field, H_0, given by

$$H_0 \sim \frac{mc}{e\hbar}\frac{\varepsilon_g^2}{\zeta} = \frac{H}{\hbar\omega_c}\frac{\varepsilon_g^2}{\zeta} \tag{7.8}$$

Thus the criterion for appreciable MB, i.e. $H \gtrsim H_0$, is equivalent to (7.1).

Another simple argument is to treat MB as the magnetic analogue of Zener breakdown (Kane 1960, Ziman 1964 p. 163), in which electrons tunnel across an energy gap in an insulator or semiconductor under the influence of an electric field. It was in fact shown by Blount (1962) that the probability of MB in a weak field is governed by the same analysis as for Zener breakdown, but with the electric force eE, replaced by the Lorentz force evH/c.

These are essentially 'low field' arguments, since the magnetic field is

334

regarded as a perturbation of a band structure determined by the lattice potential responsible for the energy gap ε_g and for the breaking up of the FS. Pippard (1962) has considered the problem in the opposite limit, treating the pseudopotential as a perturbation on a level structure determined by a *high* magnetic field, i.e. a perturbation on the completely broken-down situation. In the absence of any perturbation the free electron model applies and the energy levels are just $(n + \frac{1}{2})\hbar\omega_c$. The effect of introducing the weak perturbing potential is to broaden the originally sharp levels by an amount Δ, which Pippard shows to be given by

$$\Delta/\varepsilon_g \sim (\hbar\omega_c/\zeta)^{1/2} \tag{7.9}$$

He also shows that the probability Q of an electron *failing* to continue on the free electron trajectory is given by

$$Q \sim (\Delta/\hbar\omega_c)^2 \tag{7.10}$$

so, using (7.9) and (7.8) we find

$$Q \sim \varepsilon_g^2/\hbar\omega_c\zeta = H_0/H \tag{7.11}$$

This is just what is to be expected if (7.7) is valid in the high field limit, when Q is small and P approaches 1, since of course we must have

$$P + Q = 1 \tag{7.12}$$

Other demonstrations of (7.7) have been given by Blount (1962), who first derived it, and by various authors since (see Stark and Falicov 1967a for references), and it has been shown to be valid for intermediate fields also. The form of H_0 as given in (7.8) is only approximate. Blount's more general treatment gives

$$H_0 = \frac{\pi}{4} \frac{\varepsilon}{e\hbar} \frac{\varepsilon_g^2}{v_x v_y} \tag{7.13}$$

where v_x and v_y are the components of electron velocity perpendicular to H and respectively normal and tangential to the Brillouin zone plane. If the gap is due to straightforward Bragg reflection of the kind envisaged in fig. 7.2a, then $v_x v_y$ is just $\frac{1}{2}v_F^2 \sin 2\theta$, but (7.13) applies also to rather more general situations, such as gaps, not necessarily at zone boundaries, associated with lifting of degeneracies by spin-orbit coupling.

R. G. Chambers (1966) has shown in some detail that the ideas based on an almost free electron model still remain valid for more general models. In particular he derives a useful expression for the breakdown field H_0 which involves only the local geometry of the Fermi surface in the breakdown region. If the FS is as shown schematically in fig. 7.3, H_0 can be expressed to a good approximation as

$$H_0 = \frac{\pi \hbar c}{e} \left(\frac{k_g^3}{a+b} \right)^{1/2} \tag{7.14}$$

where $1/a$ and $1/b$ are the radii of curvature of the two FS sections at 1 and 2 and k_g is their separation. Since for the special case of almost free electrons $a + b \sim 1/k_g$ and $k_g/k_F \sim \varepsilon_g/\zeta$ (k_g is essentially the same as the Δk of (7.2)), it can easily be seen that for this case (7.14) is essentially the same as (7.8); a more quantitative comparison shows indeed that (7.14) reduces exactly to (7.13).

As already pointed out in the introduction, the importance of (7.7) and (7.8) (or the more general form (7.13)) is that MB occurs much more readily than had at first been supposed, since the H_0 of the Blount criterion is ε_g/ζ times the value originally suggested by Cohen and Falicov (1961). Even so, it is only for extremely small energy gaps that MB can be observed in reasonably modest magnetic fields. If for instance $\varepsilon_g \sim 0.1\,\text{eV}$, typical of a small gap at a zone boundary associated with a weak pseudopotential, (7.8) gives H_0 of order $100\,\text{kG}$. In fact in typical MB situations, such as the giant orbit in Mg, H_0 is found to be only a few kG, corresponding to $\varepsilon_g \sim 0.02\,\text{eV}$. As pointed out by Priestley, Falicov and Weiss (1963), this small gap may plausibly be interpreted in terms of the lifting of an accidental degeneracy due to spin-orbit coupling.

Fig. 7.3. Illustrating the breakdown region between two orbits on the FS (after Chambers 1966). Essentially this is a generalized version of fig. 7.2(b), with k_g replacing Δk; the radii of curvature at 1 and 2 are $1/a$ and $1/b$.

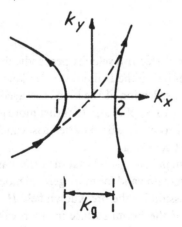

7.3. The coupled network

In order to deal with the rather intractable problem of the quantum mechanical motion of electrons in a magnetic field when MB occurs, Pippard (1962, 1965a) introduced the concept of a coupled network of electron orbits in real space. This provides a visualizable picture of the electron motion which makes possible a calculation of how MB modifies the electron energy levels (and hence the dHvA effect) and also of how it modifies the electrical resistivity in a magnetic field.

The idea is most simply illustrated by the artificial model illustrated in fig. 7.4, in which the lattice potential varies only along one direction normal to *H*. In real space the electron follows a chain of circular paths along a direction normal to *H* and also normal to the direction in which the lattice potential varies. The probability of switching from one circular path to another where they cross is determined by the probability of MB, and the chain of coupled orbits can be treated rather like an electrical network with a characteristic structure of energy levels at which the electron wave can propagate through the network.

In order to discuss the characteristics of the network it is essential first to formulate how the phase of the electron wavefunction varies along any path followed through the network. If the vector potential *A* is chosen to be

$$A = \tfrac{1}{2}H \times r$$

where *r* is measured from the centre of a particular orbit, the solution of the Schrödinger equation is a wavefunction which is appreciable only close to the track of the classical orbit, and the phase of the wavefunction varies uniformly round the orbit. The phase in fact is given simply by $eH/\hbar c$ times

Fig. 7.4. Artificial model of a chain of coupled orbits along a single direction in real space which is perpendicular to the single direction in which the lattice potential varies. Symbols such as $\alpha e^{i\chi}$, $u e^{2i\lambda}$, etc. indicate the complex amplitude of the electron wave at the various points.

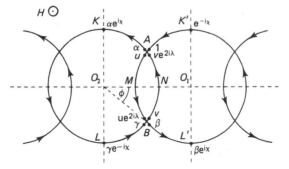

Magnetic breakdown

the area swept out by the radius vector (in real space). Thus, referring to fig. 7.5, in going along the path AMB the phase change would be $eH/\hbar c$ times the area of the sector O_1AB (note that the phase change for the full circle is consistent with the Onsager quantization condition of chapter 2 and works out to be just the dHvA phase $2\pi F/H$). As was shown by Pippard (1965a; see also Chambers 1966), this simple rule can still be maintained even when the electron switches to another circle by Bragg reflection, provided an extra phase θ is added at each such switching. This extra phase θ can also be specified by an area recipe* which we shall merely quote. It is simply that if the electron switches from circle 1 to 2 at A, θ is $eH/\hbar c$ times the area O_1AO_2 (counted positive if the sense of rotation $O_1 \rightarrow A \rightarrow O_2$ is the same as the sense of rotation within each circle).

We can now consider the conditions for electrons to propagate through the network of fig. 7.4. The analysis is somewhat similar to that of the Fabry–Perot interferometer in optics, but with two important differences: first that a wave γ enters from the bottom, as well as the wave 1 from the top and secondly that there is a periodicity condition involving a phase change ω in going from one unit cell to the next. If we suppose that a wave of unit amplitude divides into one of amplitude p after MB and one of amplitude q after Bragg reflection, it is easily shown that conservation of particles requires the phases of the two waves to be in quadrature, so that

Fig. 7.5. Illustrating the meanings, in terms of areas, of the phases θ, λ and χ of fig. 7.4 and equations (7.16)–(7.21) and also of the related phases ξ and σ. The various areas shown are just those of fig. 7.4 and the scale is such that the Greek letter (which is a phase) is $eH/\hbar c$ times the area in real space or $(\hbar c/eH)$ times the area in k-space. Note that

$$\xi = 2\lambda - \theta, \quad \sigma = 2\chi + \theta \quad \text{and} \quad 2(\xi + \sigma) = \odot = (\hbar c/eH)A_0$$

where A_0 is the k-space area of the circle.

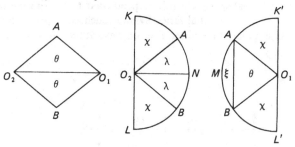

* In an earlier paper (Pippard 1964), the explicit addition of θ was omitted, but the correct answer was nevertheless obtained by including it implicitly in the recipe for the phase change round the lens orbit of the network.

338

7.3. The coupled network

we can conveniently denote their amplitudes to be ip and q respectively, with p and q both real. We then have that

$$p^2 + q^2 = 1, \qquad p^2 = P, \qquad q^2 = Q$$

and

$$p = e^{-H_0/2H}, \qquad q = (1 - e^{-H_0/H})^{1/2}$$

(7.15)

where P and Q are the probabilities (in terms of particle flux) introduced in §7.2.

The continuity conditions at A and B then give in the notation of fig. 7.4:

$$\alpha = qe^{i\theta} + ipve^{2i\lambda} \tag{7.16}$$

$$u = ip + qve^{-i\theta + 2i\lambda} \tag{7.17}$$

$$\beta = q\gamma e^{i\theta} + ipue^{2i\lambda} \tag{7.18}$$

$$v = que^{-i\theta + 2i\lambda} + ip\gamma \tag{7.19}$$

while the periodicity requirements are

$$\psi(K) = e^{i\omega}\psi(K') \quad \text{and} \quad \psi(L) = e^{i\omega}\psi(L')$$

which become

$$\alpha = e^{i(\omega - 2\chi)} \tag{7.20}$$

$$\gamma = \beta e^{i(\omega + 2\chi)} \tag{7.21}$$

Eliminating α, β, γ, u and v from (7.16) to (7.21) we find after some manipulation the condition

$$\cos \omega = [\sin(\xi + \sigma) + q^2 \sin(\xi - \sigma)]/2q \sin \xi \tag{7.22}$$

where ξ and σ are the areas indicated in fig. 7.5, such that $\xi = 2\lambda - \theta$ and $\sigma = 2\chi + \theta$.

For given energy ε, the ratio σ/ξ is a constant, say s, so (7.22) can be written as

$$\cos \omega = [\sin(1 + s)\xi + q^2 \sin(1 - s)\xi]/2q \sin \xi \tag{7.23}$$

Since $\cos \omega$ cannot lie outside the range -1 to 1, the condition (7.23) defines a spectrum of permitted values of ξ and since only trigonometrical functions of ξ enter the condition, this spectrum will have a periodic structure.

The definitions of the phases ξ, σ etc. as real space areas multiplied by $eH/\hbar c$ are equivalent to relations such as

$$\xi = (\hbar c/2eH)A_L, \qquad \xi(1 + s) = (\hbar c/2eH)A_0 \tag{7.24}$$

339

where A_L and A_0 are the k-space areas of the lens and circle for energy ε. Thus the spectrum of permitted values of ξ (the 'phase' spectrum) can be interpreted as $(\hbar c A_L / 2e)$ times the spectrum of values of $1/H$ at which the system has energy ε. This can conveniently be called a *field spectrum*. If moreover ε is the Fermi energy ζ, any regular periodicity of period $\Delta\xi$ in the ξ spectrum, as specified by (7.23), corresponds to $(\hbar c A_L / 2e)$ times a periodicity $\Delta(1/H)$ in the field spectrum, which will be a periodicity in the dHvA oscillations. The corresponding dHvA frequency F will of course be

$$F = \frac{\hbar c A_L}{2e \, \Delta\xi} = \frac{\hbar c A_0}{2e \, \Delta\xi(1 + s)} \tag{7.25}$$

where A_L and A_0 are now areas of cross-section of the Fermi surface.

The ξ spectrum can also be related to an *energy spectrum* at constant H. Thus if we measure the energy ε from an origin such that all the sines in (7.23) vanish simultaneously, and use the definition (2.21) of cyclotron mass, we find, apart from additive constants which are integral multiples of π

$$\xi = \frac{m_L c \pi \varepsilon}{e\hbar H} \quad \text{and} \quad \xi + \sigma = \frac{m_0 c \pi \varepsilon}{e\hbar H} \tag{7.26}$$

where m_L and m_0 are the cyclotron masses of the L and $\odot$ orbits (the latter in our model is just the free electron mass).* Thus (7.23) can be rewritten as

$$\cos\omega = \frac{\sin\gamma\varepsilon + q^2 \sin\gamma\varepsilon[2(m_L/m_0) - 1]}{2q \sin\gamma\varepsilon(m_L/m_0)} \tag{7.27}$$

where

$$\gamma = m_0 c \pi / e\hbar H = \pi/\hbar\omega_{co} \tag{7.28}$$

This then specifies a spectrum of permitted energies ε at constant H and it is easily seen that apart from a scaling factor, the ε spectrum is obtained from the ξ spectrum by changing the meaning of s from σ/ξ (i.e. $(A_0/A_L) - 1$) to $(m_0/m_L) - 1$ and replacing ξ by ε in (7.23).

Examination of the periodicity of the field and energy spectra for simple limiting cases helps to understand their physical meaning. Consider first the limiting case of $q \ll 1$, i.e. very little Bragg reflection or nearly complete MB. In this limit (7.23) can be satisfied only if

$$\xi + \sigma = n\pi + \delta \tag{7.29}$$

---

* In the free electron model these relations would hold for arbitrary energy, but for a real metal the cyclotron mass varies with ε, so (7.26) are valid for only a limited range of ε. Since, however, many periods of ξ (i.e. many Landau levels) occur over this limited range, this is not an important reservation.

with

$$-1 < \delta/2q \sin \xi < 1$$

This implies a set of nearly sharp levels of ξ repeating at intervals $\Delta\xi$ given by $\pi/(1 + s)$. These correspond to the dHvA frequency F_0 of the full circular orbit of area A_0, i.e. $F_0 = \hbar c A_0/2\pi e$, as we should expect in the absence of Bragg reflection. But if q is not exactly zero the levels are slightly broadened to an extent given by $4q \sin \xi$. This broadening varies periodically with a period given by $\Delta\xi = \pi$, which corresponds to the dHvA frequency F_L of the lens orbit, so the Fourier analysis of the ξ spectrum will contain the frequency F_L as well as F_0. Similarly, the energy spectrum consists of equally spaced nearly sharp levels at intervals such that $\gamma \Delta\varepsilon = \pi$, or

$$\Delta\varepsilon = e\hbar H/m_0 c = \hbar \omega_{c0} \tag{7.30}$$

i.e. just the expected cyclotron resonance levels of the circular orbit. Here again the levels are slightly broadened to an extent which repeats at intervals of $e\hbar H/m_L c$, i.e. the separation of the cyclotron resonance levels of the lens orbit.

In the opposite limit of q approaching 1, i.e. strong Bragg reflection and very little breakdown, the ξ spectrum is practically continuous, corresponding to the open orbits, but there are periodic breaks in the continuity whenever ξ or σ is close to an integral multiple of π. Detailed analysis shows that the dominant periodicity in the spectrum corresponds to the dHvA period of the lens, but the periodicity of the circular orbit is also weakly present if q is not exactly 1. Similarly the energy spectrum is dominated by the cyclotron resonance levels of the lens, but with the levels of the circle also showing up if q is not exactly 1.

Calculations by Pippard in which, for a given value of q, values of ξ are computed at equal intervals of ω (for the particular ratio of $\sigma/\xi = 11/3$), show (fig. 7.6) how the broadening of the sharp levels at $q = 0$ develops as q increases. For a particular value of $q (= 1/\sqrt{2})$ he Fourier analyzed the ξ spectrum and showed that its periodicity corresponded, as expected, to the lens and circular orbit areas as well as various combinations of them and harmonics. Before describing this spectrum in more detail, however, it is convenient to outline a rather different approach due to Falicov and Stachowiak (1966) which leads to a simpler method of deriving the dHvA spectrum. The two approaches give essentially the same amplitudes of the various frequencies in the spectrum.

As explained in §2.5, for the purpose of determining the oscillatory free

Magnetic breakdown

energy it is sufficient to determine the density of states of the system and Falicov and Stochowiak showed that this can be done without a detailed calculation of the individual states, such as Pippard's method involves. Their method is based on a theorem which relates the density of states to

Fig. 7.6. Variation with q (probability of Bragg reflection) of phase spectrum of ξ (equivalent to field spectrum) for a particular example in which $\sigma/\xi = 11/3$ (after Pippard 1962, but with different notation). A continuum of permitted values of ξ or σ occurs in the shaded regions, with gaps in between. The states in the black regions come together at $q = 1$ to produce highly degenerate levels at $\xi = n\pi$.

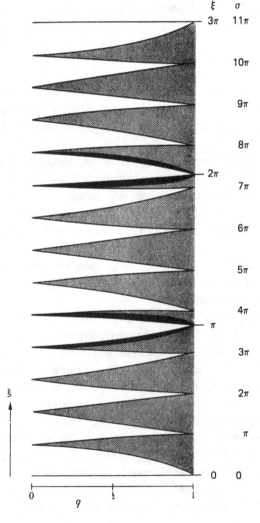

the Fourier transform of an appropriate Green's function. This Green's function corresponds essentially to the sum of semi-classical wave packets which return to some given point of the coupled network by all possible paths through the network, with amplitudes reduced according to the number of Bragg reflections and magnetic breakdowns encountered *en route*, and phases corresponding to the areas swept out. The final result is quite simple and once formulated seems almost obvious; it was indeed suggested independently on an intuitive basis by R. G. Chambers (1966)*. The result is that the oscillatory free energy is given by the sum of terms of the standard LK form (2.151) for all the possible closed orbits permitted by the network, but with the modification that the amplitude of any particular term is multiplied by a 'breakdown reduction factor', R_b given by

$$R_b = (\mathrm{i}p)^{n_1} q^{n_2} \tag{7.31}$$

where n_1 is the number of branching points in the relevant orbit at which breakdown occurs and n_2 is the number at which Bragg reflection occurs. The phase factor i is usually irrelevant, but if the orbit is a complicated one the same area may be produced by several non-equivalent paths which do not necessarily have the same phase and this must be taken into account in adding up their various contributions to the dHvA term of the relevant F in the LK formula.

In using the LK formula to calculate the contribution of any particular orbit some paths must be given extra weight because of their asymmetry, which means that they can occur in more than one version; this is taken into account by introducing a weighting factor C, which is essentially the number of times the orbit occurs per unit cell (see table 7.1). Also if any closed path is traversed more than once, say p times, its contribution is just that corresponding to the pth harmonic of the 'once round' orbit in the LK formula. The cyclotron mass to be used in the LK formula for any particular orbit can easily be shown to be proportional to the perimeter of the orbit (since this measures the time taken to complete the orbit) with the free electron mass for the basic circular orbit ⊙. However, for the pth harmonic of any orbit it is simpler to use the mass and area of the 'once round' orbit and introduce p explicitly, as in (2.151). The appropriate value of $|\mathrm{d}^2 A/\mathrm{d}\kappa^2|$ is easily shown to be $2\pi m/m_0$ for an orbit of cyclotron mass m.

---

* In Chambers' paper it is incorrectly stated that Pippard's spectrum specified by (7.22) should be regarded as an energy rather than a field spectrum, i.e. that ξ etc. should be proportional to perimeters rather than areas of orbits. Professor Chambers has kindly reexamined the question and has discovered an algebraic error in his analysis; he now confirms Pippard's original result.

The way this recipe works is illustrated in fig. 7.7 which shows some of the important orbits of the network drawn out, including the alternative versions of asymmetric orbits. The masses of the various orbits and the values of n_1, n_2 and C are collected in table 7.1. The $1/H$ dependence of $\ln (|M|H^{-1/2})$, i.e. essentially a 'Dingle plot', for a few of the dHvA frequencies at $T = 0$ is shown in fig. 7.8. In the absence of MB these would be just horizontal lines, though of course only the L orbit would then have any finite amplitude. The most important feature of the plot is in showing that as the field increases and MB becomes increasingly significant, the L oscillation amplitude dies away and the giant orbit $\odot$ oscillations become dominant.

The amplitude of the $2\odot$ area serves to illustrate the relevance of worrying about the phase factor i in (7.31). Thus because $(ip)^6 = -p^6$, the contribution of the b variant of $2\odot$ has to be subtracted instead of added to the a and c variants. The resulting total amplitude of $2\odot$ is thus proportional to $p^4(p^2 - 2q^2)^2$ and this leads to a complicated field

Table 7.1

Area	m/m_0	Harmonic	Weight factor (C)	Broken-down junctions n_1	Bragg reflected junctions n_2
L ⎫		1	1	0	2
L ⎬	0.5	2	1	0	4
L ⎭		3	1	0	6
$\odot$ ⎫		1	1	4	0
$\odot(a)$ ⎭	1	2	1	8	0
$2\odot(b)$	2	1	2	6	2
$2\odot(c)$	2	1	2	4	4
$2\odot - L$	1.5	1	1	4	2
$L + \odot$	1.5	1	2	4	2

Notes: The entries differ slightly from those in Table 2 of Stark & Falicov (1967a). Thus the areas and masses for harmonics are here given as for the fundamental, since the harmonic numbers p are to be added explicitly in (2.151); however the numbers n_1 and n_2 take into account that the pth harmonic corresponds to p circuits of the single orbit. The weight factor corresponds to the number of ways in which the particular kind of orbit can be achieved, as shown in fig. 7.7. The masses quoted assume that the angle ϕ of fig. 7.4 is $45°$.

7.3. The coupled network

dependence of amplitude, with the amplitude vanishing at the field for which $p^2 = 2q^2$ (i.e. for $H_0/H = \ln(3/2) = 0.405$); the contribution from the harmonic of $\odot$ alone which involves only the $(ip)^6$ term behaves more normally of course. However, in a more realistic situation with an appreciable lattice potential, Chambers (1966) pointed out that the various

Fig. 7.7. Some typical orbits of the system of fig. 7.4 with $\phi = 45°$ (after Stark and Falicov 1967a). The dots indicate points of Bragg reflection and the dashes points of magnetic breakdown. The frequency $2\odot$ can appear in a variety of ways. The most obvious (referred to as (a) in table 7.1) is the harmonic of $\odot$, i.e. a double passage round the orbit $\odot$; two others are (b) and (c), each of which occurs in two asymmetric versions as does $L + \odot$. Note that only one sense of rotation is allowed so combinations such as $\odot - L$ cannot occur.

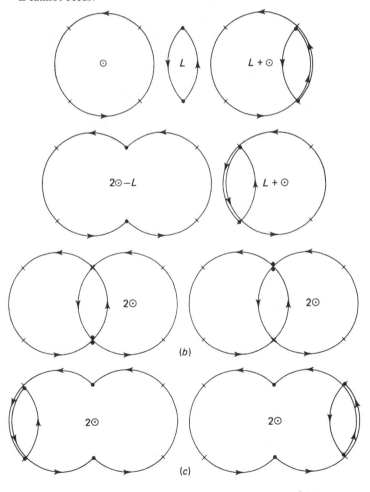

345

Magnetic breakdown

2 ⊙ orbits would not be quite exactly equal in area (see fig. 7.9). The phase relations between the a, b and c terms would therefore vary with field and complicated beats might be expected when the p^4, $4q^4$ and $4p^2q^2$ contributions were comparable.

Although the one-dimensional coupled network we have considered so far does not correspond to any real situation, we have deliberately discussed it in some detail, because it is appreciably simpler than even the simplest real situation and so serves more easily to illustrate the basic principles. We now turn to the most studied real situation, which is that of MB in Mg and Zn for the field along the hexagonal axis. As discussed in chapter 5, the free electron picture provides a fair approximation to the FS, and the coupled network, now spreading in two dimensions rather than one, is as illustrated in fig. 7.1. Pippard (1964) first considered this problem in detail by an extension of his approach for the much simpler hypothetical network, which extends in only one dimension. The problem of calculating how MB broadens the level structure is not only more complicated in that there are many more boundary conditions to satisfy (there are twelve such continuity equations, rather than the four of (7.16)–(7.19)), but there is also a difficulty of principle. In the one-dimensional real space network there

Fig. 7.8. Dingle plots for some of the orbits of table 7.1. The plots would be horizontal lines if not for MB, since we assume $T = 0$ and a perfect sample. In this idealized situation the plotted quantity, $\ln(R_bF/F_0) - \frac{3}{2}\ln(rm/m_0)$, is $\ln(|M|H^{-1/2})$ apart from a constant, where r is the harmonic number and F_0 is the frequency corresponding to the full circular orbit ⊙ (since A'' for sections defined by the intersecting spheres, varies as m/m_0).

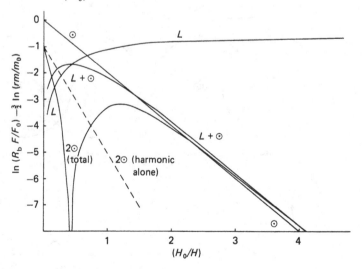

346

was no lattice periodicity to contend with in the direction of the network; thus the scale of the network (which is, of course, inversely proportional to H) is immaterial. In the two-dimensional real space network this is however, no longer true, because the lattice periodicity now extends in both dimensions and consequently it is only for special discrete values of H that points which are equivalent as regards the periodicity of the network are separated by an exact multiple of a lattice period. For other values of H, the periods are non-commensurate and equivalent points in the network are no longer equivalent in the lattice; thus the formulation of boundary conditions analogous to (7.16)–(7.19) is no longer possible.

It is easily shown that the discrete values of H for which the periodicity of the real space network matches that of the lattice are given simply by the condition that the flux through a unit cell of the two-dimensional lattice of giant orbit centres should be an integral multiple of the flux quantum $2\pi\hbar c/e$; the integer is of the same order as the value of F/H for the giant orbit. Since H cannot be varied continuously, the phase spectrum which is found (by elaborate algebra) for any one of the permitted discrete values of H cannot be translated into a field spectrum at constant energy and this greatly complicates the interpretation of the analysis. This difficulty of principle has also been discussed by W. G. Chambers (1965) in the context of a rectangular rather than a hexagonal network, and he was able to show how to calculate a phase spectrum for any value of H such that the flux is a rational fraction times $2\pi\hbar c/e$ (say N/M times, with N and M any integers). Although this is some advance, it does not really overcome the difficulty and further clarification is still needed.

Fig. 7.9. Schematic diagram of two of the orbits of fig. 7.7 ($2\odot b$ and $2\odot c$) with more realistic indication of reflection and breakdown paths. The actual paths are indicated by the full outline and other possible paths (or parts of them) are shown broken. The area of the l.h. orbit is easily seen to be $2\odot + a - b$, and that of the r.h. orbit $2\odot + 2a - 2b$, and since in general $a \neq b$, these areas are slightly different from each other and from the exact harmonic $2\odot$.

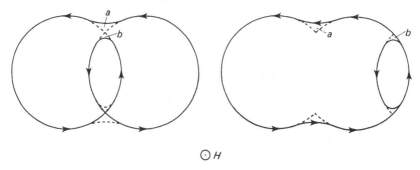

$\odot H$

Falicov and Stachowiak suggested that the difficulty of principle might be sidestepped by calculating the dHvA amplitudes from the density of states at fixed H in the same way as described above for a one-dimensional network. The calculation of the reduction factor R_b of (7.31) and the weight factor C for the real two-dimensional network proves to be little more complicated than for one-dimension. The only new features are that many more closed orbits are possible and that C may now be as high as 6. The determination of C is a little more subtle here than in the one-dimensional network; the simplest recipe for it is just $6/D$, where D is the rotational symmetry of the orbit (but D must be taken as 6 for the $\odot$ orbit). The values of C for the various orbits shown in fig. 7.10 are given in table 7.2, together with the numbers n_1 and n_2 of junctions at which breakdown and Bragg reflection occurs and the values of the cyclotron masses calculated from the

Fig. 7.10. Some possible orbits in the hexagonal network (Falicov and Stachowiak 1966). As in fig. 7.7, dots indicate points of Bragg reflection and dashes points of magnetic breakdown. The area χ is taken as a negative quantity to take account of the opposite sense of rotation of this hole orbit as compared with the electron orbits $\odot$, θ and λ. The characteristics of the various orbits are set out in table 7.2. Note that some of the orbits, though identical in area, have different rotational symmetry and therefore different weight factors C (e.g. compare (5) with (11), or (8) with (9) or with (7) and (10)).

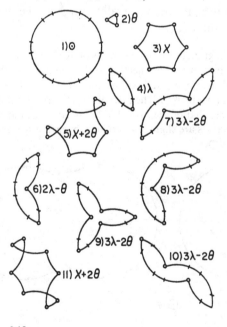

orbit perimeters. The calculation of the factor R_b and the resultant dHvA amplitude is then straightforward.

Figure 7.11 shows calculated plots of the field dependence of $|dM/dH|$ (the quantity usually measured) for a few of the important orbits in Mg (the plots for Zn would be qualitatively similar). In order to make this illustration somewhat more realistic, the temperature reduction factor $z/\sinh z$ (with $z = 1.47 \times 10^5 (m/m_0) T/H$), has been included, assuming $T = 1$ K, so that the amplitudes of high mass orbits fall off more rapidly at low fields than those of low mass (however, we ignore the spin and Dingle factors, so the actual amplitudes would be smaller than those plotted). We see that the amplitude of the giant orbit $\odot$ only becomes large for fields above about $5H_0$, so fairly large fields are needed to make it appreciable ($H_0 \sim 5.8$ kG for Mg). This is partly because there are so many (twelve) breakdown points in the orbit, but also because with a cyclotron mass of $1.4m_0$, the temperature reduction factor is very severe at 1 K until H exceeds say 30 kG. The 'hexagonal' orbit χ round the inside of the monster

Table 7.2. *Characteristics of orbits shown in fig. 7.10 (after Falicov and Stachowiak 1966)*

Orbit		m/m_0	Weight factor (C)	Broken down junctions n_1	Bragg reflected junctions n_2
1	$\odot$	1	1	12	0
2	θ	3α	2	0	3
3	χ	$1 - 6\alpha$	1	0	6
4	λ	$\frac{1}{3} + 2\alpha$	3	4	2
5	$\chi + 2\theta$	1	6	4	8
6	$2\lambda - \theta$	$\frac{2}{3} + \alpha$	6	6	3
7	$3\lambda - \theta$	1	3	8	4
8	$3\lambda - 2\theta$	1	6	8	4
9	$3\lambda - 2\theta$	1	2	6	6
10	$3\lambda - 2\theta$	1	3	8	4
11	$\chi + 2\theta$	1	3	4	8

Note: The parameter α involved in the values of m/m_0 for some of the orbits is the ratio of the arc length of each side of the triangular orbit θ (needle or cigar) to the circumference of the full circle. Appropriate values of α are 0.023 for Mg and 0.0023 for Zn; these give reasonable agreement with experimental values of ratios of masses of the different orbits. However the mass of the giant orbit $\odot$ is found to be $1.40\,m_0$ rather than m_0 (Stark 1967). Note that our weight factor C is defined otherwise than in Stark and Falicov.

is always much feebler than the breakdown orbit–at high fields because MB makes the q^6 factor so small (at $10H_0$ this factor is roughly 1/1200) and at low fields because of the temperature factor (at $H = 2H_0$, where q^6 is about 0.06, the temperature factor is 4×10^{-4} for Mg). Thus it is not too surprising that this orbit has not yet been observed experimentally.

The behaviour of the θ orbit, which is the 'needle' in Zn and the 'cigar' in Mg, provided early evidence for the reality of MB. As fig. 7.11 shows, the amplitude of this oscillation should fall off rapidly as the field increases beyond H_0. Such a fall off was indeed observed in Zn by Dhillon and Shoenberg (1955) but remained a mystery until it was interpreted as due to MB by Pippard (1964). In fact the introduction of the factor q^3 into the amplitude accounts very well for the fall away from a linear Dingle plot in Zn (see fig. 7.12) and similar agreement has also been obtained for Mg (Stark and Falicov 1967a). Falicov and Stachowiak have calculated the

Fig. 7.11. Field dependence of $|\mathrm{d}M/\mathrm{d}H|$ in Mg at 1 K for the basic orbits $\odot$, θ, χ and λ of fig. 7.10, with and without MB. The graphs ignore the Dingle and spin factors and assume the following values of the relevant parameters: $H_0 = 5.8\,\mathrm{kG}$, $m/m_0 = 1.4, 0.1, 0.86$ and 0.39, $F = 625, 2.24,$ 316 and $5.38\,\mathrm{MG}$. These are based on Stark's (1967) data, but since the χ oscillations have not been observed, the figures for m/m_0 and F were deduced by comparing the orbit perimeter and area with those of orbits of known behaviour. The broken curves (denoted by primes) indicate the behaviour in the absence of MB. The calculations ignore MI which would of course be relevant when $|\mathrm{d}M/\mathrm{d}H| \sim 1/4\pi$.

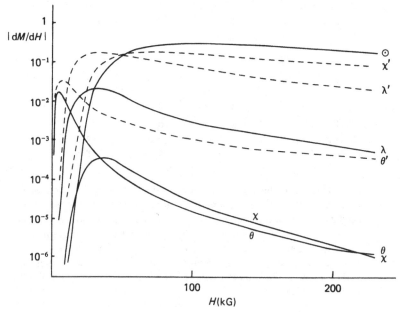

7.3. The coupled network

detailed field dependence of amplitude for other more complicated orbits (such as those shown in fig. 7.10), and as with the one-dimensional network there are several areas which can be achieved in a variety of ways, some involving a change of sign because of the i factor in $(ip)^{n_1}$; once again this can lead to complicated field dependences. As before too, further complication arises from the fact that these various 'equivalent' areas are only exactly equal in the free electron model. In reality they differ slightly and this leads to complicated beating effects.

The behaviour of the complicated dHvA spectrum of Mg has been studied experimentally by Stark (1967) and at the time of Stark and Falicov's (1967a) review it seemed that the theoretical predictions of Falicov and Stachowiak agreed well with experiment. Since then, however, experiments with extremely pure Mg have revealed striking discrepancies with the theory. Thus Eddy and Stark (1982) find not only that the dHvA spectrum contains frequencies which should be impossible on the Falicov

Fig. 7.12. Dingle plot of the 'needle' dHvA torque amplitude C in Zn with H at about $20°$ to the hexad axis; the units of C are arbitrary: $\bigcirc$ 4.2 K, $+$ 1.6 K (after Dhillon and Shoenberg 1955). This plot would be linear if not for MB, which reduces the amplitude by the factor $R_b = q^3 = (1 - e^{-H_0/H})^{3/2}$ and this is demonstrated by the nearly straight broken line on which the values of the experimental curve fall after subtracting $\log_{10} R_b$. For this purpose H_0 has been taken as 2.7 kG, the value indicated by galvanomagnetic data (see table A14.1), but the restoration to linearity is not sensitive to H_0; thus Pippard got an equally good straight line (of lower slope) with $H_0 = 6$ kG.

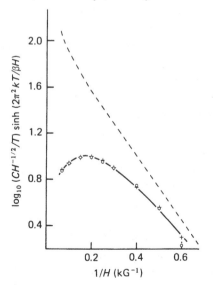

351

and Stachowiak scheme, but that the spectrum changes drastically for a small change of the field range over which the oscillations are Fourier analyzed (e.g. for a change from the range 50.11–50.25 kG to one of 50.25–50.39 kG). Moreover the field dependence of amplitude of some of the prominent oscillations is quite at variance with the theoretical prediction.

This new evidence perhaps throws doubt on the assumption that Falicov and Stachowiak's approach has really avoided the difficulty of commensurability between the real space orbit periodicity and the lattice. Indeed Stark (private communication) speculates that the commensurability may become more and more important as improved perfection of the sample increases the coherence length of the electron motion appreciably beyond a period of the real space orbit network. Perhaps the relatively simple Falicov and Stachowiak method may still prove a valid approximation in the limit of relatively poor coherence, even though it seems to fail for longer range coherence. Evidently these questions, which relate to the behaviour of almost macroscopic quantum systems, are of basic interest and require much more investigation.

Finally we should mention that we have throughout our discussion ignored one other complication. This is that the breakdown parameter H_0 may vary appreciably with κ (i.e. k_H) over the range of κ which is significant in contributing to the final amplitude. This effect proves to be negligible in the Zn needles, but appreciable in Mg and must be allowed for in a detailed calculation. Lonzarich and Holtham (1975) have shown that this effect is particularly important in the very small β-orbit frequency of Al, where the breakdown field H_0 increases rather rapidly from a value of less than 1 kG as κ increases away from the central plane ($\kappa = 0$). The reduction factor R_b which determines the de Haas–van Alphen amplitude is in this case an appropriate average of q^2 i.e. of $(1 - e^{-H_0/H})$. The effective range of κ which matters increases as $H^{1/2}$ (see §2.3.6), so, because of the rapid increase of H_0 with κ, R_b does not reduce the amplitude at high fields nearly as drastically as in Zn or Mg. The detailed numerical integration over κ shows that over a wide range of H the Dingle plot, though lowered from what it would be in the absence of MB, remains approximately straight and this was confirmed experimentally. There are probably many other situations in metals with complicated Fermi surfaces where complications of this kind are important in determining the field dependence.

7.4. Comparison of MB and MI

The problem sometimes arises of deciding whether the appearance of unusually strong harmonics or combination frequencies and of peculiar amplitude behaviour in the de Haas–van Alphen effect is to be attributed to

MI or to MB. Although it is difficult to lay down precise rules because of the many complicating factors which occur in both effects, some simple criteria are usually sufficient to decide whether MI or MB is the basic cause of the observed effects. If information on the galvanomagnetic behaviour is available, the occurrence of MB should immediately be evident through the striking effects to be discussed in the next section. If, however, the only evidence is from the dHvA effect, the following are some of the simple criteria which may serve to distinguish MB from MI:

(1) *The magnitude of the absolute amplitude.* It is only if $|4\pi dM/dH|$ becomes appreciable compared with unity that MI can be important, though as we saw in §6.3 it is really the magnitude of $|4\pi dM_0/dH|$, the amplitude before reduction by the phase smearing part of the Dingle factor, which matters, so that we may have a kind of 'hidden' MI.

(2) *The temperature variation of amplitude, particularly of combination frequencies.* The factor R_b, which determines how MB modifies the amplitude of any orbit involving MB, should be independent of temperature. For MI, however, if the observed frequency involves two basic frequencies, the amplitude is a product of functions of the separate amplitudes (see for instance (6.85)), and so falls off more rapidly with increasing temperature than it would if MB was the origin of the combination frequency.

(3) *Field dependence of amplitude.* This does not usually offer a decisive criterion in itself because, as we have seen, the field dependence is complicated by a number of effects both in MI and MB. However if the amplitude falls considerably below what might be expected from linear extrapolation of Dingle plot to high fields and to an extent which is independent of temperature, it is likely that MB rather than MI is involved, with a reduction factor involving a power of q (as for the Zn needle).

(4) *Detailed knowledge of the Fermi surface.* If such knowledge is indeed available, it is usually possible to identify where MB might occur for particular field orientations and so to have some *a priori* estimation of whether MB is likely to be the cause of any observed peculiarities. Moreover if the FS is known in detail, it is possible to predict just which of all possible combination frequencies are to be expected from MB. With MI, in principle all combination frequencies should occur and there should usually be little difference of amplitude between frequencies $F_1 \pm F_2$ if $F_2 \ll F_1$, while for MB, as can be appreciated from the examples given earlier, $F_1 - F_2$ may not be a possible combination at all. However, the recent results of Eddy and Stark (1982) suggest that this last criterion may not be reliable.

7.5. Galvanomagnetic effects and MB

We have already mentioned the striking manifestation of MB in the galvanomagnetic effects and we shall now give a brief outline of what these are and how they come about. A more detailed account with fuller references to original papers is given in the review by Stark and Falicov (1967*a*). We shall start by discussing the magnetoresistance behaviour of the hypothetical one-dimensional coupled network which serves to illustrate the principles rather more simply than the situation of any real metal.

7.5.1. One-dimensional network

For simplicity we shall consider only the limiting case where the lens orbits are relatively very small, so that the electron spends nearly all its time on the major arcs of the circles. The conductivity tensor can be calculated rather simply by the 'effective path' method (Pippard 1964, 1965a), in which the current density J produced by an electric field E is related to the average vector distance L (in real space) travelled by an electron starting at some arbitrary point on the network. For an electric field in the plane of the network it can be shown (for a compact exposition see Pippard 1969) that

$$J = \frac{nec}{\pi r H} \int_0^{2\pi} L(-E_x \sin \phi + E_y \cos \phi)\,d\phi \qquad (7.32)$$

where the starting point P of the electron is defined by the angle ϕ in fig. 7.13, r is the radius of the circle and n the number of electrons per unit volume. Note that the term in brackets is the component of E along the velocity direction at P, which is the same as that normal to the FS. Here it is assumed that the effective path L is determined entirely by the random wanderings of the electron through the coupled network rather than by electron scattering, which for simplicity we assume to be absent. An electron starting at any point P on the upper arc of a circle will terminate on average at a point with coordinates (x, y) with respect to the centre O or, using complex notation, at $Z = x + iy$, while one starting at a point P' on the lower arc will terminate at a point $-Z$. Thus since the starting point is $re^{i\phi}$, the vector L is specified as

$$L = \begin{cases} Z - re^{i\phi} & \text{for } 0 < \phi < \pi \\ -Z - re^{i\phi} & \text{for } \pi < \phi < 2\pi \end{cases} \qquad (7.33)$$

Fig. 7.13. Coupled orbits as in fig. 7.4, but only just intersecting, so that the 'lenses' can be regarded as switches.

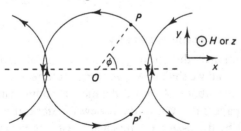

7.5. Galvanomagnetic effects

At first sight the determination of the terminal point Z would seem to be a formidable combinatorial problem, but as Pippard (1965*a*) showed, it can be derived very simply by a symmetry argument. Each small lens (i.e. double junction) can be thought of as a 'switch' which sends the electron either forward with probability T, or backward with probability S. These probabilities are related to the parameters p, q or the probabilities P, Q introduced earlier in a rather subtle way and it will be convenient to defer discussion of this relation for the moment. If the switch sends the electron starting at the point P on to the next circle to the left of O, it will inevitably reach the next switch and its terminal point will be $2r$ further to the left, i.e. at $Z - 2r$. If, however, the switch keeps it on the same circle, its terminal point will be that of an electron starting at P', i.e. $-Z$. Bearing in mind that these alternatives have probabilities T and S respectively, we have

$$Z = T(Z - 2r) - SZ$$

and since we must have

$$S + T = 1$$

we find

$$Z = -Tr/S \tag{7.34}$$

Substituting into (7.33) and (7.32) gives the current as

$$J = \frac{nec}{H}\left[\left(\frac{4T}{\pi S}E_x - E_y\right) + iE_x\right] \tag{7.35}$$

i.e. a conductivity tensor:

$$\sigma = \frac{nec}{H}\begin{pmatrix} \dfrac{4}{\pi}\dfrac{T}{S} & -1 \\ 1 & 0 \end{pmatrix} \tag{7.36}$$

or, if the conductivity tensor is inverted, a resistivity tensor

$$\rho = \frac{H}{nec}\begin{pmatrix} 0 & 1 \\ -1 & \dfrac{4}{\pi}\dfrac{T}{S} \end{pmatrix} \tag{7.37}$$

We must now return to the interpretation of the probabilities T and S. In our discussion so far, we have tacitly ignored the phases of the electron waves travelling on the main arcs of the circular orbits and considered only particle fluxes. This is usually justified in practice since phase coherence is not sufficiently preserved over the long paths involved, because of small angle scattering by dislocations or other causes. Such incoherence of course implies that the dHvA amplitude associated with the circular orbit is much reduced. However in a good quality sample there may be appreciable phase coherence over the small lens orbit. If such phase coherence does occur, amplitudes and phases must be matched at each junction and the probabilities obtained by squaring amplitudes. If, however, there is no appreciable phase coherence even over the short paths within the 'switch', we must deal with squares of amplitudes from the start. The two extreme situations are illustrated in fig. 7.14(a) and (b) in which the notations for the various amplitudes and phases in the coherent case and for the particle fluxes (i.e. squares of amplitude) in the incoherent case are indicated.

(a) coherent case

Just as in (7.16)–(7.19), at each junction the emergent complex amplitudes must be equated to the sum of the incident amplitudes multiplied by ip or q according as MB or Bragg reflection occurs. We thus have (exactly as in (7.16)–(7.19) but with $\gamma = 0$, i.e. with no feed-in at B from the left)

$$
\begin{aligned}
\alpha &= qe^{i\theta} + ipve^{2i\lambda} \\
u &= ip + qve^{-i\theta + 2i\lambda} \\
\beta &= ipue^{2i\lambda} \\
v &= que^{-i\theta + 2i\lambda}
\end{aligned}
\tag{7.38}
$$

from which we easily find (recalling that the dHvA phase ψ of the small lens orbit (i.e. $2\pi F/H$) is given by $2\xi = 4\lambda - 2\theta$)

$$
\begin{aligned}
T &= |\alpha|^2 = 2Q(1 - \cos\psi)/(1 - 2Q\cos\psi + Q^2) \\
S &= |\beta|^2 = P^2/(1 - 2Q\cos\psi + Q^2)
\end{aligned}
\tag{7.39}
$$

Fig. 7.14. Showing how a wave 1 entering a switch at A, either switches to the next orbit to the left or remains on the same orbit. In (a) the wave is coherent through the switch with the complex amplitudes shown and we have the emergent fluxes $T = |\alpha|^2$ and $S = |\beta|^2$. In (b) no coherence is maintained within the switch and the fluxes are as indicated.

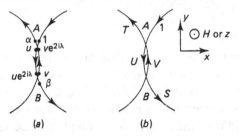

(a) (b)

7.5. Galvanomagnetic effects

For T/S, which determines the conductivity, we then have

$$T/S = 2Q(1 - \cos\psi)/P^2 \qquad (7.40)$$

which oscillates between 0 and $4Q/P^2$ as the field (and so ψ) varies.

(b) incoherent case

This can be worked out either by averaging* (7.39) over ψ or, rather more simply and directly, by the same sort of continuity argument as was used above, but working with particle fluxes instead of complex amplitudes. Referring to fig. 7.15b, we have

$$\begin{aligned} U &= P + QV \\ T &= PV + Q \\ V &= QU \\ S &= PU \end{aligned} \qquad (7.41)$$

from which we obtain

$$T = 2Q/(1 + Q) \qquad S = P/(1 + Q) \qquad T/S = 2Q/P \qquad (7.42)$$

Note that this 'incoherent' value of T/S rises less steeply as $P \to 0$ than the average over ψ of the coherent value (7.40).

The variation of the resistivity, which is proportional to HT/S, with H is shown in fig. 7.15 for both cases. From (7.11), (7.37) and (7.42) or (7.40) we find in the limit of high H that the resistivity ρ_{22} (for the current perpendicular to the network axis in real space or parallel to the network axis in k-space) saturates to the value

$$\rho_{\text{sat}} = 8H_0/\pi nec \qquad (7.43)$$

for the incoherent case, and oscillates between zero and twice this value for the coherent case. In the low field limit the resistivity increases indefinitely and we have for $H \ll H_0$

$$\rho_{22}/\rho_{\text{sat}} = \begin{matrix} (H/H_0)e^{H_0/H} & \text{for incoherence} \\ (H/H_0)e^{2H_0/H}(1 - \cos\psi) & \text{for coherence} \end{matrix} \Bigg\} \qquad (7.44)$$

This rise to infinity as $H \to 0$ is of course merely a consequence of our simplifying assumption that there is no resistivity due to electron scattering. If a relaxation time τ for scattering is introduced, the calculation becomes a good deal more complicated (Falicov and Sievert 1964) but it is not difficult to see what happens in the limits of low and high fields. At $H = 0$ electron

---

* It should be noted, however, that though the averages of (7.39) over ψ do give T and S in (7.42) correctly, the average of T/S as given by (7.40) is not the same as the ratio of the averages.

Magnetic breakdown

scattering becomes the only resistive mechanism, giving the usual resistivity $\rho_0 = m/ne^2\tau$, so we find

$$\rho_0/\rho_{\text{sat}} = \pi/8\omega_0\tau \qquad (7.45)$$

where ω_0 is the cyclotron frequency for $H = H_0$. For $H \ll H_0$ we have the situation of a metal with open orbits perpendicular to H and the resistivity rises quadratically with field, until eventually, when $H \gtrsim H_0$, the magnetic breakdown mechanism becomes dominant. For $H \gg H_0$ the small scattering contribution behaves as for closed orbits, with the resistivity having the same value as in zero field. This small contribution can be thought of as in series with the saturation value due to MB alone and we find for a finite τ

$$\rho_{\text{sat}}(\tau) = \rho_{\text{sat}}(\infty)[1 + (\pi/8\omega_0\tau)] \qquad (7.46)$$

where $\rho_{\text{sat}}(\infty)$ is as given by (7.43). If then, as can be achieved in very pure samples, $\omega_0\tau \gg 1$, the dominant resistive mechanism at high fields comes from MB rather than electron scattering. The behaviour at intermediate fields is indicated schematically in fig. 7.15, showing how the magneto-resistance regime changes over from the quadratic increase characteristic

Fig. 7.15. Plots of $\rho_{22}/\rho_{\text{sat}}$ showing behaviour of transverse magnetoresistance of the coupled orbits of fig. 7.13: (a) for incoherent limit, (b) for full coherence within the lens orbit, (c) line of mid-points of the oscillations (b). These plots assume no electron scattering ($\tau = \infty$) so that (a) and (c) grow without limit (dotted lines) as H is reduced. For a finite τ such that $\omega_0\tau \gg 1$ the behaviour of (a) and (c) at low fields is indicated schematically by the broken lines (the ordinate for $H = 0$ is too small to show on the scale of the diagram). The effect of a finite τ on the behaviour of the oscillations (b) is complicated but presumably the oscillations would fade out once $\omega_c\tau \lesssim 1$, i.e. at low enough fields. In (b), F for the lens is arbitrarily taken as $10\,H_0$.

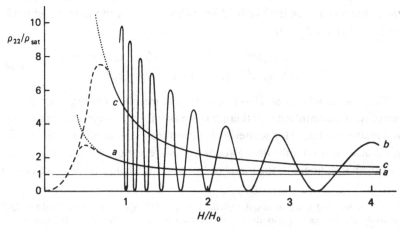

of open orbits to the saturation characteristic of an uncompensated metal with all orbits closed. If coherence is maintained round the small lens orbit, so that its switching probability oscillates with the dHvA frequency of the lens, the magnetoresistance oscillates in sympathy with the variations of the extent to which the magnetoresistance regime approaches one limit or the other.

7.5.2. Two-dimensional network with hexagonal symmetry

As can be seen from fig. 7.16, even though the one-dimensional model is only a crude model of the experimental situation, the observed field dependence of transverse magnetoresistance for Mg and Zn (with H along the hexagonal axis) is qualitatively similar to the predicted behaviour of the model, with oscillatory behaviour when the sample is good enough to maintain coherence over the small orbit. Similar behaviour occurs in Be (see fig. 6.28) though, as we saw earlier, considerably complicated by MI effects. The next step is to consider a better representation of the experimental situation and this is the two-dimensional network of fig. 7.1.

For $H \ll H_0$ there are equal numbers of 'electrons' (on the needle orbit and the inner circle of the monster) and 'holes' (on the 'hexagonal' orbit), all in closed orbits, and in the absence of MB the magnetoresistance increases as H^2. However, as MB sets in, the exact compensation is destroyed although the orbits remain closed. For $H \gg H_0$, all the carriers, i.e. those on the giant orbit and those on the inner waist of the monster, are electron-like and so the magnetoresistance saturates. The oscillations have the frequency associated with the small triangular orbit (needle in Zn or cigar in Mg and Be) and this orbit can be thought of as a switch (see fig. A14.1), rather more complicated than the lens of fig. 7.13, whose switching probabilities oscillate with the dHvA phase if there is sufficient coherence in going round the small orbit. It is therefore plausible that the behaviour of the more realistic model should be qualitatively similar to that of the simple one-dimensional model, but the detailed mechanism is somewhat different and even a simplified calculation of the two-dimensional hexagonal network is appreciably more complicated.

The details of the calculation (based on Falicov, Pippard and Sievert (FPS) 1966) are presented in Appendix 14 and we shall only summarize the main points here. Just as for the one-dimensional model, the calculation of the conductivity tensor is based on the effective path method and the effective path for an electron at any point of the network can be written down if the switching probabilities of an electron entering a 'triangular junction' are known. If complete coherence round the small orbit is assumed, the switching probabilities oscillate with the dHvA frequency of the small orbit (needle or cigar) and the magnetoresistance oscillates accordingly. The mean level of the field dependence may be calculated approximately by calculating the switching probabilities for complete lack of coherence, though it is probably more correct in principle to use the coherent model and phase average the final formula as in FPS. As discussed

Fig. 7.16. Experimental curves due to R. W. Stark of transverse magnetoresistance (ρ_{11}) and Hall effect (ρ_{12}) in (a) Zn at 1.6 K; (b), (c), Mg at 1.2 and 1.1 K respectively (reproduced in Falicov et al. 1966 and in Stark and Falicov 1967a). In (b) only the envelopes of the oscillations are shown; (c) is for a very pure sample and only ρ_{11} is shown.

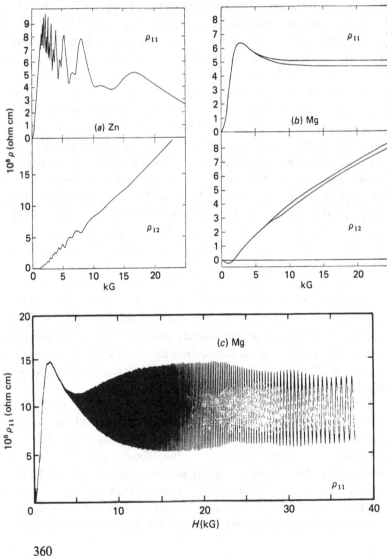

7.5. Galvanomagnetic effects

in Appendix 14, there is in practice not a great deal of difference in the final answers for the mean level of resistance, but following FPS we shall base our discussion on the result for phase averaging after the conductivity has been calculated.

Experimentally, of course, it is the resistivity rather than the conductivity tensor which is actually measured, since it is the potential (both parallel and perpendicular to the current) which is measured for a given current through the specimen. Thus to compare with experiment we need to invert the conductivity tensors derived in Appendix 14. For $H \ll H_0$, the breakdown probability P, which is $e^{-H_0/H}$, becomes negligible compared with the term a/bH (see table A14.1) and the inverse of (A14.14) gives simply

$$\rho_{11} = \rho_{22} = H^2/a,$$
$$-\rho_{21} = \rho_{12} = 0.55(bH^3/a^2)e^{-H_0/H} \tag{7.47}$$

Thus the magnetoresistance starts off quadratically (as of course was deliberately arranged in the formulation of Appendix 14, by introducing the term a/H^2 in the conductivity formula), while ρ_{12}, the Hall effect, falls off extremely rapidly below about $\frac{1}{3}H_0$. If the simpler theory leading to (A14.11) had been used, the result would have been qualitatively similar, except that the Hall effect would have fallen off even more drastically as $H \to 0$.

For $H \gg H_0$, $Q \simeq H_0/H$ and in the formula for the conductivity (A14.11) (which is the same for both the simple and the more elaborate theory), the term a/bH is less than a tenth of H_0/H (for both Zn and Mg) so we find to a fair approximation

$$\rho_{11} = \rho_{22} = H_0/b, \qquad -\rho_{21} = \rho_{12} = H/\sqrt{3b} \tag{7.48}$$

i.e. a saturating magnetoresistance and a Hall effect increasing linearly with H.

As can be seen from fig. 7.17, the general character of the non-oscillatory part of the field dependence of ρ_{11} and ρ_{12}, as computed by FPS for intermediate fields, is reasonably in accord with the experimental curves of fig. 7.16 for Zn, but the ρ_{12} theoretical curve for Mg differs appreciably from the experimental curve. No detailed analysis of the non-oscillatory galvanomagnetic effects has yet been made for Be and as mentioned in §6.9 the oscillatory magnetoresistance is complicated by magnetic interaction. It is, however, clear from comparison of the fields at which the steady magnetoresistance turns over in all three metals that H_0 for Be is several times higher than for Mg, and is perhaps of order 25 kG. The initial negative ρ_{12} observed in Mg can be qualitatively explained by supposing

361

that the holes and electrons (before MB occurs) have different values of $\omega_c\tau$, which effectively destroys the complete compensation assumed in the calculated curve. A similar change of sign of Hall effect has also been observed in samples of Zn less pure than that of fig. 7.16. Another important refinement of the theory is to take into account the variation of H_0 with κ, which as mentioned earlier, is appreciable in Mg, though not in Zn, and FPS show that agreement between theory and experiment can be somewhat improved by this refinement (cf. Al, as discussed earlier).

When we turn to theoretical predictions regarding the oscillatory part of the magnetoresistance and Hall effect, the agreement with experiment is only qualitative and there are some puzzling discrepancies. One unsatisfactory feature of the comparison between theory and experiment is that though the calculated amplitude of the oscillations in Mg agrees qualitatively with the amplitude observed in the particular sample of fig. 7.16(a), the calculation ignores the possibility of any phase smearing, i.e. assumes the Dingle temperature is zero. The sample was indeed a very good one, but it is disturbing that in a later experiment, with a still more perfect sample with an exceptionally low dislocation density, Stark (see Stark and Falicov 1967a) observed an amplitude $|\Delta\tilde{\rho}_{11}| \sim 5 \times 10^{-8}$ ohm cm (see fig. 7.16(c)) which is something like seven times higher than the calculated amplitude

Fig. 7.17. Theoretical curves of ρ_{11} and ρ_{12} in (a) Zn, (b) Mg, computed by FPS as outlined in the text; only the envelopes of the oscillations are shown in (b).

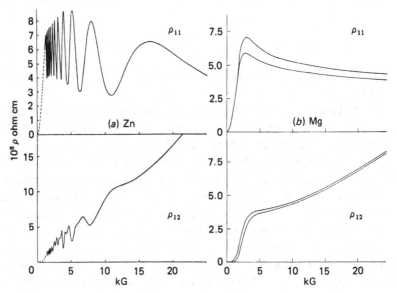

indicated in fig. 7.17. Since a finite Dingle temperature can only *reduce* the calculated amplitude, this discrepancy is not easy to explain away on the basis of the theory that has been used.

A clue to the possible origin of the discrepancy is provided by the fact that the oscillations of the sample of fig. 7.16(*c*) contain a much higher frequency component (not shown) as well as the dominant low frequency cigar oscillations. These higher frequency oscillations have just the frequency of the lens orbit (whose area is about 25 times that of the cigar) and this suggests that because of the perfection of the sample some phase coherence is maintained even round the much longer lens perimeter. Thus the sample conditions seem to be intermediate between the completely 'quantum mechanical' regime in which phase-coherence is maintained over all the orbits, and the semi-classical model used in the calculations of FPS in which phase-coherence is assumed only round the smallest orbit (the cigar). The completely quantum mechanical regime is the one discussed earlier in considering the level structure with MB, and Pippard (1964) shows how the transport properties can be regarded as determined by a new kind of quasiparticle with quasimomentum proportional to ω and energy ε related by the two-dimensional generalisation of (7.27). According to his theory, the magnetoresistance oscillations in this ideal quantum mechanical regime should be much stronger than those usually observed but, as he showed later (Pippard 1965*a*), this 'ideal' amplitude should be drastically reduced by sample imperfections. Thus it would seem plausible that the abnormally large oscillations of fig. 716(*c*) are to be interpreted as an approach to the ideal quantum regime, while the more usual lower amplitudes found with less perfect samples are characteristic of the semi-classical regime assumed in the FPS calculations.

It is disturbing, however, that some recent results of Stark (private communication) appear to conflict with this interpretation. In this recent work, the samples were even more perfect than that of fig. 7.16(*c*)), but the oscillations were actually weaker, rather than stronger. Moreover the amplitude was found to vary with the measuring current density j, increasing as j^2 and extrapolating to almost nothing for $j = 0$. Since j had not been varied in the earlier experiments, the experimental picture is still far from complete; but it does look as if existing theories cannot after all explain the magnetoresistance oscillations characteristic of the most perfect samples, in what had been provisionally thought of as approaching the ideal quantum regime. However, as with the dHvA effect, the theory ignoring or avoiding the commensurability difficulty does seem to provide a semi-quantitative explanation of the magnetoresistance oscillations in the semi-classical regime characteristic of less perfect samples.

7.5.3. The Stark quantum interferometer

Oscillations in the magnetoresistance of Mg due to a novel MB mechanism analogous to that of an optical interferometer have been extensively studied by Stark and his students (Stark and Friedberg 1974, references to later work in Stark and Reifenberger 1977; see also Sandesara and Stark 1983). The novelty is that the MB is between open orbits along which electrons travel in the *same* direction and the frequency of the oscillations is determined by the area of a loop which carries no circulating current. The principle is illustrated by the greatly oversimplified scheme of fig. 7.18 in which an electron can proceed along an open orbit by two alternative paths a or b whose relative probabilities depend on MB at the junctions J_1 and J_2.

It is easily seen that the probability of an electron (in the sense of particle flux) continuing along the open orbit path from P_1 to P_2 is given by

with
$$\left. \begin{aligned} T = |q^2 e^{i\phi_2} - p^2 e^{i\phi_1}|^2 = P^2 + Q^2 - 2PQ \cos\psi \\[8pt] \psi = \phi_2 - \phi_1 = (\hbar c/eH)A, \end{aligned} \right\} \tag{7.49}$$

where A is the k-space area of the loop and the ps and qs and Ps and Qs are as defined earlier. It is important to notice that although ψ looks exactly like a dHvA phase there should be no corresponding dHvA oscillations, because the electrons do not circulate round the loop of area A (as they did, for instance, round the small triangular orbit in the two-dimensional hexagonal network considered earlier).

We see then that the flow of electrons from P_1 to P_2 in this simplified model will oscillate in a de Haas–van Alphen-like manner as H varies and the resistivity associated with conduction along the open orbit system will

Fig. 7.18. Schematic sketch illustrating the principle of the quantum interferometer (after Stark and Friedberg 1974). The broken lines show schematically that the orbit paths a and b in k-space change with a small change of energy in such a way as to keep the area A of the loop nearly constant.

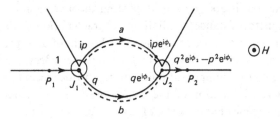

oscillate in sympathy. One important feature of these oscillations, which distinguishes them from oscillations associated with a genuine dHvA orbit, is that they should be insensitive to temperature. In a circulating orbit the area A varies with the energy ε and, as has been discussed earlier, the effect of a finite temperature is equivalent to phase smearing over the range of areas corresponding to the range of ε in the Fermi 'tail'; thus the crucial parameter determining the phase smearing and the amplitude reduction is the cyclotron mass, which is proportional to $dA/d\varepsilon$. Here, however, as indicated schematically in fig. 7.18, the paths for an energy slightly different from the Fermi energy ζ are both shifted in the same sense, leaving the area unchanged.

In the practical realization of this idea – the quantum interferometer as Stark calls it, because of its similarity to a two-beam optical interferometer – the magnetic field is applied along a $[10\bar{1}0]$ axis (perpendicular to the hexad axis) in a very perfect Mg crystal and the resisitivity is measured parallel to the $[11\bar{2}0]$ axis. The quantum inteferometer oscillations are the low frequency ones in fig. 7.19 and it can be seen that, as predicted, they are quite insensitive to temperature. The mechanism of the interferometer is in fact considerably more complicated than that of the simple model discussed above. The relevant orbits in k-space are shown in

Fig. 7.19. Quantum interference oscillations in the transverse magnetoresistance of a pure Mg crystal for $\boldsymbol{H}$ along $[10\bar{1}0]$ and current along $[11\bar{2}0]$. The upper curve is for $T = 1.5\,\text{K}$ and the lower for $T = 4.2\,\text{K}$ (Stark and Friedberg 1974).

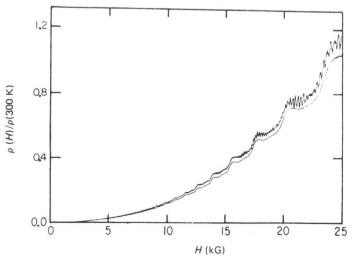

365

fig. 7.20 and there are three different kinds of MB junction (shown by black dots) with three different breakdown fields. The basic area A determining the frequency of the quantum interferometer oscillations is each of the shaded areas and it is not surprising, in view of the presence of two such areas linked by a breakdown junction, that the quantum interferometer oscillations should contain a $\cos 2\psi$ as well as a $\cos \psi$ term, where ψ is defined by A as in (7.49). The amplitude of the $\cos \psi$ term should pass through zero for the field at which $P_1 = \frac{1}{2}$, where P_1 is the breakdown probability at the entrance and exit of the quantum interferometer. This vanishing is indeed observed experimentally and measurement of the field at which it occurs provides a rather precise way of determining the breakdown field H_1. The fast oscillations of fig. 7.19, which show the

Fig. 7.20. (Stark and Friedberg 1974). (a) Electron trajectories in k-space for Mg in the repeated zone scheme for H along $[10\bar{1}0]$ and $\kappa = 0$; A, H, K and L are symmetry points in the Brillouin zone. The broken and solid FS contour lines represent trajectories on the second and first band hole sheets ('monster' and 'cap') respectively. The breakdown junctions are shown by black dots.

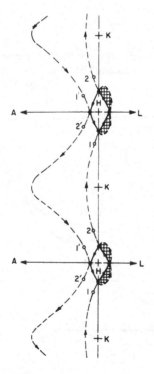

7.5. Galvanomagnetic effects

normal dHvA temperature dependence, come from the circulation of electrons round the 'cap' orbit.

The detailed working out of the properties of the quantum interferometer is quite complicated and here we shall only refer briefly to one or two general points. First it is important to realize that the observed effects are extremely sensitive to the precise orientation of the magnetic field. This comes about because of the κ dephasing; as with the dHvA effect, it is only a small range round the κ which makes the relevant area an extremum, that contributes to the observed oscillation. If the magnetic field is ever so slightly off the exact symmetry orientation, the picture of fig. 7.20 is no longer exactly periodic along the open orbit direction, either in k-space or in real space, and only for a short length of the open orbit will the area A which determines the oscillations be close to an extremum with respect to κ. As we saw in §2.3.6, the dephasing due to integration over κ is equivalent to regarding only a small range $\kappa_{\rm eff}$ of κ as contributing to the oscillations and this 'effective' range is given by

$$\kappa_{\rm eff} = (H/|\partial^2 F/\partial \kappa^2|)^{1/2} \tag{7.50}$$

If then the field is tilted by an angle δ out of the exact symmetry plane, the repeats of the area beyond a distance $\kappa_{\rm eff}/\delta$ further along the orbit will already fall out of the range of κ which effectively contributes. In particular, if

$$\delta > \kappa_{\rm eff}/K \tag{7.51}$$

where K is the repeat distance along the orbit, only a single interferometer unit (i.e. the one which is nearly an extremum with respect to κ) will contribute appreciably to the oscillations. We have, in fact, all along tacitly assumed that we need not worry about the effect of having several units in series, and this is usually justified because the critical tilt δ as given by (7.51) is so very small that it is difficult to orient the crystal sufficiently precisely. Thus for $H = 10\,{\rm kG}$, $\kappa_{\rm eff} \sim 0.7 \times 10^5\,{\rm cm}^{-1}$ and, since $K \sim 1.3 \times 10^8\,{\rm cm}^{-1}$, the critical value of δ is about 0.03°, and of course at lower fields the critical angle would be still smaller. With sufficient care a sample can be oriented to rather better than 0.01° but two other circumstances can easily undo the benefit of such a careful setting. First the sample itself may not be of quite uniform orientation (slight mosaic structure or bending) and secondly the field can never be perfectly homogeneous in direction because of the self field of the measuring current. In the conditions of Stark and Friedberg's experiments the measuring current had to be at least 0.5 A and this produced something like 2 G at the sample surface which would cause tilts of order 0.01° in an applied field of 10 kG. In fact they observed that if

367

the field was systematically tilted there was a range of tilt of something like $\pm 0.03°$ around the optimum direction, over which the oscillation amplitude was practically unchanged.

In the later experiments of Stark and Reifenberger (1977), Reifenberger (1977) and particularly Sandesara and Stark (1983), even more care was taken over sample perfection and lower measuring currents were used, and it was then found possible to establish a new regime when the field orientation was 'exactly' right. In this 'on-axis' regime new features appear in the line shape of the oscillations and detailed analysis shows that these are just what would be expected if several interferometers are acting in series, rather than only a single one.

Up to now we have tacitly assumed that there is no electron scattering, i.e. that the relaxation time τ is infinite. Evidently, if the relaxation time is much shorter than the time t_1 the electron takes to travel through the interferometer, the oscillations will disappear altogether and even if τ is appreciably longer than t_1, the oscillations will be affected. Similarly the characteristic features of the 'on-axis' situation must be destroyed if τ is much less than the time t_2 taken to go from one unit to the next, and again even if τ is appreciably greater than t_2 the oscillations will be affected. For a field of $10\,\text{kG}$, $t_1 \sim 2 \times 10^{-12}\,\text{s}$ and the detailed theory shows that the nature of the oscillations (e.g. the field dependence of the ratio of amplitudes of fundamental to harmonic) should be sensitive to the value of τ for τ as long as $10^{-10}\,\text{s}$. At $10\,\text{kG}$, $t_2 \sim 2 \times 10^{-11}\,\text{s}$, so the features of the 'on-axis' regime are sensitive to considerably larger values of τ – the detailed analysis shows that even relaxation times as long as $10^{-8}\,\text{s}$ should be detectable. This relaxation time dependence of the quantum interferometer provides a new and sensitive tool for the measurement of long relaxation times, though its application is of course limited to very special Fermi surface situations. One advantage of this method is that for a given τ the oscillations are insensitive to temperature, so that it becomes possible to study the temperature dependence of τ associated with phonon scattering, without having to disentangle temperature effects associated with the Fermi distribution. So far the only application of this method of measuring τ has been to ultra-pure Mg (Reifenberger and Stark 1975) and dilute alloys of Mg with Zn and Cd (Friedberg 1974).

8

The Dingle temperature

8.1 Introduction

As outlined in the historical introduction (chapter 1), a slight but puzzling discrepancy between the early experimental results on the de Haas–van Alphen oscillations in Bi (Shoenberg 1939) and Landau's theoretical formula was that the observed field and temperature dependences of the amplitude could not be consistently reconciled with the formula. To a fair approximation it was as if the temperature needed to fit the formula was higher than the actual temperature. An explanation of the discrepancy was suggested by Dingle (1952b) who showed (as discussed in §2.3.7.2) that if electron scattering is taken into account, the Landau levels are broadened and this leads to a reduction of amplitude very nearly the same as would be caused by a rise of temperature from the true temperature T to $T + x$. This extra temperature, x, which is needed to reconcile theory and experiment, has come to be known as the Dingle temperature and we shall refer to the amplitude reduction factor $\exp(-2\pi^2 kx/\beta H)$ as the Dingle factor. Dingle's suggestion also explained an earlier puzzling observation, which was that addition of any impurity to Bi always reduced the oscillation amplitude (Shoenberg and Uddin 1936); this would be expected in view of the increased probability of electron scattering.

For a good many years Dingle temperatures were recorded somewhat casually in studies devoted mainly to FS determinations from frequency measurements, but no systematic studies were attempted and there was little attempt to interpret such results as there were. It was only gradually realized that systematic measurements might yield valuable information about electron scattering, particularly by the foreign atoms in a pure metal host. This was emphasized (Shoenberg 1969a) at the 1968 Zürich conference* on electron mean free paths in a review of the possibilities and preliminary results. Once the necessary experimental techniques had been developed – and, as we shall see in §8.2, there are many pitfalls to be avoided – systematic results on a variety of dilute alloy systems began to

* These conference proceedings (*Physik der Kond Materie* 9, 1–210, 1969) and those of two later conferences in Oregon and Norwich, both in 1974 (*Phys. Cond. Matter* 19, 1–434, 1975) serve to demonstrate the rapid pace of development in the field of electron scattering.

pour out. By suitable deconvolution techniques it became possible to extract from the orientation dependence of the Dingle temperature how the scattering probability per foreign atom varies over the Fermi surface and there is now a considerable body of results and theory in this field, which will be reviewed in §8.3.

Much less satisfactory is the position regarding the Dingle temperatures of relatively pure samples, where the dominant mechanism is not impurity scattering. It was noticed quite early on (see for instance Shoenberg 1952a) that the Dingle temperatures x of pure samples (typically a few tenths of a degree) were much higher (by a factor of 10 to 100) than would be expected from the relation (2.136) between x and the relaxation time τ, if τ was estimated from the electrical conductivity. It also soon became apparent that x was often sensitive to the mechanical state of the sample and slight deformation in handling a sample could increase x considerably while hardly affecting the electrical conductivity. It was fairly clear that the observed high Dingle temperatures and their sample sensitivity were associated with imperfections such as dislocations and mosaic structure.

The damping effect of dislocations will be discussed in §8.4, particularly in the limiting cases of the orbit size r small or large compared with the range d over which the strain varies appreciably. For small r/d, the damping mechanism is essentially one of phase smearing due to strain dependence of the dHvA frequency. For large r/d, the damping is really one of a diffusive kind of low angle scattering, but here too a qualitative estimate of the damping can be obtained by a phase smearing argument. Such few experiments as there are in which the dislocation structure of the samples is at all reliably characterised, prove to agree only qualitatively at best with the calculations and this is not too surprising in view of the idealized models assumed and various other simplifications introduced into the calculations. One puzzle that most of the theories cannot explain is why Dingle plots are usually straight, i.e. why the damping factor has the form $\exp(-H_0/H)$, over as wide a range of H as is often observed.

The damping effect of mosaic structure, which is discussed in §8.5, is a rather more straightforward matter and calculations based on a simple phase smearing approach seem to agree reasonably well with the few available measurements on well characterised samples. An extreme form of mosaic structure is that of completely random orientation, i.e. a polycrystal. It turns out that for metals with a nearly spherical FS a small amplitude of oscillation can survive the drastic phase smearing over orientation and feeble oscillations have indeed been observed in alkali colloids and in polycrystalline Cu.

Finally we shall return briefly to the question already introduced in

§2.6.1.2 of how scattering of electrons by thermal phonons might affect the Dingle temperature. As we saw earlier, Palin (1972) found, in agreement with many-body theory but contrary to naive intuition, that for Hg there was essentially no change of Dingle temperature with rise of sample temperature. In §8.6 we shall discuss Palin's experiment and its interpretation in a little more detail and also consider evidence for specifically many-body effects from recent experiments (Elliot, Ellis and Springford 1980) in more extreme conditions.

8.2. Measurement of Dingle temperatures

Apart from certain reservations to be discussed below, the field dependence of the amplitude A_p of the pth harmonic of the de Haas–van Alphen oscillations should be given by

$$A_p = C_p T H^{-n} R_D / \sinh(\alpha p T / H) \qquad (8.1)$$

where R_D is the Dingle reduction factor, which subject to certain assumptions should have the form

$$R_D = \exp(-\alpha p x / H) \quad \text{with} \quad \alpha = 1.469(m/m_0) \times 10^5 \, \text{G K}^{-1} \quad (8.2)$$

The values of C_p and n depend on the particular method of measurement. For instance if it is M that is measured, it can be seen from (3.1) that $n = \frac{1}{2}$, while if it is (dM/dH), as in the modulation method for weak modulation, then $n = \frac{5}{2}$. If, however, the modulation is varied as H^2 to keep it always the same fraction of a dHvA period, n is again $\frac{1}{2}$ and in the torque method (see (3.5)), $n = -\frac{1}{2}$. The p dependence of C_p usually follows straightforwardly from the basic formula (2.152), though the dependence is more complicated if strong modulation is used in the field modulation technique and of course C_p also includes the spin factor which is p-dependent; we shall not, however, need to consider the p dependence of C_p explicitly here.

In favourable circumstances it is possible to measure the amplitudes of several harmonics as well as of the fundamental over a range of fields, though very often the harmonics are too weak to be measured with adequate accuracy. The usual procedure for determining x from the field dependence of A_p (which need not be measured absolutely) is to make what is often called a 'Dingle plot.' This is a plot of $\ln(A_p H^n \sinh(\alpha p T / H))$ against $1/H$, which should be linear if (8.2) is valid; the slope of the plot immediately gives $\alpha p x$. It will be noticed that in this procedure α (i.e.. m/m_0) needs to be accurately known before the plot can be made and is needed again to derive x from the slope. A variant of the procedure, valid if $\alpha p T / H$ is large enough for sinh to be replaced by $\frac{1}{2}\exp$, is to plot simply $\ln(A_p H^n)$ against $1/H$ and the slope then gives $\alpha p(x + T)$; α still needs to be known to obtain $x + T$

371

from the slope, but the linearity of the plot no longer depends on the correct choice of α. We may notice that 'spot' values of x, i.e. at a single value of H, cannot be extracted either from an absolute value of A_p or from a ratio of harmonics such as A_2/A_1, unless the spin factor is known (and also the curvature factor if A_p alone is measured) because the spin and curvature factors are included in C_p. Rather it is the other way round: if x has been determined, the absolute values of A_p and quantities such as A_2/A_1 can be used to give information about the spin and curvature factors. This point will be elaborated in chapter 9.

Although (8.2) is indeed often found to be valid, it may break down for a variety of reasons; basically, either because the underlying damping mechanism may be more complicated than we have supposed, or because of unsatisfactory experimental conditions. The most obvious check of the validity of (8.2) is the linearity of the Dingle plot over as wide a range of fields as possible and other checks are that the same value of x should be found for different harmonics and also for different temperatures. If any of these checks fail and it is certain that the failure is not due to inappropriate experimental conditions, we have an interesting situation from which we may learn something about the underlying damping mechanism responsible for the amplitude reduction factor.

We shall now review briefly the various experimental pitfalls which may falsify the significance of a Dingle plot. We assume that where the oscillations have an appreciable harmonic content a proper Fourier analysis has been made so that the measured A_ps do truly represent the harmonic amplitudes.

(1) Curved Dingle plots may come about both from MI and MB and it is important to recognize if either of these effects is relevant. Some discussion of the symptoms which may be used for diagnosing these effects has been given in §7.4 and need not be repeated here. It should be emphasized, however, that in assessing the absolute amplitude it is the 'local' value $4\pi|dM_0/dH|$, i.e. $4\pi|dM/dH|$ divided by something like the Dingle reduction factor, that matters in deciding whether or not MI is significant (see §6.6). Sometimes it has been asserted that MI was negligible because $4\pi|dM/dH|$ was small, but in fact there may have been 'hidden' MI, in the sense that 'locally' $4\pi|dM_0/dH|$ was not so small (for example, see 'note added in proof' to Bibby and Shoenberg (1979)).

(2) It is important to be sure that there is no long beat in the oscillations due to the presence of two very slightly different frequencies. An example is shown in fig. 8.7, where measurements over a limited field range at first suggested a curved Dingle plot. When the field range was sufficiently extended, however, the beat was clearly revealed and the 'true' Dingle plot for a single frequency could be conjectured and was found to be quite linear. This example will be further discussed below.

(3) Field inhomogeneity can cause a field dependent reduction factor if the field variation over the sample is comparable to a dHvA period. From (2.142) we see

8.2. Measurement of Dingle temperatures

that to avoid such an effect we must have

$$\Delta H/H \ll H/\pi F$$

where ΔH is the field variation over the sample. For $F = 5 \times 10^8$ G and $H = 5 \times 10^4$ G this implies a homogeneity of better than 1 in 10^5.

(4) Sample imperfections such as mosaic structure, dislocations and other defects are obviously undesirable if the aim is to study the properties of an ideal crystal. The effect of mosaic structure, which will be discussed in §8.5, is particularly serious away from symmetry orientations. Crude imperfections such as a bent crystal or the presence of inclusions of a different orientation may indeed be avoided by careful sample preparation and handling, but some residual defects can never be completely eliminated. As we shall see below, these residual defects often provide the main contribution to the observed Dingle reduction factor, so it will be more appropriate to discuss their effect as one of the basic mechanisms, rather than under the heading of experimental artefacts.

(5) Each particular technique of studying the dHvA effect has additional pitfalls of its own. Thus in the torque method the compliance of the system must not be high enough for torque interaction (§3.3.2 and §6.7.1.2) to be appreciable. In the impulsive field method (§3.4.1) the oscillation amplitude may be affected in a complicated way by the resonant characteristics of the detecting system and also by eddy currents induced in the sample, which can cause field inhomogeneity and heating. In the modulation method (§3.4.2.1) it is important to take proper account of the effect of modulation amplitude if comparable values of dHvA amplitude at different fields are to be obtained. It is also important to allow adequately for eddy currents, whose effect is in general field dependent through the field dependence of the magnetoresistance; indeed it is best to make their effect negligible by working at a sufficiently low frequency, even though this may involve a loss of sensitivity.

Finally it is convenient to mention here an ingenious method due to Paul and Springford (1977) of improving the precision of *relative* Dingle temperature determinations. This method is appropriate when the aim is to investigate how the Dingle temperature depends on some variable, such as orientation or stress, which can be changed *in situ*. If the constant C_p in (8.1) is expected to be sufficiently independent of the relevant variable (as for instance if the orientation of an alkali metal crystal is varied), then the Dingle plots for different values of the variable should all intersect at a single point, say of ordinate y_0, when extrapolated to $1/H = 0$. Experimentally, of course, the intercepts of the various Dingle plots will not be identical because of random experimental errors, but if all of them are 'force-fitted' to go through a single point at $1/H = 0$ chosen as a fair estimate of y_0 – this might be simply the mean of the actual intercepts – the statistical error of the individual Dingle temperatures can be appreciably reduced. The reduction comes about essentially because of the 'leverage' provided by insisting that the straight line through the experimental points lying in the range $1/H_1$ to $1/H_2$ must go through a specified point at $1/H = 0$, and the error reduction is naturally most effective when $1/H_1$ is large compared with the range $1/H_2 - 1/H_1$. The individual Dingle temperatures will of course be subject to systematic error depending on the error in estimating y_0, but it is easily shown that the *differences* of the Dingle temperatures are *not* subject to the error, so that the effect of the variable under study is indeed brought out more clearly by this procedure. However, a word of caution is in place. This is the crucial importance of the assumption that the constant C_p is independent of the variable being studied. If in fact it is not quite constant, any apparent variation of Dingle temperature revealed by the procedure may just as well be, at least in part, merely a reflection of the variation of C_p. An outline of the analysis of Paul and Springford's method is given in Appendix 15.

8.3. Dilute alloys

In experiments on dilute alloys with a view to determining probabilities of scattering by foreign atoms, attention is focussed on the *changes* of x with impurity content. The fact that even nominally pure samples have irreproducible and sometimes rather high values of x due to sample imperfections can usually be allowed for, at least roughly, by extrapolation to zero concentration. As illustrated in fig. 8.1, the results do show a reasonably linear dependence of x on concentration c (or on resistivity which often provides the most reliable measure of concentration), particu-

Fig. 8.1. Variation of Dingle temperature x with concentration of added constituent: (a) Early measurements on Sn(Hg)($\bigcirc$) and Sn(In)($+$); ρ/ρ_0 is the ratio of resistivity at 4.2 K to that at room temperature and $\rho/\rho_0 = 10^{-2}$ corresponds to roughly 0.23% by weight of either Hg or In (Shoenberg 1952a). (b) Au(Ag) $\langle 111 \rangle$ belly (Lowndes *et al.* 1973).

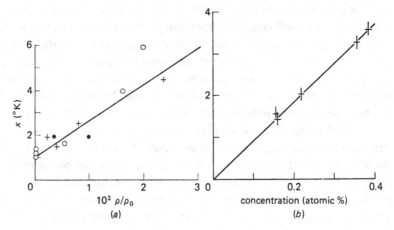

8.3. Dilute alloys

larly, as in more recent work, when care is taken to prepare samples in a reproducible way; we may notice that in favourable conditions (fig. 8.1(b)) the residual value due to imperfections is almost negligible. The linearity of the $x-c$ dependence suggests that we are dealing with the 'dilute' limit, in which the impurity atoms scatter independently of each other. Thus the slope of the $x-c$ line gives a scattering probability per unit time per impurity atom for the particular field direction concerned, if it is assumed that any contribution to x from sample imperfections is independent of c. This scattering probability is in fact an average round the extremal orbit of the FS normal to H, and we shall discuss below how it is possible to derive from such information the scattering probability at any point k on the Fermi surface.

The early experiments on dilute alloys of Sn with Hg or In (Shoenberg 1952a) were very much of an exploratory nature, but promising in showing that the relaxation time was of the same order of magnitude as enters into resistivity and as measured by other techniques. It was only a good deal later that this promise was followed up by King-Smith (1965), who measured the Dingle temperatures of a few noble metal dilute alloys. This work and later much more thorough and accurate studies mainly on the noble metals, by groups at Sussex, Leeds, NRC Ottawa and Oregon (for full bibliography see Coleridge 1980), not only confirmed the general order of magnitude agreement between dHvA and resistivity relaxation times in dilute alloys, but could be interpreted in a more fundamental way, in terms of the detailed electronic structure which had in the meantime been rather thoroughly worked out. More recently, Llewellyn, Paul, Randles and Springford (1977) have studied impurity scattering in K, where the nearly spherical FS might be expected to simplify the interpretation.

Before discussing the results on anisotropy of scattering we shall first consider in a little more detail what can be learnt from the mean value of x, when the anisotropy is not very strong (Brailsford 1966). For free electrons the distinction between the effective scattering probability $1/\tau_\rho$, which determines the electrical resistivity ρ and the $1/\tau$ which determines the Dingle temperature, is that for resistivity there is a weighting factor $(1 - \cos\theta)$ which favours a large angle θ of scattering. Thus if there is a scattering mechanism with a probability $P(\theta)\,\mathrm{d}\theta$ of scattering through an angle between θ and $\theta + \mathrm{d}\theta$, then

$$\tau/\tau_\rho = \int_0^\pi P(\theta)(1 - \cos\theta)\sin\theta\,\mathrm{d}\theta \bigg/ \int_0^\pi P(\theta)\sin\theta\,\mathrm{d}\theta \qquad (8.3)$$

Hence a comparison between τ_ρ and τ immediately gives some information about $P(\theta)$. Brailsford considers the special case of a scattering potential

$V(r) \propto e^{-qr}/r$, where $1/q$ is a range parameter (for $q \to \infty$, this becomes a δ-function). The corresponding $P(\theta)$ based on the Born approximation (which is not always valid) varies as $1/(u - \cos\theta)^2$, where $u = 1 + q^2/2k_F^2$ and substitution into (8.3) gives

$$\frac{\tau}{\tau_\rho} = (u - 1)\{\tfrac{1}{2}(u + 1)\ln[(u + 1)/(u - 1)] - 1\} \tag{8.4}$$

which increases from zero for $u = 1$ (which gives infinite weight to zero angle scattering) to 1 for $u = \infty$ (corresponding to the isotropic scattering of a δ-function-like scattering potential).

The quantitative development of this idea is essentially limited to the alkali metals and the bellies of the noble metals, for which the basic assumptions of free electron-like energy surfaces and isotropy of τ are most nearly satisfied. Before considering the experimental results, however, we must comment on a subtle point regarding the effective masses which enter into the derivations of τ and τ_ρ from x and ρ. As was pointed out in §2.6, although the electron-phonon interaction enhances the cyclotron mass which enters into the temperature damping, it does not affect the mass which enters into the Dingle factor. Thus it is the enhanced mass m which enters into the temperature damping, but the bare mass m_b that enters into the correct Dingle factor, and the masses are related by

$$m = m_b(1 + \lambda) \tag{8.5}$$

where λ is typically between 0.1 and 1. If, as is commonly done, this complication is ignored and (2.136) and (2.137) are used to determine τ, with β given by the enhanced mass m, then the 'bare' relaxation time τ_0 is related to the τ deduced in this way from x by

$$\tau_0 = \tau/(1 + \lambda) \tag{8.6}$$

This needs to be remembered if the absolute value of relaxation time is relevant and it is also of importance if λ is anisotropic. In our present context, however, where λ is nearly isotropic (see fig. 5.13) and we are interested only in the ratio of τ_0 to the bare resistivity relaxation time $\tau_{\rho 0}$, the complication is easily eliminated by introducing an 'effective' τ_ρ. The resistivity due to impurity scattering is in fact given by

$$\rho = m_b/ne^2\tau_{\rho 0} \tag{8.7}$$

so if we define our 'effective' τ_ρ by

$$\rho = m/ne^2\tau_\rho \tag{8.8}$$

376

8.3. Dilute alloys

where m is the enhanced mass of (8.5), we see that

$$\tau_{p0}/\tau_0 = \tau_p/\tau = \frac{x}{(\hbar n e^2/2\pi k m)\rho} \equiv x/x_p \qquad (8.9)$$

Thus the ratio of either bare or effective relaxation times can be equated to the ratio of the Dingle temperature x (determined in the usual way) to a 'resistive' Dingle temperatue x_p defined by (8.9), in terms of the enhanced rather than the bare mass.

Some experimental results for x per atomic % and for x/x_p are shown in table 8.1 and it can be seen that although x per atomic % varies widely, x/x_p typically lies between 1 and 2, indicating that τ/τ_p is typically between 0.5 and 1. In terms of the $P(\theta)$ in (8.3) this implies that there is some bias towards forward rather than backward scattering and, with the particular model used to derive (8.4), this would mean that q/k_F was roughly between 1 and 3. This scattering model is, however, too simple to do more than indicate that the average Dingle temperatures bear a reasonable relation to the electrical resistivities. In fact, as we shall see below, it is actually possible to calculate the resistivity on the basis of a detailed interpretation of the Dingle temperatures.

We shall now consider the anisotropy of Dingle temperature and, following Lowndes, Miller, Poulsen and Springford (1973), show how the

Table 8.1. *Estimates of x/x_p (i.e. τ_p/τ) and x per atomic %
for some dilute alloys*

Cu + (Au)[1]1.3(15) (Ge)[1]1.4(110) (Ni)[1]1.3(28) (Fe)[2]1.0(280) (Pd)[3]1.8(33)
(Pt)[3]1.2(51) (Rh)1.3(90) (Zn)2.6(18) (H)1.6(50)
Ag + (Au)[6,7]1.1(9)
Au + (Cu)[8]1.2(10) (Ag)[8]1.5(9) (Zn)[8]2.2(35) (Fe)[8]0.7(120) (Vac)[9]1.2(38)
K + (Na)[10]2.3(1) (Rb)[10]1.9(5)

References (1) Poulsen *et al.* (1974), (2) Coleridge (1972), (3) Coleridge, Templeton and Toyoda (1981), (4) Coleridge, Templeton and Vasek (1983), (5) Wampler and Lengeler (1977), (6) Springford, Stockton and Templeton (1969), (7) Brown and Myers (1972), (8) Lowndes *et al.* (1973), (9) Lengeler (1977), (10) Llewellyn *et al.* (1977).

Notes: The bracketed figures are rounded off values of x per atomic %. For the noble metals the values are for the belly (usually a mean of $\langle 100 \rangle$ and $\langle 111 \rangle$); the estimate of x_p is based on the average of cyclotron mass for $\langle 100 \rangle$ and $\langle 111 \rangle$. The values for Au(Zn) and Au(Fe) are based on only a single alloy and therefore are particularly rough. For the K alloys the figures are approximate averages over the FS. 'Vac' means vacancies.

377

anisotropy of the basic $\tau(k)$ can be derived from it. This is essentially a deconvolution problem: the scattering probability $1/\tau$ proportional to an observed x (see (2.136)) is a time weighted average of $1/\tau(k)$, (i.e. of the scattering probability at a point k) round the orbit, and we wish to extract $\tau(k)$ from a series of measurements of x for different orbits. The relevant average is

$$x = \frac{ehH}{4\pi^2 kmc} \oint \frac{\mathrm{d}t}{\tau(k(t))} \tag{8.10}$$

since the time for a complete orbit is $2\pi mc/eH$ (i.e. $2\pi/\omega_c$). Since moreover

$$\mathrm{d}t = \frac{\hbar c}{eH} \frac{\mathrm{d}k'}{v'(k)} \quad \text{and} \quad m = \frac{\hbar}{2\pi} \int \frac{\mathrm{d}k'}{v'(k)} \tag{8.11}$$

we obtain

$$x = \frac{\hbar}{2\pi k} \oint \frac{\mathrm{d}k'}{\tau(k)v'(k)} \Big/ \oint \frac{\mathrm{d}k'}{v'(k)} \tag{8.12}$$

where $\mathrm{d}k'$ is a scalar element of the orbit and $v'(k)$ is the velocity component in the plane of the orbit. If the FS and the velocity at each point on it are known in detail, (8.12) provides an explicit formulation of the relation between x and the local values $\tau(k)$*. A useful check on the computing procedure is that the denominator of (8.12) should be just $2\pi m/\hbar$.

In order to deconvolute (8.12) and obtain $\tau(k)$ from the Dingle temperatures measured at a variety of orientations, we need first to express the function $\tau(k)$ in some appropriately parametrized way. For cubic symmetry the parametrizations discussed in chapter 5, in the context of specifying Fermi surfaces, are available and it turns out that just as in the specification of a FS, the Fourier expansion formula (5.7) works best for the noble metals, while the cubic harmonic expansion (5.2) works best for the alkalis. A more basic deconvolution, in terms of a band structure parametrization, will be discussed later. If, in general, we choose some expansion of the schematic form

$$1/\tau(k) = \sum_i c_i \psi_i(k) \tag{8.13}$$

where the ψ_i are the chosen basis functions, then for any particular field

---

* For computing purposes, the integrals over k' in (8.12) can be expressed more conveniently in terms of the angle swept out by a radius vector to the orbit from an arbitrary origin (see Lowndes *et al.* 1973).

orientation (8.12) becomes

$$x = \sum_i c_i x_i \qquad (8.14)$$

with

$$x_i = \frac{\hbar}{2\pi k} \oint \frac{\psi_i(k)\,\mathrm{d}k'}{v'(k)} \bigg/ \oint \frac{\mathrm{d}k'}{v'(k)} \qquad (8.15)$$

The determination of the x_i involves a straightforward–though complicated–computation in terms of the known electronic structure. Once it has been carried out for all the field directions of interest, the determination of the c_i from the observed x values is simply a matter of solving a number of simultaneous linear equations, each of the form (8.14). Evidently at least as many values of x must be available as there are coefficients c_i to be determined and the results are more meaningful if the c_i have been determined by least squares from appreciably more values of x than the number of c_i.

Quite a reasonable representation of the anisotropy of $\tau(k)$ in the noble metals has been achieved by using only three or four c_is in a Fourier expansion similar to (5.7) and fitting the xs of six basic orbits for which F is at a turning point, so that mosaic structure effects are minimal. It was found that inclusion of values of x at other orientations, with appropriate corrections applied to remove increases of x due to mosaic structure (see §8.5), did not change the values of $\tau(k)$ by more than a few %. However the use of a greater number of terms in the expansion does modify the picture more significantly; typically, in going from three to seven terms, the values of $\tau(k)$ changed by 10% or so and even by as much as 50% at $\langle 100 \rangle$. For K (the only alkali metal so far studied), with dilute additions of Rb and Na, Llewellyn et al. (1977) found that a good fit to a cubic harmonic expansion (similar to (5.2)) could be achieved with only three harmonics and that the fit was not significantly improved if more harmonics were used.

Some typical results of these anisotropy determinations are illustrated in fig. 8.2. Not surprisingly, the anisotropy is somewhat more marked in the noble metals than in K, and we may note also that the anisotropy of τ is usually much stronger – between 10 and 100 times – than that of the radius vector of the FS from the zone centre. Although the relatively weak anisotropy of scattering in K finds no immediate qualitative interpretation, some of the general features of the anisotropy of scattering in the noble metal alloys are qualitatively plausible, as pointed out by Poulsen, Randles and Springford (1974).

Thus there are striking differences between the effects of adding homovalent, heterovalent and transition element impurities to the host

The Dingle temperature

noble metal. For instance we should expect that for addition of Au, Ge or Ni to Cu, the scattering would be strongest in those regions of the FS where the Cu wavefunction has s-, p- or d-like character respectively. For Cu(Au) the scattering is in fact fairly isotropic over most of the belly of the Fermi surface, which has s-like character, but appreciably smaller over the neck, which has p-like character. The p-like character means that the host wavefunction is strongest between the host atoms, so the addition of a homovalent impurity, whose potential differs from that of the host atom it has replaced only fairly close to the site, causes relatively less scattering.

Fig. 8.2. Contour maps of scattering probabilities for (a) Cu(Ni), (b) Cu(Ge), (c) Cu(Au) (Poulsen et al. 1974) and (d) K(Na) (Llewellyn et al. 1977). The contours are of $1/\tau$ where τ is the relaxation time deduced from the Dingle temperatures using the actual rather than the 'bare' value of m. The units are $10^{12} s^{-1}$ per atomic % in (d), 10^{13} in (a) and (c) and 10^{14} in (b).

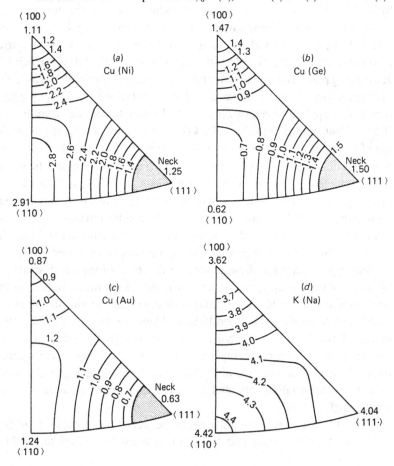

380

8.3. Dilute alloys

This was already pointed out in the early work of King-Smith (1965). For Cu(Ge) the impurity site carries a net charge, so the change of potential caused by introducing a Ge atom is much longer range than for addition of a homovalent atom. In consequence, scattering by p-like regions becomes more important and in fact the neck scatters rather more than most of the belly, and the variation of the p-like component over the belly causes a stronger anisotropy over the belly than for Cu(Au). The anisotropy for Cu(Ni) for which it is the d-like parts which scatter most, proves to be almost exactly complementary to that of Cu(Ge), in the sense that wherever the scattering is greater than the average in the one it is less in the other. Poulsen et al. suggest, indeed, that an alloy containing Ni and Ge in the ratio 5:1 should have an almost isotropic $\tau(\mathbf{k})$, and this has been confirmed by a recent study of the Hall effect (Gallagher and Greig 1981).

A complicating feature in any more detailed interpretation of the scattering anisotropy is that addition of impurity atoms introduces scattering, not only through modification of the potential around the impurity sites, but also through lattice distortion. Although the effects of lattice distortion can be partly included in the formulation of the potential scattering problem, the small angle scattering caused by the longer range components of the strain around an impurity atom cannot be easily dealt with. Fortunately, in the noble metals, this effect, though appreciable, is not large enough to matter too much in the alloys that have been studied (Watts 1973), but in an alkali metal the contribution to x of this small angle scattering by the strain field is quite comparable to the scattering by the potential of the impurity atom and as we shall see later this greatly complicates the detailed interpretation of the scattering anisotropy.

For the noble metals a remarkably detailed interpretation, not only of the anisotropy of $\tau(\mathbf{k})$, but also of the changes of FS (see §5.6) and of the electrical resistivity of the dilute alloys, can be given in terms of fundamental band structure theory. This interpretation, developed particularly by Morgan (1966) and by Coleridge, Holzwarth and Lee (1974), Holzwarth (1975) and Lee, Holzwarth and Coleridge (1976) provides a striking example of how the dHvA effect can in favourable conditions illuminate rather subtle aspects of electronic structure. We shall not attempt to give more than an outline of the main features of the theory and it should be emphasized that our outline omits any serious consideration of the reservations that should be made regarding the applicability of the theory in particular situations. Some discussion of such reservations may be found in the 1974 Oregon Conference Proceedings (*Phys. Cond. Matter* **19**, 219, 1975) and the theory is presented in detail in the review by Coleridge (1980).

As was mentioned in chapter 5, the observed FS of a pure noble metal host can be rather precisely described by a band structure calculation with the periodic potential specified in the 'muffin tin' approximation by three partial wave phase shifts $\eta_l^h (l = 0, 1, 2)$. By introducing a different set of phase shifts η_l^i, characteristic of the impurity atom, it proves possible to calculate various properties of the dilute alloy in terms of the differences

$$\Delta \eta_l = \eta_l^i - \eta_l^h \qquad (8.16)$$

and certain other parameters, which can be expressed in terms of the band structure of the host metal.

Thus for any given field direction, the value of x (or equivalently the reciprocal of τ) and the change of area ΔA for concentration c can be expressed as

$$mx/c = \frac{\hbar^2}{2\pi^2 k} \sum_l \left(\frac{1}{I_l} \frac{\partial A}{\eta_l^h} \right) \sin^2 \phi_l \qquad (8.17)$$

$$\Delta A/c = \sum_l \left(\frac{1}{I_l} \frac{\partial A}{\partial \eta_l^h} \right) \sin \phi_l \cos \phi_l \qquad (8.18)$$

where ϕ_l is the 'Friedel phase shift' defined by

$$\phi_l = \Delta \eta_l + \theta_l \qquad (8.19)$$

Here θ_l is the phase of a complex 'back scattering' coefficient and I_l is a normalizing factor. The factor $(1/I_l)(\partial A/\partial \eta_l^h)$ can be computed for any given field direction from the detailed band structure of the host metal in terms of the phase shifts η_l^h. Thus if the Friedel phase shifts are regarded as adjustable parameters their values can be determined from measurements of either x/c or $\Delta A/c$, or both, at a number of field directions. An important check on the results is that the ϕ_l should obey (or nearly obey, if the theory is not adequately valid because of lattice distortion effects) the Friedel sum rule

$$\mathcal{F} \equiv \frac{2}{\pi} \sum_l (2l + 1) \phi_l = \Delta Z \qquad (8.20)$$

where ΔZ is the valence difference between impurity and host. The importance of this sum rule is that (8.17) involves only $\sin^2 \phi_l$, and the sum rule provides some extra guidance to determine the signs of the ϕ_l.

Finally it is possible to calculate the resistivity ρ in terms of the same parameters, i.e. the ϕ_l and the parameters of the host band structure. The formulation of the calculation is rather complicated and we refer to the papers by Coleridge (1972, 1979) and Coleridge *et al.* (1974) for details.

For the theory to apply at all well it is important that there should not be too much lattice distortion and that the valence difference should not be too large ($\Delta Z \leqslant 2$ proves to be a rough criterion) and these conditions are reasonably satisfied by the noble metal alloys of table 8.2. It can be seen that for several of the alloys a single set of ϕ_l can be found which gives a good account of all three phenomena, i.e. electron scattering, change of FS area and resistivity. It should be noticed that the RBM predictions for the area changes, though qualitatively reasonable, do not fit the data nearly as well as the more sophisticated treatment. For the homovalent alloys Au(Ag) and Ag(Au) considerably better fits are obtained if different sets of the ϕ_l are fitted for the different phenomena. This is probably a consequence of the relative smallness of the alloying effects in the homovalent alloys as compared with heterovalent alloys, so that the effect of lattice distortions is relatively more serious. One nice feature of the homovalent alloy results is that when host and impurity are interchanged the ϕ_ls change sign without greatly changing their magnitudes and this encourages confidence in the physical meaningfulness of the analysis.

The relation of the ϕ_l to the more basic $\Delta\eta_l$ involves considerations of backscattering by the host lattice of electrons scattered by the impurity atom, and a detailed discussion by Lee et al. (1976) shows that in this respect too, the results make sense. It should be noted that in some of the earlier papers no distinction was made between the ϕ_l and the $\Delta\eta_l$ so that the values obtained do not really have the physical meaning attributed to them.

Notes to Table 8.2
References (1) Lowndes *et al.* (1973), (2) Coleridge *et al.* (1974), (3) Templeton and Coleridge 1975*b*, (4) Brown and Myers (1972), (5) Coleridge *et al.* (1983), (6) Coleridge *et al.* (1981), (7) Coleridge (1979).

Notes: Dr P. T. Coleridge has helped greatly in the preparation of this table and some of the entries have been either modified by him from the originally published form to bring them up to date or specially computed to bring all the information into the same form. To avoid complication, no errors are given but the experimental numbers have been rounded off so that in general the last figure is reliable to 2 or 3. It should be noted that all the frequency changes are given in terms of F_s, the free electron sphere value, rather than of the pure F value and that the experimental values have been corrected for lattice distortion. For (*a*), (*b*) and (*c*) three different fits are shown to illustrate that some improvement can be achieved in (*a*) and (*b*) by using different sets of phase shifts for the frequency and the scattering data, rather than a common set for both, while for (*c*) the improvement is only marginal. For (*d*) and (*e*) only single sets of phase shifts are shown, since these give fits well within the experimental error. The rigid band model (RBM) entries are included to show that while this model can give quite reasonable qualitative fits to the observed frequency changes, these fits are quantitatively poorer than those of the phase shifts.

Table 8.2

(a) Au(Ag)[1,2]

	ϕ_0	ϕ_1	ϕ_2	$\mathscr{F}$
Fit 1 (ΔF)	−0.22	−0.047	0.069	−0.01
Fit 2 (x)	−0.28	−0.072	0.051	−0.15
Fit 3 (Combined)	−0.26	−0.052	0.079	−0.01

	$\Delta F/cF_s$ (per at. %)			$x(m/m_0)/c$ (K per at. %)		
	meas.	fit 1	fit 3	meas.	fit 2	fit 3
B_{100}	−0.027	−0.026	−0.031	11.1	11.0	10.9
B_{111}	0.018	0.023	0.025	10.1	10.3	10.7
N	−0.011	−0.010	−0.011	0.82	0.82	0.72
R	−0.013	−0.011	−0.013	9.2	8.4	8.7
D	0.007	0.020	0.022	7.6	7.4	7.3

ρ/c ($\mu\Omega$ cm/at. %) meas.: 0.36   calc.: fit 2: 0.36   fit 3: 0.41

(b) Ag(Au)[3,4]

	ϕ_0	ϕ_1	ϕ_2	$\mathscr{F}$
Fit 1 (ΔF)	0.135	0.032	−0.050	−0.01
Fit 2 (x)	0.176	0.087	−0.085	0.01
Fit 3 (Combined)	0.238	0.047	−0.077	−0.003

	$\Delta F/cF_s$			$x(m/m_0)/c$		
	meas.	fit 1	fit 3	meas.	fit 2	fit 3
B_{100}	−0.002	−0.001	0.008	8.2	8.2	9.0
B_{111}	−0.032	−0.031	−0.038	8.0	8.1	8.8
N	0.014	0.015	0.022	1.3	1.3	0.6
R	0.005	0.009	0.008	7.3	7.2	7.0
D	−0.011	−0.008	−0.018	7.0	7.0	7.0

ρ/c meas.: 0.38   calc.: fit 2: 0.39   fit 3: 0.37

Table 8.2 (*cont.*)

(c) Cu(Zn)[5]

	ϕ_0	ϕ_1	ϕ_2	$\mathscr{F}$
Fit 1 (ΔF)	0.315	0.190	0.145	1.023
Fit 2 (x)	0.154	0.190	0.136	0.894
Fit 3 (Combined)	0.206	0.186	0.150	0.946

	$\Delta F/cF_s$				$x(m/m_0)c$		
	meas.	fit 1	fit 3	RBM($\Delta Z = 1$)	meas.	fit 2	fit 3
B_{100}	0.67	0.67	0.62	0.65	24	24	28
B_{111}	0.67	0.68	0.65	0.66	25	25	29
N	0.18	0.18	0.18	0.21	8.2	8.1	8.2
R	−0.62	−0.62	−0.60	−0.63	24	24	27
D	−0.58	−0.59	−0.56	−0.61	23	23	26

ρ/c meas.: 0.33   calc.: fit 2: 0.36   fit 3: 0.42

(d) Cu(Pd)[6]

$\phi_0 = -0.41$ $\phi_1 = -0.20$ $\phi_2 = -0.16$ $\mathscr{F} = -1.14$

	$\Delta F/cF_s$			$x(m/m_0)/c$	
	meas.	fit	RBM($\Delta Z = -1$)	meas.	fit
B_{100}	−0.72	−0.74	−0.65	47	48
B_{111}	−0.79	−0.76	−0.67	48	46
N	−0.18	−0.18	−0.21	9.3	9.1
R	0.67	0.68	0.63	40	40
D	0.61	0.64	0.61	39	38

ρ/c meas.: 0.89   calc.: 0.72

(e) Au(Ga)[7]

$\phi_0 = 0.72$ $\phi_1 = 0.36$ $\phi_2 = 0.28$ $\mathscr{F} = 2.03$

	$\Delta F/cF_s$			$x(m/m_0)/c$	
	meas.	fit	RBM($\Delta Z = 2$)	meas.	fit
B_{100}	1.30	1.27	1.41	115	115
B_{111}	1.23	1.24	1.32	110	109
N	0.30	0.28	0.35	21	20
R	−1.17	−1.16	−1.26	100	98
D	−1.04	−1.08	−1.22	87	92

ρ/c meas.: 2.15   calc.: 2.37

It must also be remembered in comparing various papers that the choice of η_1^h specifying the host band structure depends on the choice of ζ. Nevertheless the quantities $(1/I_l)(\partial A/\partial \eta_l^h)$ which enter into the analysis are found empirically to be independent of ζ as are also the ϕ_ls.

Although the theory works so well for selected noble metal alloys, it is unable to describe the scattering and its anisotropy at all adequately in the alkali alloys K(Na) and K(Rb) because of the relatively much larger effect of lattice distortion. Indeed to fit the Dingle temperature anisotropy found by Llewellyn et al. to (8.17) would require a negative value of one of the $\sin^2 \phi_l$, which is clearly nonsense. The nearest to a fit is obtained by setting $\phi_2 = \phi_3 = 0$ and choosing ϕ_1 to make the mean scattering agree with experiment, but this produces an anisotropy considerably smaller than the observed one. An alternative theory of anisotropic scattering based on lattice distortion alone was proposed by Benedek and Baratoff (1973) but this also fails to agree with experiment. Llewellyn et al. suggest that a better fit to experiment might be obtained if the two theories were combined, and a recent calculation by Molenaar, Coleridge and Lodder (1981) has gone some way in this direction.

Relatively little systematic work has yet been done on anisotropy of scattering in dilute alloys based on polyvalent metals, though frequency changes (i.e. changes of FS) have been studied in a number of systems, as discussed in §5.6. Coleridge (1980) discusses the application of pseudo-potential theory to both the scattering and the FS changes and also gives full references to such experiments as there are.

8.4. Dislocations

Most of our discussion of the damping of dHvA amplitude by imperfections in chemically pure metal crystals will be concentrated on the effect of dislocations, since most imperfections can indeed be described in terms of various combinations of dislocations. Two types of imperfection, however, need to be considered separately.

One of these is vacancies but, as can be seen from table 8.1, their effect is very similar to that of the addition of impurity atoms. This is not surprising since, just like impurity atoms, they act as rather concentrated scattering centres. To obtain the results quoted in table 8.1 Lengeler (1977) had to quench samples from a high temperature in order to reach concentrations (typically 10^{-2} atomic %) which increased the Dingle temperature by a measurable amount (of order 0.5 K). But a crystal sample prepared in the ordinary way with some annealing would have too few vacancies to contribute appreciably to the Dingle temperature. We shall therefore not need to consider vacancies further as imperfections likely to contribute to the Dingle temperature and in any case their contribution may better be considered under the heading of impurities.

The other type of imperfection needing separate consideration is mosaic structure. This can indeed be described in terms of a particular arrangement of dislocations, but is much more simply treated by regarding the sample as an assembly of small (but not too small!) grains of slightly different orientations and therefore slightly different dHvA frequencies. The problem is then a straightforward phase smearing problem which we shall discuss in §8.5. We shall see that usually the contribution of mosaic structure to the Dingle temperature is only significant away from orientations such as symmetry directions at which F has an extremum. Thus the effects of mosaic structure can usually be eliminated by working at such extremal orientations.

Returning now to the effect of dislocations *per se* we must first emphasize that our discussion will be greatly oversimplified in its assumptions about their nature, and the results quoted will be at best order of magnitude guides to real situations. Thus we shall consider mainly edge dislocations and only simple arrays of dislocations, rather than complicated ones such as loops. We shall occasionally have to bear in mind that dislocations have a sign attached to them which determines the sign of their strain field. Sometimes dislocations of opposite sign come closely paired together rather like dipoles. In that case they act as concentrated lines of scattering centres and behave like rows of vacancies without the long range strain field characteristic of well separated dislocations.

An important parameter throughout our discussion will be the 'scale length' d of the strain variation through the sample and its relation to a second important parameter; the orbit radius r. If we imagine for concreteness a square lattice array of dislocations with a density D cm^{-2}, i.e. of lattice constant $D^{-1/2}$, we may suppose

$$d = D^{-1/2} \tag{8.21}$$

Well within a distance d from a dislocation the strain field is dominated by the $1/R$ field of that dislocation alone, since the fields of the neighbouring dislocations, which will be as often plus as minus, tend to cancel out. Beyond the distance d, the strain field of the next nearest dislocation to the point concerned takes over and so on.

Another important scale length is b, the magnitude of the Burgers vector, which is of atomic dimensions, typically 3×10^{-8} cm (see table 8.3). As explained in Appendix 16, b determines the magnitude of the strain field around the dislocation. Thus the strain at distance R from the dislocation

Table 8.3. *Estimates of important parameters for Cu and Bi*

	$10^8 b$(cm)	$10^{-8} k$(cm^{-1})	$10^{-8} F$(G)	$d \ln F/ds$	$\mu = \pi bk(d \ln F/ds)$	$H_0/D^{1/2}$(G cm)
Cu (belly)	2.5	1.4	5.8	0.29	3.1	26
Cu (neck)	2.5	0.27	0.22	2.9	5.9	10
Bi ($H\|$bin.)	4.5	6.5×10^{-3}	1.4×10^{-4}	~150	~14	~0.6

Notes: The various symbols are defined in the text. The estimates of $d \ln F/ds$ are based on the discussion of Appendix 16 and are really little more than orders of magnitude, especially for Bi.

is of order $b/2\pi R$ for $R > b$. Closer in than $R \sim b$ the concept of strain loses its meaning and b can be regarded as a 'cut-off' distance in calculations which involve integrations over R.

The strain in the immediate neighbourhood of a dislocation (i.e. for $R \sim b$) is of the same order of magnitude as that of a vacancy or an impurity atom and so contributes to the Dingle temperature in much the same way as a line of such point imperfections. This contribution, however, proves to be unimportant simply because the dislocation density D is in practice too small. It is easily seen that a density D of lines is roughly equivalent to an concentration c of point scattering centres given by

$$c \sim D/n^{2/3}, \tag{8.22}$$

where n is the total number of atoms per cm^3. If we take $D = 10^{10} \, cm^{-2}$, a high value for a typical sample, and $n \sim 10^{23}$ we find $c \sim 5 \times 10^{-6}$, i.e. 5×10^{-4} atomic %, which (cf. table 8.1) would contribute only 0.1 K or less to the Dingle temperature and perhaps $5 \times 10^{-3} \, \mu\Omega \, cm$ to the residual resistivity. The order of magnitude of this rough estimate is confirmed by direct measurement (Basinski and Saimoto 1967) of the residual resistivity due to dislocations which gives about $10^{-3} \, \mu\Omega \, cm$ for $D = 10^{10} \, cm^{-2}$. In fact the measured residual resistivity provides a simple measure of any appreciable point scattering contribution to the Dingle temperature from dislocations and, if it should be appreciable, it can be estimated and subtracted from the total observed x to give a residue caused by the long-range strains of the dislocations alone.

The novelty of the long-range strain field (scale length d as in (8.21)) is that it can cause only very low angle scattering (since the electron wavelength is much less than d) and so contributes negligibly to the resistance, though it can contribute quite appreciably to the Dingle temperature even for relatively low dislocation densities D. We shall now show how the dHvA damping factor can be roughly estimated in two extreme regimes: (a) small orbits: $r \ll d$ and (b) large orbits: $r \gg d$. The situation for an intermediate regime $r \sim d$ can be still more roughly estimated by interpolating between the extreme regimes.

8.4.1. Small orbits: $r \ll d$

If the orbit is small enough in relation to d it can be considered as lying in a region of almost constant strain and therefore constant dHvA frequency F. The damping can be thought of as essentially coming from phase smearing due to the differences of F between one region and another. This is the problem already discussed in §2.3.7, where it was shown that the amplitude damping factor depends on the distribution function of ΔF, i.e.

of phase, through the sample, which depends in turn on the distribution of strain s through the sample.

If we suppose that for strain s

$$\Delta F = s\,\mathrm{d}F/\mathrm{d}s \tag{8.23}$$

then (2.138), which gives the dHvA phase shift ϕ, becomes (for the pth harmonic)

$$\phi = 2\pi ps(\mathrm{d}F/\mathrm{d}s)/H \tag{8.24}$$

We need not for the present worry about the precise definition of s, which is discussed in Appendix 16. If the distribution function of s through the sample is $D(s/\sigma)$, where σ is some characteristic strain* such that $D(s/\sigma)$ falls off rapidly once $|s|$ exceeds σ, the corresponding distribution of phase can then be expressed as $D(\phi H/pH_0)$, where

$$H_0 = 2\pi\sigma(\mathrm{d}F/\mathrm{d}s) \tag{8.25}$$

and we see that pH_0/H plays the role of the scaling parameter λ of (2.118). The reduction factor due to the phase smearing is then

$$R_\mathrm{D} = |f(pH_0/H)|/f(0) \tag{8.26}$$

where $f(\lambda)$ is the Fourier transform with respect to λ of $D(z)$ (see (2.122)). Two strain distributions which are simple to deal with mathematically are the Lorentzian

$$D(s/\sigma) \propto 1/[1 + (s/\sigma)^2] \tag{8.27}$$

and the Gaussian

$$D(s/\sigma) \propto \mathrm{e}^{-s^2/\sigma^2} \tag{8.28}$$

(we need not worry about normalization factors, which are automatically looked after by the denominator in (8.26)). The corresponding reduction factors are

$$R_\mathrm{D} = \mathrm{e}^{-pH_0/H} \tag{8.29}$$

for the Lorentzian, and

$$R_\mathrm{D} = \mathrm{e}^{-p^2 H_0^2/4H^2} \tag{8.30}$$

for the Gaussian. The Lorentzian result (8.29) becomes, of course, the

---

* As in §2.3.7 we are assuming that the distribution function is characterized by only a single 'scale' parameter σ. This is an oversimplification and in the more detailed theories of §8.4.3, where more than one parameter is introduced, the Fourier transform formulation becomes less simple.

8.4. Dislocations

standard Dingle factor $\exp(-2\pi^2 pk_B x/\beta H)$, if we put for the Dingle temperature

$$x = \beta H_0/2\pi^2 k_B = \beta\sigma(\mathrm{d}F/\mathrm{d}s)/\pi k_B \tag{8.31}$$

In most of our discussion it will however be more convenient to specify the reduction factor by H_0 rather than by x.

As we shall see below, the H dependence of (8.29) seems to fit experiments surprisingly well, suggesting that if the basic assumption of $r \ll d$ is indeed valid, the distribution of strains must be Lorentzian or close to it. We shall see also that this result is not at all well understood in terms of what might be expected of real dislocations. The order of magnitude of σ (and hence of H_0 if $\mathrm{d}F/\mathrm{d}s$ is known) may be estimated roughly in terms of the dislocation density D from the known strain fields of dislocations and it is shown in Appendix 16 that

$$\sigma \sim bD^{1/2} \sim b/d \tag{8.32}$$

should be a reasonable estimate, where b is the Burgers vector. Thus from (8.25) we find

$$H_0 \sim 2\pi b D^{1/2}(\mathrm{d}F/\mathrm{d}s) = 2\pi(b/d)(\mathrm{d}F/\mathrm{d}s) \tag{8.33}$$

It is also sometimes convenient to express H_0 in terms of a dimensionless parameter μ characteristic of the material and defined as

$$\mu = \pi b k(\mathrm{d}\ln F/\mathrm{d}s) \tag{8.34}$$

where k is the k-space orbit radius and it is easily seen that

$$H_0 = 2F\mu D^{1/2}/k = 2F\mu/kd \tag{8.35}$$

The significance of μ is that it provides a link between H_0/H and r/d (which determines whether the orbit is 'small' or 'large'). Thus from (2.4) and (2.7) it follows that

$$H = 2F/rk \tag{8.36}$$

and we find

$$H_0/H = \mu r/d \tag{8.37}$$

Some rough estimates of μ based on the discussion of Appendix 16 are given in table 8.3 and it can be seen that μ has not very different values for widely different situations, and the assumption $\mu \sim 10$ should not be too far wrong. This implies that we can never get far into the 'large orbit' regime ($r/d \gg 1$) because the reduction factor R_D of (8.29) would make the amplitude too small to observe.

The Dingle temperature

The prediction of (8.33) or (8.35) is that H_0, and hence the Dingle temperature x, should vary as $D^{1/2}$ and some rough estimates of $H_0/D^{1/2}$ are included in table 8.3. However, we shall postpone any comparison with experiment until we have discussed the large orbit regime and presented some account of more detailed calculations.

8.4.2. Large orbits: $r \gg d$

In this regime the electron encounters many regions of varying strain (i.e. varying F) in the course of an orbit and we can estimate the net change of phase (as compared with that of a constant F) by an argument due to Watts (1974). We shall however present only a rather crudely simplified version of the argument, in which the actual continuous variation of phase (i.e. of strain) round the orbit is replaced by a series of N equal abrupt changes $\pm \Delta$ occurring at equal intervals, which we take to be d (i.e. $D^{-1/2}$). For the pth harmonic we have to consider p circuits of the orbit rather than 1. The essential point of the argument is that the total phase change Φ in the p circuits must have a Gaussian distribution of the form

$$D(\Phi) \propto e^{-\Phi^2/\lambda^2} \tag{8.38}$$

but λ^2 varies as $1/H$ so, rather surprisingly, the final reduction factor $e^{-1/4\lambda^2}$ (see table 2.2) still has the Dingle form $e^{-H_0'/H}$. The parameter H_0' is easily calculated.

We first note that the number N of abrupt changes in p circuits is

$$N = 2\pi pr/d = (2\pi p/\mu)(H_0/H) \tag{8.39}$$

(using (8.37)). The magnitude of Δ corresponds to a fraction $1/N$ of the phase change that would have occurred if the strain change had continued unchanged for the whole of the p circuits, i.e.

$$\Delta = (2\pi p\sigma/NH)(\mathrm{d}F/\mathrm{d}s) = pH_0/NH = \mu/2\pi \tag{8.40}$$

The total phase change Φ is

$$\Phi = \pm \Delta \pm \Delta \pm \Delta \pm \cdots \text{ to } N \text{ terms} \tag{8.41}$$

and as shown in A17.1, the parameter λ of its Gaussian distribution is given by*

$$\lambda^2 = 2N\Delta^2 = (p\mu/\pi)(H_0/H) \tag{8.42}$$

---

* However, the derivation requires Δ, i.e. $\mu/2\pi$, to be small, while actually $\mu/2\pi \sim 1$. Hopefully this does not affect the qualitative conclusions too seriously.

8.4. Dislocations

so that the reduction factor is

$$R_D = \exp(-\tfrac{1}{4}\lambda^2) = \exp[-(p\mu/4\pi)(H_0/H)] \tag{8.43}$$

Thus not only does the reduction factor in the large orbit regime have the same Dingle form as for a Lorentzian distribution in the small orbit regime, but since $\mu/4\pi$ is typically of order 1, the coefficient of $1/H$, and therefore the Dingle temperature, is not very different from what it is in the small orbit regime.

8.4.3. More detailed calculations

In this section we shall outline attempts by Watts (1972, 1974, 1977) and Chang and Higgins (CH) (1975) to develop the rather hand waving arguments of the previous section in a more quantitative way. These calculations take account of the strain field of an array of dislocations in more realistic detail and also formulate the problem in such a way that the whole range of fields from 'small' to 'large' orbit regimes is embraced by the same formulation. Although this basic formulation is rather general, somewhat special assumptions and simplifications have to be made in applying it: Watts assumes a random array of edge dislocations normal to the plane of the orbit, while CH assume a regular array which simulates the condition of some specially prepared samples. In both calculations mathematical simplifications are made in order to obtain explicit results and it is possible that some features of the results are due to these simplifications, rather than genuine properties of the model.

The formulation of both calculations is based on phase smearing and is essentially the same as we have already used in the limits of small and large orbits.* Formally, the reduction factor R_D is given by

$$R_D = \frac{1}{S}\left|\int e^{i\Delta\psi(x_0, y_0)}\,dS\right| \tag{8.44}$$

where dS is an area element of a suitably large area S of the orbit plane, and $\Delta\psi(x_0, y_0)$ is the average phase round an orbit of radius r centred on (x_0, y_0). This average is (for a circular orbit)

$$\Delta\psi(x_0, y_0) = \frac{1}{2\pi}\int_0^{2\pi} \delta\psi\,(x_0 + r\cos\phi, y_0 + r\sin\phi)\,d\phi \tag{8.45}$$

where $\delta\psi(x, y)$ is the phase shift at (x, y) defined for the pth harmonic in the

---

* A theoretical justification for this formulation has been given by Ginzburg and Maksimov (1977).

usual way as

$$\delta\psi(x, y) = 2\pi p \, \delta F(x, y)/H \tag{8.46}$$

Evidently if we are in the small orbit regime, the integrand in (8.45) is practically constant so that there is no distinction between $\Delta\psi(x_0, y_0)$ and $\delta\psi(x_0, y_0)$. If a distribution function of δF, i.e. of $\Delta\psi$, is introduced into (8.44) this leads to essentially the results we obtained earlier. Our hand waving argument in the large orbit limit amounts to a crude evaluation of (8.45) followed by assuming a Gaussian distribution of the phases in (8.44) with a breadth which is field dependent.

CH do not introduce a distribution function, but use explicit forms of the strain field, and carry through numerical calculations of the integrals for particular values of the relevant parameters chosen to match their experimental conditions. One example they give to illustrate how the calculated damping varies with field (for the Cu neck and $D = 1.5 \times 10^7 \, \mathrm{cm}^{-2}$) shows a strange feature corresponding to disappearance of amplitude, i.e. a blowing up of the effective Dingle temperature, at a critical field of about $16 \, \mathrm{kG}$ (for which $r \sim \frac{1}{2}D^{-1/2}$). This may well be a consequence of some of the simplifications in their calculation, such as the assumption of a strictly periodic array and the assumption that all the dislocations have the same sign. But it makes the degree of validity of their predictions, even well away from the critical field, somewhat uncertain. Some quantitative comparison of their calculations with experiment will be discussed in §8.4.5.

Watts in his most recent calculations (1977) considers a random array of edge dislocations (but all normal to the plane of the orbit) and evaluates (8.44) and (8.45) analytically in a rather general way. He first shows that the strain field of any edge dislocation line which threads an orbit contributes exactly nothing to the integral (8.45), while that of a line which passes outside the orbit and is perpendicular to the plane of the orbit contributes exactly $\delta\psi(x_0, y_0)$, i.e. just the effect of its strain at the centre of the orbit. Thus as far as any particular orbit is concerned, the relevant change of phase $\Delta\psi$ is the sum of the effects of the strains at the orbit centres of all dislocation lines outside the orbit.

By invoking a general statistical theorem and assuming that the dislocations are completely randomly disposed (which may not be quite valid) he is eventually able to obtain a closed expression for the phase smearing reduction factor R_D, without explicitly determining the strain distribution.* His result for the fundamental ($p = 1$), expressed in our

---

* The result (8.47) of course implies a certain strain distribution, which was considered explicitly in an earlier paper (Watts 1972), though under slightly different assumptions.

8.4. *Dislocations*

notation, for $\mu/2\pi \lesssim 1$ is approximately

$$\ln R_D = -\frac{1}{8\pi}(H_0/H)^2 \ln[1 + (\mu v H/H_0)] \tag{8.47}$$

where μ is as defined in (8.34), and

$$v = t/d \tag{8.48}$$

The extra parameter v has been introduced to take account of the 'correlation' distance t which is such that the strain at (x_0, y_0) due to dislocations further away than t from the orbit can be ignored. It is not very clear how large v should be made; in our hand waving arguments we had effectively put $v \sim 1$, but as we shall see later, it has to be made a good deal higher (~ 10) to fit experimental data. We notice that in this limit of $\mu/2\pi \lesssim 1$, only the single parameter μv enters into the relation between $\ln R_D$ and H_0/H. For $\mu/2\pi \lesssim 1$ and fields so low that $H_0/H \gg \mu v$, (8.47) approaches the asymptotic limit

$$\ln R_D = -\frac{\mu v}{8\pi}\left(\frac{H_0}{H} - \frac{1}{2}\mu v\right) \tag{8.49}$$

Another limiting case is again for $H_0/H \gg \mu v$ (i.e. $r/d \gg v$) but valid for any value of μ such that $\mu \ll [(2\pi/v)(H_0/H)]^{1/2}$. We then have

$$\ln R_D = -2\pi v^2\left\{\left(\frac{H_0}{\mu v H} + \frac{1}{2}\right)[1 - J_0(\mu/2\pi)] - (\mu/4\pi)J_1(\mu/2\pi)\right\} \tag{8.50}$$

We note that the two parameters μ and v now enter separately. This formula reduces, as it should, to (8.49) if $(\mu/2\pi) \ll 1$. For $\mu > 1$ and $H_0/H \lesssim 1$ a more elaborate formula must be used which we shall not quote, but in fact (8.47) is still a fair approximation to the more elaborate result even for $\mu/2\pi = 1.75$, the highest value we shall need to discuss.

The general behaviour of Watts' formula over a wide range of H_0/H is illustrated in fig. 8.3 and over the more limited range of practical interest in fig. 8.6. As with the CH calculations, the Dingle plots are appreciably curved in the small orbit regime and only straighten out at such low fields that the amplitude would be too small to observe. For comparison the Dingle plot for our simple handwaving arguments, i.e.

$$\ln R_D = -H_0/H \tag{8.51}$$

is also shown.

Before comparing these theoretical predictions with experiment we shall first point out some reservations that need to be made about the applicability of idealized models, such as have been assumed in the detailed calculations.

8.4.4. Some reservations

First of all, our discussion so far has assumed a homogeneous sample, in the sense that exactly the same pattern of dislocations goes all the way through, with the same density D of the same kind of dislocations and the same distribution function of strain throughout. Except perhaps in very specially prepared samples (such as those of CH), this is a gross oversimplification. In fact the usual situation is that some parts of the sample are heavily dislocated while others may be almost perfect over macroscopic dimensions. A possible way of dealing with this is to think of the sample as made up of many regions in each of which the factor R_D is reasonably uniform and then to average over the sample. If ideally R_D has the same type of H

Fig. 8.3. Examples of Dingle plots over a wide range of H_0/H according to Watts' (1977) theory; (a) $\mu = 4.1$, (b) $\mu = 11$ and $\nu = 8$ for both. These parameters are found to give a fair fit to the experimental data of Chang and Higgins (1975) for Cu belly and neck respectively, as shown in fig. 8.6. The appropriate values of H_0 corresponding to these values of μ would be about 130 kG and 70 kG for belly and neck respectively. Only the small region of the diagram down to $\ln R_D \sim -10$ is of any practical interest, since the amplitudes for smaller fields would be unobservably small, but the continuation illustrates how the asymptotic limit of the formula (broken lines) is approached. The simple formula $\ln R_D = -H_0/H$ is also shown (light lines) for comparison.

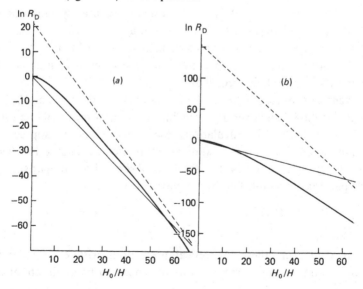

8.4. Dislocations

dependence throughout, say $e^{-H_0/H}$, we could usefully think of a distribution function say $D'(H_0)$ of the parameter H_0 which determines the local damping. We should then find for the overall reduction factor

$$R_D = \int D'(H_0) e^{-H_0/H} \, dH_0 \bigg/ \int D'(H_0) \, dH_0 \qquad (8.52)$$

It is hardly worth developing the mathematical formalism of this idea (involving Laplace transforms) because it is unlikely that the function D' has any useful general form in practice and moreover because the assumption of $e^{-H_0/H}$ for the local reduction is already rather a special one.

However, (8.52) may be used to illustrate (again in an idealized way) some of the situations which may occur to complicate the interpretation of experiments:

(a) *'The curate's egg'–good in parts:* suppose a fraction f has H_0 throughout, while the rest, $(1 - f)$, is perfect ($H_0 = 0$). The reduction factor for the whole sample will be

$$R_D = (1 - f) + f e^{-H_0/H} \qquad (8.53)$$

The plot of $\ln R_D$ versus $1/H$ will no longer be straight; if f is small we have approximately

$$\ln R_D = -f (1 - e^{-H_0/H})$$

the plot of which against $1/H$ would curve appreciably once $H \lesssim H_0$.

(b) *Converse of the curate's egg–bad in parts:* we suppose a fraction f is so bad that it contributes no amplitude at all while $(1 - f)$ has a reasonable H_0. This gives

$$\ln R_D = \ln(1 - f) - H_0/H \qquad (8.54)$$

The Dingle plot would be linear and its slope would correctly give the H_0 of the reasonable part, but the intercept would give a falsified estimate of the absolute amplitude.

(c) *Top hat distribution:* suppose H_0 is uniformly distribution betwen H_{01} and H_{02}, then (8.52) becomes

$$R_D = \frac{1}{H_{01} - H_{02}} \int_{H_{02}}^{H_{01}} e^{-H_0/H} \, dH_0 = \frac{H}{H_{01} - H_{02}} (e^{-H_{02}/H} - e^{-H_{01}/H})$$

$$(8.55)$$

For H larger than both H_{01} and H_{02} we approach a linear Dingle plot, corresponding to the mean of H_{01} and H_{02}, but when $H \sim H_{01} - H_{02}$ the Dingle plot becomes curved and at sufficiently small H its slope will be given by the smaller of H_{01} and H_{02}.

Another complication is that we have throughout assumed the distribution function of strains to be symmetrical about zero, so that the phase of the oscillations is unaffected by the amplitude reduction (see §2.3.7). In fact in an inhomogeneous sample it is unlikely that this assumption will be locally valid, and in consequence a formulation of the type (8.52) is too simple, since it ignores possible variations of the average local phase which would accompany variations of the local H_0.

Finally it should be emphasized again that all the models on which calculations have been made are oversimplified and some of the simplifications in working out the consequences of the models may not be altogether valid. How far such simplifications may be responsible for particular features predicted by the calculations is difficult to say.

8.4.5. Experimental evidence

We shall first review a number of experimental approaches to deciding whether a Dingle plot falls in the small or large orbit regime and then present what little experimental evidence there is for the validity of the theoretical calculations we have outlined above.

8.4.5.1. Which regime?

(a) The most obvious test is actually to determine the dislocation density D and to compare $D^{-1/2}$ (which in our simple-minded approach gives d, the 'scale' of the strain variation) with the orbit radius r (as specified by $r = 6.6 \times 10^{-8} k/H$). The field H for which $r = D^{-1/2}$ is shown in table 8.4 for various values of D for Cu (neck and belly) and for Bi (H parallel to binary axis). In fact only a few crystals used in dHvA studies have so far been examined with a view to determining D by etch pit counts. Coleridge and Watts (CW) (1971b) and Terwilliger and Higgins (TH) (1973) studied Cu crystals with D in both studies ranging from about 10^5 to $6 \times 10^7 \, \text{cm}^{-2}$. CH also studied Cu, with rather smaller values of D ($1.4 \times 10^7 \, \text{cm}^{-2}$ or less), but with a technique of sample preparation which produced a much more regular and well characterized array of dislocations. The only other study in which D has been estimated was that of Barklie and Shoenberg (1975 and unpublished) on Bi (with H parallel to the bi-

Table 8.4   *Comparison of $d = D^{-1/2}$ and orbit radius r*

$D(\text{cm}^{-2})$	10^6	10^7	10^8	10^9	10^{10}
Cu (belly)	9	29	90	290	900
Cu (neck)	1.8	5.6	18	56	180
Bi ($H \parallel$ bin.)	0.04	0.14	0.43	1.4	4.3

Notes: The table gives the values of H in kG for which $r = d = D^{-1/2}$ ($r = 6.6 \times 10^{-8} k/H$ with k as in table 8.3)

nary axis): the distribution of dislocations was very variable over different regions of the sample but D was probably somewhere between 10^6 and 10^7 cm^{-2}. The contrast between the fair regularity of distribution in Chang and Higgins' specially prepared samples and the rather wild irregularity in those of Barklie and Shoenberg is illustrated in fig. 8.4.

We shall discuss the results of these experiments more fully below but for the moment we wish only to point out that with the values of D indicated above, table 8.4 suggests that for the Cu neck (lowest field used about 25 kG) and Bi (lowest field used 130 G) the experiments were entirely or almost entirely in the small orbit regime. For the Cu belly, however, for which the lowest fields used were also about 25 kG, the regime for the higher values of D may at the lower fields have been a bit beyond the small orbit regime and approaching (but not fully reaching) the large orbit regime. However we should not be too dogmatic, because the criterion $r = d$ (or in Watts' theory $r = vd$) may be misleading: it could be that it is $2r$ or even $2\pi r$ which should be compared with d or $\frac{1}{2}d$, and in that case the small orbit regime, even for the Cu necks and Bi, might apply fully only at the upper end of the field range.

(b) If no direct information is available about the dislocation density D, an indirect but simple pointer is to look at the magnitude of the reduction factor actually observed. Provided the Dingle plot is not too strongly

Fig. 8.4. Etch pit micrographs of (a) a specially prepared $\langle 111 \rangle$ surface of Cu (CH 1975), (b) a cleavage plane in a Bi crystal similar to those used by Barklie and Shoenberg (1975).

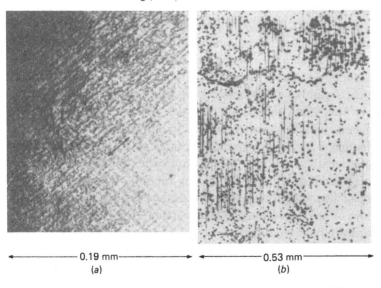

$\longleftarrow$————— 0.19 mm —————$\longrightarrow$ $\longleftarrow$————— 0.53 mm —————$\longrightarrow$

(a)   (b)

curved it can be extrapolated back to $1/H = 0$ without too much uncertainty and $\ln R_D$ can be immediately estimated for any particular value of $1/H$. If now we identify $-\ln R_D$ with H_0/H (which makes sense if the Dingle plot is straight) and use (8.37), we can estimate the value of r/d and decide which regime we are in. With the values of μ of table 8.3, this method gives estimates of r/d roughly comparable to those of the direct method (a) above, in the few cases where experimental evidence is available.

(c) As mentioned earlier, the reduction factor $R_D(p)$ for the pth harmonic at field H should be the same as that for the fundamental (i.e. $R_D(1)$) at field H/p, *provided* that the orbit regime is sufficiently similar at both fields. In itself this does not tell us *which* regime is the one, but if there is other evidence that at some particular field we are probably in the small orbit regime, then the behaviour of the pth harmonic at p times this field could provide confirmatory evidence. A special case is if the Dingle plot is linear, i.e. if $-\ln R_D$ is proportional to $1/H$; we then expect the plot of $-\ln R_D(p)$ to have p times the slope of that of $-\ln R_D(1)$. As we shall see below, the experimental evidence on the behaviour of harmonics is somewhat confused.

(d) As discussed in §6.5 the effect of magnetic interaction on the amplitude and harmonic content of dHvA oscillations is different according as the Dingle reduction factor comes from 'microscopic' causes (e.g. impurity scattering and probably dislocations in the large orbit regime) or from phase smearing. The assumption that the sample is essentially perfect over regions large compared with the orbit size, i.e. that we have the small orbit regime, seems to be confirmed by various experiments in fields above 50 kG on the noble metals (Shoenberg 1976, Bibby and Shoenberg 1979) and on Na (Perz and Shoenberg 1976). It would be useful to repeat experiments of this kind on samples for which dislocation densities were known so that there was independent evidence of the nature of the regime.

(e) If it is possible to reach the quantum limit in practicable fields, as for Bi with H parallel to the binary axis, the detailed line shape of the last one or two dHvA oscillations provides evidence on the nature of the mechanism which reduces the amplitude at lower fields. For these last few oscillations it is more appropriate to think of line shape rather than of Fourier analysis into harmonics, because even in the presence of a little temperature and scattering broadening and some phase smearing the line shape is still a recognizable distortion of the ideal cusped oscillation in the absence of any broadening (as illustrated in fig. 8.7). The effect of phase smearing is then best calculated by a convolution of the phase distribution function with the ideal oscillation curve, and in principle it should be possible to determine the distribution function, or at least its spread, by

400

comparison of trial convolutions with the experimental line shape. If the phase smearing is determined in this way at the quantum limit, it is immediately possible to predict the behaviour of the amplitude at lower fields, where the phase smearing has killed all but the fundamental and one or two harmonics. If the reduction factors at the lower fields are still behaving in accordance with prediction, we can be reasonably sure that over the whole range the orbit size is in the same regime. Some experimental results on Bi (see below) do in fact appear to suggest that down to fields as low as 150 G the orbit regime is the same as at the quantum limit (about 1.5×10^4 G), where the orbit size is certainly small compared with the scale of dislocation separations.

(f) The giant quantum oscillations in ultrasonic attenuation (see §4.6) are even more 'spiky' than the dHvA oscillations, being δ-functions in the absence of broadening, so for good samples at low temperatures the spike-like character is still apparent and its broadening can be studied, even as far as the tenth oscillation below the quantum limit. According to the theory of §4.6, if a Dingle temperature (say of order 0.5 K) is due to 'microscopic' causes, e.g. electron scattering by impurities and perhaps scattering by dislocations in the large orbit regime, the GQO should be so much broadened as to lose its spiky character altogether. If therefore the spiky character is still evident, we can conclude that electron scattering makes only a minute contribution (of order 0.05 K or less) to the observed x. As we saw in §4.6, the peak broadening over and above the thermal broadening can be separated into contributions from inhomogeneity (due to phase smearing in the small orbit regime), and from 'microscopic' causes, the latter contribution being identifiable by its characteristic asymmetry. In principle, comparison of the phase smearing part with the amplitude decay at low fields should reveal to what extent the same broadening parameter is operative at high and low fields. The available evidence (see §4.6) suggests that the broadening parameter (i.e. the Dingle temperature that should be observed in the dHvA effect) may be much the same at low fields as it is near the quantum limit, but further experiments are needed, designed to test this point more precisely. Ideally such experiments should include dHvA as well as GQO studies on the same sample.

8.4.5.2. Experiments on Cu

As already mentioned, the various experiments (CW, TH and CH) differed in the techniques of sample preparation and consequently in the pattern of dislocation lines and the uniformity of this pattern through the sample. If for the moment we ignore these differences and also the slight curvatures which show up in some of the Dingle plots of CH, we can compare the experimental results with the prediction (8.33) or (8.35) based on our rather handwaving arguments, assuming that we are indeed in the small orbit regime.

The first check is as regards the dependence of H_0 (i.e. in the simple theory, the slope of the plot of $-\ln R_D$ against $1/H$) on D. If (8.35) were correct we should have a $D^{1/2}$ dependence but, as can be seen from fig. 8.5, even though the scatter of the experimental points is considerable, the results would appear to vary more nearly as D than $D^{1/2}$.* If we ignore this discrepancy and apply (8.35) at about the middle of the experimental range, say at $D = 4 \times 10^7 \text{cm}^{-2}$, we can compare the predicted and observed values of H_0. We find (see table 8.3) for the neck a predicted value $H_{0N} \sim 60 \text{kG}$, compared with the observed value of about 100kG, while

Fig. 8.5. Variation of x (average Dingle temperature) with D, the density of dislocation lines, for Cu belly and neck; note that H_0 (in G) is $2.0 \times 10^5 x$ for belly and $6.8 \times 10^4 x$ for neck. The points are experimental ($\bullet$, $\blacktriangle$ CH, two different samples, $\triangle$ TH, $\square$ CW). The curves are theoretical; *dotted*: according to the simple theory which gives $x = 1.28 \times 10^{-4} D^{1/2}$ for belly and $1.48 \times 10^{-4} D^{1/2}$ for neck; *broken*: based on Watts (1977) with $v = 8$ and $\mu = 4.1$ (belly) and 11.0 (neck); *chain*: based on CH neck only). Some recent measurements (Basinski *et al.* 1981, unpublished) give $x \sim 1 \text{K}$ (belly) and $\sim 3 \text{K}$ (neck) for $D \sim 2 \times 10^9 \text{cm}^{-2}$, but D could be estimated only very roughly.

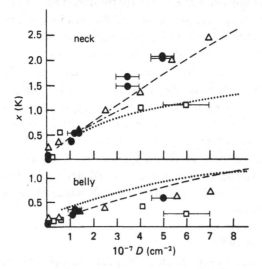

---

* However some recent experiments by Basinski, Howie, Lonzarich and Sigfusson (private communication 1981) with more heavily dislocated crystals ($D \sim 2 \times 10^9 \text{cm}^{-2}$) suggest a variation *slower* than as $D^{1/2}$. It is possible that the estimates of D are less reliable than they seem and other factors characterizing the dislocations may be different in different experiments.

8.4. Dislocations

for the belly the predicted value is $H_{0B} \sim 160\,\text{kG}$ and the observed value is about $100\,\text{kG}$. It is encouraging that the simple theory gives reasonable orders of magnitude, though, of course, because of the possible discrepancy between the observed and predicted dependence on D, the ratios of predicted to observed values of H_0 would be reduced for $D > 4 \times 10^7\,\text{cm}^{-2}$ and increased for $D < 4 \times 10^7\,\text{cm}^{-2}$. The fact that the predictions for neck and belly differ from experiment in opposite directions need not worry us too much, since the calculations in Appendix 16 of the appropriate values of $d\ln F/ds$ are based on an idealised model, the shortcomings of which might well affect neck and belly differently.

In Watts' theory, the Dingle plot is curved and the mean slope is no longer H_0, as in the simple theory. If values of x are deduced from the mean slopes of the Dingle plots over the relevant field range on the basis of the Watts theory, using the parameters chosen to fit the Dingle plot curvature at one particular value of D ($1.5 \times 10^7\,\text{cm}^{-2}$) as discussed below, the D dependence of x shown in fig. 8.5 is obtained and this agrees marginally better with the experimental points than the prediction of the simple theory. The CH theory involves rather elaborate numerical integrations and has been carried through only for the neck; although approximate agreement with the data could be obtained as shown in fig. 8.5, the values of the parameters needed to produce this agreement were not very plausible.

In our handwaving arguments H_0 was merely a scaling parameter in the sense that $-\ln R_D$ was expected to be a function of H_0/H, in particular a linear function if the strain distribution was Lorentzian, but the argument provided no guide as to what the distribution function should be. The more detailed theories outlined above do indicate the form of the field dependence of $-\ln R_D$ and as already mentioned predict that the Dingle plot should be appreciably non-linear in the small orbit regime. The experimental evidence is, however, not very satisfactory. In the earlier experiments of CW and TH the Dingle plots appear to be convincingly straight, but in those of CH on rather better characterised samples, the Dingle plots of both belly and neck were occasionally appreciably curved (see fig. 8.6). Possibly samples with a more irregular arrangement of dislocations than those of CH somehow straighten out the curvature characteristic of a more regular sample, but it is difficult to see why.

The curved Dingle plots of CH can be fitted nearly quantitatively by either of the more detailed theoretical calculations discussed above, though both fits involved somewhat implausible assumptions; only the Watts fit is shown in fig. 8.6. Two unsatisfactory features of the CH theory are that the parameters used for specifying the strain dependence of F have been given rather unrealistic values and that the calculated Dingle temperature (and hence also $-\ln R_D$) diverges at about $16\,\text{kG}$, though this is outside the range of the experimental points. In Watts' calculation the adjustable parameters are essentially the value of μ, which determines H_0, and v, the ratio of the 'correlation length' to $D^{-1/2}$ (H_0/μ is $(2F/k)D^{1/2}$ and so is not adjustable if D is known). A fairly reasonable fit to both the neck and belly Dingle plots can be obtained by choosing $v = 8$ and $\mu = 4.1$ for the belly and 11.0 for the neck. The fact that these values of μ are rather higher than the objective estimates of table 8.3 (3.1 and 5.9 respectively) need not worry us too much, since these estimates are based on rather idealized models. However, the choice of such a high value of v as 8, which is necessary to avoid excessive curvature in the Dingle plots, is not easy to understand. Another worrying feature is that Watts' theory assumes a random array of dislocations and so would not appear to be suitable to apply to the regular arrays of the CH samples, though perhaps the regularity was not sufficiently perfect to destroy the random assumption. But of course even more worrying is the fact that both calculations predict curved Dingle plots, while in most experiments no appreciable curvature has been observed.

CH present some data on the behaviour of the second harmonic amplitude of the neck for one of their samples, (unfortunately not the one which appeared to show departures from a linear Dingle plot for the fundamental). If the harmonic amplitude (after removing the temperature dependent factor and a factor $1/\sqrt{2}$) is M'_2 and that of the fundamental is

Fig. 8.6. Dingle plots for (a) belly and (b) neck of Cu $\langle 111 \rangle$. The points are experimental (CH), but with the zero of the ordinates shifted to agree approximately with the Watts' theoretical curves (broken lines). The parameters of the theoretical curves are $v = 8$ and $\mu = 4.1$ (belly) and 11 (neck) (as in fig. 8.5); this choice is somewhat of a compromise, but a high value of v is needed to avoid too curved a plot.

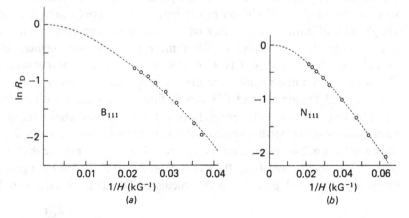

M'_1 we should have

$$M'_2/M'_1 = ce^{-\alpha x/H} \tag{8.56}$$

where c is the ratio G_2/G_1 of the spin factors and α is as defined in (3.1). Thus if c is reliably known, x may be determined from M'_2/M'_1 at any one field H. In this way CH found a value of x which varied appreciably and linearly with H and infer that this confirms their theoretical prediction of curved Dingle plots. However this method of deducing x is sensitive to any errors in the process of reducing the raw data to the form (8.56) and further experiments on harmonic amplitudes in a variety of samples would be well worth while.

Finally we should mention one other interesting result of the CH experiments. This was their demonstration that the four $\langle 111 \rangle$ necks had quite different amplitude reduction factors, because of the different angles between the various $\langle 111 \rangle$ directions and the direction of the dislocation lines. The $\langle 111 \rangle$ axis most nearly parallel to the dislocations (which were actually along [121], at about 20° to [111]) had the lowest Dingle temperature (0.35 K) while that perpendicular to the dislocation lines had the highest Dingle temperature (0.79 K). This considerable anisotropy, which is in qualitative agreement with the detailed calculations of CH, provides useful confirmation that the amplitude reduction is indeed caused by the dislocations.

8.4.5.3. Experiments on Bi

For the field parallel to the binary axis, the cyclotron mass is exceptionally small ($m/m_0 = 0.0091$) and at a temperature of about 0.6 K the oscillations can be followed all the way from the quantum limit at about 14 kG, down to fields as low as 130 G. This provides an opportunity of studying the phase smearing mechanism over an unusually wide range – of order 100 to 1 – in field. An attempt to exploit this opportunity was made by Barklie and Shoenberg (BS) (1975), but the experiments were unfortunately never completed and some of the published provisional conclusions later proved to be ill-founded. Perhaps the brief account of these experiments which follows in which some of the results are reinterpreted will stimulate others to continue a potentially fruitful line of research. As mentioned earlier, simultaneous measurements of the giant quantum oscillations would also be very informative.

The advantage of the binary orientation is that two of the three frequencies associated with the three electron ellipsoids are low and should be exactly equal for this orientation, while the third electron frequency and the hole frequency are both much higher. Consequently, except at the highest fields, only the oscillation pattern of a single low frequency (and its harmonics) should be observed, thus simplifying analysis of the amplitudes. In the preliminary experiments the oscillations were followed only down to about 400 G and the pronounced non-linearity of the Dingle plot (left-hand part of fig. 8.7) was attributed to a Gaussian distribution of strains. However when, somewhat later, the measurements were extended down to about 130 G it became obvious that the curvature was due to a very long beat between the two nominally equal low frequencies. In fact these frequencies are exactly equal only if the orientation is exactly along the binary axis and a very small misalignment (of order 0.3°)–such as may well have occurred–would have been sufficient to produce the observed pattern. If this explanation of the anomalous Dingle plot is accepted it is not difficult to work out a correction for the misalignment (see Appendix 18) and, as can be seen from fig. 8.7, the corrected Dingle plot becomes rather accurately linear. The slope of the Dingle plot gives $H_0 = 500$ G ($x = 0.37$ K) and from the very rough estimate of the dislocation density: $D \sim 10^6 - 10^7$ cm^{-2} we find $H_0/D^{1/2} \sim$ 0.16 – 0.5 G cm, which is indeed comparable to the very roughly predicted value 0.6 (table 8.3). We also see that since even at the lowest field H_0/H is only about 4, we should be still in the small orbit regime, since r/d should be only 1/14 of H_0/H, i.e. $r/d \sim 0.3$. This is also just about consistent with

Fig. 8.7. Dingle plot for Bi with binary axis nearly along H; $T = 0.57$ K (BS 1975, unpublished). The dots are the directly measured values of ln [amplitude $\times$ $H^{5/2}$ sinh(z/z)] and indicate a beat between two frequencies which would be exactly equal if H was exactly along the binary axis. Values of ln R_D appropriate to the exact binary orientation, as calculated by the analysis of Appendix 18, and corrected for harmonic content are marked ⊙ and their linear extrapolation to $1/H = 0$ is taken as the zero of ln R_D. The slope of the linear plot gives $H_0 = 500$ G corresponding to $x = 0.37$ K.

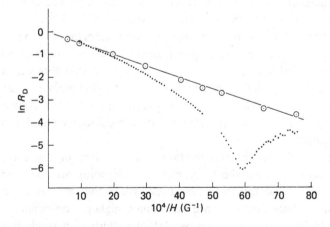

8.4. Dislocations

the direct comparison of $r = 3.3 \times 10^{-4}$ cm (at 130 G) and $d = D^{-1/2}$, which is between 3 and 10×10^{-4} cm.

Further evidence for the applicability of the exponential reduction factor, characteristic of a Lorentzian distribution in the small orbit regime, comes from harmonic analysis of oscillations of dM/dH at relatively high fields, such as shown in fig. 8.8. After making appropriate adjustments for the varying envelope width through a single oscillation, the graphs of oscillations 8 and 13 were digitized and Fourier analyzed; the harmonic amplitudes A_p and phases are given in table 8.5. The analysis of these results is rather sensitive to what is assumed for the spin factor $\cos(\frac{1}{2}\pi pgm/m_0)$. Experiments to be discussed in chapter 9 show that this factor can be written as $\cos p(\pi + \chi)$, where χ is a small angle; the evidence is that $\chi = 16.9°$, with an uncertainty of order $1°$. If we substitute into the LK formula for dM/dH we find that the pth harmonic should vary with p as

$$\exp(-pH_0/H)p^{1/2}(\cos p\chi/\sinh pz)\sin(2\pi pF/H + \tfrac{1}{4}\pi) \qquad (8.57)$$

where $z = 2\pi^2 kT/\beta H$ and we have assumed an exponential Dingle factor. Thus if the observed amplitude is A_p (in arbitrary units), we should have

$$A_p' \equiv A_p(\sinh pz)/(p^{1/2}\cos p\chi) \propto \exp(-pH_0/H) \qquad (8.58)$$

and H_0 can be determined from the ratios A_p'/A_{p-1}' which are independent of the proportionality constant. The fact that the values of H_0 determined in this way are not very different from the value of 500 G found from the

Fig. 8.8. Line shapes of oscillations numbers 8 and 13 as traced from recorder chart for Bi with H along binary axis (counting from the quantum limit oscillation as 1); the envelope curve is indicated by the broken lines. dM/dH is in arbitrary units, but dM/dH per unit height is about 1.27 times greater for 13 than for 8; the H scales, as indicated, are also different. The mean H and T were about 1.75 kG, 0.58 K for 8 and 1.06 kG, 0.59 K for 13 (BS, unpublished 1974).

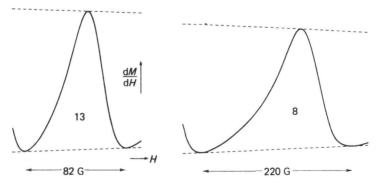

Table 8.5. *Fourier analysis of the two oscillations of fig. 8.8*

p	A_p (degrees)	ϕ_p (degrees)	$p\phi_1 - \phi_p$ (degrees)	$A'_p = \dfrac{A_p \sinh pz}{p^{1/2} \cos p\chi}$	$-\ln(A'_p/A'_{p-1}) = H_0/H$	$H_0(G)$
Oscillation No. 8 H = 1750 G						
1	4.54	−43.9	—	2.22	—	—
2	1.85	−133.5	45.7	1.63	0.31(0.33)	540(570)
3	0.73	−218.7	87.0	1.22	0.29(0.33)	510(570)
4	0.25	63.3	121.1	0.97	0.23(0.31)	400(550)
Oscillation No. 13 H = 1060 G						
1	5.60	−58.6	—	4.90	—	—
2	1.56	−160.3	43.1	2.90	0.52(0.54)	560(570)
3	0.43	−254.1	78.3	1.86	0.45(0.48)	470(510)

Notes: A_p is the amplitude of the pth harmonic in arbitrary units and ϕ_p is its phase measured from an arbitrary origin; χ is taken to be 16.9°. The values shown in brackets are those obtained if χ is taken as 16.0° rather than 16.9°. The calculated values of $p\phi_1 - \phi_p$ shown above ignore a number of small effects such as slight inadequacies of the LK formula at low quantum numbers (e.g. variation of Fermi energy ζ with H), slight MI, the slight effect of the beat due to the orientation being slightly off the binary and possible systematic phase shifts due to mosaic structure, though this last effect should be negligible (see table 8.6). In as far as these effects can be estimated, however, they tend to worsen, rather than improve, the agreement of $p\phi_1 - \phi_p$ with (8.59) by up to 5° or so.

8.4. Dislocations

Dingle plot of fundamental amplitudes is encouraging. We need not be too worried by the relatively small inconsistencies, since the answers are so sensitive to the assumed value of χ (reducing χ to $16.0°$ improves the consistency appreciably, but increases the mean value of H_0). Moreover the allowance made for the variable envelope width of the oscillations may introduce appreciable errors, particularly in the higher harmonics. The phases provide further evidence that the Fourier analysis is reasonably reliable. The observed phases ϕ_p as determined in this experiment were with reference to an arbitrary origin, but it can easily be seen (see (9.15)) that if (8.57) is valid, we should have

$$p\phi_1 - \phi_p = (p - 1)\pi/4 \tag{8.59}$$

and this relation appears to be not too badly satisfied (see table 8.5), bearing in mind the uncertainties of the analysis.

Evidence about the phase smearing at still higher fields can be obtained by examining the line shape of the last oscillation* before the quantum limit (fig. 8.9). Here it is no longer profitable to think in terms of harmonic

Fig. 8.9. The last oscillation of Bi for H along the binary axis at $T = 0.6\,\text{K}$ (BS 1975). The higher frequency oscillations come from other parts of the FS. The broken curve is an idealized version of how the oscillation would appear at $T = 0$ for a perfect sample in the absence of the high frequency oscillations. The presence of two cliffs rather than one is due to spin-splitting (see §9.5.1).

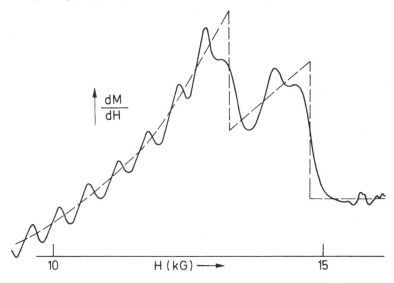

---

* In principle the $n = 2$ or 3 oscillation line shapes could also be used, but the clearest evidence comes from $n = 1$.

analysis, but instead directly in terms of the convolution of an 'ideal curve' at $T = 0$, with the effect of temperature and of phase smearing (whether due to electron scattering or sample inhomogeneity). The experimental curve is complicated by the presence of higher frequency oscillations (due to the holes and the third electron ellipsoid) and also by the spin splitting, but some idea of the form of the ideal line shape can be obtained by sketching a free-hand curve through the mean level of the fast oscillations, and with sharp 'cliffs' of equal heights at the positions where the Landau levels for the two spin directions abruptly empty as the field is raised. The effects of temperature and phase smearing should of course be most pronounced at these cliffs and hardly appreciable over the more gentle approaches to the cliff.

If the form of the ideal line shape expressed in terms of the phase $\psi = 2\pi F/H$ is $v(\psi)$, then phase smearing over a distribution function $D(\phi/\lambda)$ will modify $v(\psi)$ to

$$v'(\psi) = \int v(\psi + \phi)D(\phi/\lambda)\,\mathrm{d}(\phi/\lambda) \tag{8.60}$$

where it is assumed that D is correctly normalized. This is equivalent to a convolution of the ideal line shape $u(H)$ expressed in terms of H, to give

$$u'(H) = \int u(H + h)D\left(\frac{2\pi Fh}{H^2\lambda}\right)\mathrm{d}\left(\frac{2\pi Fh}{H^2\lambda}\right) \tag{8.61}$$

The appropriate forms of D and λ, as shown in §2.3.7 are:

$$D_1(z) = 1/2(1 + \cosh z), \qquad \lambda_1 = 2\pi kT/\beta H \tag{8.62}$$

for temperature smearing and

$$D_2(z) = 1/\pi(1 + z^2), \qquad \lambda_2 = H_0/H = 2\pi^2 kx/\beta H \tag{8.63}$$

for Lorentzian phase smearing, if $e^{-H_0/H}$ is the amplitude reduction factor of the fundamental at low fields. The order in which the convolutions are carried out is of course immaterial. If the distribution were Gaussian rather than Lorentzian (8.63) would be replaced by

$$D_2'(z) = \pi^{-1/2}e^{-z^2}, \qquad \lambda_2 = H_0/H, \tag{8.64}$$

with an amplitude reduction factor of $e^{-H_0^2/4H^2}$ at low fields.

A few convolutions were computed*, first for the temperature smearing

---

* In these computations, since the main concern was with the region round the last cliff in fig. 8.9, (8.61) was somewhat simplified by putting $F/H = 1$: this simplification should not have caused any great errors.

(at the experimental value of T, i.e. 0.6 K) and then with either the Lorentzian or the Gaussian functions for various trial values of H_0. Unfortunately the presence of the high frequency oscillations makes it difficult to say which distribution gives the best overall fit to the experimental line shape. The Lorentzian distribution, strongly suggested by the linearity of the Dingle plot at lower fields, is best fitted with an H_0 of about 300 G. This is rather less than the 500 G estimated from the slope of the Dingle plot, but the high frequency oscillations make estimation of the true experimental broadening of the cliffs rather uncertain and there may well be no real discrepancy. Thus, apart from this possibility that the broadening parameter may be slightly less at the highest fields, we see that essentially the same broadening parameter $H_0 \sim 500$ G is valid over a range of something like 100 to 1 in field.

In the above discussion the 'ideal' curve in the quantum limit was based essentially on a free-hand sketch; a rather better version can be obtained on the basis of band structure theory. Rode and Lowndes (1977) have carried out convolutions based on this approach to estimate Dingle temperatures for both pure and alloyed Bi. They too were able to show that the same Dingle temperature gave a fair fit both at high and low fields.

8.5. Mosaic structure

A sample which is nominally a single crystal, in reality usually consists of a large number of grains, each of which is a perfect crystal but whose orientations spread over a range around the mean characterising the whole sample. The damping effect of this mosaic structure, first considered by Shoenberg (1962) and Priestley (1962), ranges from very slight, at orientations such that F is extremal, to devastating at orientations where F varies rapidly with angle. In this section we shall calculate the reduction factor due to mosaic structure in various simple situations and compare its field and orientation dependence with the rather meagre experimental evidence available. We assume throughout that the grain size is large compared to the orbit radius. An extreme example of mosaic structure is that of a polycrystal, in which all orientations are equally probable and we shall show that sometimes an appreciable oscillatory amplitude may survive the averaging over orientations. In the various calculations we shall usually assume somewhat arbitrarily that the spread of orientations in the mosaic structure has a Gaussian distribution. What little experimental evidence there is tends to support this assumption, but the exact form of the distribution is not important for estimating the order of magnitude of the reduction factor in terms of the 'spread' of the distribution. If, of course, a reliable determination of the real distribution has been made by special

X-ray techniques, it should be possible to make a more objective numerical calculation of the reduction factor.

8.5.1. Orientation with *F* extremal

If the FS consists of a single sheet, then F is extremal with respect to angle at symmetry directions (such as $\langle 100 \rangle$, $\langle 111 \rangle$ and $\langle 110 \rangle$ in a cubic crystal); F may also be extremal at non-symmetry directions, as for instance in Cu at $16°$ to $\langle 100 \rangle$ in a (110) zone, where there is an absolute minimum, or at $12°$ to $\langle 100 \rangle$ in a (100) zone, where there is a saddle point. However, if the FS consists of several symmetry related sheets, such as the three ellipsoids in Bi, a symmetry direction may be one where several branches of the F spectrum cross, but there is no extremum of F.

Since we shall usually be concerned with only small angular ranges (beyond which dephasing effectively kills the oscillations) we may without appreciable error use rectangular angular coordinates (α, β) to specify the departure of the relevant axis of a grain from the mean orientation at $(0, 0)$ which is assumed to be an extremum of F. A Gaussian spread of orientation may then be specified by $\exp[-(\alpha^2 + \beta^2)/u^2]\,d\alpha\,d\beta$; the r.m.s. spread (i.e. $\langle \alpha^2 + \beta^2 \rangle^{1/2}$) is then u. It should be noted, however, that this distribution implies that all directions on a cone of given semi-angle round $(0, 0)$ are equally probable, which may not necessarily be true. Since F is an extremum at $(0, 0)$, the phase departure $\Delta\psi$ at (α, β) may be expressed as

$$\Delta\psi = a\alpha^2 + b\beta^2 \qquad (8.65)$$

if the (α, β) axes are appropriately chosen. For a and b of the same sign, $(0, 0)$ is either an absolute minimum or maximum, while if they are of opposite sign it is a saddlepoint. The values of a and b are, of course, just π/H times $\partial^2 F/\partial\alpha^2$ and $\partial^2 F/\partial\beta^2$ respectively for the fundamental, or p times as much for the pth harmonic. The reduction factor R_M (M for mosaic) for the fundamental is then given by

$$R_M = \left| \int_{-\infty}^{\infty} \int_{-\infty}^{\infty} \exp[i(a\alpha^2 + b\beta^2)]\exp[-(\alpha^2 + \beta^2)/u^2]\,d\alpha\,d\beta \right|$$

$$\div \int_{-\infty}^{\infty} \int_{-\infty}^{\infty} \exp[-(\alpha^2 + \beta^2)/u^2]\,d\alpha\,d\beta \qquad (8.66)$$

Provided $u \ll 1$, so that the integrands are negligible when α and β become too large for the planar representation to be accurate, no appreciable error is made by using the planar system and by extending the limits of integration to $\pm\infty$. The double integrals are immediately separable and

8.5. Mosaic structure

some simple algebra leads to the results

$$R_M = (1 + a^2 u^4)^{-1/4}(1 + b^2 u^4)^{-1/4}$$
$$= [1 + (\pi F_1'' u^2/H)^2]^{-1/4}[1 + (\pi F_2'' u^2/H)^2]^{-1/4} \qquad (8.67)$$

where $F_1'' = \partial^2 F/\partial \alpha^2$ and $F_2'' = \partial^2 F/\partial \beta^2$. There is also a phase shift δ_1 of the resultant oscillation given by

$$\delta_1 = \tfrac{1}{2}[\tan^{-1}(\pi F_1'' u^2/H) + \tan^{-1}(\pi F_2'' u^2/H)] \qquad (8.68)$$

An important special case is for $a = b$, which is appropriate for instance at $\langle 100 \rangle$ or $\langle 111 \rangle$ (but not $\langle 110 \rangle$) in a cubic crystal with a single sheet of FS. We then have

$$\left.\begin{array}{l} R_M = (1 + \gamma^2)^{-1/2}, \qquad \delta_1 = \tan^{-1}\gamma \\ \text{where} \quad \gamma = \pi F'' u^2/H \end{array}\right\} \qquad (8.69)$$

The origin of the phase shift is, of course, that all departures from the extremal orientation change F in the same sense.

Some values of F'' are given in table 8.6 together with the corresponding values of $\pi F'' u^2/H$, i.e. γ, for appropriate values of H and assuming $u = 3 \times 10^{-3}(0.2°)$ as typical (though with care much smaller mosaic spreads can be achieved). It can be seen that usually the factor R_M at an extremal orientation differs little from 1, except for surfaces with very sharp contours, as in the nearly free electron-like second zone surface of Al (see fig. 5.15). Priestley (1962) was in fact unable to observe the predicted $\langle 100 \rangle$ frequency even at 150 kG and the mosaic reduction factor of order 1/30 may well have been the cause (for this orientation the amplitude of a perfect crystal would in any case be small, because of the smallness of $|d^2 A/d\kappa^2|^{-1/2}$).

It should be noticed that although R_M is usually close to 1 for an extremal orientation, the phase shift δ_1 may be quite appreciable. As CH point out, the presence of such a phase shift can be detected if γ is large enough by accurate measurement of the *relative* phase shift $2\delta_1 - \delta_2$ between the fundamental and the $p = 2$ harmonic. Since for circular symmetry ($a = b$), $\delta_2 = \tan^{-1} 2\gamma$, we find (omitting the 45° contribution from the phase constant in the LK formula)

$$2\delta_1 - \delta_2 = 2\tan^{-1}\gamma - \tan^{-1} 2\gamma = \tan^{-1}[2\gamma^3/(1 + 3\gamma^2)] \qquad (8.70)$$

This is just appreciable for the $\langle 111 \rangle$ belly in Cu (with the conditions assumed in table 8.6, $2\delta_1 - \delta_2 \sim 4°$), though barely observable for the Cu neck ($\sim 0.1°$). Even though $2\delta_1 - \delta_2$ may be too small to detect, δ_1 itself is much larger ($\sim 3°$ for the Cu neck and $\sim 20°$ for $\langle 111 \rangle$ belly) and it may be important to consider it in measurements of absolute phase.

Table 8.6. *Parameters relevant in estimating effect of mosaic structure*

		$10^{-8}F''(G)$	$\gamma = |(\pi/H)F''u^2|$	R_M
Cu	B $\langle 100 \rangle$	-2.3	0.13	0.992
	B $\langle 111 \rangle$	-6.7	0.38	0.94
	'Dip' 12° from $\langle 100 \rangle$ in (100) zone	2.9, -0.07	0.16, 0.004	0.994
	'Dip' 16° from $\langle 100 \rangle$ in (110) zone	3.1, 0.25	0.18, 0.014	0.992
	N $\langle 111 \rangle$	0.9	0.05	0.999
	D $\langle 110 \rangle$	12.3, 6.4	0.70, 0.36	0.88
	R $\langle 100 \rangle$	11.5	0.65	0.84
K	$\langle 100 \rangle$	-0.06	0.003	
	$\langle 111 \rangle$	-0.02	0.001	
	$\langle 110 \rangle$	0.01, 0.006	0.0006, 0.0004	
				$1 - 5 \times 10^{-6}$
Na	$\langle 100 \rangle$	-0.01	0.0006	
	$\langle 111 \rangle$	-0.06	0.003	
	$\langle 110 \rangle$	-0.01, 0.03	0.0006, 0.002	
Al	second-zone surface			
	$\langle 111 \rangle$, $\langle 110 \rangle$	~ 50	~ 5	~ 0.2
	$\langle 100 \rangle$	~ 300	~ 30	~ 0.03
Bi	$H \parallel$ bin	~ 0.0001	$\gamma \sim 0.0001$	~ 0.998
		$F'_{\text{max}} \sim 10^4 \, G$	$\pi F'u/H \sim 5 \times 10^{-2}$	

Notes: F'' is defined as $\mathrm{d}^2F/\mathrm{d}\theta^2$ and u is the r.m.s. mosaic spread; the entries are only rough estimates based on published tables or curves of the orientation dependence of F; at best they are reliable to 10%, but those marked $\sim$ are little more than orders of magnitude. The double entries for the Cu 'dips' refer first to variation in and second across the zone; in the (110) zone the dip is an absolute minimum but in the (100) zone it is a saddle point. The other double entries are all at $\langle 110 \rangle$ and the first entry is in the (100) zone, the second in (110). The value of R_M is calculated by (8.67) or (8.69) as appropriate, except for Bi where (8.76) is used since the two relevant ellipsoids of the FS have a finite value of F' at the binary orientation. The values of u and H are chosen to match the conditions of typical experiments, though u may be much smaller for specially perfect samples. For Cu, K and Na, the values are $u = 3 \times 10^{-3}$, $H = 5 \times 10^4 G$; for Al, $u = 7 \times 10^{-3}$, $H = 1.4 \times 10^5 G$ and for Bi, $u = 3 \times 10^{-3}$, $H = 2 \times 10^3 G$.

8.5.2. Orientation off extremal

Consider the special case of circular symmetry round the extremal orientation (i.e. $a = b$ in (8.65)) but for an orientation at a small angle ϕ to the extremal; α and β are now measured from this off-extremal orientation and the reduction factor becomes

$$R_M = \left| \int_{-\infty}^{\infty} \int_{-\infty}^{\infty} \exp[i(2a\alpha\phi + a(\alpha^2 + \beta^2))] \exp[-(\alpha^2 + \beta^2)/u^2] \, d\alpha d\beta \right|$$

$$\div \int_{-\infty}^{\infty} \int_{\to\infty}^{\infty} \exp[-(\alpha^2 + \beta^2)/u^2] \, d\alpha d\beta \tag{8.71}$$

The double integrals are again separable and we eventually obtain (applying the trick of 'completing the square' to the quadratic in α which appears in the exponent in the numerator)

$$R_M = (1 + \gamma^2)^{-1/2} \exp\left(-\frac{\phi^2}{u^2} \frac{\gamma^2}{1 + \gamma^2}\right) \tag{8.72}$$

where γ is as defined in (8.69), and also a phase shift given by

$$\delta_1 = \tan^{-1}\gamma - \frac{\gamma^3(\phi^2/u^2)}{1 + \gamma^2} \tag{8.73}$$

It is not difficult to generalize (8.72) and (8.73) to extend beyond the parabolic region for which $F' = \phi F''$ and also to take account of the possibility $a \neq b$, but we shall not write down the resulting rather cumbersome expressions. CH give the answers for the former generalization, but still assume $a = b$, which is unlikely to be realistic beyond the parabolic region.

The phase shift (8.73) may become quite large if ϕ/u is large (e.g. for $\phi = 4°$ and $u = 0.2°$) even though γ itself is small, and this makes significant absolute phase determination almost imposssible except at directions of extremal F and for some small pieces of FS, for which γ is exceptionally small. The relative phase difference between fundamental and harmonic is given by

$$2\delta_1 - \delta_2 = \tan^{-1}[2\gamma^3/(1 + 3\gamma^2)] + 6\gamma^3\phi^2/u^2(1 + \gamma^2)(1 + 4\gamma^2) \tag{8.74}$$

The reduction factor (8.72) simplifies in two limiting cases. If, as for the second zone surface of Al, $\gamma \gg 1$, it becomes approximately

$$R_M = (1/\gamma)e^{-\phi^2/u^2} \tag{8.75}$$

and we see that the oscillations have appreciable amplitude for only a minute range of angles–of order u–around the mean orientation. The other limiting case is for γ very small when (8.72) reduces to

$$R_M = \exp[-(\pi F'u/H)^2]$$

(8.76)

(since in our approximation $F' = \phi F''$). This result can of course be obtained much more directly by starting with the assumption that the contour lines of F are straight and run normal to the α axis and that we can neglect the quadratic term in the expansion of F in powers of α around the off-extremal orientation. This simpler approach is however equivalent to assuming $F'' = 0$, which would imply zero phase shift, rather than the small shift given by (8.73) for the small but non-zero value of γ (i.e. of F'').

We may note here that had we assumed a Lorentzian rather than a Gaussian distribution, only this 'linear' limiting case is easily calculated. As can be easily seen from (2.140), the reduction factor for $p = 1$ is then given by

$$R_M = \exp(-2\pi F'u/H)$$

(8.77)

if the distribution is such that the volume of grains having α between α and $\alpha + d\alpha$ is proportional to $1/(1 + \alpha^2/u^2)$.

The experimental evidence on the damping of various off-extremal oscillations in Cu (CW, CH) is scanty but on the whole supports the predictions based on a Gaussian distribution. Thus $\ln R_M$ at given H varies linearly with ϕ^2 and the slope is reasonably consistent with the mosaic spread u as roughly estimated by X-rays. CH also observed non-linearity of Dingle plots consistent with (8.76) and obtained qualitative confirmation of the predicted relative phase shift (8.74).

Apart from occasional qualitative observations which have supported the theory in explaining why some oscillations are particularly feeble or unobservable (e.g. Priestley 1962), the only other semi-quantitative verification has been in the interpretation of experiments on dilute alloys of the noble metals (see particularly Miller, Poulsen and Springford 1972). In the determination of the anisotropy of impurity scattering, as outlined in §8.3, corrections for mosaic structure had to be made before the Dingle temperatures could be reliably ascribed to impurity scattering alone. It was found that data taken at off-extremal orientations did not agree well with a parametrized scheme for the impurity scattering contribution based on data from extremal directions alone (where the mosaic correction could safely be ignored). However, if the off-extremal data were corrected in accordance with (8.76), assuming the same value of u for all orientations of

416

a given sample, the consistency of the scheme could be greatly improved. Eventually, by successive approximations, a more accurate scheme could be determined consistently based on all the data rather than on the extremal orientations alone. This provides slightly indirect support for the validity of the Gaussian assumption. With modern X-ray techniques it should be possible to check the form of the statistical distribution of orientations directly, but this does not appear to have been done except for rough estimation of the mosaic spread u.

8.5.3. A limit to the reduction factor

In all our discussion of amplitude reduction factors so far, we have assumed that the phase smearing is characterized by a *continuous* probability distribution. For the mosaic structure problem, however, there is only a finite number, say N, of grains, each of which has a definite phase and, if N is small enough, this may invalidate the result calculated from a continuous distribution. This consideration is likely to be less relevant for the problem of phase smearing due to variable strains, since the strain field of dislocations is really continuous, even though it may be nearly constant over regions large compared with orbit size.

Intuitively it would seem that if H is reduced so much that phase differences between grains become very large, all phase correlation would be lost and the amplitude reduction factor would level out at $1/\sqrt{N}$ rather than continuing to decrease with field as in (8.76) or (8.77). This intuitive suggestion (originally mentioned by Shoenberg and Stiles 1964) is in fact confirmed by statistical analysis, though, not surprisingly, the statistical fluctuations as between samples with the same N become comparable with the amplitude itself. This analysis (see Appendix 17.2) shows that if R is the reduction factor for an infinite population (i.e. for a truly continuous distribution of phases), then for a population N the expectation value of the true reduction factor R' is given by

$$\langle R'^2 \rangle = \frac{1}{N} + \frac{N-1}{N} R^2 \simeq \frac{1 + NR^2}{N} \tag{8.78}$$

We notice that if $R = 1$, corresponding to no phase smearing, $R' = R$, as we should expect, but if $R < 1/\sqrt{N}$ (e.g. for a sufficiently low field or a sufficiently broad distribution), the first term in (8.78) becomes dominant. With good measurement techniques it should in favourable circumstances be possible to see oscillations with R as low as 10^{-3}, (as for instance would occur for a Gaussian distribution with $u = 3 \times 10^{-3}$ and $F' = 10^7$ G at a field of about $40 \, \text{kG}$ according to (8.76)). Thus a sample with

10^6 grains would already have its true reduction factor R', $\sqrt{2}$ times the calculated R. It would be interesting to look for this effect experimentally, i.e. to look for a flattening out of the Dingle plot at low fields in samples of relatively large grain size, but some possible complications should be mentioned.

One is that the square root of the variance of the actual reduction factor becomes comparable to the expectation value given by (8.78), when the $1/N$ term is comparable to R^2 in (8.78). As shown in Appendix 17.3, the ratio of the variance of R'^2 to the square of the expectation value of R'^2 is given approximately (when $R \ll 1$, $N \gg 1$, but $NR^2 \lesssim 1$) by

$$\frac{V(R'^2)}{\langle R'^2 \rangle} = \frac{1 + 2NR^2}{(1 + NR^2)^2} \tag{8.79}$$

This ratio approaches 1 as NR^2 is reduced below 1, which implies that the variations of observed amplitude between samples with the same N and in otherwise identical conditions should be comparable with the amplitude itself.

Another possible complication is that we have tacitly assumed that the electron orbits are always small compared with the grain size. This is probably nothing to worry about since for a 1 mm sample the typical grain size for $N = 10^6$ would be of order 10^{-3} cm, which is still 20 times larger than a Cu neck orbit size in the conditions envisaged. No doubt a 'large orbit' calculation could be made if the grains are too small (cf. §8.4.2), but this will not be attempted here.

Finally it should be mentioned that if a high frequency modulation method is used to detect the dHvA oscillations (as in the experiments of Shoenberg and Stiles (1964) on alkali metals) the effective N may be considerably reduced because of skin effect (by something like the ratio of the skin depth to the sample size).

8.5.4. Polycrystal as extreme example of mosaic structure

A sample containing grains of random orientations spread over the whole solid angle (i.e. what is usually called a 'polycrystal', in contrast to a 'single crystal') would not be expected to show any great dHvA amplitude because of the enormous phase smearing over the random orientations. Recent experiments, however, have demonstrated that the residual amplitude, though small, may be quite detectable and we shall discuss two examples where an effect has indeed been observed.

The first example came about in rather a curious way. While testing their equipment at the University of Waterloo, J.J. Grodski and A.E. Dixon

8.5. Mosaic structure

(private communication in 1977)* were astonished to observe feeble dHvA oscillations at about 100 kG even though there was no sample in their pick-up coil. The oscillations had the characteristic frequency of the Cu belly and showed a long beat pattern. Eventually they were able to demonstrate that the oscillations must be coming from the polycrystalline Cu wire of the pick-up coil and, as we shall now show, a simple analysis based on treating a polycrystal as an extreme case of mosaic structure is able to account plausibly for their observations. The essential point of the analysis is that only those grains whose orientations lie close to extremal ones with respect to the field will be sufficiently coherent in phase to survive.

With this idea in mind we start with a modified form of (8.66) and examine what would be the amplitude for orientations distributed uniformly over a *wide range* around an extremal orientation. The wide range is achieved effectively by making u so large that the factor $\exp[-(\alpha^2 + \beta^2)/u^2]$ can be put equal to 1. This, however, requires modification of the normalization factor in (8.66), which is valid only if u is small enough, and it is easily seen that in the limit of equal probabilities for all orientations, the denominator must simply be replaced by 4π. Thus the contribution to the amplitude from those grains with orientations centred on an extremal orientation $(0,0)$ is simply

$$A = \frac{A'}{4\pi}\left|\int_{-\infty}^{\infty}\int_{-\infty}^{\infty} e^{i(a\alpha^2 + b\beta^2)}\, d\alpha\, d\beta\right| \tag{8.80}$$

where A' is the amplitude of oscillation of a single crystal of orientation $(0,0)$. Here of course we are supposing that only a small range of angles (α, β) will contribute effectively to (8.80) because of the rapid dephasing that occurs beyond $a\alpha^2 + b\beta^2 \sim \pi$. In other words we are assuming that the approximation of planar coordinates (α, β) is still permissible over the range which matters, and reference to table 8.6 shows that this is reasonable, except perhaps for the alkali metals. We are also ignoring the fact that if we go far enough away from $(0,0)$ we shall reach other extremal orientations. The contributions of such other extremal orientations to the oscillations are very simply dealt with by repeating the analysis with a new centre of coordinates for each of the extrema over the 4π of solid angle that has been included in the normalization factor.

Equation (8.80) reduces to

$$A = A'/(4|ab|^{1/2}) = A'H/(4\pi|F_1''F_2''|^{1/2}) \tag{8.81}$$

---

* I am grateful to Dr Grodski for telling me about these observations and stimulating me to develop the analysis presented here. No account of either the observations or the analysis has been published before.

419

and it is easy to show that there is a change of phase from the ψ of the single crystal oscillation to $\psi + \phi$ for the polycrystal where

$\phi = \frac{1}{2}\pi$ if a, b both $+$ve (absolute minimum)

$\phi = -\frac{1}{2}\pi$ if a, b both $-$ve (absolute maximum)   (8.82)

$\phi = 0$ if a, b of opposite signs (saddlepoint)

This phase change, however, is probably of only academic interest since it would be difficult to detect. If we now examine the FS of Cu, the reduction factors for the various belly extrema should be roughly as shown in table 8.7. After taking into account the number of repetitions of each orientation over the 4π solid angle it is evident that the two 'dip' orientations are likely to be the strongest candidates for survival. In fact the single crystal amplitudes at these orientations are particularly strong, so that their predominance in a polycrystal is even stronger than appears from the table. The frequencies of these dips differ by about 0.31%, so we should expect

Table 8.7. *Reduction factors for various oscillations in a polycrystal of Cu and Na*

Orbit	R for single orbit	Number of repeats in 4π	R(net)
Cu			
B $\langle 100 \rangle$	0.35×10^{-4}	6	2×10^{-4}
B dip 12° from $\langle 100 \rangle$ in (100) zone	1.8×10^{-4}	24	42×10^{-4}
B dip 16° from $\langle 100 \rangle$ in (110) zone	0.9×10^{-4}	24	22×10^{-4}
B $\langle 111 \rangle$	0.12×10^{-4}	8	1×10^{-4}
N $\langle 111 \rangle$	1.0×10^{-4}	8	8×10^{-4}
Na			
$\langle 100 \rangle$	80	6	~ 0.05
$\langle 110 \rangle$	50	12	~ 0.06
$\langle 111 \rangle$	13	8	~ 0.01

Note: R is the factor by which the LK amplitude for a single crystal should be multiplied to give the contribution to the amplitude of a randomly oriented polycrystal coming from orbits of a particular type. The entries are based on the estimates of F'' in table 8.6 and on (8.81), assuming $H = 10^5$ G. The contributions from other extremal orbits such as R $\langle 100 \rangle$ and D $\langle 110 \rangle$ are small (comparable to that of B $\langle 111 \rangle$) and have been ignored. The estimates of R for Na are rough not only because the F'' values are rough, but also because (8.81) can at best be only qualitatively valid.

beats in the observed oscillations with something like 320 oscillations per beat, and this was just about what was observed (the beat period could not be very precisely established since little more than a single beat period could be observed before the oscillations faded out).

Oscillations in an assembly of randomly oriented small crystals have also been observed in Li and Na (Randles and Springford 1973, 1976 (RS), see fig. 5.6) as already mentioned in §5.3.1. Because the Na FS is so nearly spherical, the various extremal orientations differ very little in phase (at 100 kG by only about 5π between $\langle 111 \rangle$ and either $\langle 110 \rangle$ or $\langle 100 \rangle$, which differ from each other by less than π). Thus we could hardly expect our simple analysis (in which complete dephasing is assumed over a considerable range between extrema) to apply more than very roughly. Even so, the observed beat frequency of about 3.1×10^5 G is fairly close to the difference, about 2.5×10^5 G, of the $\langle 111 \rangle$ frequency and either the $\langle 100 \rangle$ or $\langle 110 \rangle$ frequencies. It is, at first sight, a little surprising that the observed beats should be as deep as they are, since it would seem that the $\langle 111 \rangle$ frequency should be something like five times more severely reduced than the predominant $\langle 100 \rangle$ and $\langle 110 \rangle$. However it could be that there is a much longer beat between these two nearly equal frequencies (the uncertainty in their difference as measured by Lee (1966) is several times larger than the estimate of the difference) and that the whole of the experimental field range happened to lie towards the minimum of this long beat.

A more realistic calculation of the expected form of the random assembly oscillations was made by RS based on a parametrized (cubic harmonic) description of the FS, from which they computed the actual frequency distribution function for an array of random orientations, and hence the detailed form of the oscillations as a function of field. As can be seen from fig. 5.6, the computed amplitude variations show a close similarity to those actually observed. RS computed reduction factors for Na of order 0.2 (private communication from Dr Springford) which are reasonably consistent with the very rough estimates based on (8.81) which suggest that the reduction factor should be of order 0.1 (see table 8.7). RS quote only a rough order of magnitude of the observed absolute amplitude at 1.2 K ($|M| \sim 2 \times 10^{-4}$ G) and this is only about 2×10^{-3} times the single crystal amplitude in comparable conditions ($|M| \sim 10^{-1}$ G, see Perz and Shoenberg 1976), i.e. rather appreciably less than the estimate of 0.1. This discrepancy may merely indicate that the individual colloid particles were of much poorer quality (i.e. higher Dingle temperature x) than the single crystal sample of Perz and Shoenberg, and indeed an increase in x of 1 K would nearly bridge the gap.

421

The amplitude of the Li colloid oscillations was about 20 times smaller than for Na and the beat pattern more complicated and of something like 10 to 15 times higher frequency than that of the Na colloid. These facts are consistent with a considerably more distorted FS. As discussed in §5.3.1, RS were able to get a semi-quantitative description of the FS of Li by a trial and error procedure of comparing the beat patterns calculated for trial models with those actually observed.

8.6. Scattering of electrons by phonons (see also §2.6.1.2)

On the basis of the independent particle theory developed in chapter 2 we should expect to find a Dingle temperature associated with any form of electron scattering and in particular with scattering by phonons. However it is only in metals with exceptionally low Debye temperatures that such an effect—an increase of Dingle temperature with temperature—should be appreciable at temperatures for which the dHvA amplitude is still observable. The most promising metal is Hg, for which estimates based on the temperature dependence of thermal and electrical conductivity suggest that this thermal contribution to x should already become appreciable at 4 K, and by 17 K be as large as several degrees (the variation should be roughly as T^3). Such an effect might be expected to show itself both in direct determinations of x by conventional Dingle plots at various temperatures and in departures from linearity of plots of $\ln(A/T)$ against T at fixed H. It came as a surprise when the experiment was carried out by Palin (1970, 1972), that neither effect could be observed and the results at fields up to about 80 kG and temperatures up to 17 K were quite consistent with a Dingle temperature which did not change appreciably with temperature.

As already discussed in §2.6.1.2, this negative result is in fact just what would be expected from the many-body treatment of the electron–phonon interaction (Engelsberg and Simpson 1970) due to a subtle interplay of the real and imaginary parts of the self-energy. Before considering the more extreme conditions in which appreciable deviations from the independent particle treatment (i.e. from the LK formula) are predicted by many-body theory, we should mention an alternative proposal for interpreting Palin's negative result without appeal to many-body theory.

Gantmakher (1972) suggested that the absence of an appreciable thermal contribution might be merely a consequence of the fact that at sufficiently low T a phonon has too small a momentum to scatter an electron from one Landau tube to another. He showed that this would make the probability of scattering by phonons vary as $HT^{3/2}$ rather than T^3, and this would leave the *slope* of the Dingle plot independent of T, as found by Palin. It would however also imply that the plot of $\ln(A/T)$ against T should be

8.6. Scattering of electrons by phonons

slightly non-linear and that the cyclotron mass deduced from its average slope should increase appreciably with H, while Palin found good linearity and a mass almost independent of H (apart from the possible slight increase discussed below). A further doubt about Gantmakher's explanation is that it should apply only at temperatures so low that

$$kT/\hbar\omega_c \ll (ms^2/kT)^{1/2} \tag{8.83}$$

where s is the velocity of sound, and this is satisfied only below about 4 K even at the highest fields used. However it is only above 4 K that the absence of appreciable variation of x with T need occasion any surprise. Thus, although Gantmakher's comments on scattering by phonons at high fields and low temperatures are interesting in principle, it is unlikely that they are more than slightly relevant to the explanation of Palin's negative result.

As outlined in §2.6.1.2, the many-body theory not only explains Palin's results, but indicates the conditions in which characteristic many-body deviations from the independent particle predictions should occur. It turns out that the best conditions for showing up such deviations are at the highest fields and lowest temperatures and that appreciable effects could be

Fig. 8.10. Dingle plots for fundamental ($r = 1$) and harmonic ($r = 2$) of the β-oscillations of an Hg crystal at 2.2 K (Elliott et al. 1980). The predicted deviations from the independent particle LK theory are indicated by the broken lines and agree qualitatively with the $r = 1$ and perhaps the $r = 2$ points. The slight difference of the intercepts of the $r = 1$ and 2 plots is consistent with the known slight departures of ($\cos rS/\cos S$) from 1, but the $r = 2$ line has a slope which is only 10% higher than that of $r = 1$ instead of the predicted twice.

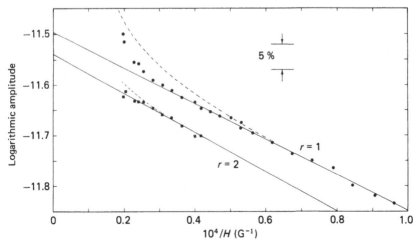

expected in a number of metals other than Hg (see Mueller and Myron 1976 for details).

In fact Palin's experiments did give a hint of the sort of deviation predicted, in the form of a slight apparent increase (by 2 or 3%) in the cyclotron mass as the field was raised from 40 to 90 kG, but this was hardly more than the experimental uncertainty. More recently Elliott, Ellis and Springford (1978, 1980) have obtained more positive evidence in rather delicate experiments, again on Hg. Their results are illustrated in fig. 8.10, which shows that the deviations from a linear Dingle plot, though small ($\sim 5\%$ in amplitude) are probably larger than the experimental uncertainty and are very comparable to the theoretical prediction. A puzzling feature of the results, however, is that the Dingle plot of the higher harmonics does not relate as it should to the plot of the fundamental in the lower field region where the conventional theory should still apply. If it is supposed that in this region the Dingle plots are linear and we are in the 'small orbit' regime throughout, the plots for the harmonics should have slopes varying as the harmonic number r; in fact the slopes for $r = 1, 2, 3$ are as 1 to 1.1 to 1.5. The fact that the reduction factors are close to 1 (of order 0.7 between $1/H = 0$ and the lowest field) make it likely that the small orbit regime applies throughout (see §8.4.4.1), so it is difficult to understand the inconsistent behaviour of the harmonics.

As we have seen earlier, the whole theory of the Dingle factor is not in a very happy state and indeed deviations from linearity of Dingle plots are to be expected for a variety of reasons. It is therefore perhaps a little risky to assume that the observed deviation upwards at high field is due to the predicted many-body effect rather than to some cause which has not yet been tracked down. Against this it should be said that most deviations from linear plots curve the opposite way to that observed here and that excellent agreement with the LK theory was found as regards the harmonic phases which suggests that there was no serious systematic error in the technique used. Evidently it would be of considerable interest to continue these experiments in the search for further confirmation of the many-body theory.

9

Phase and spin-splitting

9.1. Introduction

As was already mentioned in §2.3.7.4 the spin degeneracy of the conduction electrons is lifted in a magnetic field, leading to an energy difference between spin-up and spin-down electrons given by

$$\Delta \varepsilon = \frac{1}{2} g \beta_0 H = \frac{1}{2} g \frac{m}{m_0} \beta H \qquad (9.1)$$

(see 2.145). For free electrons, $g = 2$ and the spin splitting coincides with the Landau level separation (both are just $\beta_0 H$). In real metals, however, spin-orbit coupling and many-body interactions can modify the g-factor considerably. Spin-orbit coupling is particularly important in small pieces of Fermi surface which are associated with electrons undergoing Bragg reflection at zone boundaries, while electron–electron many-body interactions are usually more important in larger pieces of FS. Measurement of the spin-splitting, i.e. of the g-factor, is thus of some interest in contributing to the detailed theoretical understanding of electrons in metals.

The spin-splitting (9.1) can in favourable circumstances be directly observed and measured as a splitting of the magnetic oscillations near the quantum limit. More usually, however, the oscillations are of high quantum number, with only a feeble harmonic content. The superposition of the spin-up and spin-down oscillations then no longer gives the appearance of 'split' oscillations but leads to modification of the amplitudes of the fundamental and the weak harmonics by factors $\cos(\frac{1}{2} p \pi g m / m_0)$ (see 2.148). Thus measurements of the amplitudes can in principle be used to determine g-factors, and various practical methods based on this idea will be outlined below.

Unfortunately, as will be explained in §9.2, both the direct method of observing spin-split oscillations near the quantum limit and the indirect methods based on amplitudes suffer from an ambiguity, in as far as the results are consistent with not just a single g value but a whole series of values. If in the direct method the structure of the oscillations can be followed all the way to the quantum limit, the ambiguity can be completely or nearly completely resolved, but this is rarely possible because of the high

425

magnetic fields required. With the indirect methods, the ambiguity can be reduced (though not eliminated) if the absolute phase of the oscillations is determined and it is convenient therefore to discuss the determination of absolute phase before that of the spin-splitting. The determination of absolute phase is also of interest in its own right as a check on the validity of the LK formula and hence the validity of the quantization condition which lies at the heart of the whole theory. Departures from the LK prediction may occur if complications such as magnetic interaction, magnetic breakdown, spin-dependent impurity scattering or sample inhomogeneity are relevant and observation of such departures can sometimes provide a useful diagnostic tool.

In our presentation we shall first discuss the intrinsic ambiguity in the determination of g-factors and the relevance of absolute phase determination to reducing such ambiguity, then the experimental methods used and the results obtained in such phase determinations, followed by the main theme of the chapter, the determination of g-factors. Finally we shall consider briefly some special situations in which the standard theory needs modification because the spin-splitting does not follow the simple field proportionality rule (9.1) or because the amplitudes of the spin-up and spin-down oscillations are for various reasons unequal. Such situations occur for instance in metals alloyed with very small amounts of transition metals and, in an extreme form, in ferromagnetics.

9.2. Ambiguity of g-factor determination

The ambiguity is perhaps most easily appreciated by examination of a diagram such as fig. 9.1, which indicates the 'positions' of the oscillations on a $1/H$ scale. These are the values of $1/H$ for which the Landau tubes just part company with the FS and are given by

$$F/H = n + \gamma \pm \tfrac{1}{2}S \qquad (9.2)$$

where γ should normally be close to $\tfrac{1}{2}$ (we retain the more general form for convenience of later discussion but in fig. 9.1, γ is assumed to be $\tfrac{1}{2}$) and the spin-splitting parameter S is defined as

$$S = \tfrac{1}{2}gm/m_0 \qquad (9.3)$$

as in (2.148). The diagram can also be regarded as an energy level diagram in which the main levels are βH apart and the splitting is $\pm\tfrac{1}{4}g(m/m_0)\beta H$; however in general, (except for a parabolic band) β varies from the bottom of the band to the Fermi level, so as an energy level diagram, fig. 9.1 is only schematic.

426

9.2. Ambiguity of g-factor determination

The significant feature of the diagram is that apart from the first one or two oscillations near the quantum limit the *same* apparent splitting can be produced by a *variety* of g values. These are such that

$$g = \frac{2r}{(m/m_0)} \pm g_0 \quad \text{or} \quad S = r \pm S_0 \tag{9.4}$$

where g_0 is the smallest value of g and

$$S_0 = \tfrac{1}{2} g_0 m / m_0 \quad (S_0 < \tfrac{1}{2}) \tag{9.5}$$

The sign of g is also indeterminate but it is natural to suppose it is positive, i.e. that the energy is lower when the spin is in the same, rather than the opposite direction as the field and we shall assume this in what follows. As

Fig. 9.1. Illustrating ambiguity of the relation between the apparent and real spin-splitting. The diagram shows short vertical full lines at the values of F/H such that $F/H = n + \tfrac{1}{2} \pm \tfrac{1}{2} S$, where $S = r \pm S_0$ (n and r are integers). Here $S_0 = 0.2$ and (a), (b), (c), (d), (e), (f) correspond to $S = S_0$, $1 - S_0$, $1 + S_0$, $2 - S_0$, $2 + S_0$ and $3 - S_0$ respectively. All of these show the same apparent splitting except that at small F/H (b), (d) and (e) show an unsplit level and (f) shows two unsplit levels. The crossed out lines correspond to negative values of F/H and so cannot occur.

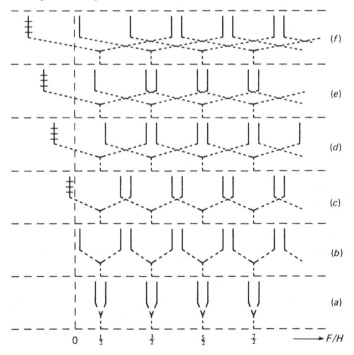

427

already mentioned, the spin-splitting is in favourable conditions directly visible in the oscillations. In ideal conditions (perfect sample and $T = 0$) the positions given by (9.2) correspond to those of the cusps of the M oscillations, of the spikes in the giant quantum oscillations of attenuation of ultrasound or of the resistance peaks in magnetic breakdown galvanomagnetic oscillations.

If the splitting can indeed be directly seen (as in fig. 9.6), it is a fair approximation to suppose that the positions of the observed peaks are given by the ideal positions (9.2), and if γ is known, it is possible to decide from the field dependence of the peak positions whether r in (9.4) is odd or even, thus, as it were, halving the ambiguity. In fact (9.2) is equivalent to

$$F/H = n' - \tfrac{1}{4} \pm \tfrac{1}{4} + \gamma \pm \tfrac{1}{2}S_0 \qquad (9.6)$$

where n' is an integer and the first $\pm$ is plus for r even and minus for r odd. If then a plot is made of the peak $(1/H)$ values against successive integers, the two intercepts* at $1/H = 0$ on the integer axis will give either $(\tfrac{1}{2} - \gamma) \mp \tfrac{1}{2}S_0$ if r is odd or $(1 - \gamma) \mp \tfrac{1}{2}S_0$ if r is even. If $\gamma = \tfrac{1}{2}$ these can be easily distinguished but if the mean of the intercepts is not close to 0 or $\tfrac{1}{2}$ and there is no *a priori* knowledge of γ, the ambiguity between odd and even r remains.

If high enough fields are available to approach close to the quantum limit, further information may come from the positions and structures of the last one or two oscillations. As can be seen from fig. 9.1, the fields at which the last oscillations should come and whether they are single or double, are quite different as between the various alternative values of g. In fact, as we shall see from the examples to be discussed in more detail below, the interpretation may be complicated not only by departures of γ from $\tfrac{1}{2}$, but also by field dependence of the Fermi energy at high fields, which upsets the even spacing of the peaks. However, with evidence from other experiments and theoretical guidance from band structure considerations, the choice between g values can usually be greatly narrowed down.

The distinction between even and odd r in (9.4) is essentially a phase difference of π between the oscillations in the situation of (a), (d) or (e) and one such as (b) or (c), and this distinction finds its counterpart in the determination of g from measurement of the factors $\cos p\pi S$ in the fundamental and harmonic amplitudes of the oscillations, when the spin-splitting is no longer directly observable.

---

* Here and in what follows it is important to realize that by intercept we usually mean the fractional part above the next lowest integer. However the same 'lowest' integer must be used for both the split positions, so if the intercept is a small positive fraction for one spin it may be slightly negative for the other.

The LK formula for M_p, the pth harmonic of the oscillatory magnetization* can be expressed compactly as

$$M_p = a_p \sin\left(\frac{2\pi p F}{H} + \phi_p\right) \tag{9.7}$$

where a_p is positive and contains the factor $|\cos p\pi S|$ and

$$\frac{\phi_p}{2\pi} = \frac{1}{4} \pm \frac{1}{4} - \gamma p \pm \frac{1}{8} \tag{9.8}$$

The first $\pm$ is chosen to be the same as the sign of $\cos p\pi S$ while the second $\pm$ indicates whether the extremal area is minimum ($+$) or maximum ($-$); γ is the Onsager phase factor, normally expected to be close to $\frac{1}{2}$. Any integer may of course be added to or subtracted from the r.h.s. of (9.8), and we shall normally assume that an appropriate integer has been included to bring it into the range 0 to $+1$. We note that for free electrons with $g = 2$, $m/m_0 = 1$, $\gamma = \frac{1}{2}$ and a maximum extremal area, (9.8) reduces to $\frac{3}{8}$ for all p.

We see from (9.8) that if ϕ_p can be determined absolutely, the sign of the spin factor $\cos p\pi S$ is unambiguously determined, provided it is known whether the extremal area of the FS is a maximum or a minimum, and provided the value of ϕ_p is indeed compatible with the theoretically based assumption $\gamma = \frac{1}{2}$. If it is *not* compatible, (9.8) merely determines two possible values of γ, one for each sign of the spin factor. We need to consider the sign of the spin factor only for the fundamental ($p = 1$), since it is easy to show that no extra resolution of the ambiguity is provided by the higher harmonics. If the sign of $\cos \pi S$ has been determined, the possible values of g compatible with the magnitude of the cosine are again given by (9.4), with r restricted to even values if the sign is plus or to odd values of r if minus. Thus measurement of ϕ_1 and hence determination of the sign of the spin factor can effectively halve the ambiguity of determination of g.

9.3. Phase determination: methods

The problem of determining ϕ_p as defined in (9.8) is essentially one of measuring absolute values of H sufficiently precisely. We shall consider only the fundamental ($p = 1$) since, as will be shown later, the absolute phases for the harmonics are easily determined once ϕ_1 (which we shall call simply ϕ in what follows) is known. The easily identified features of the oscillations of M are the zeros, maxima and minima and if the oscillations

---

* We discuss only the dHvA effect, but the generalization to other oscillatory effects or to observations of derivatives of M is straightforward.

are nearly simply harmonic, and described by (9.7) with $p = 1$, they occur at

$$F/H = -(\phi/2\pi) + n + \tfrac{1}{4} \pm \tfrac{1}{4}$$

for a zero ($\pm$ of same sign as that of dM/dH at zero M) and

$$F/H = -(\phi/2\pi) + n \pm \tfrac{1}{4}$$

for a maximum of M ($+$) or minimum ($-$). $\left.\begin{array}{c} \\ \\ \\ \\ \end{array}\right\}$ (9.9)

It is useful to note here that the mean position of the spin-slit pair of cusps of the 'ideal' oscillations as specified by (9.6) can be expressed by (9.8) as

$$F/H = -(\phi/2\pi) + n + \tfrac{1}{2} \pm \tfrac{1}{8} \qquad (9.10)$$

according as the extremal area of the FS is a minimum ($+$) or maximum ($-$) and independently of the parity of r in (9.4). This value of F/H is either $\tfrac{3}{8}$ or $\tfrac{1}{8}$ greater than that for the simple harmonic oscillation maximum given by (9.9).

To determine ϕ from the fundamental oscillations, the obvious procedure is to plot a series of values of $1/H$ at which any particular feature occurs against a series of consecutive integers n. We should then find a straight line of slope $1/F$ and the fractional part of the intercept at $1/H = 0$ on the n axis determines $\phi/2\pi$ in accordance with (9.9). In principle the zeros can be rather more accurately determined than the maxima or minima and this can be important if the highest accuracy is essential. If the oscillations are appreciably non-simple harmonic the error due to the harmonics can be reduced by averaging between the values of ϕ as determined from the zeros corresponding to positive and negative dM/dH.

It is not difficult to estimate the precision of determination of ϕ by this linear plot procedure if we know the uncertainty $\Delta H/H$ of determining the fields at which the zeros occur, the number of readings, N, (assumed to be equally spaced), and the highest and lowest values of H used, say H_1 and H_2 respectively. From the statistical formulae (A15.3) it can be shown that the uncertainty in $\phi/2\pi$ is

$$\Delta(\phi/2\pi) = \frac{4}{(3N)^{1/2}} \frac{F}{H_2} \frac{[1 + (H_2/H_1) + (H_2/H_1)^2]}{1 - (H_2/H_1)} \frac{\Delta H}{H} \qquad (9.11)$$

if $\Delta H/H$ is assumed to be independent of H (this may not be valid in practice but it can be shown that even if it supposed that it is ΔH which is independent of H the answer is not greatly different). For a typical high phase (e.g. for a noble metal belly), we might have $H_2/H_1 = \tfrac{1}{2}$ and $F/H_2 = 2 \times 10^4$, so for $N = 10$ say, we find

$$\Delta(\phi/2\pi) \sim 5 \times 10^4 (\Delta H/H)$$

Thus if ϕ is to be determined to better than 20°, $\Delta H/H$ must not exceed something like 1 part in 10^6. Even if we were to make $N = 2 \times 10^4$, i.e. to measure $1/H$ at *every* zero (which would be almost prohibitively slow and tedious), $\Delta H/H$ would need to be a few parts in 10^5 or so. The only practical method of achieving such precision is to measure H by n.m.r. and so it is not surprising that absolute phase determination in metals with a large FS had to wait for the introduction of this technique into dHvA studies (cf. Coleridge and Templeton (1972)). We shall come back to the practical realization of the method a little later.

For the much lower phases typical of a small Fermi surface, the requirement of precision of field measurement is less stringent, and absolute phases for Bi and a few other low frequency oscillations were determined at quite an early stage (Shoenberg 1939, 1952a, Dhillon and Shoenberg 1955). In the experiment on Bi (see fig. 9.2), H_2/H_1 was about 0.2, F/H_2 was

Fig. 9.2. An early example of phase determination (Shoenberg 1939). The values of $1/H$ at which the dHvA oscillations of Bi (such as those of fig. 1.3) cross the mean line are plotted against successive integers r (careful account of signs was taken in assigning the parity of r with respect to the initial point of each series of points). The various series of points refer to various crystal orientations. The slope of any line determines the dHvA frequency F and the intercept on the r axis gives the phase. At that time (before the effect of spin was understood) it was thought that the oscillations should be of the form $-\sin(2\pi F/H - \pi/4)$, but the experimental result appears to give π (with an accuracy of order $\pi/4$) instead of $\pi/4$. The correct value was later found to be close to $3\pi/4$.

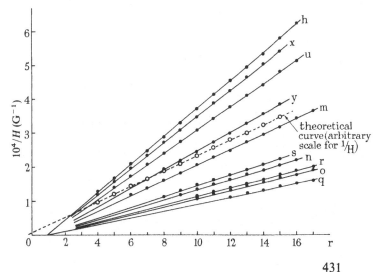

Phase and spin-splitting

Fig. 9.3. Illustrating CT's (1972) method of absolute phase and frequency determination: (a) recorder trace of Ag $\langle 110 \rangle$ dog's bone d^2M/dH^2 oscillations at about $27\,\text{kG}$ and $1.1\,\text{K}$ ($F = 2.01323 \times 10^8\,\text{G}$); the field interval of each oscillation is about $3\,\text{G}$ and the time interval across the picture is about $5\,\text{min}$, with complete stoppages over the horizontal portions where an n.m.r. reading is taken. The oscillations are of the form $a'\cos(2\pi F/H + \phi + \pi/2)$ (cf. (9.7)) with a' positive and ψ is defined by $F/H + \phi/2\pi + \frac{1}{4} = r + \psi$, where r is an integer (note that our notation is different from that of CT: their $\phi_0 = \phi/2\pi + \frac{1}{4}$). (b) analysis of recordings such as (a); the fraction f is determined as explained in the text for a good trial value F_0 of F and $\psi - f$ (increased by a unit at appropriate places) is plotted against $1/H$. The intercept gives $\phi/2\pi + \frac{1}{4}$ and the slope gives $F - F_0$ (in this example $= 5.0 \times 10^4\,\text{G}$). The broken line shows a replot with the fs calculated from a new trial value $F'_0 = F_0 + 5 \times 10^4$.

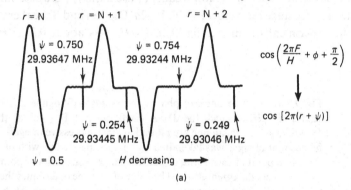

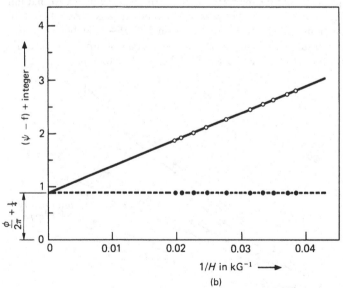

432

about 10 and N was also about 10, which gives

$$\Delta(\phi/2\pi) \simeq 10\Delta H/H$$

Thus, even using the relatively crude methods of those days, for which $\Delta H/H$ was of order 1%, ϕ could be determined to within 40° or so. With better current control and measurement technique, such as is available today, it would be easy to improve the accuracy by a factor of ten or more, even without the use of n.m.r.

The straightforward linear plot method can be used for frequencies a good deal higher than that of Bi if the field is measured with n.m.r. precision, but as F/H becomes larger, say beyond a few thousand, a practical difficulty appears. This is that the numbers n in (9.9) or (9.10) become very large and if the range from H_2 to H_1 to be covered is such that $(1 - H_2/H_1)$ is not too small (as is necessary if $\Delta(\phi/2\pi)$ in (9.11) is to be kept small), a very large number of oscillations, i.e. as many as a thousand or more, must be swept and reliably counted, even though accurate field measurements need be made at only a few of the oscillations. An ingenious method of avoiding the need to keep accurate count of very large numbers of consecutive oscillations was devised by Coleridge and Templeton (CT) (1972), based on a principle somewhat similar to the method of exact fractions in optical interferometry.

At some particular field, a few oscillations are drawn out very slowly to establish the local envelope. The field sweep is stopped altogether at several successive oscillation zeros and an n.m.r. reading taken at each such zero. The zeros are identified as having ordinates half-way between the maxima and the minima and even if the stopping points are not exactly at the half-way points it is easy to make appropriate small corrections based on the actual stopping positions. These accurately determined values of H of the oscillations zeros must then satisfy (9.9), with n of order 10^4 for a large FS (for simplicity we shall suppose it is only the zeros with dM/dH negative which are measured). The value of F can be estimated to a few parts in 10^4 from accurate measurements of the fields at the start and finish of a series of a few hundred oscillations and this precision is sufficient to establish the values of n to within a few integers if F/H is of order 10^4. Suppose this estimate of F is F_0 and that

$$F_0/H = n + m + f \tag{9.12}$$

where m would be a small integer if F_0 has been well estimated and f is the residual fraction. The values of m and f should not change appreciably over a few consecutive zeros. Subtracting (9.9) from (9.12) we have

$$(F_0 - F)/H = m + f + \phi/2\pi \tag{9.13}$$

If now this procedure is repeated at a number of fields separated from each other by many oscillations, which can be rapidly swept through without keeping count, we should find a linear relation between the values of f (to which 1 must be added or subtracted whenever m change by 1) and the values of $1/H$ at the various stopping places. The slope of this linear plot immediately gives $(F_0 - F)$ and the intercept gives $-\phi/2\pi$. As a check on the method, it can be repeated using a revised estimate, say F_0', of F, based on the value of $(F_0 - F)$ found from the first linear plot. This revised estimate F_0' should be within a few parts in 10^6 of the true value of F, so that m in (9.13) would vanish and the value of f would be constant, equal to $-\phi/2\pi$ over the whole range of fields, i.e. the linear plot of f against $1/H$ should be horizontal with the same intercept as the previous plot. An example illustrating the procedure is shown in fig. 9.3.

In all the above discussion we have for simplicity supposed that we have an 'ideal' situation in which the oscillations consist of only a single frequency. If the only other frequencies are high harmonics the ideal situation can be approximately achieved either by working at a temperature high enough to kill the harmonics or using a level of modulation which suppresses the dominant harmonic*. If the oscillations have a complicated frequency spectrum, Fourier analysis becomes essential and absolute phase determination of the separate frequency components becomes more complicated and inevitably somewhat less precise.

The absolute phases of the harmonics of a fundamental frequency can be fairly easily determined if the absolute phase of the fundamental is already known. If the oscillations are Fourier analyzed relative to some local origin (which need not be precisely known) it is straightforward to convert the relative phases of the harmonics into absolute phases. Suppose for instance a field H_0 is chosen as origin and a few oscillations round H_0 are Fourier analyzed to give the pth harmonic of M as

$$M_p = a_p \sin\left[2\pi pF\left(\frac{1}{H} - \frac{1}{H_0}\right) + \phi_p'\right] \tag{9.14}$$

Comparing with (9.7), we see that

$$\phi_p' = \phi_p + 2\pi pF/H_0$$

---

* It should be noted that the use of high modulation amplitude does not affect the basic phase to be measured, provided the induced eddy currents do not introduce a Shubnikov-de Haas contribution (of a different basic phase) to the observed oscillations. If detection is at a multiple of the modulation frequency ω, say $n\omega$, the basic phase of the output signal will be that of $d^n M/dH^n$.

so that

$$\phi'_p - p\phi'_1 = \phi_p - p\phi_1 \qquad (9.15)$$

This result (which we have already used in another context (§8.4.5.3)) enables all the ϕ_p to be determined absolutely provided ϕ_1 (i.e. ϕ) is known.

9.4. Phase determination: some results

As already mentioned, the first phase determination was made for Bi (Shoenberg 1939) and it was noted as a slightly worrying puzzle that the $p = 1$ phases for a variety of orientations came out roughly π different from the prediction of Landau's formula.* The rather more accurate later measurements of Dhillon and Shoenberg (1955) confirmed this discrepancy and revealed moreover that all the odd harmonics showed a similar discrepancy. As can be seen from fig. 9.4, the experimental oscillations agree well with a synthetic curve calculated on the assumption that Landau's

Fig. 9.4. Torque oscillations in Bi near the quantum limit with H at 12° to bisectrix axis in trigonal-bisectrix plane (Dhillon and Shoenberg 1955); C is the torque per unit volume. The open ($T \sim 1.5$ K) and black ($T \sim 1.0$ K) circles are experimental and illustrate the insensitivity to temperature at high fields; the crosses are calculated from a Fourier synthesis assuming a factor $(-1)^p$ must be added to Landau's formula for the thermodynamic potential and allowing for the quadrature term (usually neglected) in differentiating the thermodynamic potential.

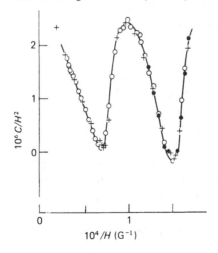

* The discrepancy was originally given as $\frac{3}{4}\pi$ with an uncertainty of about $\frac{1}{4}\pi$ (see fig. 9.2).

435

formula needed correction by a factor $(-1)^p$ for the pth harmonic (and also embodying an appreciable correction to take account of the fact that on approach to the quantum light the assumption $F/H \gg 1$ is no longer valid). The origin of the discrepancy is, of course, that Landau's treatment did not take account of spin. In fact as was first pointed out by Cohen and Blount (1960) and will be discussed later, there is a strong spin-orbit coupling effect which makes g very high and close to the value $2/(m/m_0)$; thus $\cos[\frac{1}{2}p\pi g(m/m_0)]$ (i.e. $\cos p\pi S$) is indeed close to $\cos p\pi$ or $(-1)^p$.

A few other early results on phases of relatively low frequency oscillations in Ga, graphite and Zn (Shoenberg 1952a, Berlincourt and Steele 1954, Dhillon and Shoenberg 1955) suggested some discrepancies with the theoretical predictions. The most clearly demonstrated discrepancy was for the Zn needle oscillations for which the phase $\phi/2\pi$ of the fundamental (as defined by (9.7)) came out as 0.58. Bearing in mind that the extremal area of the needle is a maximum and using (9.8), this implies that γ must be either 0.3 or 0.8 (according as the spin factor is negative or positive respectively), rather than the expected $\frac{1}{2}$; thus without further evidence, the sign of the spin-factor is left undecided. Since the torque method requires an off-extremal orientation, it might at first sight seem that this discrepancy was a consequence of a phase shift due to mosaic structure (see (8.73)) but realistic estimates make this improbable. Moreover, later experiments by quite different techniques, in which the orientation was exactly along the hexad axis and better samples were used, confirmed the discrepancy, as will be elaborated below (Stark 1964, O'Sullivan and Schirber 1967). The discrepancy is probably associated with magnetic breakdown and interband effects, which might modify the value of γ appreciably (Roth 1966), though the theory has been only partially worked out. We shall return to this question in discussing results on the g-factor below (see §9.6.1).

Most of the absolute phase results for relatively high frequencies have come from the Ottawa group, based on the technique of Coleridge and Templeton (1972) outlined above or variants of it. The results for the noble metals are summarized in table 9.1 and it can be seen that they do indeed agree well with the theoretical predictions based on $\gamma = \frac{1}{2}$. For the major extrema (i.e. belly, rosette and dog's bone) the spin factors for $p = 1$ would be expected to have the same sign as in the free electron model, i.e. negative, and the experimentally determined phases agree well with this expectation, i.e. with (9.8) if the first $\pm$ is chosen to be negative and the second is chosen to agree with the known information about the FS. For the necks, however, there is no *a priori* knowledge of the signs of the spin factors and the experimentally determined absolute phases are useful in deciding these signs and so reducing the ambiguity of the g values derived from the oscillation amplitudes; this will be elaborated in §9.5.2.

The fact that the experimental and theoretical values of $\phi/2\pi$ agree to within the experimental uncertainty of typically ± 0.05 or less indicates that γ cannot differ from $\frac{1}{2}$ by more than this. The deviations from $\frac{1}{2}$ predicted by Roth's theory are however far smaller and much higher precision would be needed to reveal them. Similar absolute phase measurements have been made for Al (Coleridge and Holtham 1977), Cd (Coleridge and Templeton 1971), K (Templeton 1972) and Rb (Gaertner and Templeton 1977). All the results point to $\gamma = \frac{1}{2}$ with the same sort of precision as in the noble metals.

We have already shown that for Bi (see table 8.5) the observed relative phases of the harmonics are indeed as predicted by the LK formula. An illustration of the use of relative phases in diagnosing effects which modify the LK formula is provided by the results for Na (Perz and Shoenberg 1976). In Na the harmonics due to MI are comparable in amplitude to

Table 9.1. *Absolute phases for the noble metals (CT 1972)*

		B_{111}	B_{100}	R_{100}	D_{110}	N_{111}	
Cu	exp.	0.64	0.35	0.57	0.62	0.12	0.12
	theory	0.625	0.375	0.625	0.625	0.625($-$)	0.125($-$)
						0.125($+$)	0.625($+$)
Ag	exp.	0.34	0.37	0.65	0.63	0.12	—
	theory	0.375	0.375	0.625	0.625	0.625($-$)	0.125($-$)
						0.125($+$)	0.625($+$)
Au	exp.	0.61	0.42	0.61	0.63	0.13	0.59 ± 0.06
	theory	0.625	0.375	0.625	0.625	0.625($-$)	0.125($-$)
						0.125($+$)	0.625($+$)

Notes: The entries are values of $\phi/2\pi$ except for the last column which gives $\phi_2/2\pi$ for N_{111}. The experimental errors are typically ± 0.05 for B, R and D, but for N only ± 0.01, except where noted. It should be noted that our definition of phase differs from that of Coleridge and Templeton; the relation between our ϕ and their ϕ_0 is $\phi/2\pi = \phi_0 - \frac{1}{4}$. It is assumed that the spin factor is negative for all the B, R and D orbits. For the neck the assumed sign of the spin factor is entered in brackets.

those predicted by the LK formula, so the relative phases vary appreciably as the field changes. As can be seen from fig. 9.5, the field variation of the experimentally determined relative phases agrees well with theoretical prediction (see §6.7.2).

Fig. 9.5. Relative phases of harmonics in Na oscillations at 1.2 K (Perz and Shoenberg 1976). The points are experimental and the curves are theoretical for two trial values of $\cos \pi S$ (0.38 and 0.42) and according as the 'old' or 'new' treatment of MI is used in the calculation (broken and full lines respectively). At low fields the harmonics begin to be dominated by MI, but as the field increases the LK harmonics become more comparable. For MI alone $\phi_2 - 2\phi_1$, $\phi_3 - 3\phi_1$ and $\phi_4 - 4\phi_1$ should be 270°, 180° and 90° respectively.

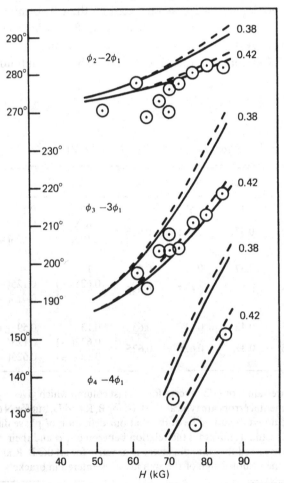

9.5. Determination of the spin-splitting (g) factor

9.5.1. Direct observation of spin-splitting

As was pointed out in §9.1, the spin-splitting of the energy levels in a magnetic field can in favourable circumstances be directly observed as a splitting of the oscillations. The circumstances are most favourable for oscillations of low quantum number, i.e. close to the quantum limit, and for oscillations of as 'peaky' a line shape as possible. The giant quantum oscillations of ultrasonic attenuation (fig. 4.4), ΔT (fig. 4.1), dM/dH (fig. 8.9), d^2M/dH^2 (fig. 9.6) and the magnetic breakdown resistivity oscillations (fig. 7.16), all show this direct splitting as we approach the quantum limit, provided the sample is good enough and the temperature low enough. The essential point is that the oscillations should be rich in harmonics and that the smearing of phase due to temperature, electron scattering and imperfections should not exceed the phase separation of the up and down spins. The phase separation between up and down spins is independent of H (it is $\pi g m/m_0$), while the phase smearing due to the other causes varies as $1/H$ (it is of order $2\pi^2 k(T + x)/\beta H$), so any observable splitting of the oscillations is rapidly masked as H decreases. Once $2\pi^2 k(T + x)/\beta H$ appreciably exceeds 1 the higher harmonics are more and more damped out and 'direct' splitting can no longer be observed. The effect of the spin-splitting is then only through the interference between two nearly simple harmonic oscillations, i.e. it is reflected only in the amplitude of the resultant oscillations, as will be discussed in §9.5.2. If the spin-splitting can be directly observed, the derivation of possible values of the g-factor is straightforward, though the resolution of the ambiguity may be rather subtle, and we shall now discuss some results in detail.

The directly visible spin-splitting in the GQO of Ga was shown in fig. 4.9 and the field dependence of the mean peak positions provides perhaps the simplest example of the principles discussed in §9.2. The separation in F/H of the two peaks, i.e. S_0, is 0.031 (± 0.002) and $m/m_0 = 0.066$, so (9.4) gives (in round numbers), the following possible values of S and g:

$$S = r \pm 0.031, \quad g = 30r \pm 1, \quad \text{i.e.} = 1, 29, 31, 59, 61 \text{ etc.}$$

However from the data of Shapira (1964) or Shapira and Lax (1965) a linear plot of the mean F/H for the split peaks against consecutive integers is found to give an intercept $\frac{1}{2}$ on the number axis (with an uncertainty of ± 0.1) and (9.6) therefore implies (with the usual convention that 1 can be added to keep γ a positive fraction)

$$-\gamma + \tfrac{1}{4} \mp \tfrac{1}{4} = \tfrac{1}{2} \quad \text{(minus for } r \text{ even, plus for } r \text{ odd)}$$

The obvious interpretation is that $\gamma = \frac{1}{2}$, as suggested by theory, and therefore that r is even, which means that the odd r values above may be ruled out, and the ambiguity is thus halved, i.e.

$$g = 1, 59, 61, 119, 121 \text{ etc.}$$

This reduction in ambiguity appears to have been overlooked in the published papers. Some further guidance is provided by band structure theory (see §9.6.1) which suggests that $g = 1$ is probably the right answer.

Our next example is Bi where the direct spin-splitting has been observed by a number of techniques. Some of the spin-split oscillations have already been mentioned earlier (see figs 4.1, 4.4 and 8.9) and we refer to the review by Edelman (1976) for a detailed discussion of the orientation dependence of the splitting. Here we shall consider only the special case of H along the binary axis, for which the split oscillations can be followed right up to the quantum limit. Measurements of the splitting in the various experiments agree in showing that $S_0 \simeq 0.1$ (Smith, Baraff and Rowell (1964) give 0.07, Edelman (1976) gives 0.09 ± 0.01); moreover since, as mentioned earlier, the absolute phase measurements indicate the spin-splitting is nearly equal to the Landau level splitting, it is clear that only *odd* numbers r should be considered in (9.4) i.e.

$$S = 1 \pm S_0 \qquad 3 \pm S_0 \text{ etc.}$$

corresponding to (b), (c), (f) etc. in fig. 9.1. However $3 \pm S_0$ and higher values of S can be ruled out, because they would imply the presence of several unsplit peaks at fields beyond the last 'doublet' of peaks and this is not observed. It should be noticed that the situation in Bi differs in one important respect from that assumed in fig. 9.1 (b and c). This is that the Fermi energy ζ becomes quite strongly field dependent as the quantum limit is approached in order to keep the number of electrons equal to the number of holes ('charge neutrality') and this works out in such a way that the last Landau tube never empties as the field is raised (i.e. F *increases* with H and F/H can never reach the requisite low value). Consequently the lowest level of fig. 9.1 can never correspond to an oscillation peak.

A decision between $1 + S_0$ and $1 - S_0$ is suggested by the orientation dependence of S_0. If the field is rotated away from the binary axis in the binary-trigonal plane, S_0 is found to decrease and to pass through zero at about 65° from the binary axis. The natural interpretation is that the true S decreases smoothly from a value of 1.1 at the binary axis to a value

9.5. Determination of the g-factor

below 1 beyond the 65° orientation, with a minimum of about 0.35 near the trigonal axis. We could suppose alternatively that S *increased* from 0.9 through 1 at 65° to a maximum value 1.65 near the trigonal axis, but band structure considerations make this unlikely. If then we suppose that $S = 1.1$ is the correct choice, we find from (9.3), since $m/m_0 \simeq 0.009$, that $g \simeq 240$. We shall discuss the theoretical significance of this result in §9.6.1.

Another example is the needle in Zn, for which direct spin-splitting has been observed in dHvA oscillations of d^2M/dH^2 (O'Sullivan and Schirber 1967)[*], in galvanomagnetic oscillations in magnetic breakdown conditions (Stark 1964) and ultrasonic attenuation (Myers and Bosnell 1966). All three experiments give essentially similar results for g and we shall discuss only the d^2M/dH^2 oscillations, shown in fig. 9.6. It can be seen that there are two clearly separated peaks A and B in each oscillation and their positions can without much error be identified with the 'ideal' positions defined in the discussion of §9.2 (see (9.2) and (9.6)). The mean F/H for an A and a B peak could at worst (if the peak was a simple harmonic maximum) differ from the 'ideal' value by $\frac{1}{8}$ according to (9.9) and (9.10), (the minus sign must be chosen in (9.10) since the extremal area is a maximum).

Direct measurement of the average of the separation AB for the various oscillations showed that the S_0 of (9.5) was 0.36 (cf. 0.34 in Stark's (1964) experiments) and, since $m/m_0 = 0.0075$, this immediately implies (see (9.4)) that

$$S = r \pm 0.36 \quad \text{and} \quad g = 267(r \pm 0.36)$$

Fig. 9.6. Oscillations of d^2M/dH^2 for the Zn needle with H along the hexad axis and $T = 1.2\,\text{K}$, obtained by field modulation with 2ω detection (after O'Sullivan and Schirber 1967).

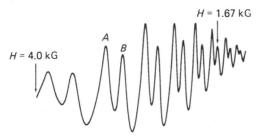

[*] The torque oscillations observed by Dhillon and Shoenberg (1955) did not show split peaks because the sample was of poor quality and because in any case the harmonic content is lower (by a factor p^2 for the pth harmonic) in torque oscillations than in oscillations of d^2M/dH^2.

441

Phase and spin-splitting

Thus for $r = 0, 1, 2$, etc., we find (in round numbers)

$$S = 0.36, 0.64, 1.36, 1.64, 2.36 \text{ etc.} \quad g = 96, 170, 360, 440, 630 \text{ etc.}$$

corresponding qualitatively to cases (a) to (e) of fig. 9.1. However the situation is complicated by the result mentioned earlier, that γ is appreciably different from $\frac{1}{2}$. If (9.2) and (9.4) are applied to the spin-split peaks of O'Sullivan and Schirber (see their fig. 4), γ is found to be 0.32 or 0.82, which is reasonably consistent with the 0.3 or 0.8 rather less precisely determined by Dhillon and Shoenberg. With these two possible values of γ the values of F/H at which peaks should occur are as shown in fig. 9.7, which is essentially a modified version of fig. 9.1. Because the electrons in the needle represent only a minute fraction of the total number of carriers in the Zn FS, the assumption of a constant Fermi energy (and so a constant F) even in the quantum limit is quite valid here, so that fig. 9.7 implies equal spacings of $1/H$, in contrast to the situation in Bi, where F increases with H at high fields, so that peaks of very low F/H may not occur. It can be seen that the possible values of g consistent with the observed peaks now fall into two groups:

$$g = 96, 440, 630 \text{ etc.} \quad \text{if} \quad \gamma = 0.32 \text{ (even } r)*$$

and

$$g = 170, 360 \text{ etc.} \quad \text{if} \quad \gamma = 0.82 \text{ (odd } r)$$

Fig. 9.7. Positions of spin-split peaks in Zn. The lowest observed values of F/H (O'Sullivan and Schirber 1967) are shown by arrows. The values of F/H such that $F/H = n + \gamma \pm \frac{1}{2}S$ ($S = r \pm 0.36$) are indicated for $\gamma = 0.32$ and 0.82 and for $r = 0, 1, 2$. Only the schemes a, d, e, B, C (indicated by heavier lines) are consistent with experiment.

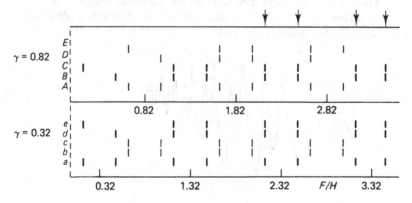

* The 440, 630 and still higher possibilities (for $r > 2$) were ruled out by Stark because they would produce negative values of F/H for low n and the minus choice in (9.2).

442

but further guidance is needed to choose between the various possibilities. Such guidance should be provided if the experiment could be extended to considerably higher fields. Thus the presence of a peak near $F/H = 0.50$ ($H = 31\,\text{kG}$) would rule out $g = 360$ and 630, while one near $F/H = 0.14$ ($H = 110\,\text{kG}$) would rule out 170 and 440. Thus if both these peaks were found $g = 96$ would be unambiguously indicated. This crucial experiment remains to be done, though it may in practice prove difficult because magnetic breakdown would greatly reduce amplitudes at such high fields.

O'Sullivan and Schirber (1967) used an ingenious alternative indirect method to decide between the possible g values and their results suggest that $g = 170$ or 440 is in fact the correct choice. They point out that the cyclotron masses of the two spin-split states should differ appreciably (by an amount proportional to H/F: by about 4% at $5\,\text{kG}$) so that the amplitudes of the spin-up and spin-down oscillations should become appreciably different at high temperatures and high fields. Their estimate suggests that above about $4\,\text{K}$ the oscillations (which at such temperatures are already nearly simple harmonic) begin to be dominated by the low mass state, which is the higher energy state of the spin-split pair. Now from their measurements on the pressure dependence of the oscillations they find that the F/H separation of the pair, i.e. S_0 as defined in (9.4), decreases with increasing pressure. Since the c/a ratio of Zn decreases with pressure but increases with temperature and there is good reason for supposing that the effects of temperature and pressure are mainly due to the consequent changes of c/a (see §5.5), it follows that S_0 should *increase* with rising temperature. But if it is only the low mass oscillation which has appreciable amplitude at high temperatures, the observed peak position will move to *lower* F/H (relative to the mean of the two peak positions) for the $g = 170$ or 440 possibilities, but to *higher* F/H for the others, as T is increased (see fig. 9.9). In the former case the observed phase* ϕ will *increase* as T increases, in the latter it will *decrease*. The experiments of Berlincourt and Steele (1954) mentioned earlier not only revealed an appreciable decrease of F from 1.57 to about $0.8 \times 10^4\,\text{G}$ (see §5.5), but also an increase of $\phi/2\pi$ by about $\frac{1}{4}$ as T increases from 4 to $60\,\text{K}$. The sense of the phase change suggests that it is the $g = 170$ or 440 possibilities which must be chosen, and detailed calculations by O'Sullivan and Schirber confirm that

---

* Some care is needed in relating our phase ϕ which was defined in the context of simple harmonic oscillations (nearly appropriate for the Zn torque oscillations, particularly at high T) with the phase Φ used by O'Sullivan and Schirber, defined as the intercept on the integer axes of a plot of F/H for peak positions. If the peaks are simply maxima of simple harmonic oscillations the relation is $\Phi = \phi/2\pi - \frac{1}{4}$, but if the peaks are 'ideal' cusps, while ϕ still relates to the fundamental component alone $\Phi = \phi/2\pi - \frac{3}{8}$ (see (9.10)). Berlincourt and Steele's phase δ is again differently defined; in fact it differs by π from our ϕ.

on this assumption the magnitude of the phase change agrees well with that observed experimentally. Band structure considerations to be discussed later (§9.6.1) prove to be reasonably consistent with the choice of $g = 170$.

Direct observation of spin-splitting has also been studied in Sb (Hill and Vanderkooy 1978), but because spin-splitting oscillations of several frequencies are superimposed, the resulting pattern is complicated and we shall not discuss its interpretation beyond mentioning that it leads to values of S between 0.2 and 0.5 (varying with orientation and as between holes and electrons) and of g between 4 and 6.

9.5.2. g-factor from amplitudes of fundamental and harmonics

We shall now describe a number of different approaches to determining $\cos \pi S$ and hence the g-factors; the results on the alkalis and noble-metals are summarized in table 9.3 (p. 459). Which is the most reliable, or indeed which can be used at all, depends on the particular metal and orbit concerned, but in general it is useful to apply as many of these methods as possible for the same metal and orbit in order to check the validity both of the LK theory and of the theory of MI which is often involved. It seems that both theories are in fact sufficiently valid at the level of experimental precision achieved so far, but this precision is not very high (of order a few per cent) and it would be desirable to repeat some of these checks at an improved level of precision.

9.5.2.1. The harmonic ratio (HR) method

As already pointed out, the direct observation of spin-splitting is no longer possible if the harmonic content of the oscillations becomes too small. However, even if the harmonic content is too small to split the oscillations into distinct peaks it can still be revealed by Fourier analysis, and measurements of a harmonic amplitude relative to the fundamental is one possible way of determining g. This harmonic ratio (HR) method was first exploited by Shoenberg and Vuillemin (1966) to measure g for the gold neck and later systematically applied by Randles (1972), Knecht (1975) and others.

As can be seen from the LK formula (2.152) the ratio of the $p = 2$ harmonic amplitude to that of the fundamental ($p = 1$) in M is given by

$$\frac{a_2}{a_1} = \frac{1}{2\sqrt{2}} \left| \frac{\cos 2\pi S}{\cos \pi S} \right| \frac{\exp(-2\pi^2 kx/\beta H)}{\cosh(2\pi^2 kT/\beta H)} \tag{9.16}$$

(if it is dM/dH or a higher derivative which is observed, the r.h.s. of (9.16)

444

requires multiplication by 2 or an appropriate power of 2). Thus the value of $|\cos 2\pi S/\cos \pi S|$ can be extracted from measurement of the harmonic ratio, provided the Dingle temperature x and β ($= e\hbar/mc$) are both known. An implicit assumption is that the damping of the pth harmonic due to phase smearing has indeed the form $\exp(-2\pi^2 pkx/\beta H)$. As discussed in chapter 8, this may not always be valid, but its validity can of course be checked by appropriate logarithmic plots, and the plot for $p = 1$ is in any case necessary to determine x. The great merits of this HR method are that only relative amplitudes need be measured, so that the difficulties of calibration are avoided, the sample volume need not be known (a considerable advantage in dealing with alkalis), and perhaps most important, the awkward curvature factor $|A''|^{-1/2}$ (which we shall denote by C) is not involved in the ratio. This last point can be important if the FS is complex or not very precisely determined, so that C cannot be reliably estimated.

The extraction of $|\cos \pi S|$ from the ratio $|\cos 2\pi S/\cos \pi S|$ is not quite straightforward unless the ratio is greater than 1. If the ratio is less than 1, two possible values of $|\cos \pi S|$ are consistent with a given value of the ratio but, as explained in Appendix 19, this ambiguity can be resolved if the relative phase of the harmonic is measured or if higher harmonic amplitudes can be measured. Of course, the determination of $|\cos \pi S|$ still leaves ambiguity in S and hence g, as discussed in §9.2, where we saw that the ambiguity can be halved if the absolute phase of the fundamental is known.

We have discussed the HR method in the context of the $p = 2$ harmonic because usually it is only this harmonic that can be measured with reasonable accuracy. If, however, higher harmonics are strong enough, they can also be used to advantage. An ingenious variant of the HR method developed by Gold and Schmor (1976) and Gold and Van Schyndel (1981) and used to determine g for lead, makes use of the $p = 3$ harmonic. The idea is that the combination $a_2^2/|a_1 a_3|$ is basically determined by $(\cos 2\pi S)^2/|\cos \pi S \cos 3\pi S|$ and is independent of the Dingle temperature, so that in principle amplitude measurements at a single field should be sufficient to determine S. In practice, however, as already mentioned, it is essential to check that the amplitudes are indeed damped as $\exp(-2\pi^2 pkx/\beta H)$, so elimination of the Dingle temperature is not as much of an advantage as might appear at first sight.

We must now consider a complication built into any form of the HR method and which we have up to now ignored. This is that essentially the same conditions (i.e. low T and high H) which favour the appearance of appreciable harmonics in the LK formula (LK harmonics as we shall call them), also favour appreciable harmonics due to magnetic interaction (MI harmonics). The MI harmonics prove to be comparable to, or larger than, the LK harmonics for oscillations associated with a large FS area (e.g. noble metal bellies or alkalis), but become relatively less significant for smaller FS areas (roughly in proportion to $F^2/|A''|^{1/2}$), and are usually no longer important for $F < 10^6$ G.

If conditions of field and temperature can be chosen so that the LK harmonics are strong enough to measure accurately while the MI harmonics are not overwhelming, the two are easily separated because they have different phases relative to the fundamental. For the LK harmonic (see (9.8)) $\phi_2 - 2\phi_1$ is $\mp\frac{1}{4}\pi$ if the spin factor is negative or $\pm\frac{3}{4}\pi$ if it is positive (the upper sign for a minimum area, the lower for a maximum) but is π for the MI harmonic in the weak limit (see (6.18)). Here we refer to the phases as defined in (9.8), where the oscillation variable is $1/H$. In chapter 6 the variable was taken as h, the difference of H from a fixed field H_0, and it can easily be shown that this effectively changes $\phi_2 - 2\phi_1$ by π. It is important to remember too that the above values of $\phi_2 - 2\phi_1$ require changing by $\frac{1}{2}\pi$ if the oscillations are in dM/dH rather than M. An example of how this separation of LK and MI harmonics works in practice is shown in fig. 9.8 (based on Knecht's (1975) results for K).

It will be seen that this type of analysis not only gives the LK harmonic ratio from which the g-factor is derived, but also the MI harmonic ratio, which gives an immediate estimate of the absolute value of $|dM/dH|$ without having to know any calibration constants or the sample volume. We shall return to this absolute determination later and show how it too can be used to derive the g-factor.

The HR method is no longer straightforward if $|4\pi\,dM/dH|$, and hence the MI harmonics, become too large in conditions for which the LK harmonics are strong enough to measure. Both the amplitudes and phases of the observed harmonics then become complicated functions of the true LK amplitudes and to extract the true harmonic ratios becomes a matter of trial and error. Rough preliminary values of $\cos\pi S$ and of the Dingle x can be obtained by ignoring the complications and these can be used as trial values ($\cos\pi S$, of course, determines all the harmonic factors $\cos\pi S$). With these trial values, the LK formula provides a trial set of amplitudes for the fundamental and the LK harmonics as they would be without MI. These can then be fed into the Phillips and Gold iteration scheme (see §6.7.2) for calculating the MI modified* amplitudes and phases which would be

---

* As discussed in chapter 6, if the Dingle x is predominantly due to phase smearing the effect of MI in detail is different according as the phase smearing is introduced into the calculation before or after MI ('old' and 'new' treatments).

observed if the trial assumptions were valid. Comparison of the field dependence of the predicted trial amplitudes and phases with those actually observed then indicates how the trial assumptions need to be modified. This trial and error procedure was found necessary in Perz and Shoenberg's (1976) experiments on Na, for which (on account of the higher F) MI is rather more important than for the other alkalis (the phase comparison is shown in fig. 9.5). For still stronger MI, as in the belly oscillations of the noble metals in conditions such that the LK harmonics are appreciable, the harmonic content is entirely dominated by MI (except in Cu) and it becomes impossible to extract any information about the LK harmonics.

As already mentioned in §6.10, another way of dealing with the MI difficulty, successfully used by Gold and Van Schyndel (1981) to study the g-factor for the γ oscillations in Pb, is to feed back a signal proportional to M into the magnetic field sensed by the sample. If the strength of feedback is suitably adjusted, the field sensed by the electrons in the sample (which without feedback would be the B in the sample) can be brought back to

Fig. 9.8. Illustrating the HR, MI and AA methods (after Knecht 1975). The sample (with $n \sim 0.1$) is of K at 0.55 K and the oscillations (period ~ 39 G at 84 kG) were recorded by the field modulation method. Fourier analysis gives (absolutely) $4\pi \, dM/dH = 0.141 \sin kh + 0.0161 \sin(2kh + 281.3°)$ where h is measured from an appropriate zero. The harmonic term can be expressed in terms of an MI contribution $0.0189 \sin(2kh + 270°)$ and an LK contribution $0.0045 \sin(2kh + 45°)$. The value of $|\cos(\pi S)|$ can now be estimated as: (a) 0.60 by the HR method from the ratio $0.0045/0.141$, (b) 0.67 by the MI method from the absolute magnitude 0.149 of $4\pi |dM/dH|$ inferred from the ratio $0.0189/0.141$ and (c) 0.64 from the directly estimated absolute magnitude 0.141 of $4\pi |dM/dH|$.

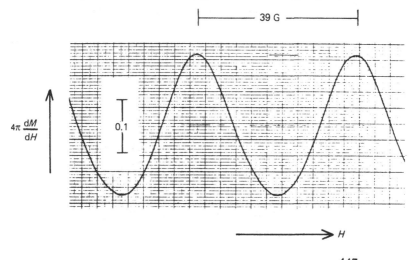

H_e, the field of the magnet, so that MI is effectively eliminated, provided it is not too strong, and provided that phase smearing is not too severe (see §6.10). Once MI is eliminated, the LK harmonics can be studied on their own and the HR method applied to determine g. A useful by-product of this approach is that when the feedback adjustment has been correctly made (as diagnosed by the minimization of MI side bands in the frequency spectrum around the higher α frequency), the current in the feedback coil effectively measures $4\pi(1 - n)M$, where $4\pi n$ is the sample demagnetizing coefficient. Thus once again the complication of MI can be put to good use in measuring the absolute amplitude of M without the need to know the sample volume or the calibration of the detecting equipment (though of course, here it is necessary to know the field constant of the feedback coil).

9.5.2.2. The spin-zero (SZ) method

For certain orbits in certain metals there are particular orientations for which the spin factor, by coincidence as it were, exactly vanishes and this manifests itself as a vanishing of the fundamental amplitude (or occasionally, less obviously, as the vanishing of a particular harmonic amplitude). If such a vanishing occurs (we consider only the usual case of the fundamental) it follows immediately that

$$\cos \pi S \equiv \cos\left(\frac{\pi}{2}g\frac{m}{m_0}\right) = 0 \tag{9.17}$$

and hence that

$$S = r + \tfrac{1}{2} \quad \text{or} \quad g = (2r + 1)/(m/m_0) \tag{9.18}$$

where r is any integer. It should be noted that the degree of ambiguity is the same as if the sign of a non-vanishing $\cos \pi S$ had been determined (since no sign can be attached to a zero!).

Such spin-zeros were first observed by Joseph and Thorsen (1964a) at orientations on either side of $\langle 111 \rangle$ in the neck oscillations of Cu (see fig. 9.9) and have since also been seen for particular orientations in the necks of Ag and Au (Joseph and Thorsen 1964b) and in the belly oscillations of Cu (Joseph, Thorsen, Gertner and Valby 1966, Halse 1969). In Cu, the neck mass is close to $\tfrac{1}{2}m_0$, so if g is close to 2, S is close to $\tfrac{1}{2}$, while the belly mass is close to $\tfrac{3}{2}m_0$, so if g is close to 2, S is close to $\tfrac{3}{2}$. Even at orientations well away from the spin zeros, the value of $\cos \pi S$ for Cu is rather small, typically of order 0.1 to 0.2 for both neck and belly. This implies that $\cos 2\pi S$ is close to -1, and that the second harmonic is unusually strong relative to the fundamental over a wide range of orientations. This is reflected in the appearance of the oscillations (see fig. 9.10) which show splitting directly in the form of two separate maxima.* Spin-zeros have also been found in a number of other metals, such as Sb (Windmiller 1966) and Pt and Pd (Windmiller, Ketterson and Hörnfeldt 1970).

9.5.2.3. The absolute amplitude (AA) method

With the development of improved measuring techniques and more precise knowledge of Fermi surfaces, it was gradually realized that g-factors could be determined from measured absolute amplitudes even when the LK harmonics were negligibly small. This possibility was first systematically exploited by Knecht (1975) who developed the technique described in

Fig. 9.9. Spin zeros in the neck oscillations of Cu (Joseph and Thorsen 1964a). The oscillations were observed by the torque method and the vanishing of amplitude at about 13° either side of ⟨111⟩ in the (110) rotation zone demonstrates the spin zeros, i.e. the vanishing of $\cos(\pi S)$ at these orientations. The vanishing at ⟨111⟩ is of course merely because F has a minimum there with respect to orientation. The rapid falling off beyond 20° to insignificant values is probably partly because of poor sample quality; the neck ceases to exist at about 25° on either side of ⟨111⟩.

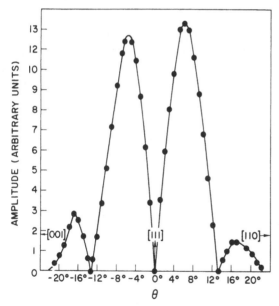

* The fact that the separation of the two maxima is 0.37 of a period, while S as estimated by Randles (1972) using the HR method at this orientation is about 0.44, is probably due to insufficient 'peakiness' of the separate spin-up and spin-down oscillations. This tends to bring the two maxima in the superposed curve somewhat closer together than they would be if drawn separately. The Zn oscillations discussed earlier were rather 'peakier' and the measurement of S by the separation of maxima was probably more reliable.

§3.4.2.1 of calibrating the pick-up coil by using it as a modulating coil and observing a Bessel zero with oscillations of known F. To extract the spin factor $\cos \pi S$ from the observed absolute amplitude it is essential to know the curvature factor C (i.e. $|A''|^{-1/2}$) in the LK formula as well, of course, as F and (m/m_0). For the alkali metals which Knecht studied, the Fermi surfaces are nearly spherical so that $C \simeq (2\pi)^{-1/2}$; the small departures of C from $(2\pi)^{-1/2}$ due to the slight departures from sphericity can be estimated and allowed for (these are of order 0.1% for Na, 1% for K, 5% for Rb, but up to 20% for Cs). The values of $\cos \pi S$ obtained by this AA method for the alkalis agreed well with those obtained by the HR method, thus confirming that the LK formula is reliable absolutely as well as relatively.

This confirmation encouraged Bibby and Shoenberg (BS) (1979) to use the AA method for the noble metal bellies where the HR method fails because of excessive MI, as we saw above. All that was necessary to reduce MI was to raise the temperature sufficiently to reduce $4\pi|dM/dH|$ well below 1 and thus make MI either negligible or amenable to correction. The value of the curvature factor could be computed from the specifications of the FS which, as mentioned earlier, are now known with high precision (Bibby *et al.* 1979). In fact the computations showed that the values of C obtained from the various quite different formulae which specify the Fermi surfaces equally well, were remarkably consistent for most orientations, though there were discrepancies of order 10% around $\langle 111 \rangle$ between the parametrized band structure (KKR) and Fourier representations, especially for Au (possibly due to neglect of relativistic effects in the KKR description).

A valuable check on the whole procedure is that for Cu there are many orientations where the belly amplitude vanishes (these lie on a few discrete

Fig. 9.10. Torque oscillations of the neck in Cu at 6.1° to $\langle 111 \rangle$ in a (110) zone at 1.2 K and about 67 kG. The evident presence of a strong second harmonic is a consequence of the proximity of a spin zero (after Shoenberg and Vanderkooy 1970).

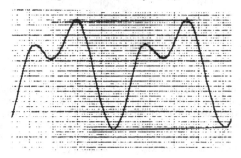

lines in a stereographic projection), so that the SZ method can be applied to give reliable S (and hence g) values at a variety of orientations. This enabled Randles (1972) to work out an interpolation scheme and predict the S values at other orientations. In fact Bibby and Shoenberg's AA results agreed remarkably well with Randles' interpolated values, even when it was $\cos \pi S$ which was compared, rather than the much less sensitive S. This agreement gives confidence that the AA results for Ag and Au are also reliable.

9.5.2.4. The magnetic interaction (MI) method

This is really a variant of the AA method, in which the absolute amplitude is determined from the strength of the MI harmonics or from the strength of the feedback field in the Gold and Van Schyndel (1981) technique. This has the merit, of course, that no calibration of the equipment is required and the sample volume need not be known. However, the values of C (i.e. $|A''|^{-1/2}$), F, (m/m_0) and x are still required to derive $\cos \pi S$ from the absolute amplitude. If MI is not too strong and the LK harmonics are also weak, the MI harmonic component is easily separated from the LK component by careful measurement of the relative phase of the observed harmonic (as in the example of fig. 9.8). Allowance must, of course, be made for the shape of the sample: as explained in §6.4.2, if the demagnetizing coefficient is $4\pi n$, the ratio of the $p = 2$ MI harmonic amplitude to that of the fundamental in dM/dH_e is given in the weak MI limit by

$$(a_2/a_1)_{MI} = 4\pi |dM/dH|(1 - n) \qquad (9.19)$$

so that the absolute value of $|dM/dH|$ can be determined once $(a_2/a_1)_{MI}$ has been separated out. If, however, MI becomes stronger, the simple additivity of the a_{2MI} and a_{2LK} components, each with its appropriate phase, begins to break down and the higher order terms in the Phillips and Gold scheme begin to be significant. In that case the trial and error procedure outlined in §9.5.2.1 can still be used; this is really a combination of the HR and MI methods, rather than either of them separately.

9.6 Discussion of g value results

As mentioned earlier, the main mechanisms responsible for departures from the free electron value $g = 2$ are spin-orbit coupling and electron–electron interactions, and usually one of these is so dominant that we need not consider the other. Very high g values, such as we have already discussed in Zn and Bi (typically of order $1/(m/m_0)$), are characteristic of the spin-orbit coupling mechanism in the context of very small orbits located in k-space at places where there is a small energy gap to a neighbouring band. The spin-orbit mechanism depends on Z^4, where Z is the atomic number of the metal and is therefore most effective for heavy metals; this is sometimes a useful criterion in comparing results for a series of similar metals with different Z, e.g. in going from Cu to Au for the noble metals.

Another criterion is to compare the g value with that obtained by spin resonance. If spin-orbit coupling is the only relevant mechanism the two values should be the same, but not if it is electron–electron interaction. This is because electron–electron interaction, although relevant to the energy separation of spin-up and spin-down states, produces an 'effective' magnetic field which is always in the same direction as the instantaneous spin magnetic moment, and so can exert no couple on it which will affect the resonant frequency. Unfortunately there are two difficulties in applying this criterion. First, electron spin resonance is a difficult experiment in metals because of eddy current difficulties, so that as yet resonance has been observéd in relatively few metals and secondly there is the problem of the anisotropy of g. Spin resonance can measure only an average of g over the FS and it is not always obvious how to compare this average with strongly orientation dependent g values deduced from dHvA oscillations. Nevertheless we shall see that spin resonance can often provide useful information.

Finally, before discussing the results for individual metals, a general point about accuracy should be made. Since, as we saw in §9.2, the value of S ($=\frac{1}{2}gm/m_0$) is of the form $r \pm S_0$ where r is an integer and $S_0 < \frac{1}{2}$, the relative precision of S, and hence of g, may be higher than that of what is directly deduced from experiment. This of course assumes that the value of r is reliably known from other evidence (such as comparison with g values obtained from quite different experiments or with theoretical estimates). For instance in Na, K and Rb, $S = 2 - S_0$, so that $\delta S/S$ is at least three times smaller than $\delta S_0/S_0$ (for the actual values of S_0 in fact five or more times smaller). Moreover S_0 itself is derived from either $\cos \pi S_0$ (in the AA method) or from $\cos 2\pi S/\cos \pi S$ (in the HR method) so that $\delta S_0/S_0$ depends in a complicated way on the relative accuracy of the actually measured quantity. In the AA method, if S_0 is close to $\frac{1}{2}$, quite a large error in the measured $\cos \pi S_0$ would produce a much smaller error in S_0; if, however, S_0 is close to 0, it is the other way round. Similarly in the HR method, if S_0 is close to $\frac{1}{4}$, $\cos 2\pi S_0$ is small and again quite large errors in the observed harmonic ratio correspond to much smaller error in S_0. In the spin-zero method, of course, S_0 is exactly $\frac{1}{2}$ and the only error in g comes from the error in m/m_0. As examples of how these considerations work, we may mention that for Na, K and Rb and for spin-zero orientations in Cu, the g values determined from the dHvA effect are probably accurate to better than $\frac{1}{2}\%$, the accuracy being limited mainly by the uncertainty of m/m_0.

9.6. Discussion of g value results

Such accuracies are much higher than those with which absolute or relative amplitudes are usually measured (typically a few %).

9.6.1. Spin-orbit splitting

The problem of deriving spin-orbit splitting in metals from basic band structure theory has been reviewed by Yafet (1963) (though mainly in the context of interpreting spin resonance experiments). We shall not attempt any detailed exposition of the rather complicated theory involved, but give only some relatively simple arguments which help to understand the orders of magnitude of the observed effects.

For small pieces of FS with extremal orbits involving a number of Bragg reflections, Pippard (1968, 1969) has given a semi-quantitative argument which explains why high g values are observed, and sets an upper limit to what is to be expected. In moving through the periodic electric field **grad** V of the lattice with velocity v, the electron spin 'feels' a magnetic field $(v \times \mathbf{grad}\ V)/c$ and the interaction of its spin magnetic moment μ_B (i.e. a Bohr magneton) with this field contributes a term

$$\varepsilon = \pm \frac{1}{2}\frac{\mu_B}{c}|v \times \mathbf{grad}\ V| \tag{9.20}$$

to the energy, which is $\pm$ according as the spin is antiparallel or parallel to the direction of the magnetic field. The factor $\frac{1}{2}$ arises from the relativistic origin of the spin moment of an electron, but is not important for the present qualitative argument. For this spin-orbit interaction the appropriate magnitudes of V are rather heavily weighted by regions close to the atomic nuclei and so can be much larger–as much as several 10^4 times larger–than the small pseudopotential V' which determines the band structure. Similarly the relevant v may be somewhat larger than the ordinary orbital velocity–perhaps ten times larger. If for simplicity we suppose the lattice potential varies only in one dimension (cf. fig. 7.4) as $\cos(x/a)$, where a is the lattice parameter, the spin-orbit energy (9.20) varies as $\sin(x/a)$; substitution of numerical values shows that its magnitude (i.e. $\frac{1}{2}\mu_B v|V|/ca$) may become comparable to that of the pseudopotential energy (i.e. $e|V'|$) and exceptionally it might become the dominant contribution.

If it were indeed dominant–and this can be thought of as representing an upper limit–the change of phase from cos to sin is in a sense equivalent to a shift of the lattice planes by one quarter of a lattice spacing in one direction for one spin direction and in the opposite direction for the other. Correspondingly the phase change on Bragg reflection would differ by π between the states of opposite spin. Thus in an orbit which requires several Bragg reflections for its realization (e.g. three for the Zn needle) the Onsager

453

condition is different for the two spins by a phase difference of as many times π as there are Bragg reflections, say $l\pi$. Since a phase difference of 2π corresponds to going from one Landau level to the next, this implies that the spin splitting will be $\frac{1}{2}l$ times the Landau level separation, i.e. $S = \frac{1}{2}l$. If, of course, the extreme condition $\varepsilon \gg e|V'|$ does not hold, the spin-splitting would be reduced, so in general only an upper limit is set to S or g:

$$S \leqslant \tfrac{1}{2}l \qquad g \leqslant l/(m/m_0) \tag{9.21}$$

This kind of argument should apply particularly well to the Zn needle, since the energy gap which separates it from the monster is thought to be due entirely to spin-orbit coupling. Here $l = 3$, so we might expect that S should be close to 1.5 ($g \simeq 400$), which would correspond to a spin-zero. A spin-zero is in fact not observed and, as we saw in §9.5.1, the evidence points rather to $S = 0.64$ ($g = 170$); this is consistent with the inequality (9.21) but the rather large difference between 1.5 and 0.64 suggests that in fact the ε of (9.20) is not much larger than the pseudopotential contribution. A discussion of the Zn situation based on the detailed band structure has been given by Van Dyke, McClure and Doar (1970) and points to $g = 170$ as the probable answer, but not quite unambiguously.

For Bi the band structure implies that there are two Bragg reflections in the construction of the electron ellipsoids of the FS and this would imply that $S \leqslant 1$. The experimental value of S is 1.1 (see §9.5.1), and this appears to contradict the idea that $S = \frac{1}{2}l$ represents an upper limit, but once again the argument is probably oversimplified. A calculation by Cohen and Blount (1960) based on a slightly idealized model of the band structure suggests that S is exactly 1.

In general we should expect the value of S to approach the upper limit $\frac{1}{2}l$ of Pippard's argument when the spin-orbit splitting for the state in the atomic spectrum which is similar in character to that of the metal wave function is large compared with the energy gap in the metal band structure. This is certainly so with Bi, for which the atomic spin-splitting is nearly 2 eV and the energy gap only about 0.02 eV. For Zn the atomic spin-splitting is 7×10^{-2} eV and the energy gap about 2×10^{-2} eV, so the situation seems less extreme. Without more detailed information about the band structure of Ga than is readily available, it is difficult to assess the applicability of the Pippard argument to the spin-splitting observed in the GQO, but we notice that $S = 2 - S_0$ (one of the possibilities left open in §9.5.1) would be just consistent with Pippard's argument for four Bragg reflections. It is, however, unlikely that the spin-orbit effect should be so extreme in Ga and $S = S_0$ (i.e. $g \simeq 1$) seems the more probable answer. Gold and Van Schyndel use the Pippard criterion for restricting the choice of g values consistent with their data on the γ-oscillations in Pb to $g = 0.70$ or 6.44 and give reasons for preferring the smaller value.

For the various metals just discussed, clear evidence from conduction electron spin resonance is available only for Bi (Smith, Galt and Merritt 1960). This agrees well with the assumption that the spin-splitting is nearly equal to the Landau level separation over most of the electron Fermi surface.

For orbits on larger sections of the FS, where g is much closer to 2, a useful guide to the magnitude of the g-factor (as far as spin-orbit coupling is concerned) has been suggested by Dupree and Holland (1967). This is based on the tight binding approximation and sets an upper limit to the departure δg of g from the free electron value $g_0 = 2.0023$. This upper limit is given by

$$|\delta g|_{max} = \frac{4l^2}{(2l+1)} \frac{\Lambda}{\Delta E} \tag{9.22}$$

Table 9.2. *Maximum spin-orbit deviation of g from g_0* (*Dupree and Holland 1967*)

	l	Λ(eV)	ΔE(eV)	theory $|g - g_0|_{max}$	spin res. $g - g_0$	dHvA $g - g_0$
Cu	1	0.03	1.9	0.02	0.03	0.18B
	2	0.25		0.42		−0.10N
Ag	1	0.11	4.0	0.04	−0.02	0.28B
	2	0.55		0.44		−0.08N
Au	1	0.47	2.4	0.26	0, 0.26	0.04B
	2	1.52		2.0		−0.8N
Na	1	0.002	2.3	0.0012	0.0007	0.64
K	1	0.007	1.2	0.008	0.0026	0.80
Rb	1	0.029	1.3	0.03	0.0039	0.83

Notes: Λ is the relevant atomic spin-orbit splitting for an orbit of azimuthal quantum number l and ΔE is the band gap to the nearest state of appropriate symmetry type; $g_0 = 2.0023$. The table is based on (9.22) and is a combination of information given by Dupree and Holland (1967), Randles (1972) and Knecht (1975); two spin resonance measurements for Au are quoted by Randles but there is some doubt about the reliability of the higher figure. In the dHvA entries B refers to belly (mean of $\langle 100 \rangle$ and $\langle 111 \rangle$ values) and N to $\langle 111 \rangle$ neck.

where Λ is the largest atomic spin-orbit splitting for atomic orbitals contributing to the FS wavefunction, l the corresponding azimuthal quantum number and ΔE the band gap to the nearest state of appropriate symmetry type. Estimates of $|\delta g_{\text{max}}|$ for the alkalis and the noble metals are shown in table 9.2 and compared with spin resonance and dHvA values. It is immediately obvious that though these estimates are indeed higher than the δg values obtained from spin resonance in the alkalis and perhaps the noble metals, they do not explain the much larger δg obtained from the dHvA effect, except perhaps as regards the noble metal neck data. The reason for this is, as already mentioned, the importance of electron–electron interaction, which does not show up in spin resonance experiments but is dominant in determining δg for the alkalis and the noble metal bellies. The observed spin resonance values of g are averages over the whole FS and since for the noble metals the FS is mostly 'belly' like, it can be inferred that spin-orbit interaction contributes little to the observed belly g values. Spin-orbit interaction may, however, be the major effect for the neck orbits, which are predominantly p-like in character, though electron–electron interaction may be significant there too.

9.6.2. Many-body interactions

Once again we shall not attempt to present the basic theory, which goes beyond the scope of this book, but merely summarize the key results of the theory which are relevant to the interpretation of the experimental data. For the general background of the many-body theory involved, Pines and Nozières (1966) and Platzmann and Wolf (1973) may be consulted.

As has already been pointed out in §2.6, the LK formula retains its validity in the presence of both electron–electron (EE) and electron–phonon (EP) interactions, provided the conditions are not too extreme, but the parameters entering into the formula have to be appropriately modified or 'renormalized', This renormalization can be expressed within the framework of the Landau theory of Fermi liquids and for an isotropic metal the interactions are determined by the coefficients A_0, A_1, etc. and B_0, B_1, etc. of the Legendre expansion of the spin-symmetric and spin-antisymmetric parts, respectively, of the Landau scattering function. All the results quoted below are to some extent approximations based on assuming that the As and Bs are small.

For the g-factor, Kaplan and Glasser (1969) suggest that the renormalization is simply

$$g = g_{\text{s}}/(1 + B_0) \tag{9.23}$$

where g_s is the g value that would be measured by spin resonance, i.e. including spin-orbit coupling but not many-body effects. The coefficient B_0 is determined by both EE and EP effects and can be expressed in an obvious notation as

$$(1 + B_0) = (1 + B_0^{\mathrm{EE}})(1 + B_0^{\mathrm{EP}}) \tag{9.24}$$

As we shall see below, typically B_0 is small and negative (-0.29 for K), though B_0^{EE} and B_0^{EP} are somewhat larger in magnitude and of opposite sign (B_0^{EP} is positive). The value of B_0 as derived from the dHvA effect proves to agree well with theoretical estimates and estimates from other experiments.

The cyclotron mass m is renormalized as

$$m = m_{\mathrm{b}}(1 + A_0^{\mathrm{EP}}) = m_{\mathrm{c}}(1 + A_0^{\mathrm{EP}})(1 + A_1^{\mathrm{EE}}), \tag{9.25}$$

where m_{c} is the 'crystalline' mass as determined by a band structure calculation which ignores both EE and EP interaction, and m_{b} is the 'bare' mass which is modified by EE but not EP interactions. It can be seen that A_0^{EP} is identical with the λ defined by (2.189). Usually A_1^{EE} is rather small (typically <0.1), so that m_{b} and m_{c} are nearly equal, while A_0^{EP} is rather larger (typically between 0.1 and 1).

One other important relation is that

$$B_0^{\mathrm{EP}} = A_0^{\mathrm{EP}} \tag{9.26}$$

This implies that the combination $g(m/m_0)$ which enters into the dHvA amplitudes depends only on the EE interaction and *not* on the EP interaction. Thus

$$S = \frac{1}{2} g(m/m_0) = \frac{1}{2} g_s \frac{m_{\mathrm{c}}(1 + A_1^{\mathrm{EE}})}{m_0(1 + B_0^{\mathrm{EE}})} \tag{9.27}$$

The g-factor is also related to the spin susceptibility χ by

$$\chi/\chi_0 = g\left(\frac{g_s}{g_0^2}\right)\left(\frac{m}{m_0}\right) \simeq S \tag{9.28}$$

where χ_0 is the spin susceptibility for a free electron gas of the same electron density. The approximate equality with S is valid for the alkalis and noble metals since both g_s and g_0 are very close to 2.

In comparing the data with the theory outlined above, it should be emphasized that the theory is not intended to be more than approximate and that it has been developed only for an isotropic metal. The latter consideration means that its application to the noble metals can at best be only qualitative because of the quite appreciable anisotropy. In order to make comparisons, we somewhat arbitrarily suppose that a mean of the properties for the $\langle 111 \rangle$ and $\langle 100 \rangle$ bellies represents something like the behaviour of an isotropic metal with the same electronic density.

The g values (or the equivalent B_0 or χ/χ_0 values) as given by the dHvA effect may be compared with estimates based on other experiments. The most significant comparison is with experiments on a new kind of spin wave phenomenon discovered by Schultz and Dunifer (1967). These spin waves, which show up as resonances when a slab of metal is exposed to high frequency electromagnetic radiation in the presence of a high magnetic field, are essentially a many-body effect in the sense that they would not occur in a non-interacting electron system and the parameter B_0 can be determined directly from their dispersion relation. As can be seen from table 9.3, the value of B_0 obtained in this way is in excellent agreement with the dHvA value for K, but for Na and Rb can only be said to agree if it is supposed that experimental errors are rather larger than stated. It is of course possible that the discrepancies for Na and Rb are real and a consequence of the theory used to derive B_0 from the respective experiments being too approximate. These spin waves have not as yet been observed in the noble metals. For convenience the comparison of B_0 values is also presented as a comparison between values of χ/χ_0 as obtained from the dHvA value of S and from the spin wave B_0 (using (9.23) and (9.28)).

These values of χ/χ_0 may also be compared with estimates based on (*a*) the Knight shift, (*b*) the amplitude of the conduction electron spin resonance and (*c*) the directly measured magnetic susceptibility, with due allowance for contributions other than the spin susceptibility. These estimates are inevitably rather rough–typically the uncertainty is of order 10%, but, as can be seen from table 9.3, they do agree broadly with the considerably more accurate dHvA values. This agreement incidentally confirms that we have correctly resolved the ambiguity of g values, i.e. we have correctly chosen the integer in (9.4).

There have been many attempts, at varying levels of sophistication, to calculate the many-body effects which determine the g values, either by way of the Landau A and B coefficients, or more fundamentally in terms of the detailed band structures. In table 9.3 we give the values of χ/χ_0 as calculated by Vosko, Perdew and MacDonald (1975) (see also MacDonald and Vosko 1976) for the alkali metals and it can be seen that they agree reasonably well with the experimental values, especially those from the dHvA effect. Various other theoretical estimates quoted by Knecht (1975) are not very different.

Another way of presenting the data is to isolate a quantity containing

Table 9.3. *Estimates of B_0 and χ/χ_0 from g values for the alkalis and noble metals*

Metal	g(dHvA)	g_s	$-B_0$ dHvA	$-B_0$ spin wave	m/m_0	χ/χ_0 dHvA	χ/χ_0 spin wave	χ/χ_0 Knight shift	χ/χ_0 spin res. ampl.	χ/χ_0 direct	χ/χ_0 Vosko theory	χ/χ_0 crude theory
Na	2.64(2)	2.00	0.241(7)	0.215(30)	1.24(1)	1.63	1.58		1.7	1.62	1.62	1.64
K	2.80(1)	2.00	0.286(3)	0.285(20)	1.217(2)	1.70	1.70	1.6		1.70	1.79	1.71
Rb	2.83(5)	2.00	0.295(12)	0.21(5)	1.22(2)	1.72	1.54	1.6		1.93	1.78	1.66
Cu	2.18(5)	2.03	0.07(2)		1.36(2)	1.50						1.41
Ag	2.28(15)	1.98	0.13(6)		0.93(1)	1.05						1.05
Au	2.04(18)	2.00	0.02(8)		1.10(4)	1.13						1.08

Notes: The experimental g values are those of Perz and Shoenberg (1976; Na), Knecht (1975; K, Rb; his data for Cs have been omitted because their interpretation was rather uncertain), Bibby and Shoenberg (1979; Cu, Ag, Au); the sources for the other entries are given in these papers. The values of B_0 are derived from (9.23). The figures in brackets indicate the uncertainties in the last significant figures of the entries; for the noble metals the entries are means between belly values at $\langle 100 \rangle$ and $\langle 111 \rangle$ and the uncertainty is dominated by the actual anisotropy. The first two columns of the χ/χ_0 entries are based on (9.28) and so give essentially the same information as the $-B_0$ columns, but in a form more convenient for comparison with the other estimates. The Vosko theory refers to Vosko et al. (1975); the crude theory column is based on linear extrapolation of (9.33) beyond the low r_s^* for which it should be valid.

459

only terms due to EE interaction, which might be expected to depend only on the electron density; this dependence could then be compared with theoretical prediction. Such a quantity is

$$X = gm/g_s m_c = (\chi/\chi_0)(g_0/g_s)^2(m_0/m_c) \tag{9.29}$$

since from (9.23) to (9.28) it follows that

$$X = (1 + A_1^{EE})/(1 + B_0^{EE}) \tag{9.30}$$

Thus X, which can be obtained from the dHvA estimate of g, the spin resonance value g_s and the band structure mass m_c, is a function of purely EE interaction and so for an isotropiç metal should depend only on the electron density, or equivalently, on the average interelectronic separation, conventionally denoted by r_s and defined by

$$r_s = n^{-1/3}/a_0 \tag{9.31}$$

where n is the electron density and a_0 is the Bohr radius. Unfortunately this argument it too naive because it ignores the screening effect of the ion cores. This difficulty can be dealt with to some extent by treating the screening as equivalent to that of a medium of dielectric constant ε and it can be shown that X in the metal should be the same function of r_s^*,

Table 9.4. *Parameters determining X and r_s^* (after BS 1979)*

Metal	gm/m_0	m_c/m_0	r_s	ε	r_s^*	X
Na	3.26(1)	1.02	3.93	1.09	3.68	1.60(2)
K	3.40(1)	1.01	4.86	1.15, 1.21	4.16(12)	1.68(2)
Rb	3.45(2)	0.99	5.20	1.25, 1.26	4.10(4)	1.74(3)
Cu	2.96(4)	1.26	2.66	6.0, 7.0	0.52(4)	1.16(2)
Ag	2.12(13)	0.93	3.00	2.3, 3.8	0.9(2)	1.15(7)
Au	2.25(18)	1.01	3.00	7.0, 7.8	0.41(2)	1.11(9)

Notes: Since it is the combination gm/m_0 which comes out of the dHvA measurements, its relative uncertainty is usually slightly smaller than that of the g values in table 9.3 (which include the uncertainties of m). Where two entries for ε appear for the same metal, these are results of different experiments (full references in BS); the mean is used in calculating r_s^*. The errors in r_s^* reflect only the discrepancies between the two entries and may in fact be much greater.

9.6. Discussion of g value results

as it would be of r_s in a pure electron gas, where

$$r_s^* = (r_s/\varepsilon)(m_c/m_0) \tag{9.32}$$

Values of ε can be estimated from optical data by suitable extrapolation to infinite frequency, though the reliability of the procedure is far from certain. A plot of X against r_s^* is shown in fig. 9.11 and it can be seen that within the rather large uncertainties, the points for the noble metals and the alkalis can indeed be regarded as lying on a single curve, which moreover is reasonably in agreement with theory. There have been a number of calculations of the theoretical form of the curve which do not differ greatly from each other where they overlap. We have chosen that of Hamman and Overhauser (1966) for comparison with the experimental points mainly because it gives explicit results over the whole of the relevant range. Because the values of ε are high for the noble metals (7 for Cu and Au) the noble metal points would have fallen far to the right of the curve if the plot had been against r_s rather than r_s^*. Moreover, because of some uncertainty in the true meaning of ε and how it should be extracted from the optical

Fig. 9.11. Plot of X (see (9.30)) against r_s^* (see (9.31) and (9.32) and table 9.4) for the noble metals and alkalis (BS 1979). The experimental points are from left to right for Au, Cu, Ag, Na, Rb and K, with rough indications of the uncertainties. The full line is theoretical (Hamman and Overhauser 1966) and the broken line indicates the asymptotic form (9.33) of the theory for low r_s^*.

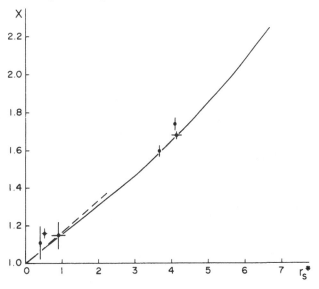

data, the horizontal error bars should probably be larger than shown; those shown reflect only the differences in ε as determined from different experiments.

The theoretical form of the variation of X with r_s^* should take a particularly simple form for low r_s^*, where the first term of an expansion in powers of r_s^* should be valid. This form (see for instance Pines and Nozières 1966, p. 308) is

$$X = 1 + 0.166r_s^* \tag{9.33}$$

Rather surprisingly, the detailed calculations for $r_s^* > 1$ give values of X not greatly differing from the simple linear formula (8.33) and in table 9.3 under the heading 'crude theory' we have given the values of (χ/χ_0) obtained by applying (9.29) to the values of X given by (9.33). It can be seen that they agree with experiment no worse than the more sophisticated theories.

Finally we must comment on what at first sight appears a paradoxical feature of the theory. This is that EE interaction increases with increasing r_s^* (i.e. with decreasing electron density) rather than the reverse as naive consideration might suggest. The reason for this is that the EE interaction energy should not be thought of absolutely but relative to the kinetic energy. The interaction is electrostatic and so varies as $1/r$ while the kinetic energy (essentially the Fermi energy) varies as $1/r^2$; thus as the electrons get closer together, the *relative* importance of the electrostatic interaction should indeed get smaller.

9.7. Anomalous situations

We have up to now tacitly assumed (*a*) that the oscillations associated with spin-up and spin-down electrons have equal amplitudes and (*b*) that the spin-splitting of the energy levels is proportional to H. In this section we shall consider what happens if these assumptions fail, so that the effect of spin can no longer be described by the simple constant factors $\cos(p\pi S)$ in the harmonic amplitudes. Failure of these assumptions was first discovered in dilute alloys containing magnetic impurities (Coleridge and Templeton 1968; Coleridge, Scott and Templeton 1972). On the one hand scattering by the magnetic impurity is in general spin dependent so that the amplitudes for spin-up and spin-down oscillations may be appreciably different, and on the other there is an exchange splitting which is approximately independent of H, so that the effective g-factor becomes field dependent. Study of these effects has proved to be a useful tool for studying the interaction of magnetic impurity atoms with the metallic electrons, but we shall give only a brief outline of the possibilities and the results obtained; a more detailed review has been given by Higgins and Lowndes (1980).

If we go back to our derivation in §2.3.7.4 of the spin factor in the amplitude, but suppose that the phase difference ϕ $(=2\pi S)$ between the spin-up and spin-down oscillations is field dependent and also that their amplitudes may be different, say $\frac{1}{2}A(1+a)$ and $\frac{1}{2}A(1-a)$, the superposition gives an oscillation

$$M = A\left\{\left(\frac{1+a}{2}\right)\sin\left(\psi + \frac{\phi}{2}\right) + \left(\frac{1-a}{2}\right)\sin\left(\psi - \frac{\phi}{2}\right)\right\} \quad (9.34)$$

where as usual $\psi = (2\pi F/H) \pm \frac{1}{4}\pi$ and this can be expressed as

$$
\left.
\begin{aligned}
M &= A\left(\cos^2\frac{\phi}{2} + a^2\sin^2\frac{\phi}{2}\right)^{1/2}\sin(\psi + \theta) \\
\text{where} \quad & \\
\tan\theta &= a\tan\frac{\phi}{2}
\end{aligned}
\right\} \quad (9.35)
$$

In the context of dilute alloys with a magnetic impurity, a field dependence of ϕ can occur through an exchange splitting superimposed on the usual splitting $\frac{1}{2}g(m/m_0)\beta H$. If this exchange interaction is antiferromagnetic (as is found experimentally for the system to be discussed below) the phase difference ϕ can be conveniently expressed as

$$\phi = 2\pi\left\{\frac{1}{2}\left(g - \frac{H_{ex}}{H}\right)\frac{m}{m_0}\right\} \quad (9.36)$$

where H_{ex} is related to the exchange interaction energy Δ_{ex} (i.e. the exchange splitting) by

$$\tfrac{1}{2}(m/m_0)\beta H_{ex} = \Delta_{ex} \quad (9.37)$$

A particularly simple situation arises if there is no appreciable difference in the amplitudes of the spin-up and spin-down oscillations, i.e. $a = 0$, for then (9.35) reduces to

$$M = A\cos\left(\frac{\pi}{2}\left(g - \frac{H_{ex}}{H}\right)\frac{m}{m_0}\right)\sin\psi \quad (9.38)$$

i.e. there is no phase change, but the spin cosine factor is field dependent.

This is in fact the situation for dilute alloys of Cr in Cu and leads to striking novel features in the dHvA oscillations (Coleridge *et al.* 1972). Since, for pure Cu, gm/m_0 is close to 3 for belly oscillations and close to 1 for neck oscillations, it is possible to produce a spin-zero simply by varying H

until $(g - H_{ex}/H)m/m_0$ is exactly 3 or 1. This is illustrated in fig. 9.12, and it can be seen that the experimental field dependence can be fitted very well to (9.38) if H_{ex} is suitably chosen. Another demonstration of the same effect is to look at the oscillations as the sample is rotated in a fixed field. It is then found that the spin-zero occurs at an orientation which is field dependent and different from that of pure Cu (see fig. 9.13).

The value of H_{ex} (i.e. of Δ_{ex}) for a given orientation as determined from such experiments is of some theoretical interest. To a crude approximation we should expect

$$\Delta_{ex} = c|J|S \tag{9.39}$$

where c is the concentration of magnetic impurity, J is the exchange integral between the magnetic impurity and the conduction electrons, and S is the spin quantum number of the impurity ($\frac{3}{2}$ for Cr). In this way the important quantity J which enters into the Kondo effect can be determined. In fact the detailed theory is a good deal more complicated (see for instance

Fig. 9.12. Field dependence of amplitude of $\langle 111 \rangle$ belly oscillation at about 1 K in Cu + 34 ppm Cr. The points are experimental (field modulation technique) and the curve is fitted to (9.38) by suitable choice of H_{ex} (Coleridge *et al.* 1972).

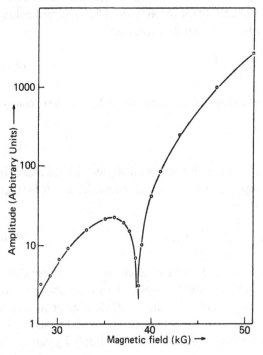

Simpson and Paton 1971) and Δ_{ex} may be appreciably field and temperature dependent. Analysis of the experimental data proves to be consistent with the more detailed theory and indicates values of J in agreement with the evidence from the Kondo effect.

The existence of sharp spin zeros in the Cu–Cr alloys, i.e. the absence of spin dependent scattering, is apparently rather a lucky accident. In general, the scattering is appreciably spin dependent so that $a \neq 0$ and spin-zeros can no longer occur. The extraction of basic information from the oscillations is then more difficult. For sufficiently dilute alloys the spin-zero is replaced by a minimum amplitude, when $\phi = \pi$ in (9.35), and both a and H_{ex} can still be extracted from (9.35) and (9.36), but the precision deteriorates, since A also varies rapidly with field and orientation, and thus makes the location of the field or orientation for $\phi = \pi$ less precise. Indeed the overall amplitude no longer shows a minimum at all if the variation of A is too dominant.

If conditions are such that the LK harmonics are strong enough to be measured, but MI is not too strong, it is possible–at least in principle–to study both the field dependence of g and the spin-dependent scattering without the necessity of being near a spin-zero. This approach was pioneered by Alles, Higgins and Lowndes (1973) (see also Alles and Higgins 1974) and further developed by Chung, Lowndes, and Hendel (1978). Because of the restriction that the MI harmonics must not be too strong in conditions where the LK harmonics are measurable, the method cannot be applied to the noble metal bellies, except perhaps for Cu close to a spin-zero, but it has been applied to the neck oscillations of Cu and Au to which magnetic impurities have been added.

Fig. 9.13. Spin zeros for neck oscillations in pure Cu (upper trace) and Cu + 15 ppm Cr (lower trace). The oscillations are produced by rotating the sample in a fixed field of 50 kG in a (110) zone at about 1 K and are observed by the field modulation technique; the angles are measured from $\langle 111 \rangle$ (Coleridge et al. 1972).

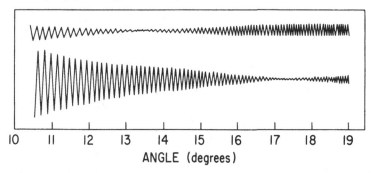

ANGLE (degrees)

Phase and spin-splitting

Suppose first that MI is completely negligible, then the first three harmonics of M can be written as

$$
\left.
\begin{aligned}
M_1 &= A_1 \left[z \sin\left(\psi + \frac{\phi}{2} \right) + z' \sin\left(\psi - \frac{\phi}{2} \right) \right] \\
&= A_1 (z^2 + z'^2 + 2zz' \cos \phi)^{1/2} \sin(\psi + \theta_1) \\
M_2 &= A_2 \left[z^2 \sin\left(2\psi \mp \frac{\pi}{4} + \phi \right) + z'^2 \sin\left(2\psi \mp \frac{\pi}{4} - \phi \right) \right] \\
&= A_2 (z^4 + z'^4 + 2z^2 z'^2 \cos 2\phi)^{1/2} \sin\left(2\psi \mp \frac{\pi}{4} + \theta_2 \right) \\
M_3 &= A_3 \left[z^3 \sin\left(3\psi \mp \frac{\pi}{2} + \frac{3\phi}{2} \right) + z'^3 \sin\left(3\psi \mp \frac{\pi}{2} - \frac{3\phi}{2} \right) \right] \\
&= A_3 (z^6 + z'^6 + 2z^3 z'^3 \cos 3\phi)^{1/2} \sin\left(3\psi \mp \frac{\pi}{2} + \theta_3 \right)
\end{aligned}
\right\} \quad (9.40)
$$

where z and z' are the Dingle factors of the two components and

$$
\tan \theta_1 = \frac{z - z'}{z + z'} \tan \frac{\phi}{2}, \quad \tan \theta_2 = \frac{z^2 - z'^2}{z^2 + z'^2} \tan \phi,
$$

$$
\tan \theta_3 = \frac{z^3 - z'^3}{z^3 + z'^3} \tan \frac{3\phi}{2} \quad (9.41)
$$

Here ψ is again defined as $(2\pi F/H) \pm \frac{1}{4}\pi$, but the formulation (9.34) to describe the spin-dependent amplitudes is no longer convenient for dealing with several harmonics; the upper sign of the $\mp$ in (9.40) and the $\pm$ in ψ is to be used for a minimum FS area and the lower for a maximum. In the formulation it has been assumed that the Dingle factors for the two spin components in the pth harmonic have the standard form

$$
z^p = \exp(-p\alpha x/H) \quad \text{and} \quad z'^p = \exp(-p\alpha x'/H) \quad (9.42)
$$

and it should be emphasised that the analysis becomes unreliable if, as may well happen, this assumption is not valid. We also suppose that the mass m is independent of spin. If the value of m is known, the ratios A_2/A_1 and A_3/A_1 at given H and T can be calculated from the LK formula. Thus from the measured amplitude ratios $|M_2|/|M_1|$, $|M_3|/|M_1|$ and the relative phases $\theta_2 - 2\theta_1$ and $\theta_3 - 3\theta_1$ we have, at any given H and T, four relations for determining the three unknowns z, z' (i.e. x and x') and ϕ, with some redundancy for checking the consistency of the relations. The variations of z, z' and ϕ with H and T provide further checks on the validity of the initial assumptions. If MI is significant, but not overwhelming, it can be taken into account by something like the Phillips and Gold iteration scheme and the appropriate modifications of (9.40) and (9.41) can be written down. Although the resulting equations look complicated in detail they are still perfectly amenable to solution by computer techniques.

In Alles and Higgins' (1974) experiment on the neck in Cu + 93 ppm Fe, the third harmonic was too weak to use, so only two of the four relations were available, but by a semi-empirical procedure based on the field dependence of the measured quantities they were able to estimate x, x' and ϕ. Although x and x' were indeed appreciably different (roughly 1.1 K and 1.4 K), ϕ was found to be a constant, rather than showing the expected $1/H$ dependence of (9.36). This is essentially just a shift of g downwards (by about 0.13) from the value in pure Cu, but they rather artificially apply (9.36) and deduce an H_{ex} which is proportional to H. Their results expressed as per unit concentration do not agree well with the rather rough estimates of Coleridge *et al.* who found a g-shift something like twice as large. A possible cause of the discrepancy may be that Alles and Higgins' technique involved fairly high frequency field modulation and they assumed the exact truth of eddy current theory in the skin effect limit for extracting the harmonics of M from the raw data. However, as has been discussed by Knecht *et al.* (1977), this assumption may be not reliable because of sample inhomogeneity and this might cause appreciable errors in determining the characteristics of a weak harmonic.

In the applications to Au as the host, the third harmonic was strong enough to be used, but MI corrections were also necessary. Some of the preliminary results are outlined in the review of Higgins and Lowndes (1980) and indicate that the technique, though complicated, has considerable possibilities for studying magnetic impurities. We shall mention only two examples of results obtained so far. For addition of Fe to Au (Chung, Lowndes, Hendel and Rode 1976), a value of H_{ex} was found which varied only from about 10 kG at $H = 40$ kG to something like 13 kG at $H = 75$ kG. Over the same range of H, the Dingle temperatures are essentially constant but differ by a factor of 3 ($x = 0.3$ K, $x' = 0.9$ K). The value of the exchange integral J estimated from H_{ex} at the highest field agrees well with estimates based on bulk magnetization and resistivity. For addition of Co (Chung and Lowndes 1977) the results suggest that although the mean x varies linearly with concentration c, the difference $x - x'$ varies more like c^2. This behaviour is taken as indicating the increasingly important effect of Co pairs as c is increased. At low concentrations scattering is almost entirely by isolated Co atoms which are believed to carry no magnetic moment, so that the scattering is not spin dependent, but at higher concentrations Co pairs which do carry a local moment and so scatter opposite spins differently, begin to contribute appreciably.

Appendix 1

Ellipsoidal surfaces of constant energy

A.1.1. Parabolic band

Ellipsoidal surfaces of constant energy provide illustrations of the general formulae which are one step more realistic than the spherical surface of the free electron model, particularly in their ability to represent the anisotropy of real crystals. Moreover, since such surfaces sometimes provide a good approximation to the constant energy surfaces of real metals (e.g. in Bi and parts of the band structure of many polyvalent metals), the formulae we shall obtain below can be of practical use.

If the relation $\varepsilon(\mathbf{k})$ is a quadratic one, i.e. if the band is parabolic, the general ellipsoidal energy surface referred to rectangular axes related to the crystal axes can be expressed as

$$\sum \alpha_{ij} k_i k_j = 2m_0 \varepsilon/\hbar^2 \tag{A1.1}$$

where the suffixes denote 1, 2, 3 or x, y, z and $\alpha_{ij} = \alpha_{ji}$. In this form the α_{ij} are dimensionless and we would have for instance $\alpha_{ij} = \delta_{ij}$ for a spherical surface. It should be noticed that (A1.1) will not in general satisfy the requirements of crystal symmetry unless there are restrictions on the α_{ij} or there are several other surfaces related to (A1.1) is such a way as to satisfy the crystal symmetry. For instance, if the α_{ij} are arbitrary twenty four ellipsoids are required for the assembly to have cubic symmetry (obtained by permuting x, y, z and the signs of the k components appropriately). Further multiplication may be required if the individual ellipsoids are not centred on appropriate symmetry points in the Brillouin zone.

If the field H has direction cosines (v_1, v_2, v_3) w.r.t. the (k_x, k_y, k_z) axes, it is not difficult to show* that the area a of cross-section of the ellipsoid (A1.1) by a plane normal to H and at distance κ from the origin is

$$a(\varepsilon, \kappa) = \frac{2\pi m_0 \varepsilon}{\hbar^2 W^{1/2}} \left(1 - \frac{\kappa^2}{\kappa_0^2} \right) \tag{A1.2}$$

---

* The most direct way is to transform (A1.1) to axes (X, Y, Z) such that Z has direction cosines (v_1, v_2, v_3) w.r.t. the k_x, k_y, k_z axes. The elliptical section of this surface by $Z = \kappa$ can then be written down and its area calculated.

468

where

$$W = \left. \begin{array}{l} v_1^2(\alpha_{22}\alpha_{23} - \alpha_{23}^2) + 2v_2v_3(\alpha_{12}\alpha_{13} - \alpha_{23}\alpha_{11}) \\ + \text{ similar terms obtained by cyclic change} \\ \text{of suffices} \end{array} \right\} \qquad \text{(A1.3)}$$

and κ_0 is the value of κ for the tangent plane normal to H. It is not difficult to show that

$$\kappa_0^2 = \frac{2m_0\varepsilon}{\hbar^2} \frac{W}{D} \qquad \text{(A1.4)}$$

where D is the determinant

$$D = \begin{vmatrix} \alpha_{11} & \alpha_{12} & \alpha_{13} \\ \alpha_{12} & \alpha_{22} & \alpha_{23} \\ \alpha_{13} & \alpha_{23} & \alpha_{33} \end{vmatrix} \qquad \text{(A1.5)}$$

We may note that the volume v of the ellipsoid and the number N of electrons in a sample of volume v are related to D by

$$v = \frac{4\pi}{3D^{1/2}}\left(\frac{2m_0\varepsilon}{\hbar^2}\right)^{3/2} \qquad N = vV/4\pi^3 \qquad \text{(A1.6)}$$

and that D is invariant w.r.t. transformations of the axes of reference, as it must be to give unique values for v and N.

From (A1.2) we can immediately deduce that the cyclotron mass m is independent of both ε and κ and is given by

$$m = \frac{\hbar^2}{2\pi}\left(\frac{\partial a}{\partial \varepsilon}\right)_\kappa = \frac{m_0}{W^{1/2}} \qquad \text{(A1.7)}$$

so that (A1.2) can be expressed as

$$a(\varepsilon, \kappa) = a(\varepsilon, 0)(1 - \kappa^2/\kappa_0^2) \qquad \text{(A1.8)}$$

where

$$a(\varepsilon, 0) = 2\pi m\varepsilon/\hbar^2$$

It is also instructive to calculate the helical motion in real space of the quasiparticle whose velocity is given by (2.6). We shall not give the calculation in detail, but the idea is first to transform the ellipsoid (A1.1) to (X, Y, Z) axes where Z is in the H direction, so that Z is the same as κ or k_H (see footnote on p. 468), then to use the relation (2.5) between the velocity components v_x, v_y normal to H and the corresponding components X, Y of k to obtain differential equations for the time variation of X and Y (at

469

Appendix 1

constant Z of course). The solution has the form

$$\left.\begin{array}{l} X = A\cos\omega_c t + \alpha Z \\ Y = B\cos(\omega_c t - \phi) + \beta Z \end{array}\right\} \tag{A1.9}$$

where the constants A, B, α, β and ϕ are related to the parameters of the constant energy surface (as transformed to (X, Y, Z) axes) in a rather complicated way, while ω_c is the cyclotron frequency given by

$$\omega_c = eH/mc \tag{A1.10}$$

The velocities in real space are then given by

$$\hbar v_x = \partial\varepsilon/\partial X, \qquad \hbar v_y = \partial\varepsilon/\partial Y, \qquad \hbar v_z = \partial\varepsilon/\partial Z$$

and it can be shown that the motion projected on the (x, y) plane is, as it should be, round the elliptical section of the constant energy surface by the plane $Z = \kappa$, but scaled by the factor η of (2.9) and turned through 90°. The v_z component consists of a steady part and a periodic part and can be compactly expressed as

$$v_z = \frac{\hbar\kappa}{m'}\left[1 + \left(\frac{\kappa_0^2}{\kappa^2} - 1\right)^{1/2}\left(\frac{\kappa_0^2}{Z_0^2} - 1\right)^{1/2}\cos\omega_c t\right] \tag{A1.11}$$

where m' is a 'longitudinal mass' defined by

$$\frac{m'}{m_0} = \frac{W}{D} = \frac{\hbar^2\kappa_0^2}{2m_0\varepsilon} \tag{A1.12}$$

κ_0 is as defined by (A1.4) and Z_0 is the radius of the ellipsoid in the H direction (see fig. A1.1). Explicitly, Z_0 is given by

$$Z_0^2 = \frac{2m_0\varepsilon}{\hbar^2}\bigg/\sum\alpha_{ij}v_i v_j \tag{A1.13}$$

Note that the origin of t in (A1.11) is not the same as in (A1.9).

It can be seen that the periodic part of v_z vanishes for $\kappa = \kappa_0$, when the elliptic cross-section normal to H shrinks to a point, and also for H is along a principal axis of the ellipsoid, when $Z_0 = \kappa_0$. Although the steady part of v_z vanishes for $\kappa = 0$, the periodic part in general does not.

If (A1.11) is integrated to give the actual displacement z in the field direction, we obtain the pitch P of the helical motion as the displacement during one cyclotron period. This is

$$P = \frac{2\pi\hbar mc\kappa}{m'eH} = \frac{2\pi ch}{eH}\frac{D}{W^{3/2}}\kappa \tag{A1.14}$$

470

Ellipsoidal surfaces of constant energy

This result can of course also be obtained (and much more simply) from (A1.2) by applying the general formula (2.13) for P.

We now consider the quantization of the motion and find from (2.30), using (A1.2), (A1.4), (A1.7) and (A1.12)

$$a(\varepsilon, \kappa) = \frac{2\pi m\varepsilon}{\hbar^2} - \frac{\pi \kappa^2 m}{m'} = (r + \tfrac{1}{2}) \frac{2\pi e H}{ch} \qquad (A1.15)$$

Thus the quantized energy can be expressed explicitly as

$$\varepsilon = (r + \tfrac{1}{2})\beta H + \hbar^2 \kappa^2 / 2m' \qquad \textbf{(A1.16)}$$

where

$$\beta = e\hbar/mc \qquad (A1.17)$$

We note that (A1.16) has the same form as for a free electron gas except that the free electron mass must be replaced by different masses in the two terms: by the cyclotron mass m in the discrete term and by the longitudinal mass m' in the continuous term. This is a useful result since it permits ready adaptation of any result for the free electron (spherical) model to an anisotropic ellipsoidal model, by merely changing from m_0 to m or m' at the appropriate places. The values of m and m' can of course be easily derived by calculating the determinant D and the value of W for the particular field direction.

The form of the Landau tubes can also be worked out by eliminating the energy ε between (A1.15) and the equation of the ellipsoid expressed in the (X, Y, Z) axes (where Z is the same as κ), say

$$Q(X, Y, Z) = 2m_0\varepsilon/\hbar^2 \qquad (A1.18)$$

where Q is the appropriate quadratic function. The equation of the rth tube is

$$Q(X, Y, Z) = \left(\frac{2m_0}{\hbar^2}\right)(r + \tfrac{1}{2})\beta H + \frac{DZ^2}{W}$$

which reduces to the form

$$a(X - pZ)^2 + b(Y - qZ)^2 + 2f(X - pZ)(Y - qZ)$$
$$= \left(\frac{2m_0}{\hbar^2}\right)(r + \tfrac{1}{2})\beta H$$

where p and q are expressible in terms of the coefficients in Q. This is just a cylinder of elliptic cross-section with its axis having direction cosines proportional to $(p, q, 1)$, i.e. at angle $\theta = \tan^{-1}(p^2 + q^2)^{1/2}$ to the Z axis.

This result finds a simple geometrical interpretation as illustrated in fig. A1.1. The cross-section of the constant energy surface ε' in the plane $Z = \kappa_0$ (the tangent plane to the surface ε) is an ellipse identical to the central cross section of the surface ε, while its centre is evidently located at the point of contact $C = (X_0, Y_0, \kappa_0)$. The Landau tube is defined by the two ellipses and so its axis has direction cosines proportional to (X_0, Y_0, κ_0), i.e. it is the line of points of contact of the concentric ellipsoids. It is not difficult (though a little tedious) to show that $(p, q, 1)$ are indeed proportional to (X_0, Y_0, κ_0) so that the detailed analysis does confirm the simple geometrical argument.

For the dHvA effect it is the extremal area of the surface of energy ζ, the Fermi energy, which determines the dHvA frequency F (see (2.111)). Thus

$$F = (c\hbar/2\pi e)a(\zeta, 0) = mc\zeta/e\hbar = \zeta/\beta \qquad (\text{A1.19})$$

Fig. A1.1. Illustrating notation of the text and indicating construction of a Landau tube. The two ellipsoidal surfaces for constant energies ε and ε' are such that $a(\varepsilon, 0) = a(\varepsilon', \kappa_0) = (r + \frac{1}{2})\beta H$, where $a(\varepsilon, \kappa)$ is the area of the cross-section of the surface ε at the plane $Z = \kappa$. The Landau tube is the elliptic cylinder joining the two sections indicated; its axis is shown by the arrow.

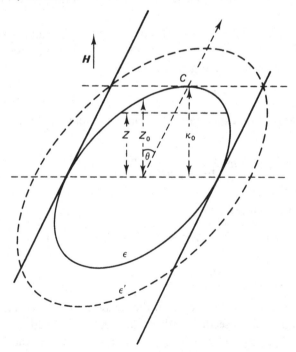

Ellipsoidal surfaces of constant energy

An important property of F in this model is that it has a particularly simple orientation dependence. As is evident from (A1.2),

$$1/F^2 \propto W$$

and if we imagine the field direction (v_1, v_2, v_3) rotated in some arbitrary plane, the vs must be expressible as

$$v_1 = l_1 \cos\psi + m_1 \sin\psi, \quad v_2 = l_2 \cos\psi + m_2 \sin\psi,$$

$$v_3 = l_3 \cos\psi + m_3 \sin\psi$$

where ψ is the angle of rotation from some direction (l_1, l_2, l_3) and (m_1, m_2, m_3) are the direction cosines of a line perpendicular to (l_1, l_2, l_3) and in the plane of rotation. Thus, from the definition (A1.3) of W, we must have quite generally

$$\frac{1}{F^2} = A \cos^2\psi + B \sin^2\psi + C \cos\psi \sin\psi$$

$$= A' \cos 2\psi + B' \sin 2\psi + C'$$

$$\text{(A1.20)}$$

where A, B, C and $A' B' C'$ can be expressed in terms of the parameters of the ellipsoid and the direction cosines (l_1, l_2, l_3) and (m_1, m_2, m_3). If the plane of rotation of H is normal to a principal axis of the ellipsoid and ψ is measured from one of the other principal axes, (A1.20) simplifies, in as far as C and B' vanish. The relation (A1.20) provides a useful criterion for deciding whether or not a Fermi surface is ellipsoidal.

By suitable choices of the signs of the α_{ij}s in (A1.1) much of the above analysis can be adapted to apply to hyperboloids as well as to ellipsoids. We shall mention only one special case which has a practical application. The FS of the noble metals for orientations around $\langle 111 \rangle$ can be well approximated to a hyperboloid of revolution

$$\rho(k_x^2 + k_y^2) - k_z^2 = c \tag{A1.21}$$

where k_z is along $\langle 111 \rangle$ and we find

$$F^2(0)/F^2(\psi) = \cos^2\psi - \frac{1}{\rho}\sin^2\psi \tag{A1.21}$$

where ψ is angle between H and $\langle 111 \rangle$. Another useful result is that

$$\partial^2 A/\partial k_H^2 = 2\pi \Big/ \left[\rho\left(\cos^2\psi - \frac{1}{\rho}\sin^2\psi \right) \right]^{3/2} \tag{A1.22}$$

Appendix 1

A1.2. Non-parabolic band

We must now consider how the analysis is modified if the energy surfaces are still ellipsoidal but the band is not parabolic, i.e. ε is *not* a quadratic function of k. We must then replace (A1.1) by the more general relation

$$\sum \alpha_{ij} k_i k_j = \frac{2m_0}{\hbar^2} f(\varepsilon) \tag{A1.23}$$

(here $f(\varepsilon)$ is to be formulated to have the same dimensions as ε).

Where no differentiation is involved (as for instance in (A1.2), (A1.4) or (A1.6)) the only change involved is to replace ε by $f(\varepsilon)$ on the r.h.s. We note that the central area is now given by

$$a(\varepsilon, 0) = 2\pi m_0 f(\varepsilon)/\hbar^2 W^{1/2} \tag{A1.24}$$

so that (A1.19) is modified to

$$F = \frac{m_b c}{e\hbar} f(\zeta) = f(\zeta)/\beta_b \tag{A1.25}$$

where $\beta_b = e\hbar/m_b c$ and m_b is now used to denote the constant mass defined by (A1.7), i.e. $m_0/W^{1/2}$

The actual cyclotron mass $m(\varepsilon)$ is no longer independent of ε. It is in fact

$$m(\varepsilon) = \frac{\hbar^2}{2\pi} \left(\frac{\partial a}{\partial \varepsilon}\right)_\kappa = m_b f'(\varepsilon) \tag{A1.26}$$

and correspondingly

$$\beta(\varepsilon) = \beta_b / f'(\varepsilon) \tag{A1.27}$$

The suffix b is used to denote 'bottom of the band' since we should usually expect the band to become parabolic and $f'(\varepsilon)$ to approach 1 close to $\varepsilon = 0$. It should be noticed that $F\beta_b$, $F\beta(\zeta)$, F/m_b and $F/m(\zeta)$ are all independent of orientation: this provides a useful test for checking that the surfaces are ellipsoidal.

The quantized energy levels can in general be expressed only in implicit form and it is easily shown that (A1.16) must be replaced by

$$f(\varepsilon) = (r + \tfrac{1}{2})\beta_b H + \hbar^2 \kappa^2 / 2m_b' \tag{A1.28}$$

where m_b' is the longitudinal mass at the bottom of the band (i.e. $m_0 \, W/D$ as given by (A1.12)). The Landau tubes are independent of the form of $f(\varepsilon)$ since they are obtained by eliminating $f(\varepsilon)$ between the ellipsoid equation and the energy level relation.

The separation of adjacent energy levels ($\Delta r = 1$) is given in the limit of

474

high r by

$$\Delta \varepsilon = \beta_b H / f'(\varepsilon) = \beta(\varepsilon) H = \hbar \omega_c(\varepsilon) \tag{A1.29}$$

where $\omega_c = eH/m(\varepsilon)c$. Thus $\Delta\varepsilon$ varies with ε and is determined by the actual cyclotron mass $m(\varepsilon)$ rather than by the constant m_b. The result (A1.29), though exact for a parabolic band (for which $f'(\varepsilon) = 1$), becomes slightly imprecise for the general case since it does not specify at which ε the value of $f'(\varepsilon)$ should be taken; the lack of precision grows as r gets smaller.

A1.3. The Lax model

These results may be illustrated by a particular example, the Lax model (Brown, Mavroides and Lax 1963), which provides a useful approximation to the electron energy surfaces in Bi. In this model

$$f(\varepsilon) = \varepsilon(1 + \varepsilon/\varepsilon_g) \tag{A1.30}$$

where ε_g is the small energy gap separating the filled valence band and the electron conduction band.

The energy levels are given by

$$\varepsilon\left(1 + \frac{\varepsilon}{\varepsilon_g}\right) = (r + \tfrac{1}{2} \pm \tfrac{1}{2})\beta_b H + \frac{\hbar^2 \kappa^2}{2m_b'} \tag{A1.31}$$

which can be solved to give ε explicitly as

$$\varepsilon = -\frac{\varepsilon_g}{2} + \left\{ \frac{\varepsilon_g^2}{4} + \varepsilon_g \left[(r + \tfrac{1}{2} \pm \tfrac{1}{2})\beta_b H + \frac{\hbar^2 \kappa^2}{2m_b'} \right] \right\}^{1/2} \tag{A1.32}$$

for the states in the conduction band. The $\pm$ term represents the contribution of electron spin.

Since $f'(\varepsilon) = 1 + 2\varepsilon/\varepsilon_g$, $m(\varepsilon)$ and $\beta(\varepsilon)$ are given by

$$m(\varepsilon) = m_b(1 + 2\varepsilon/\varepsilon_g) \quad \text{and} \quad \beta(\varepsilon) = \beta_b/(1 + 2\varepsilon/\varepsilon_g) \tag{A1.33}$$

We also have from (A1.29) that

$$F = \zeta(1 + \zeta/\varepsilon_g)/\beta_b = \frac{\zeta\,(1 + \zeta/\varepsilon_g)}{\beta\,(1 + 2\zeta/\varepsilon_g)} \tag{A1.34}$$

where, as usual β means $\beta(\zeta)$. In specifying the energy levels (A1.32), the parameters β_b and m_b' at any particular orientation are of course determined if the FS is known, but it is sometimes more convenient to express β_b in terms of the known values of F, ζ and ε_g by means of (A1.34). The longitudinal mass m_b' requires in addition knowledge of the volume v of the FS or equivalently, the number of electrons N filling the FS, since

$N = vV/4\pi^3$. It is easily shown from (A1.6) with ε replaced by $f(\varepsilon)$ and the definitions of m_b and m_b' together with (A1.34) that

$$m_b' = m_0 \frac{W}{D} = \left(\frac{3\pi^2\hbar^2 cN}{2eFV}\right)^2 \bigg/ 2\zeta(1 + \zeta/\varepsilon_g) \qquad \text{(A1.35)}$$

Appendix 2

Note on thermodynamics, illustrated by free electron model example

In various presentations of the theory, the use of Ω and differentiation at constant ζ is either taken for granted as obvious (as in Lifshitz and Kosevich 1955) or else not quite correctly justified (e.g. Gold 1968 (see pp. 55–8) or Chambers 1966 (see equation 22), in which the right answer is reached by cancelling errors). The point involved is a little subtle and so is spelt out in some detail in this appendix for the benefit of those (like the author) who are not theoreticians.

As stated in the text, the usual free energy F and the thermodynamic potential Ω are defined as

$$F = E - TS \tag{A2.1}$$

and

$$\Omega = F - N\zeta \tag{A2.2}$$

where ζ is the chemical potential, N the number of particles in the system and the other symbols have their usual meanings. For simplicity we shall ignore changes of volume and other strains and we shall also ignore the vector nature of magnetization M and magnetic field H (this is equivalent to considering only the component of M in the direction of H; the more general case involves no new principles but requires a more elaborate notation). We then find, by differentiating and applying the second law

$$dF = -S\,dT - M\,dH + \zeta\,dN \tag{A2.3}$$

$$d\Omega = -S\,dT - M\,dH - N\,d\zeta \tag{A2.4}$$

Thus for constant T (and constant volume or strain) we have

$$M = -(\partial F/\partial H)_N = -(\partial \Omega/\partial H)_\zeta \tag{A2.5}$$

$$\zeta = (\partial F/\partial N)_H, \qquad N = -(\partial \Omega/\partial \zeta)_H \tag{A2.6}$$

Since T will be constant throughout, the suffix T need not be written, though of course it should be understood. We can use either form of (A2.5) to get M, but since it is ζ rather than N which appears explicitly in the simplest forms of the statistical mechanics expressions for Ω and F, it is much simpler to use $(\partial\Omega/\partial H)$. If (perversely) we want to use $(\partial F/\partial H)_N$ the correct answer can still be obtained, but the calculation is more complicated. Starting from (A2.2), we have

$$-\left(\frac{\partial F}{\partial H}\right)_N = M = -\left(\frac{\partial \Omega}{\partial H}\right)_\zeta = -\left(\frac{\partial F}{\partial H}\right)_\zeta + \zeta\left(\frac{\partial N}{\partial H}\right)_\zeta \tag{A2.7}$$

Thus if F is expressed in terms of ζ we must evaluate the second term on the right as well as $-(\partial F/\partial H)_\zeta$ and this may be appreciably more complicated than simply starting from Ω.

This may be illustrated by the concrete example of a free electron gas without spin at $T = 0$. For any Fermi–Dirac system at $T = 0$ we have from (2.39) and (2.43)

$$F = \sum_{\varepsilon \leqslant \zeta} \varepsilon \tag{A2.8}$$

$$\Omega = \sum_{\varepsilon \leqslant \zeta} (\varepsilon - \zeta) \tag{A2.9}$$

(where levels are degenerate the term ε or $\varepsilon - \zeta$ is repeated as many times as the degeneracy and where ε varies continuously with the state variable, summation is replaced by integration). For our particular example (see Appendix 1)

$$\varepsilon = (r + \tfrac{1}{2})\beta H + \hbar^2 \kappa^2/2m \tag{A2.10}$$

as in (2.31) (for simplicity in writing we omit the zero suffices here). Thus taking account of the degeneracy factor D as given by (2.37) we have

$$F = \frac{eHV}{2\pi^2 c\hbar} \sum_r \int_{-\kappa_0(r)}^{\kappa_0(r)} [(r + \tfrac{1}{2})\beta H + \hbar^2\kappa^2/2m]\,\mathrm{d}\kappa \tag{A2.11}$$

$$\Omega = \frac{eHV}{2\pi^2 c\hbar} \sum_r \int_{-\kappa_0(r)}^{\kappa_0(r)} [(r + \tfrac{1}{2})\beta H + \hbar^2\kappa^2/2m - \zeta]\,\mathrm{d}\kappa \tag{A2.12}$$

where $\kappa_0(r)$ is given by (A2.10) with $\varepsilon = \zeta$, i.e. it is half the occupied length of the rth Landau tube (here a right circular cylinder) and the summation is taken as far as the highest occupied Landau tube. The number N is simply the number of states within the Fermi sphere, i.e.

$$N = \frac{eHV}{2\pi^2 c\hbar} \sum_r \int_{-\kappa_0(r)}^{\kappa_0(r)} \mathrm{d}\kappa = \frac{eHV}{\pi^2 c\hbar} \sum_r \kappa_0(r) \tag{A2.13}$$

(Note that (A2.13) is also derivable from (A2.6) and (A2.12).) Carrying out the integrations and substituting for $\kappa_0(r)$, we find

$$F = \frac{eHV}{3\pi^2 ch}\left(\frac{2m}{\hbar^2}\right)^{1/2} \sum_{r=0}^{n} [\zeta - (r + \tfrac{1}{2})\beta H]^{1/2}[\zeta + (2r + 1)\beta H]$$

$$(A2.14)$$

$$\Omega = -\frac{2eHV}{3\pi^2 ch}\left(\frac{2m}{\hbar^2}\right)^{1/2} \sum_{r=0}^{n} [\zeta - (r + \tfrac{1}{2})\beta H]^{3/2} \qquad (A2.15)$$

$$N = \frac{eHV}{\pi^2 ch}\left(\frac{2m}{\hbar^2}\right)^{1/2} \sum_{r=0}^{n} [\zeta - (r + \tfrac{1}{2})\beta H]^{1/2} \qquad (A2.16)$$

(n is the highest value of r for which $\zeta \geqslant (r + \tfrac{1}{2})\beta H$).

Thus we have F and Ω as *explicit* functions of H and ζ, but only as *implicit* functions of H and N, through (A2.16). The use of Ω in (A2.15) gives M immediately as

$$M = -\left(\frac{\partial \Omega}{\partial H}\right)_{\zeta}$$

$$= \frac{2}{3}\frac{eV}{\pi^2 ch}\left(\frac{2m}{\hbar^2}\right)^{1/2} \sum_{r=0}^{n} [\zeta - (r + \tfrac{1}{2})\beta H]^{1/2}[\zeta - \tfrac{5}{2}(r + \tfrac{1}{2})\beta H] \quad (A2.17)$$

or

$$M/N\beta = (\beta H/\zeta)^{3/2} \sum_{r=0}^{n} \left[\frac{\zeta}{\beta H} - (r + \tfrac{1}{2})\right]\left[\frac{\zeta}{\beta H} - \tfrac{5}{2}(r + \tfrac{1}{2})\right]$$

$$(A2.18)$$

If, however, we set out from F and use (A2.7), considerably more algebra is required to reach the same result. It should be noticed that even though M is obtained by differentiating Ω at constant ζ, the result (A2.18) carries no implication that ζ is independent of H; the only restraint on the behaviour of ζ is that it must be consistent with that of N through (A2.16). If ζ is kept constant, then N oscillates with H, while if N is kept constant ζ oscillates with H. These remarks apply quite generally and are worked out more fully in the main text.

For the special case of a free electron gas it is not difficult to work out the magnetization if N rather than ζ is kept constant as H varies. If ζ_0 is the Fermi energy for N electrons at $H = 0$ (see (2.23) and (2.24)) it is easily shown that the condition (A2.16) can be expressed as

$$(\zeta_0/\beta H)^{3/2} = \frac{3}{2}\sum_{r=0}^{n} ((\zeta/\beta H) - (r + \tfrac{1}{2}))^{1/2} \qquad (A2.19)$$

This together with (A2.18) makes it possible to compute $M/N\beta$ as a function of $\zeta_0/\beta H$ (which is directly proportional to $1/H$) rather than of $\zeta/\beta H$. In fact for $\zeta/\beta H \gg 1$, ζ_0 and ζ are almost exactly equal, since the r.h.s. of (2.19) approximates to $(\zeta/\beta H)^{3/2}$ by the Euler–Maclaurin theorem, so that it is only for low values of $\zeta_0/\beta H$ that the distinction between constant N and constant ζ matters. The extent to which it does matter is illustrated in fig. 2.14 (p. 69).

It is convenient to note here two generalizations of the result (A2.18): First it is not difficult to show that it is valid for *any* parabolic band, whatever the specification of the ellipsoidal constant energy surfaces. Although the m in (A2.10) etc. must be replaced by the longitudinal mass m' of (A1.12) and β is given by (A1.7) and (A1.17), the extra parameters W and D which specify the ellipsoid are absorbed in the expressions for N and ζ in such a way as to leave (A2.18) unchanged.

The second generalization is to take account of spin. If it is assumed that the spin moment is just a Bohr magneton so that the spin-splitting is exactly equal to the Landau level splitting (this however is true only for the free electron model and not for any arbitrary parabolic band), the levels of (A2.10) are modified by adding $\pm\frac{1}{2}\beta H$ to (A2.10). It is easily shown that (A2.18) is then modified to

$$M/N\beta = \tfrac{1}{2} + (\beta H/\zeta)^{3/2} \sum_{r=1}^{n} \left(\frac{\zeta}{\beta H} - r\right)^{1/2} \left(\frac{\zeta}{\beta H} - \frac{5}{2}r\right) \qquad \text{(A2.20)}$$

Appendix 3

Calculation by the Poisson summation formula

The Poisson formula is easily obtained from the Fourier analysis of an infinite series of equally spaced δ-functions. Such a series can be represented as a function of a continuous variable x in the form

$$\sum_{r=-\infty}^{\infty} \delta[x - (r + \gamma)]$$

where r is an integer and γ is a phase factor which we shall choose to be $\frac{1}{2}$, to suit our particular application. The Fourier analysis of this periodic function is given by

$$\sum_{r=-\infty}^{\infty} \delta[x - (r + \tfrac{1}{2})] = 1 + 2 \sum_{p=1}^{\infty} \cos[2\pi p(x - \tfrac{1}{2})] \qquad (A3.1)$$

If we now multiply inside the summation by any function $f(x)$ and integrate we find

$$\int_0^{\infty} \sum_{r=-\infty}^{\infty} f(x)\delta[x - (r + \tfrac{1}{2})]\, dx$$

$$= \int_0^{\infty} f(x)\,dx + 2 \int_0^{\infty} \sum_{p=1}^{\infty} f(x)\cos[2\pi p(x - \tfrac{1}{2})]\, dx$$

$$(A3.2)$$

But the left hand side is just

$$\sum_{r=0}^{\infty} f(r + \tfrac{1}{2})$$

so we have

$$\sum_{r=0}^{\infty} f(r + \tfrac{1}{2}) = \int_0^{\infty} f(x)\,dx + 2 \int_0^{\infty} \sum_{p=1}^{\infty} f(x)\cos[2\pi p(x - \tfrac{1}{2})]\, dx$$

$$(A3.3)$$

This is the Poisson summation formula and we now apply it to calculate

$$D \sum_{r=0}^{n} (\varepsilon_r - \zeta)$$

Appendix 3

(see (2.45)). To apply the formula we must define $f(x)$ as $(\varepsilon_r - \zeta)$ up to $r = n$ (the highest filled level) and $f(x) = 0$ for $r > n$. The continuous variable x and its value X for $\varepsilon = \zeta$ are defined just as in (2.48) and (2.49) and we find

$$\delta\Omega/D = \int_0^X (\varepsilon(x) - \zeta)\,dx + 2\int_0^X \sum_{p=1}^\infty (\varepsilon(x) - \zeta)\cos[2\pi p(x - \tfrac{1}{2})]\,dx$$

(A3.4)

Integration by parts twice over to evaluate the second integral gives

$$\delta\Omega = \left(\frac{eHV\,\delta\kappa}{2\pi^2 c\hbar}\right)\Bigg\{ \int_0^X [\varepsilon(x) - \zeta]\,dx$$

$$+ \left([\varepsilon(x) - \zeta]\sum_{p=1}^\infty \frac{\sin[2\pi p(x - \tfrac{1}{2})]}{\pi p}\right)_0^X$$

$$+ \left(\frac{\beta(x)H}{2}\sum_{p=1}^\infty \frac{\cos[2\pi p(x - \tfrac{1}{2})]}{\pi^2 p^2}\right)_0^X$$

$$- \int_0^X \frac{H}{2}\frac{d\beta(x)}{dx}\sum_{r=1}^\infty \frac{\cos[2\pi p(x - \tfrac{1}{2})]}{\pi^2 p^2}\,dx\Bigg\}$$

The second line on the right vanishes at both limits (since $\varepsilon(X) = \zeta$) and the third can be evaluated at the lower limit by means of the relation

$$\sum_{p=1}^\infty \frac{\cos 2\pi p\theta}{\pi^2 p^2} = \theta^2 - \theta + \tfrac{1}{6}$$

valid for $0 \leqslant \theta \leqslant 1$. The last integral can again be reduced by integration by parts and its contribution can be seen to be smaller than that of the main periodic term by a factor of order $(d\ln\beta(x)/dx)_{x=X}$; this is in general of order $1/X$ or $1/n$ and so can be ignored to our order of approximation (for a parabolic band this factor is of course exactly zero). We thus obtain finally

$$\delta\Omega = \left(\frac{eHV\,\delta\kappa}{2\pi^2 c\hbar}\right)\Bigg\{ \int_0^X [\varepsilon(x) - \zeta]\,dx + \tfrac{1}{24}\beta(0)H$$

$$+ \frac{\beta H}{2}\sum \frac{\cos[2\pi p(X - \tfrac{1}{2})]}{\pi^2 p^2}\Bigg\}$$

(A3.5)

which is identical with the Fourier analysed form of (2.55) obtained by the Euler–Maclaurin method. Indeed, a closer examination of the two calculations shows that essentially they differ only in that the Fourier analysis is put in at the start in the Poisson approach and at the end in the Euler–Maclaurin one.

Appendix 4

The steady susceptibility

A4.1. The spin susceptibility

The first term in the expression (2.55) for the two-dimensional free energy $\delta\Omega$ at $T = 0$ *without* spin can be expressed in a more immediately meaningful way if we recall the definition (2.48) of x. Its contribution after integration over κ is

$$\Omega_s = \frac{V}{4\pi^3} \int d\kappa \int_0^a (\varepsilon(a, \kappa) - \zeta) \, da = \frac{V}{4\pi^3} \int (\varepsilon(k) - \zeta) \, dv \qquad (A4.1)$$

Here dv is an element of volume of k-space and the integration is to be extended over the whole volume of the Fermi surface. It is now evident that Ω_s is just the value of the free energy in the absence of field. Another useful expression of Ω_s is obtained by integrating by parts, giving

$$\Omega_s = -\frac{V}{4\pi^3} \int_0^\zeta v(\varepsilon) \, d\varepsilon \qquad (A4.2)$$

where $v(\varepsilon)$ is the volume contained by the surface of energy ε.

If now the spin is taken into account, the two-fold degeneracy of each state is lifted and for the simplest case, that in which the spin-splitting is the same all through the band, the two states originally of energy $\varepsilon(k)$ have energies given by

$$\varepsilon_\pm(k) = \varepsilon(k) \pm \tfrac{1}{2}\mu H \qquad (A4.3)$$

where μ is a constant magnetic moment, usually denoted by $\tfrac{1}{2}g\beta_0$ (g is the spin-splitting factor, independent of ε and β_0 is the double Bohr magneton). The states are now filled up until $\varepsilon_+ = \zeta$ for the antiparallel spins and $\varepsilon_- = \zeta$ for the parallel spins. Thus (A4.2) is replaced by

$$\Omega_s = -\frac{V}{8\pi^3} \left[\int_0^{\zeta - \frac{1}{2}\mu H} v(\varepsilon) \, d\varepsilon + \int_0^{\zeta + \frac{1}{2}\mu H} v(\varepsilon) \, d\varepsilon \right] \qquad (A4.4)$$

If we use the notation

$$W(\zeta) = \int_0^\zeta v(\varepsilon) \, d\varepsilon \qquad (A4.5)$$

483

we find on expanding in a Taylor series up to terms in H^2

$$\Omega_s - \Omega_{s0} = -\frac{V}{32\pi^3}\frac{\partial^2 W}{\partial \zeta^2}\mu^2 H^2 \tag{A4.6}$$

where Ω_{s0} is $-VW(\zeta)/4\pi^3$, i.e. the value of Ω_s for $H = 0$. Now

$$\partial^2 W/\partial \zeta^2 = (d\upsilon/d\varepsilon)_{\varepsilon=\zeta} = 4\pi^3 \mathscr{D}(\zeta)/V \tag{A4.7}$$

where $\mathscr{D}(\varepsilon)$ is the density of states,* so (A4.6) becomes

$$\Omega_s - \Omega_{s0} = -\tfrac{1}{8}\mathscr{D}(\zeta)\mu^2 H^2$$

which is $-\tfrac{3}{16}(N\mu^2 H^2)/\zeta$ for a parabolic band. The corresponding para-magnetic spin moment is

$$M_s = -\partial\Omega_s/\partial H = \tfrac{1}{4}\mathscr{D}(\zeta)\mu^2 H \tag{A4.8}$$

which is $\tfrac{3}{8}(N\mu^2 H)/\zeta$ for a parabolic band. This is, of course, the well known Pauli spin paramagnetism.

The above derivation is equivalent to that given by Lifshitz and Kosevich (1955), but simplified by assuming $T = 0$; the result is in fact a good approximation as long as $kT \ll \zeta$. It is also essentially the standard argument for deriving the Pauli spin paramagnetism in which no account is taken of the Landau orbital quantization. It should be emphasized that this standard argument, i.e. based on the quasi-continuum of states implied in (A4.1), is justified only by the procedure of separating out the various contributions from the expression (2.55) which takes the orbital quantization into account. Of the two other contributions in (2.55), the one leads to the steady diamagnetism, as discussed below in §A4.2, and is unaffected to our order of approximation by the introduction of spin, while the other is the oscillatory contribution whose amplitude is reduced by the introduction of spin as discussed in detail in §2.3.7.4.

We must now consider how (A4.3) should be generalized if the spin splitting varies through the band and may moreover be asymmetric. We shall suppose for simplicity that the parameters involved are functions only of the energy ε; if they vary over a surface of constant ε it is the mean values which are implied. Thus instead of (A4.3) we write

$$\varepsilon_\pm(\mathbf{k}) = \varepsilon(\mathbf{k}) \pm \tfrac{1}{2}\mu(\varepsilon)H + \lambda(\varepsilon)H^2 \tag{A4.9}$$

* The density of states defined here counts the two spin states per cell in k-space as distinct states. Lifshitz and Kosevich (1955) define a density of states $\rho(\varepsilon)$ which is half of our $N(\varepsilon)$.

Some care is now needed in reformulating (A4.4) in two respects. First it is not enough to replace the limits of integration simply by $\zeta \mp \frac{1}{2}\mu(\zeta)H - \lambda(\zeta)H^2$ since it is really $\mu(\zeta \mp \frac{1}{2}\mu H)$ which is needed, thus to a sufficient approximation the limits must be modified to $\zeta \mp \frac{1}{2}\mu(\zeta)H + \frac{1}{4}\mu(\zeta)(\partial\mu/\partial\zeta)H^2 - \lambda(\zeta)H^2$. The correction for λ is omitted, since it would lead to terms in H^3 and H^4. Secondly in the integration by parts which leads from integrals over v to integrals over ε, (A4.4) is valid only if μH is a constant. The appropriate modification taking both these points into account is

$$\Omega_s = -\frac{V}{8\pi^3}\Bigg\{\int_0^{\zeta - \frac{1}{2}\mu(\zeta)H + \frac{1}{4}\mu(\zeta)[\partial\mu(\zeta)/\partial\zeta]H^2 - \lambda(\zeta)H^2} v(\varepsilon)$$

$$\times \left[1 + \frac{H}{2}\frac{\partial\mu(\varepsilon)}{\partial\varepsilon} + H^2\frac{\partial\lambda(\varepsilon)}{d\varepsilon}\right]d\varepsilon$$

$$+ \int_0^{\zeta + \frac{1}{2}\mu(\zeta)H + \frac{1}{4}\mu(\zeta)[\partial\mu(\zeta)/\partial\zeta]H^2 - \lambda(\zeta)H^2} v(\varepsilon)$$

$$\times \left[1 - \frac{H}{2}\frac{\partial\mu(\varepsilon)}{\partial\varepsilon} + H^2\frac{\partial\lambda(\varepsilon)}{\partial\varepsilon}\right]d\varepsilon. \qquad (A4.10)$$

Retaining only terms up to H^2 we find the terms in $\partial\mu/\partial\varepsilon$ cancel out and the term in $\partial\lambda/\partial\varepsilon$ also cancels after an integration by parts. Finally (A4.10) reduces to

$$\Omega_s - \Omega_{s0} = -\left[\frac{1}{8}\mathscr{D}(\zeta)\mu^2(\zeta) - \int_0^\zeta \mathscr{D}(\varepsilon)\lambda(\varepsilon)d\varepsilon\right]H^2 \qquad (A4.11)$$

and

$$M_s = \left[\frac{1}{4}\mathscr{D}(\zeta)\mu^2(\zeta) - 2\int_0^\zeta \mathscr{D}(\varepsilon)\lambda(\varepsilon)\,d\varepsilon\right]H \qquad \textbf{(A4.12)}$$

As would be expected by elementary reasoning, the up and down spins cancel except near the Fermi surface, as far as the symmetric part of the spin splitting (i.e. $\pm\frac{1}{4}\mu H$) is concerned, and the only novelty comes from the asymmetric part λH^2, whose contribution comes from all the electrons, rather than only those at the FS. An example of how this works out will be given in §A4.3. It should be explained that we have throughout ignored many-body interactions; as is outlined in §9.6.2 these considerably modify the theory of the spin susceptibility.

Appendix 4

A4.2. The steady diamagnetic susceptibility

The second term in (2.55) gives a contribution $d\Omega_D$ to $d\Omega$ and this leads to a steady diamagnetism. This contribution when integrated over κ gives*

$$\Omega_D = \frac{eH^2V}{48\pi^2ch} \int \beta(0)\,d\kappa \qquad\qquad\text{(A4.13)}$$

where the integration is over the full range of κ and

$$\beta(0) = 2\pi e/[ch(\partial a/\partial\varepsilon)_{\kappa \text{ at } a=0}] \qquad\qquad\text{(A4.14)}$$

Correspondingly there is a diamagnetic moment given by

$$M_D = -\frac{eHV}{24\pi^2ch} \int \beta(0)\,d\kappa \qquad\qquad\text{(A4.15)}$$

For a parabolic band (and in particular for free electrons), $\beta(0)$ is independent of κ and is simply $e\hbar/mc$ with m as defined in (A1.7); the limits of integration are given by $\pm\kappa_0$, where κ_0 is as defined by (A1.4) for $\varepsilon = \zeta$, and using the relations in Appendix 1 for a general ellipsoid, we easily find

$$\Omega_D = \frac{1}{16}\frac{N\beta^2H^2}{\zeta} \quad\text{and}\quad M_D = -\frac{1}{8}\frac{N\beta^2H}{\zeta} \qquad\qquad\textbf{(A4.16)}$$

This is exactly the correct result. For free electrons μ in (A4.8) and β in (A4.16) are identical (both just β_0, the double Bohr magneton) so we have the well known relation that $M_D = -\frac{1}{3}M_S$.

If, however, the band is not parabolic, (A4.13) does *not* lead to the correct answer as first given by Peierls (1933a) on the basis of a more fundamental calculation. Thus, as we shall see below, the correct result involves only the behaviour of electrons at the FS while (A4.13) involves electrons of all energies. The source of the discrepancy lies in the assumption that the phase γ is exactly $\frac{1}{2}$. This assumption is adequate for discussing the dHvA oscillations to the approximation we have aimed at throughout (i.e. ignoring terms of order $1/n$ times those retained), but a better approximation is essential to yield the correct steady diamagnetism. This problem was first thoroughly discussed by Roth (1966) and we shall use her result for the next higher approximation to γ as our starting point. If we ignore complications such as singularities in constant energy surfaces (e.g. self interesting orbits for certain ε and κ) and interband effects, her result can be

---

* It is easily shown that the introduction of spin only introduces terms of order higher than H^2, so we shall ignore spin throughout this section.

486

The steady susceptibility

expressed in our notation as

$$\gamma - \frac{1}{2} = \frac{eH}{48\pi c h} \frac{\partial}{\partial \varepsilon} \oint \frac{dk'}{(\partial \varepsilon / \partial k'_n)} \left[\frac{\partial^2 \varepsilon}{\partial k_x^2} \frac{\partial^2 \varepsilon}{\partial k_y^2} - \left(\frac{\partial^2 \varepsilon}{\partial k_x \partial k_y} \right)^2 \right]$$

$$\equiv H \frac{\partial G}{\partial \varepsilon} (\varepsilon, \kappa) \qquad\qquad (A4.17)$$

where the integration is to be taken round the k' orbit defined by the intersection of the surface of energy ε by the $k_x k_y$ plane (H is in the k_z or κ direction) and dk' is an element of the k' orbit; k'_n is the component normal to the orbit in this plane (thus $\partial \varepsilon / \partial k'_n = \hbar v'_n$ where v'_n is the normal component of velocity in the plane). All the derivatives are taken at constant κ.

For a parabolic band the expression in square brackets is a constant independent of ε, while $\oint dk'/(\partial \varepsilon / \partial k'_n)$ is just $(\partial a / \partial \varepsilon)_\kappa$, which is $2\pi m / \hbar^2$ and this too is independent of ε. Thus $\gamma - \frac{1}{2}$ vanishes exactly, as of course we know it should from the detailed solution of the Schrödinger equation for the motion of free electrons (or of quasiparticles appropriate to a parabolic band) in a magnetic field. In general, however the difference $\gamma - \frac{1}{2}$ does not vanish. Some idea of its order of magnitude may be obtained by treating the square bracket as a constant, of order $\hbar^4 / m^2$, where m is some effective mass. If this mass is identified with the cyclotron mass and this is now considered to be a function of ε, (A4.17) gives

$$\gamma - \frac{1}{2} \sim \frac{H}{24} \frac{\partial}{\partial \varepsilon} \left(\frac{e\hbar}{mc} \right)_\kappa = \frac{H}{24} (\partial \beta / \partial \varepsilon)_\kappa \qquad\qquad (A4.18)$$

This quantity can be estimated for the example given in Appendix 1 and, using (A1.33) with $\varepsilon_g / 2\zeta \sim 0.3$, it can be easily shown to be less than about $0.03/n$ for the level of quantum number n. This is of course a very crude estimate, but it shows that the approximation of ignoring $\gamma - \frac{1}{2}$ is quite consistent with other approximations in which terms proportional to $1/n$ were ignored in the calculations of the dHvA oscillations. It cannot, however, be ignored in calculating the steady diamagnetism, because for our two-dimensional slab at $T = 0$ the steady term is itself only of order $1/n$ times the oscillatory term.

We shall now recalculate the two-dimensional thermodynamical potential starting from equation (2.47), but taking into account that γ is given by (A4.17). The continuous variable x now corresponds to $r + \gamma$, and γ is itself a function of x so that

$$dr = dx - (d\gamma/dx) dx$$

Thus (2.52) must be revised to

$$\frac{\delta\Omega}{D} = \int_{\gamma(0)}^{n+\gamma} [\varepsilon(x) - \zeta]\,dx - \int_{\gamma(0)}^{n+\gamma} [\varepsilon(x) - \zeta]\frac{\partial\gamma(x)}{\partial x}\,dx$$

$$+ \tfrac{1}{2}[\varepsilon(n+\gamma) - \zeta] + \tfrac{1}{2}[\varepsilon(\gamma(0)) - \zeta]$$

$$+ \tfrac{1}{12}\left\{\frac{\partial\varepsilon(n+\gamma)}{\partial x} - \frac{\partial\varepsilon(\gamma(0))}{\partial x}\right\} \qquad (A4.19)$$

Here we use the notation γ for the value of γ near the Fermi level and $\gamma(0)$ for the value near the bottom of the 2-D band. The *variable* γ is denoted by $\gamma(x)$ or $\gamma(\varepsilon)$, as appropriate. We have also assumed in the last term on the r.h.s. that $d\gamma/dx \ll 1$, which is certainly justified since it can be shown to be of order $1/n^2$. Integrating the second integral by parts, it becomes (with its minus sign)

$$[\zeta - \varepsilon(n+\gamma)][\gamma - \gamma(0)] + \int_{\varepsilon(\gamma(0))}^{\varepsilon(n+\gamma)} d\varepsilon[\gamma(\varepsilon) - \gamma(0)] \qquad (A4.20)$$

and following the same procedure as was used in going from (2.52) to (2.53) (A4.19) becomes (denoting $\varepsilon(0)$ by ε_0)

$$\frac{\delta\Omega}{D} = \int_0^x [\varepsilon(x) - \zeta]\,dx + \delta(\tfrac{1}{2}\beta H\delta) - \gamma(0)[\varepsilon_0 + \tfrac{1}{2}\beta(0)\gamma(0)H - \zeta]$$

$$+ \int_{\varepsilon_0}^{\zeta} [\gamma(\varepsilon) - \gamma(0)]\,d\varepsilon - \tfrac{1}{2}\beta H\delta + \tfrac{1}{2}(\varepsilon_0 - \zeta) + \tfrac{1}{2}\beta(0)\gamma(0)H$$

$$+ \tfrac{1}{12}[\beta H - \beta(0)H] \qquad (A4.21)$$

Here the second and third terms on the r.h.s. come from the adjustment of the limits of the first integral and the first term of (A4.20) is cancelled by the adjustment of the upper limit of the integral in (A4.20).

We now make the assumption that ε varies quadratically with k_x and k_y around ε_0, which implies $\gamma(0) = \tfrac{1}{2}$ and we substitute (A4.17) for $\gamma(\varepsilon) - \tfrac{1}{2}$; (A4.21) then reduces to

$$\delta\Omega = \frac{eHV\,d\kappa}{2\pi^2 c\hbar}\left\{\int_0^x [\varepsilon(x) - \zeta]\,dx + \tfrac{1}{2}\beta H(\delta^2 - \delta + \tfrac{1}{6})\right.$$

$$\left. + \tfrac{1}{24}\beta(0)H + H[G(\zeta,\kappa) - G(\varepsilon_0,\kappa)]\right\} \qquad (A4.22)$$

Thus the only difference from (2.55) is the appearance of the last term, which clearly modifies the steady diamagnetism, but leaves the oscillatory part of

$\delta\Omega$ (i.e. the second term of (A4.22)) as it was. The part of the thermodynamic potential proportional to H^2 (i.e. that which gives the steady diamagnetism) is

$$
\delta\Omega_D = \frac{eH^2 V \, d\kappa}{2\pi^2 c\hbar} \left\{ \frac{\beta(0)}{24} \right.
$$

$$
+ \frac{e}{48\pi c\hbar} \oint_{\varepsilon=\zeta} \frac{dk'}{(\partial\varepsilon/\partial k'_n)} \left[\frac{\partial^2\varepsilon}{\partial k_x^2} \frac{\partial^2\varepsilon}{\partial k_y^2} - \left(\frac{\partial^2\varepsilon}{\partial k_x \partial k_y} \right)^2 \right]
$$

$$
\left. - \frac{e}{48\pi c\hbar} \oint_{\varepsilon=\varepsilon_0} \frac{dk'}{(\partial\varepsilon/\partial k'_n)} \left[\frac{\partial^2\varepsilon}{\partial k_x^2} \frac{\partial^2 c}{\partial k_y^2} - \left(\frac{\partial^2\varepsilon}{\partial k_x \partial k_y} \right)^2 \right] \right\} \quad \text{(A4.23)}
$$

It is easily shown by reference to the properties of a general quadratic $\varepsilon(k)$ relation (as outlined in Appendix 1) that for a parabolic band, as is appropriate round ε_0, the last term in the curly brackets of (A4.23) reduces to $-\beta(0)/24$ and so exactly cancels the first term, leaving the middle term, which involves only properties at the Fermi surface. Integrating over κ, we find

$$
\Omega_D = \frac{e^2 H^2}{96\pi^3 c^2 \hbar^2} \int d\kappa \oint_{\varepsilon=\zeta,\kappa} \frac{dk'}{(\partial\varepsilon/\partial k'_n)} \left[\frac{\partial^2\varepsilon}{\partial k_x^2} \frac{\partial^2\varepsilon}{\partial k_y^2} - \left(\frac{\partial^2\varepsilon}{\partial k_x \partial k_y} \right)^2 \right]
$$

$$
\text{(A4.24)}
$$

Simple geometry shows that this is the same as

$$
\Omega_D = \frac{e^2 H^2}{96\pi^3 c^2 \hbar^2} \int_{FS} \frac{dS}{(\partial\varepsilon/\partial k_n)} \left[\frac{\partial^2\varepsilon}{\partial k_x^2} \frac{\partial^2\varepsilon}{\partial k_y^2} - \left(\frac{\partial^2\varepsilon}{\partial k_x \partial k_y} \right)^2 \right] \quad \text{(A4.25)}
$$

where dS is an element of area of the FS and k_n refers to the normal to the FS. This agrees with Peierls' formula (1933a) for the special case $T = 0$ and under the various simplifying assumptions made in formulating (A4.17); other derivations of the formula have been given by Wilson (1953b) and Roth (1966). The main reason for giving yet another derivation here is to demonstrate just how the slight H and ε dependence of γ eliminates the awkward feature of the calculation with constant γ, i.e. that the steady diamagnetism appears to depend on the form of $\varepsilon(\kappa)$ below the FS. With the slightly variable γ we find, as we should, that the steady diamagnetism depends only on the band structure at the FS.

Finally it should be emphasized that we have considered here only the simplest case, the contribution of an isolated band of quasiparticles with an arbitrary dispersion. A realistic calculation of the steady susceptibility is much more complicated, involving interaction effects between bands, contributions from filled bands and a number of other effects. A detailed calculation was first made by Hebborn and Sondheimer (1960) and the problem is also discussed in detail by Roth (1966), where full references to earlier work may be found.

A4.3 Steady susceptibility in the Lax model

The various formulae of the last two sections may be illustrated by the Lax model, which involves ellipsoidal energy surfaces but a non-parabolic band. Although this model may be somewhat artificial, it does describe the behaviour of bismuth reasonably well and has the merit that it permits explicit calculations to be made. In this model (see Appendix 1) the constant energy surfaces are specified by

$$\sum \alpha_{ij} k_i k_j = \frac{2m_0}{\hbar^2} \left(\varepsilon + \frac{\varepsilon^2}{\varepsilon_g} \right) \tag{A4.26}$$

and the energy levels are given by (A1.32), i.e.

$$\varepsilon_{\pm} = -\frac{\varepsilon_g}{2} + \left\{ \frac{\varepsilon_g^2}{4} + \varepsilon_g \left[(r + \tfrac{1}{2} \pm \tfrac{1}{2}) \beta_b H + \frac{\hbar^2 \kappa^2}{2m_b'} \right] \right\}^{1/2} \tag{A4.27}$$

(the parameters β_b and m_b' are defined in Appendix 1). For $r \gg 1$ the extra $\pm$ term can be treated as small and expansion in a Taylor series gives essentially (A4.9) with the coefficients μ and λ specified by

$$\mu(\varepsilon) = \beta(\varepsilon), \qquad \lambda(\varepsilon) = -\tfrac{1}{4} \beta(\varepsilon)/\varepsilon_g \left(1 + \frac{2\varepsilon}{\varepsilon_g} \right) \tag{A4.28}$$

where, as in (A1.33)

$$\beta(\varepsilon) = \beta_b/(1 + 2\varepsilon/\varepsilon_g)$$

and $\beta(\varepsilon)H$ is the Landau level splitting (and also the spin-splitting) of a level of energy ε; $\beta(\varepsilon)$ depends of course on the direction of H in the manner explained in Appendix 1, but we shall not be explicitly concerned with this orientation dependence here. The density of states $\mathscr{D}(\varepsilon)$ is easily shown to be

$$\mathscr{D}(\varepsilon) = \frac{3}{2} \frac{N}{\zeta} \frac{\varepsilon^{1/2}}{\zeta^{1/2}} \frac{(1 + \varepsilon/\varepsilon_g)^{1/2}}{(1 + \zeta/\varepsilon_g)^{3/2}} \left(1 + \frac{2\varepsilon}{\varepsilon_g} \right) \tag{A4.29}$$

where N is the total number of electrons within the FS, given by

$$N \equiv \int_0^{\zeta} \mathscr{D}(\varepsilon) \, d\varepsilon = \frac{V}{3\pi^2 D^{1/2}} \left(\frac{2m_0}{\hbar^2} \right)^{3/2} \zeta^{3/2} \left(1 + \frac{\zeta}{\varepsilon_g} \right)^{3/2} \tag{A4.30}$$

The steady susceptibility

If these parameters are substituted into (A4.12), we find after some algebra

$$M_s = \frac{3}{8} \frac{N\beta^2 H}{\zeta} \frac{(1 + 2\zeta/\varepsilon_g)}{(1 + \zeta/\varepsilon_g)}$$

$$\times \left[1 - 1 + \frac{(1 + 2\zeta/\varepsilon_g)}{2(\zeta/\varepsilon_g)^{1/2}(1 + \zeta/\varepsilon_g)^{1/2}} \cosh^{-1}\left(1 + \frac{2\zeta}{\varepsilon_g}\right) \right]$$

(A4.31)

where the first term in the square brackets comes from the first term in (A4.12), i.e. from the standard formula for the Pauli paramagnetism, and the other two from the second term in (A4.12). The net result is

$$M_s = \frac{3}{16} \frac{N\beta^2 H}{\zeta} \frac{(1 + 2\zeta/\varepsilon_g)^2}{(\zeta/\varepsilon_g)^{1/2}(1 + \zeta/\varepsilon_g)^{3/2}} \cosh^{-1}\left(1 + \frac{2\zeta}{\varepsilon_g}\right) \quad \text{(A4.32)}$$

For the electron ellipsoids in Bi, $\zeta/\varepsilon_g \simeq 2.2$ and if this is substituted in (A4.31) we find that the contribution of the second two terms, coming from λH^2, i.e. from the asymmetry of the spin-splitting, is about 40% more than the first term alone, i.e. the total M_s is about 2.4 times the result of the standard formula which ignores asymmetrical splitting. A useful check on (A4.31) or (A4.32) is to verify that as $\varepsilon_g \to \infty$, i.e. as $\zeta/\varepsilon_g \to 0$, the results approach those of the parabolic model. This is indeed so, since $\cosh^{-1}(1 + x) \to (2x)^{1/2}$ when x is small, so that (A4.32) does reduce to the parabolic band result (A4.8) when $\zeta/\varepsilon_g \to 0$.

We next evaluate the diamagnetic susceptibility in the Lax model. We first consider only the $\beta(0)H/24$ term in (A4.22), i.e. what would be obtained if $\gamma = \frac{1}{2}$ was exact. The moment $\beta(0)$ is given by (A1.33) for the value of ε which makes the area $a(\varepsilon, \kappa) = 0$; this is $\varepsilon_0(\kappa)$ as defined in the previous section and is given by

$$\varepsilon_0 + \frac{\varepsilon_0^2}{\varepsilon_g} = \frac{\hbar^2 \kappa^2}{2m_b'}$$

(A4.33)

The total contribution to Ω_D is

$$\Omega_D = \frac{eH^2 V}{2\pi chr} \int_0^{\kappa(\zeta)} \beta(0)\,d\kappa$$

(A4.34)

where $\kappa(\zeta)$ is given by (A4.33) for $\varepsilon_0 = \zeta$. Transforming the variable of integration from κ to ε_0 by means of (A4.33) and using relations from Appendix 1, we find after some algebra that

$$\Omega_D = -\tfrac{1}{3}(\Omega_s - \Omega_{s0}) \quad \text{and} \quad M_D = -\tfrac{1}{3}M_s \quad \text{(A4.35)}$$

491

i.e. the same relative result as for a parabolic band; this is of course not a general result, but a specific feature of the Lax model. The net susceptibility is then paramagnetic and given by two-thirds of the spin susceptibility alone. It is reassuring that this result for the Lax model agrees exactly with a formula quoted by McClure and Shoenberg (1976)* and calculated by a different method on the assumption that $\gamma = \frac{1}{2}$.

Finally we evaluate the steady diamagnetism as calculated with the Peierls formula, i.e. allowing for the slight variation of γ. The calculation is simplified by noting that for ellipsoidal surfaces the last bracket inside the integral of (A4.24) or (A4.25) is a function only of κ. Thus as far as the integration round the contour $\varepsilon = \zeta$ for given κ is concerned this bracket is just a constant of the form $A + B\kappa^2$ and (A4.24) can be expressed as

$$\Omega_D = \frac{e^2 H^2}{96\pi^3 c^2 \hbar^2} \int_{-\kappa 0}^{\kappa 0} d\kappa (\partial a/\partial \varepsilon)_{\varepsilon=\zeta} \left[\frac{\partial^2 \varepsilon}{\partial k_x^2} \frac{\partial^2 \varepsilon}{\partial k_y^2} - \left(\frac{\partial^2 \varepsilon}{\partial k_x \partial k_y} \right)^2 \right]$$

involving a simple integration over κ. The final result can be conveniently expressed as

$$\Omega_D = \frac{N\beta^2 H^2}{16\zeta} \frac{\left[1 + \frac{4}{3} \left(\frac{\zeta}{\varepsilon_g} + \frac{\zeta^2}{\varepsilon_g^2} \right) \right]}{\left(1 + \frac{\zeta}{\varepsilon_g} \right) \left(1 + \frac{2\zeta}{\varepsilon_g} \right)} \tag{A4.36}$$

Correspondingly

$$M_D = -\frac{N\beta^2 H}{8\zeta} \frac{\left[1 + \frac{4}{3} \left(\frac{\zeta}{\varepsilon_g} + \frac{\zeta^2}{\varepsilon_g^2} \right) \right]}{\left(1 + \frac{\zeta}{\varepsilon_g} \right) \left(1 + \frac{2\zeta}{\varepsilon_g} \right)} \tag{A4.37}$$

The first factor in both these expressions is the result for a parabolic band, which is achieved if $\zeta/\varepsilon_g \to 0$. For $\zeta/\varepsilon_g = 2.2$, (A4.37) gives 0.6 times the parabolic band value and only 0.25 the value obtained by ignoring the variation of γ. Once again it should be emphasized that M_D as given by (A4.37) together with M_s as given by (A4.32) do not by any means account for all the steady susceptibility; for instance in the Lax model, the states in the filled band, whose energies are given by (A4.27) with a minus instead of a plus sign outside the curly bracket, contribute a strong diamagnetic susceptibility (see for instance McClure and Shoenberg (1976)).

---

* It should be noted that the N_0 of equation (12) of that paper is the total for three ellipsoids and therefore three times the N defined here; note too that $\sinh^{-1} x = \cosh^{-1}(1 + x^2)^{1/2}$.

Appendix 5
Cornu spiral and related topics

A5.1. The Cornu spiral

If we put a finite limit v rather than ∞ in the integrals of (2.107) over κ we have

$$I'_p \equiv (-1)^p (\tfrac{1}{2} p X'')^{1/2} I_p$$

$$= C(v) \cos \psi \mp S(v) \sin \psi$$

$$= \{[C(v)]^2 + [S(v)]^2\}^{1/2} \cos\left[\psi \pm \tan^{-1} \frac{S(v)}{C(v)} \right] \qquad \text{(A5.1)}$$

where we have put

$$2\pi p X_0 = 2\pi p F/H = \psi, \qquad \tfrac{1}{2} u^2 = p X'' \kappa^2 \qquad \text{(A5.2)}$$

and $C(v)$, $S(v)$ are the Fresnel integrals

$$C(v) = \int_0^v \cos \frac{\pi u^2}{2} \, du, \qquad S(v) = \int_0^v \sin \frac{\pi u^2}{2} \, du \qquad \text{(A5.3)}$$

The variation of I'_p with v can be illustrated graphically by plotting $S(v)$ against $C(v)$. This produces the well known Cornu spiral, shown in fig. A5.1, and it is immediately obvious that for given v the amplitude of I'_p regarded as a function of ψ is simply the length of the chord of the spiral from the origin to the point v (it is easily shown that v is in fact the curved length of the spiral from the origin), while the phase angle is the inclination of the chord to the $C(v)$ axis. Evidently, as $v \to \infty$, the length of the chord approaches the limit $1/\sqrt{2}$ and the phase angle approaches $45°$, leading to the standard formula (2.109) in which v was taken as ∞. For a closed convex Fermi surface the limit of integration v is of course finite, and we now investigate what sort of error is involved in the approximation $v = \infty$, by considering the example of an ellipsoid. For an ellipsoid the relation

$$X = X_0 - \tfrac{1}{2} X''(0) \kappa^2$$

is exact, so the formulation (A5.1) should be particularly applicable if the

493

limit v is taken to correspond to the limiting value κ_0 of κ and the lower alternative sign is chosen. This limiting value is given by (2.105) (it is the value of κ which makes $X = 0$) and we see that

$$\tfrac{1}{2}\pi v^2 = \pi p X'' \kappa_0^2 = 2\pi p X_0 = 2\pi p F/H = \psi \qquad (A5.4)$$

Thus for an ellipsoid the limit v of the integration is not simply a constant but linked to the variable ψ of the oscillations; in particular v becomes of order 2π as we approach the quantum limit $F/H = 1$. This introduces a subtlety into the use of the Cornu spiral in as far as the true amplitude is no longer exactly the same as the $[C^2(v) + S^2(v)]^{1/2}$ suggested by (A5.1). In fact, for sufficiently large v, the Cornu spiral is to a fair approximation a circle of radius $1/\pi v$ centred on the point $C = S = \tfrac{1}{2}$ (see fig. A5.2), so that with v specified by (A5.4)

$$C = \frac{1}{2} + \frac{1}{\pi v}\sin\psi, \qquad S = \frac{1}{2} - \frac{1}{\pi v}\cos\psi \qquad (A5.5)$$

Fig. A5.1. The Cornu spiral: the functions $C(v)$ and $S(v)$ are defined by (A5.3) and v is the length from the origin along the spiral to any particular point (values of v are marked). The length of the chord from the origin gives the amplitude of the oscillation (A5.1) and the inclination of the chord to the $C(v)$ axis gives the value of the phase angle $\tan^{-1}(S(v)/C(v))$.

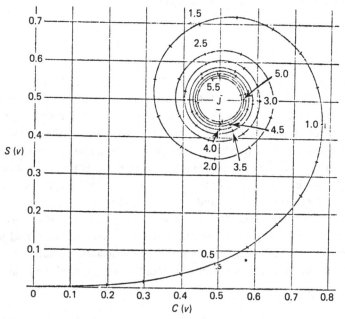

Cornu spiral and related topics

Even for pX_0 (i.e. pF/H) as low as 1 (i.e. $v = 2$) this approximation is good to about 1% and, if it is used, we find the correcting terms cancel out in I'_p and we obtain from (A5.1) with the lower sign exactly the standard answer

$$I'_p = \frac{1}{\sqrt{2}} \cos\left(\frac{2\pi pF}{H} - \frac{\pi}{4}\right) \tag{A5.6}$$

This is also evident from the geometric construction of fig. A5.2.
 A still better approximation is

$$\left.\begin{array}{l} C(v) = \tfrac{1}{2} + f(v)\sin\psi - g(v)\cos\psi \\ S(v) = \tfrac{1}{2} - f(v)\cos\psi - g(v)\sin\psi \end{array}\right\} \tag{A5.7}$$

Fig. A5.2. For large values of v it is a fair approximation to treat the Cornu spiral locally as a circle centred on $J \equiv (\tfrac{1}{2}, \tfrac{1}{2})$ and of radius $1/\pi v$. In this approximation the coordinates of a point such as P are evidently $(\tfrac{1}{2} + (1/\pi v)\sin\theta, \tfrac{1}{2} - (1/\pi v)\cos\theta)$, where θ is defined as the inclination of the tangent to the C axis, and is given by $\theta = \tfrac{1}{2}\pi v^2$ (more precisely $\tan\theta = \tan\tfrac{1}{2}\pi v^2$). For the special point P corresponding to the $\kappa(=\kappa_0)$ of the tangent plane to the ellipsoidal FS it is shown in the text that $\tfrac{1}{2}\pi v^2 = 2\pi pF/H = \psi$, so θ can be identified with ψ and the construction shows that $C(v)\cos\psi + S(v)\sin\psi = (1/\sqrt{2})\cos(\psi - 45°)$.

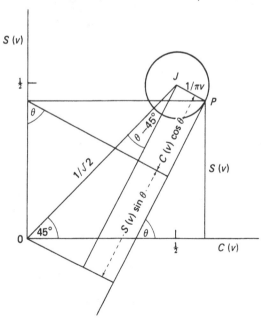

with $f(v)$ and $g(v)$ given in the form of series expansions in $1/v$, as

$$
\left.
\begin{aligned}
f(v) &= \frac{1}{\pi v} - \frac{3}{\pi^3 v^5} + \cdots \\
g(v) &= \frac{1}{\pi^2 v^3} - \cdots
\end{aligned}
\right\}
\tag{A5.8}
$$

If this is used, the $f(v)$ corrections in the first equation of (A5.1) cancel exactly and we are left with

$$
I'_p = \frac{1}{\sqrt{2}} \cos\left(\frac{2\pi pF}{H} - \frac{\pi}{4}\right) - \frac{1}{8\pi^2 (pF/H)^{3/2}} - \cdots
\tag{A5.9}
$$

Thus it is only when we are very close to the quantum limit (i.e. $F/H = 1$) for the fundamental ($p = 1$) that any appreciable correction is required. This correction is non-periodic in $1/H$ but even for $F/H = 1$ is only of order 2% of the amplitude given by the standard formula. Since in any case we have in the development of the theory ignored terms F/H times smaller than those retained, we should not expect our results to be better than approximate when F/H approaches 1.

Probably the above analysis still applies qualitatively even if the FS is not ellipsoidal, provided it is closed and everywhere convex. If the FS has a minimum rather than a maximum extremum or if it has several extrema (alternating maxima and minima), the above analysis no longer applies (there is no longer a clearly defined limit to the integration over κ, moreover if the extremum is a minimum the opposite sign must be chosen in (A5.1)). Each such case must be considered on its merits and usually there is no alternative to numerical integration to determine how the standard formula must be corrected.

A5.2. Nearly cylindrical FS

A nearly cylindrical FS provides a nice example of a situation in which the standard formula may require considerable modification. The significant feature of such a surface is that the curvature factor X'' becomes small and higher order terms in the expansion of X about its extremum X_0 need to be considered. The failure of the standard formula here is not because of the finite limits of integration but because of the presence of the higher order terms. A simple model of this situation has been discussed by Shoenberg and Templeton (ST) (1973), and has some relevance to the FS of copper for certain orientations of H. In this model it is assumed that the only even order terms in κ need be considered and terms up to κ^4 are included, i.e.

$$
X = X_0 \pm \tfrac{1}{2} X'' \kappa^2 \pm \tfrac{1}{24} X^{IV} \kappa^4
\tag{A5.10}
$$

Cornu spiral and related topics

where

$$X'' = |(\partial^2 X/\partial\kappa^2)_{\kappa=0}|, \qquad X^{IV} = |(\partial^4 X/\partial\kappa^4)_{\kappa=0}| \qquad (A5.11)$$

The $\pm$ sign in the κ^2 term is chosen, as before, according as X (i.e. the area of the FS cross-section) is a minimum or a maximum at $\kappa = 0$. If the same sign is chosen for the κ^4 term as for the κ^2 term, the variation of X continues monotonically as κ is increased, but if the opposite sign is chosen, the variation turns around for a value of κ given by

$$\kappa^2 = 6X''/X^{IV} \qquad (A5.12)$$

The problem reduces to replacing C and S in (A5.1) by C' and S' defined as

$$C' = \int_0^\infty \cos[\tfrac{1}{2}\pi u^2(1 \pm \alpha u^2)]\, du$$

and

$$S' = \int_0^\infty \sin[\tfrac{1}{2}\pi u^2(1 \pm \alpha u^2)]\, du \qquad (A5.13)$$

where

$$\alpha = X^{IV}/24p(X'')^2 \qquad (A5.14)$$

and the $+$ or $-$ sign is chosen according as $\partial^4 X/\partial\kappa^4$ and $\partial^2 X/\partial\kappa^2$ have the same or opposite signs respectively. Explicit expressions for C' and S' are given in standard books of tables (e.g. Gradshteyn and Ryzhik 1965, p. 398) and these assume relatively simple limiting forms when $\alpha \ll 1$ or $\alpha \gg 1$.

The dependence on α of the amplitude and phase of the resulting oscillations is shown in figs. A5.3 and A5.4, where it is supposed that α varies through change of X'' with fixed X^{IV} (this is not too unrealistic for the case of copper if the orientation of the magnetic field is varied through the critical orientation where X'' vanishes). If the minus sign is chosen in (A5.13) (i.e. corresponding to minus values of the abscissae in the graphs), the variation of X with κ goes through a minimum (or maximum) on either side of the central maximum (or minimum) and the resulting amplitude can be interpreted in detail as the resultant of two oscillating terms of slightly different frequencies, one from the central extremum and one from the two identical non-central extrema. For a given value of H these two oscillations get in and out of step as X'' (and $1/\alpha$) increases away from zero and this accounts for the oscillating character of the left-hand part of fig. A5.3. An interesting feature is that the largest amplitude does not come for $X'' = 0$ (i.e. $1/\alpha = 0$) but for a small finite value of X'' such that the two interfering

Fig. A5.3. Variation of amplitude as the orientation passes through that for which $X'' = 0$ ($X \propto F$). The plotted quantity is $I = \alpha^{1/4}(C'^2 + S'^2)^{1/2}$ against $s\alpha^{-1/2}$, where s is the sign chosen in (A5.13) and α is as defined by (A5.14); $s\alpha^{-1/2}$ varies with orientation (approximately linearly over a small range) and passes through zero when $X'' = 0$. If X^{IV} were constant over the range, I would be exactly proportional to the dHvA amplitude (ST 1973).

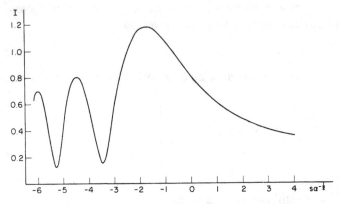

Fig. A5.4. Variation of phase ϕ as the orientation passes through that for which $X'' = 0$; ϕ is defined as $\tan^{-1}(S'/C')$ and is plotted against $s\alpha^{-1/2}$. The right-hand curve is plotted on an expanded scale of ϕ (ST 1973).

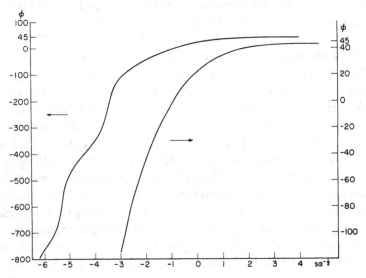

terms first have the same phase*. Once the signs of $\partial^2 X/\partial\kappa^2$ and $\partial^4 X/\partial\kappa^4$ have become the same, there is only a central extremum and the amplitude falls off smoothly as X'' increases (right-hand side of the graph). The resultant phase of the oscillations can be discussed in a similar way and it should be noted that for X'' exactly zero the $\pi/4$ phase of the standard formula is modified to $\pi/8$. Fig. A5.5 shows some experimental data for a crystal of Cu + 10 ppm Fe in which the variation with α is achieved by varying the orientation and it can be seen that the amplitude variation is qualitatively similar to that of fig. A5.3. The assumptions of the simple model used in the calculation are however too crude to give detailed agreement with experiment; a more detailed discussion is given in the published paper.

Fig. A5.5. Oscillations at $H = 50.5\,\text{kG}$ and $T = 1.2\,\text{K}$ of d^2M/dH^2 as the orientation of a crystal of Cu + 10 ppm Fe is varied through the critical orientation for which $X'' = 0$ (at about 23.3° from $\langle 100 \rangle$ in a (110) plane). This is essentially an expanded version of the region around D in fig. 5.8; the addition of a little Fe has killed the spin zero which in pure Cu introduces an extra complication. The oscillatory variation of the amplitude to the left of the critical point and the smooth variation to the right is qualitatively similar to the prediction of the idealized model shown in fig. A5.3 (Coleridge *et al.* 1972 quoted in ST 1973).

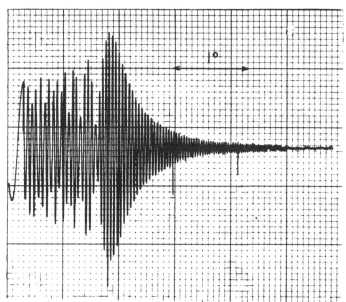

---

* This is in fact only approximately true, because the simple interpretation of two quite separate oscillatory terms (each of which can be evaluated independently by integrating to $v = \infty$) begins to break down when the frequencies become too close, and also because the amplitudes of the separate oscillatory terms vary with α in a way which pushes the maximum towards a lower value of X'' than the phase argument alone would indicate.

Appendix 6

Electron–phonon interaction

A6.1. Properties of the self energy function

The self energy $\Sigma(y)$, where $y = \varepsilon - \zeta$, is given by

$$\Sigma(y) \equiv \Delta(y) - i\Gamma(y)$$

$$= \underset{\eta=0}{\mathrm{Lt}} \int_0^\infty \alpha^2(v)F(v)\,dv \int_{-\infty}^\infty \left[\frac{1 - f(x) + g(v)}{y - x - v + i\eta} + \frac{f(x) + g(v)}{y - x + v + i\eta} \right] dx$$

$$(A6.1)$$

where $f(x)$ and $g(v)$ are the Fermi and Bose functions respectively:

$$f(x) = 1/(e^{x/kT} + 1), \qquad g(v) = 1/(e^{v/kT} - 1) \qquad (A6.2)$$

$\alpha(v)$ is a function characteristic of the electron–phonon interaction and $F(v)$ is the phonon spectrum; v is an energy given by $\hbar\omega$ if ω is the actual angular frequency.

The real and imaginary parts Δ and Γ as extracted from (A6.1) are given by

$$\Delta(y) = \int_0^\infty \alpha^2(v)F(v)\,dv\,P \int_{-\infty}^\infty \left[\frac{1 - f(x) + g(v)}{y - x - v} + \frac{f(x) + g(v)}{y - x + v} \right] dx$$

$$(A6.3)$$

where P denotes Cauchy principal value, and

$$\Gamma(y) = \pi \int_0^\infty \alpha^2(v)F(v)[2g(v) + f(v + y) + f(v - y)]\,dv \qquad (A6.4)$$

(the relevant terms from the inner integral of (A6.1) are effectively $-i\pi$ times $[1 - f(x) + g(v)]\delta(x - y + v)$ and $[f(x) + g(v)]\delta(x - y - v)$, and $1 - f(x) = f(-x)$). The functions Δ and Γ are related by what is essentially the Kramers–Kronig relation which can be expressed as

$$\Delta(y) = -\frac{P}{\pi} \int_0^\infty \frac{\Gamma(y + v) - \Gamma(y - v)}{v}\,dv \qquad (A6.5)$$

EP interaction

For purposes of computation it is simplest to evaluate Γ from (A6.4) (using data from superconducting tunnelling experiments to determine $\alpha^2 F$) and then to derive Δ from (A6.5). A graph of $\alpha^2 F$ for Hg is shown in fig. A6.1 and graphs of Δ and Γ are shown in fig. 2.16.

We shall also be concerned with the self energy for an imaginary argument (say iu). This is given by essentially the same formula as for a real argument, i.e. (A6.1) with y replaced by iu and with the $i\eta$ omitted. It is not difficult to show that $\Sigma(iu)$ is pure imaginary and so it is convenient to define $Z(iu)$ as $i\Sigma(iu)$. We then have

$$Z(u) \equiv i\Sigma(iu)$$

$$= \int_0^\infty \alpha^2(v)F(v)\,dv \int_{-\infty}^\infty \left[\frac{1 - f(x) + g(v)}{iu - x - v} + \frac{f(x) + g(v)}{iu - x + v}\right]dv \qquad \text{(A6.6)}$$

The value of the inner integral can be evaluated explicitly and the answer for the particular argument $u = (2l + 1)\pi kT$ which proves to be relevant will be given below (see (A6.17)). We can also express $Z(u)$ in terms of Γ by

Fig. A6.1. Plot of $\alpha^2(v)F(v)$ (which is a dimensionless characteristic of the electron–phonon interaction) against v, the phonon energy for Hg (Elliott *et al.* 1980). Note that $kT = 1\,\text{meV}$ for $T = 11.6\,\text{K}$.

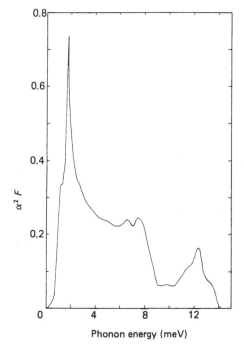

Phonon energy (meV)

means of the Kramers–Kronig relation or by some simple manipulations of (A6.8) (and using the relation $1 - f(x) = f(-x)$). We find

$$Z(u) = \frac{2u}{\pi} \int_0^\infty \frac{\Gamma(z)\,dz}{z^2 + u^2} \tag{A6.7}$$

Before embarking on a discussion of the integral in (2.184) it is useful to derive some simple limiting properties of Δ and Γ. It is convenient to define a 'characteristic' frequency v_0 such that both for $v \ll v_0$ and for $v \gg v_0$, $\alpha^2(v)F(v)$ is negligibly small; v_0 is evidently comparable to $k\Theta$ where Θ is the Debye temperature (about 60 K for Hg). As can be seen from fig. A6.1, v_0 for Hg could be taken as a few meV ($k\Theta$ is about 5 meV).

First we note from (A6.4) that

$$\Gamma(-y) = \Gamma(y) \tag{A6.8}$$

i.e. Γ is an even function of y, and (A6.5) then shows that

$$\Delta(-y) = -\Delta(y) \tag{A6.9}$$

i.e. Δ is an odd function of y, and in particular

$$\Delta(0) = 0$$

Next we calculate the slope of $\Delta(y)$ for $kT \ll v_0$ and $y \ll v_0$. We then have from (A6.5)

$$\Delta(y) = \frac{-2y}{\pi} \int_0^\infty \frac{\Gamma'(v)}{v}\,dv$$

and since $-f'(v)$ for $kT \ll v$ is practically a δ-function, differentiation of (A6.4) gives

$$\Gamma'(x) = \pi\alpha^2(x)F(x)$$

so that

$$\Delta(y) = -2y \int_0^\infty \frac{\alpha^2(v)F(v)}{v}\,dv \tag{A6.10}$$

This is exactly the linear form (2.185), with λ defined as

$$\lambda = 2 \int_0^\infty \frac{\alpha^2(v)F(v)}{v}\,dv \tag{A6.11}$$

and it is this value of λ which determines the mass enhancement factor $(1 + \lambda)$.

Simple limiting forms can also be given for $\Gamma(y)$ at both low and high T. For $kT \ll v_0$ and $y \ll v_0$ all the terms in the square brackets of (A6.4) are exponentially small so

$$\Gamma(y) \simeq 0 \tag{A6.12}$$

while for $kT \gg v_0$ and any y, $2g(v) = 2kT/v$, which is much larger than the other two terms in (A6.4), so we obtain

$$\Gamma(y) = 2\pi kT \int_0^\infty \frac{\alpha^2(v)F(v)}{v} \, dv = \pi kT\lambda \tag{A6.13}$$

Since $\Gamma(y)$ is in this limit almost independent of y, it follows from (A6.5) that

$$\Delta(y) \simeq 0 \tag{A6.14}$$

A6.2. The integral in (2.184)

For simplicity in writing, we shall omit the harmonic number p in what follows and carry out the analysis for $p = 1$; the result at any stage for higher values of p is immediately given by substituting H/p for H. The integral in (2.184) which determines the effect of electron–phonon interaction on the temperature reduction factor is then

$$I = \int_{-\infty}^\infty \exp\left[\frac{2\pi i}{\beta H}(y - \Delta(y) + i\Gamma(y))\right] dy/(e^{y/kT} + 1) \tag{A6.15}$$

From the limiting forms of Δ and Γ, we have already shown in the main text, without explicitly evaluating the integral, that in two limiting cases (a) $\beta H \ll v_0$, $kT \ll v_0$ and (b) $kT > v_0$, the result of EP interaction is in effect to enhance the cyclotron mass (i.e. to reduce β) by the factor $(1 + \lambda)$, with λ given by (A6.11). We shall now show that this simple result is still valid to a good approximation over a much wider range of H and T. We shall not present the mathematics of the integration (which can be achieved either by contour integration or by expressing the factor $\exp\{(-2\pi i/\beta H)(\Delta(y) - i\Gamma(y))\}$ as an infinite Taylor expansion in powers of y and then integrating term by term) but merely quote the answer. This is

$$I = \frac{2\pi kT}{i} \sum_{l=0}^\infty \exp\left\{\frac{-2\pi}{\beta H}[\pi kT(2l + 1) + Z(\pi kT(2l + 1), T)]\right\} \tag{A6.16}$$

It should be stressed that the quantity in square brackets is entirely real and also that $Z(\pi kT(2l + 1), T)$ is a function of T in two ways, both because its argument is $\pi kT(2l + 1)$ and because the form of the function $Z(y)$ depends on T through the Fermi and Bose functions; the rather awkward notation $Z(\pi kT(2l + 1), T)$ is intended to draw attention to this double dependence.

Appendix 6

The value of $Z(\pi kT(2l + 1), T)$ can be explicitly expressed by evaluating the internal integral in (A6.6) and again we merely state the answer

$$Z(\pi kT(2l + 1), T) = 2\pi kT \int_0^\infty \frac{dv\alpha^2(v)F(v)}{v}$$

$$\times \left\{1 + 2\sum_{q=1}^{l}\left[1 + \left(\frac{2\pi qkT}{v}\right)^2\right]^{-1}\right\} \quad \text{(A6.17)}$$

Here, of course, both kinds of T dependence have been taken into account.

We can now see more clearly how far the limiting cases (a) and (b) above can be extended to meet each other without causing serious deviation from the simple mass enhancement formulation. First of all we note that if for all relevant values of l the curly bracket in (A6.17) is approximately $(2l + 1)$ (i.e. effectively if $(2\pi lkT/v_0)^2 \ll 1$), then (A6.17) reduces to

$$Z(\pi kT(2l + 1), T) = \pi kT\lambda(2l + 1) \quad \text{(A6.18)}$$

(see (A6.11)), and (A6.16) becomes to a good approximation

$$I = \frac{2\pi kT}{i}\sum_{l=0}^{\infty}\exp[-2\pi^2 kT(2l + 1)(1 + \lambda)/\beta H]$$

$$= \frac{2\pi kT}{i}\frac{\exp[-2\pi^2 kT(1 + \lambda)/\beta H]}{(1 - \exp[-4\pi^2 kT(1 + \lambda)/\beta H])}$$

$$= \frac{\pi kT}{i\sinh\left[\dfrac{2\pi^2 kT(1 + \lambda)}{\beta H}\right]} \quad \text{(A6.19)}$$

This is exactly the standard LK result with the mass enhanced by $(1 + \lambda)$.

At first sight it might seem that this situation would occur if $kT \ll v_0$ but in fact the condition $\beta H < v_0$ is more essential. This can be appreciated by noting that as l rises the damping factor $\exp[-2\pi^2 kT(2l + 1)/\beta H]$ in (A6.16) becomes more and more severe, so that the highest l that contributes anything appreciable to the series is given by something like

$$4\pi^2 kT(2l + 1) \sim 3\beta H \quad \text{or} \quad 2\pi kT \sim \beta H/2(2l + 1)$$

For this value of l, the curly bracket in (A6.17) becomes

$$1 + \frac{2}{1 + \left[\dfrac{\beta H}{(4l + 2)v}\right]^2} + \frac{2}{1 + \left[\dfrac{2\beta H}{(4l + 2)v}\right]^2} \cdots + \frac{2}{1 + \left[\dfrac{l\beta H}{(4l + 2)v}\right]^2} \quad \text{(A6.20)}$$

and if $\beta H < v_0$, the sum of (A6.20) is indeed close to $2l + 1$ for all the v

504

that matter and the simple mass enhancement holds. To make the last term in (A6.20) differ from 2 by less than say 10%, we must have roughly $\beta H/4v_0 < 0.3$, so that in fact $\beta H \gtrsim v_0$ is already enough to guarantee negligible departure from the simple mass enhancement. If we take v_0 as say 5 meV, this means (with $m/m_0 = 0.06$, i.e. *before* enhancement) $H <$ say 25 kG.

As the temperature rises, the conditions on H become less stringent as can be seen from the limit of large T. If in fact $4\pi^2 kT/\beta H > 3$ only the first term ($l = 0$) in (A6.16) is appreciable (the next would be only 1/20 as large) and thus only Z for $l = 0$ (which is $\pi kT\lambda$) need be considered in (A6.16) giving the result

$$I = \frac{2\pi kT}{i} \exp[-2\pi^2 kT(1 + \lambda)/\beta H] \qquad (A6.21)$$

which is an excellent approximation to the general result (A6.19) since here $4\pi^2 kT/\beta H > 3$. This high temperature limit is such that for $T/H(\text{kG}) > 0.18$ we should expect the departure from the amplitude corresponding to simple mass enhancement to be $< 5\%$.

A helpful way of presenting the results of these calculations is to plot contours of given percentage deviation on an (H, T) chart. Such a chart is shown in fig. A6.2 with contours computed by Elliott *et al.* (1980) on the

Fig. A6.2. Field and temperature dependence for Hg of deviations from the LK formula due to electron–phonon interaction. The curves are contours of constant u, where e^u is the ratio of the many-body theory amplitude to that predicted by the LK formula (assuming a phonon-enhanced mass). Since u is small it is, to a good approximation, the same as the fractional increase of amplitude.

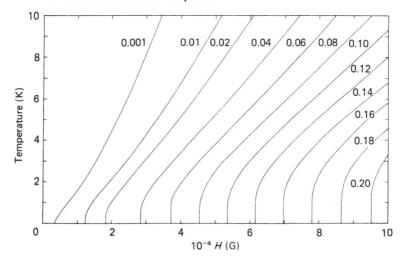

Appendix 6

basis of the $\alpha^2 F$ spectrum of fig. A6.1, and it can be seen that our limiting considerations fit in quite reasonably. The main moral of this chart is in emphasising that, even for the most extreme conditions, the deviations from the conventional formula with enhanced mass amount to only 10 or 20% in amplitude. Such a deviation is not very striking on a logarithmic plot and, as is discussed further in chapter 8, considerable precision is required to obtain convincing evidence of it.

Finally we should point out that any deviation is always an *increase* of amplitude. This can be seen from the form of (A6.16) and (A6.20). Thus if H is 'large' rather than 'small' some of the terms in (A6.20) will be less than for small H and consequently the contribution of Z in (A6.16) will be *less* than $(2l + 1)\pi k T$ for given l, which implies a *larger* amplitude. The extreme case is if *all* the terms in (A6.20), except the first, become negligible, so that Z is simply $\lambda\pi kT$ rather than $(2l + 1)\lambda\pi kT$; (A6.16) then reduces to

$$I = \frac{\pi kT}{i} \frac{\exp(-2\pi^2 \lambda kT/\beta H)}{\sinh(2\pi^2 kT/\beta H)} \tag{A6.22}$$

The amplitude is therefore larger than for full mass enhancement by a factor

$$\exp(-2\pi^2 \lambda kT/\beta H)\frac{\sinh[2\pi^2 kT(1 + \lambda)/\beta H]}{\sinh(2\pi^2 kT/\beta H)} \tag{A6.23}$$

This is the largest possible deviation: it is greatest for $2\pi^2 kT(1 + \lambda)/\beta H \ll 1$, when the factor is $(1 + \lambda)$, but for $2\pi^2 kT/\beta H \gtrsim 1$ it rapidly approaches 1 (it is 1.15 for $2\pi^2 kT/\beta H = 1$ and $\lambda = 1.6$).

Appendix 7

Numerical estimates of $|M/H|$ and $|dM/dH|$

A7.1. Estimates for a 3-D system

The amplitudes of both (M/H) and dM/dH (see (3.1)–(3.3)) show maxima as functions of H at a field H_0 given by the implicit equation

$$\frac{\alpha x}{H_0} + \frac{\alpha T}{H_0}\coth\frac{\alpha T}{H_0} = n \tag{A7.1}$$

where $n = 3/2$ or $5/2$ for $|M/H|$ and $|dM/dH|$ respectively, and it is convenient to note here the values of H_0 and the corresponding maximum amplitudes for two fairly extreme situations: $x = 0$ (a 'perfect' sample) and $x = T = 1$ K (typical of a 'poor' sample). We find

$\underline{x = 0}$

$$\alpha T/H_0 = 1.288 \quad (n = 3/2) \quad \text{or} \quad 2.464 \quad (n = 5/2) \tag{A7.2}$$

$$|M/H|_{max} = 4.03 \times 10^{-14}(2\pi/A'')^{1/2}GFT^{-1/2}(m/m_0)^{-3/2} \tag{A7.3}$$
$$(H_0 = 1.14T(m/m_0) \times 10^5\,\text{G})$$

$$|dM/dH|_{max} = 3.23 \times 10^{-18}(2\pi/A'')^{1/2}GFT^{-3/2}(m/m_0)^{-5/2} \tag{A7.4}$$
$$(H_0 = 5.96T(m/m_0) \times 10^4\,\text{G})$$

$\underline{x = T = 1\,\text{K}}$

$$\alpha/H_0 = 0.437\,(n = 3/2) \quad \text{or} \quad 1.116\,(n = 5/2) \tag{A7.5}$$

$$|M/H|_{max} = 1.91 \times 10^{-14}(2\pi/A'')^{1/2}GF(m/m_0)^{3/2} \tag{A7.6}$$
$$(H_0 = 3.36\,(m/m_0) \times 10^5\text{G})$$

$$|dM/dH|_{max} = 6.25 \times 10^{-19}(2\pi A'')^{1/2}GF^2(m/m_0)^{-5/2} \tag{A7.7}$$
$$(H_0 = 1.32\,(m/m_0) \times 10^5\text{G})$$

Appendix 7

Equations (A7.3)–(A7.7) provide a useful guide to the highest susceptibility amplitudes that can be achieved as H is varied but they should be used with care for two reasons. First the optimum field H_0 may prove larger than is actually available or else it may come close to or beyond the quantum limit $H = F$. The highest amplitude is then given by substituting in (3.1) or (3.2) the highest available field or a field given by say $F/3$ (whichever is lower). Secondly the observed output of the experiment may not be simply proportional to M/H or dM/dH, but rather to H or H^2 times the susceptibility according to the particular experimental method used. This means that the condition for maximum output signal may be different from that of maximum susceptibility, or even that there is no maximum condition, but that the largest signal occurs at the largest field that can be used. It is therefore helpful to calculate not only the maximum amplitudes given by (A7.3)–(A7.7) for some typical metals, but also to calculate the amplitudes given by (3.1) and (3.2) for one or two fixed fields.

The results of such calculations are shown in table A7.1 for two fairly extreme examples of large and small Fermi surfaces, namely (a) a hypothetical free electron-like metal with electron density similar to that of the noble metals, and (b) bismuth with the field oriented along a binary axis. For (a) one of the chosen fixed fields is $H = 10^5$ G, which is as high as is available in most laboratories, and the other is $H = 2 \times 10^4$ G, which is typical of the highest field available in an ordinary iron electromagnet. For (b) the fixed field is chosen as 5×10^3 G, which is roughly one third of F and about the highest field for which the formulae are still valid. The effect of temperature is illustrated by some results at 5 K for (a) and at 5 K and 20 K for (b). Since, for the binary orientation in bismuth, two of the three ellipsoids present identical extremal areas, all numerical values are double those indicated by the formulae (the third ellipsoid has a much higher frequency and may be ignored here). The figures in table A7.1 all refer to $M_{\parallel}$, and the corresponding figures for $M_{\perp}/H$ or $dM_{\perp}/dH$ are given by (2.114), i.e. by multiplication of the entries in table A7.1 by appropriate values of $(1/F)(dF/d\theta)$. The orders of magnitude of this anisotropy factor are indicated in table A7.2 for some typical situations, though it should be emphasized that the factor varies considerably with orientation and in detail from metal to metal.

A7.2. Estimates for a 2-D system

To illustrate orders of magnitude we consider specifically a silicon inversion layer, which as discussed in §2.3.4 should behave like a 2-D metal with a magnetic moment given by (2.61) or (2.80) at $T = 0$ with no scattering and no spin effects. The spin effects are probably complicated in detail as a result

508

Table A7.1. *Susceptibility amplitudes*

(a) free electron metal: $F = 5 \times 10^8 \text{G}, (2\pi/A'')^{1/2} = 1, |G| = 1, (m/m_0) = 1$

$T(\text{K})$	$x(\text{K})$	$10^{-5} \times H(\text{G})$	$10^5 \times |M/H|$	$|dM/dH|$
1	0	1.1 for $|M/H|_{\max}$, 0.6 for $|dM/dH|_{\max}$ 2.0		0.80
		1	1.8	0.63
		0.2	0.06	0.09
1	1	3.4 for $|M/H|_{\max}$, 1.3 for $|dM/dH|_{\max}$ 1.0		0.16
		1	0.4	0.14
		0.2	4×10^{-5}	6×10^{-5}
5	0	1	0.03	0.008
5	1	1	0.006	0.002

(b) bismuth, H along binary axis: $F = 1.4 \times 10^4 \text{G}, (2\pi/A'')^{1/2} = 11, |G| = 1,$ $(m/m_0) = 10^{-2}$

$T(\text{K})$	$x(\text{K})$	$10^{-3} \times H(\text{G})$	$10^5 \times |M/H|$	$10^4 \times |dM/dH|$
1	0	1.1 for $|M/H|_{\max}$, 0.6 for $|dM/dH|_{\max}$ 1.2		14
		5	0.8	1.3
1	1	3.4 for $|M/H|_{\max}$, 1.3 for $|dM/dH|_{\max}$ 0.6		2.7
		5	0.6	1.0
5	0	5	0.6	1.0
20	0	5	0.03	0.04

Note: Since the purpose of the table is mainly to give an idea of orders of magnitude, the entries are given to only one or two significant figures. M denotes oscillatory magnetic moment per unit volume.

Table A7.2. *Orders of magnitude of* $(1/F)(dF/d\theta)$

Metal	Typical value of $(1/F)(dF/d\theta)$
Hypothetical with spherical FS	0
Alkalis	10^{-3}(for Na and K) to 5×10^{-2}(Cs)
Noble metals	5×10^{-2}
Bismuth	1
Small pieces of Fermi surface in polyvalent metals	1

of field dependent many-body EE interactions, but we shall simply ignore them since they are unlikely to alter the order of magnitude of the oscillation amplitude. We also ignore effects associated with the multi-valley band structure degeneracy of Si. The temperature and scattering reduction factors should be similar to those for a 3-D metal, so we find for the fundamental of the oscillatory moment per unit area which we denote by $\mathcal{M}$ to avoid confusion with magnetic moment M per unit volume

$$\mathcal{M} = R_T R_D (n\beta/\pi) \sin(2\pi F/H) \tag{A7.8}$$

with

$$F = \frac{\pi c\hbar}{e} n = 2.06 \times 10^{-7} n,$$

$$R_T = \frac{2\pi^2 kT/\beta H}{\sinh(2\pi kT/\beta H)},$$

$$R_D = \exp\left(-\frac{\pi mc}{eH\tau}\right) \tag{A7.9}$$

Here n is the number of electrons per unit area of the sample and F is expressed in G. The value of n usually lies in the range of 10^{12}–$10^{13}\,\mathrm{cm}^{-2}$, so

$$F = 2 \times 10^5 \text{ to } 2 \times 10^6 \text{ G.}$$

The mass is typically given by $m/m_0 = 0.2$, i.e. $\beta \sim 10^{-19}$; the relaxation time τ varies with the carrier concentration n; it has a maximum of about 2.3×10^{-12} s for $n = 10^{12}\,\mathrm{cm}^{-2}$ and falls to about 2.3×10^{-3} s for $n = 10^{13}\,\mathrm{cm}^{-2}$.

With these numbers we find the values of $|\mathcal{M}|$ shown in table A7.3. It can

Table A7.3. *Values of $|\mathcal{M}|$ and $|d\mathcal{M}/dH|$ for 2-D sample*

H (kG)	R_T	R_D	$|\mathcal{M}| \times 10^8$ (G cm)	$|d\mathcal{M}/dH| \times 10^{11}$ (cm)
30	0.8	$(10^{12})0.6$ $(10^{13})0.005$	1.4 0.1	2.0 1.5
50	0.92	$(10^{12})0.7$ $(10^{13})0.04$	1.9 1.1	1.0 5.6
80	0.97	$(10^{12})0.8$ $(10^{13})0.14$	2.3 4.0	0.5 8.1

Note: The two values given for each field are for $n = 10^{12}$ and $10^{13}\,\mathrm{cm}^{-2}$ respectively. R_T is for $T = 1.2$ K. $\mathcal{M}$ is the dipole moment per unit area.

be seen that at best $|\mathcal{M}|$ is only a few times $10^{-8}\,\mathrm{G\,cm}$ and this is not very much above the lower limit of what can be detected by present techniques, even exploiting the possibility of modulating n by means of the gate voltage and using the special trick of fabricating a sample of large perimeter. Note that the sample area S, which can be made as large as $0.1\,\mathrm{cm}^2$ is not directly relevant in determining the optimum signal conditions.

Appendix 8

Calibration of field modulation system for finite sample

As explained in the text, in order to calibrate the system so that voltage signals can be translated into sample magnetization, we need to know the parameters η_1 and η_2 which specify the field per unit current at the sample position in the pick-up and modulation coils respectively. Effectively η_1 determines the coupling constant c in (3.27) while η_2 determines the modulation amplitude h_0 (and hence λ) if the current amplitude i_0 is known, so that the amplitude A of a simple harmonic dHvA oscillation can be determined from measurement of the amplitude $|v_k|$ of the kth harmonic in the voltage signal if η_1 and η_2 are known. In Knecht's method η_1 and η_2 are determined by using each coil in turn as the modulation coil and measuring the current amplitude i_0 required to make the signal from the other coil vanish (Bessel zero). If $k\omega$ detection is used and λ_0 is such that $J_k(\lambda_0) = 0$, then, as in (3.37)

$$\eta_1 \quad \text{or} \quad \eta_2 = \lambda_0 H^2 / 2\pi F i_0 \qquad (A8.1)$$

For simplicity we shall suppose in what follows that $k = 1$ (for which $\lambda_0 = 3.83$) but the analysis can be generalized to any k.

The purpose of this appendix is to discuss how the analysis is modified if the specimen cannot be regarded as 'small', i.e. if η_1 and η_2 vary appreciably over the sample. If we use the notation $\eta_1(P)$ and $\eta_2(P)$ to denote the values at some point P in the sample, it is evident from (3.27) that the sin ωt voltage amplitude in coil 1 for current i_2 in coil 2 will be:

$$v_1 = -2A\omega \int \eta_1(P) J_1 \left[\frac{2\pi F}{H^2} \eta_2(P) i_2 \right] d\tau \qquad (A8.2)$$

and if the coils are interchanged it will be

$$v_2 = -2A\omega \int \eta_2(P) J_1 \left[\frac{2\pi F}{H^2} \eta_1(P) i_1 \right] d\tau \qquad (A8.3)$$

where $d\tau$ is a volume element of the sample and the integration is to be taken over the sample. If v_1 is made to vanish for current i_{02} in coil 2 and the sample is not *too* big, J_1 must be small all over the sample so that its value at any point P is given to an adequate approximation by

$$J_1\left[\frac{2\pi F}{H^2}\eta_2(P)i_{02}\right] = \left(\frac{\mathrm{d}J_1(\lambda)}{\mathrm{d}\lambda}\right)_{\lambda=\lambda_0}\left[\frac{2\pi F}{H^2}\eta_2(P)i_{02} - \lambda_0\right] \quad (A8.4)$$

and similarly for the coils interchanged. Thus v_1 will vanish if

$$\frac{2\pi F}{H^2}i_{02}\int \eta_1(P)\eta_2(P)\,\mathrm{d}\tau = \lambda_0\int \eta_1(P)\,\mathrm{d}\tau \quad (A8.5)$$

and similarly v_2 will vanish for current i_{01} in coil 1, when

$$\frac{2\pi F}{H^2}i_{01}\int \eta_1(P)\eta_2(P)\,\mathrm{d}\tau = \lambda_0\int \eta_2(P)\,\mathrm{d}\tau \quad (A8.6)$$

In normal use we would measure v_1 for some current i_2 and wish to deduce the dHvA amplitude A absolutely from (A8.2). We consider two cases: (a) *small* modulation i_2, for which (A8.2) becomes

$$v_1 = -\frac{2\pi F}{H^2}i_2\omega A\int \eta_1(P)\eta_2(P)\,\mathrm{d}\tau \quad (A8.7)$$

and (b) modulation i_2 to give the Bessel maximum. At the maximum the relatively small variations of η_2 over the sample will produce only second order effects and to sufficient approximation $J_1(2\pi F/H^2)\eta_2(P)i_2$ can be replaced by $J_{1\,\text{max}}$, i.e.

$$v_1 = -2A\omega J_{1\,\text{max}}\int \eta_1(P)\,\mathrm{d}\tau \quad (A8.8)$$

Thus calibration is achieved for (a) if we know the value of $I = \int \eta_1(P)\eta_2(P)\,\mathrm{d}\tau$, and for (b) if we know the value of $I_1 = \int \eta_2(P)\,\mathrm{d}\tau$; we shall also use the notation $I_2 = \int \eta_2(P)\,\mathrm{d}\tau$.

From (A8.5) and (A8.6) we see that $1/i_{01}$ and $1/i_{02}$ give I/I_2 and I/I_1 respectively, i.e.

$$K_2 \equiv \left(\frac{\lambda_0 H^2}{2\pi F}\right)\frac{1}{i_{02}} = \frac{I}{I_1} \quad \text{and} \quad K_1 \equiv \left(\frac{\lambda_0 H^2}{2\pi F}\right)\frac{1}{i_{02}} = \frac{I}{I_2} \quad (A8.9)$$

and we note that if the sample reduces to a point $K_2 = I/I_1 = \eta_2$ and $K_1 = I/I_2 = \eta_1$ so that the equations (A8.9) assume the form of (3.38).

If η_{10} and η_{20} are appropriate average values of $\eta_1(P)$ and $\eta_2(P)$, we can write

$$\eta_1(P) = \eta_{10}[1 + \delta_1(P)], \qquad \eta_2(P) = \eta_{20}[1 + \delta_2(P)] \quad (A8.10)$$

Provided $\delta_1(P)$ and $\delta_2(P)$ do not exceed say 0.1, which allows a sample size of dimensions comparable to the smaller or the two coil sizes, we shall now show that I_1, I_2 and I are given by K_1V, K_2V and K_1K_2V to an accuracy of better than 1% (V is the sample volume). These are exactly the values for a 'point' sample (for which $\delta_1(P) = \delta_2(P) = 0$), so no appreciable correction is necessary to the calibration procedure even though η_1 and η_2 may vary by 10% over the sample.

Let

$$\left.\int \delta_1(P)\,d\tau = D_1, \qquad \int \delta_2(P)\,d\tau = D_2\right\}$$

and

$$\left.\int \delta_1(P)\delta_2(P)\,d\tau = D_{12}\right.$$

$$\text{(A8.11)}$$

then

$$\left. I_1 = \eta_{10}(V + D_1), \qquad I_2 = \eta_{20}(V + D_2)\right\}$$

and

$$I = \eta_{10}\eta_{20}(V + D_1 + D_2 + D_{12})$$

$$\text{(A8.12)}$$

Since we are assuming $\delta_1(P)$ and $\delta_2(P) < 0.1$

$$D_1/V < 0.1, \quad D_2/V < 0.1 \quad \text{and} \quad D_{12}/V < 0.01$$

and, indeed, usually $<$ can be replaced by $\ll$.

Thus to an accuracy of better than 1% (and usually much better)

$$K_1 \equiv \frac{I}{I_2} = \eta_{10}\left(1 + \frac{D_1}{V} + \frac{D_{12}}{V} - \frac{D_2^2}{V^2}\right) = \frac{I_1}{V} \qquad \text{(A8.13)}$$

i.e. $\qquad I_1 = K_1V, \quad$ and similarly $\quad I_2 = K_2V, \; I = K_1K_2V \qquad \text{(A8.14)}$

which proves our proposition that measurement of i_{01} and i_{02} determines the calibration, exactly as for a point sample. For any particular geometry it is of course always possible to estimate the very small correction terms arising from D_{12}/V, D_1^2/V^2 and D_2^2/V^2. The reason why a relatively large sample can be treated as if it were a point is, of course, that the modifications of coupling between sample and coils due to finite sample size affect both the calibration procedure and the signal in actual use in almost equal fashions.

Appendix 9

Magnitude of magnetothermal oscillations

We shall consider only the fundamental in the oscillations (i.e. $p = 1$) and write (4.7) as

$$|\widetilde{\Delta T}| = -zf'(z)\frac{H^2}{2\pi Fc}|\widetilde{M}_0| \tag{A9.1}$$

The temperature variation of c is given by

$$c = aT + bT^3 \tag{A9.2}$$

but usually only the aT term is important in the relevant range of temperature, so that substituting $z = 1.47 \times 10^5 (T/H)(m/m_0)$, (A9.1) becomes

$$|\widetilde{\Delta T}| = -1.47 \times 10^5 \frac{f'(z)H(m/m_0)}{2\pi Fa}|\widetilde{M}_0| \tag{A9.3}$$

The highest value of $H|\widetilde{M}_0|$ comes at the highest available H, say H_0, and for this highest H, $|\widetilde{\Delta T}|$ will be largest for the value of z which gives a maximum value of $|f'(z)|$. This value of z is 1.6, (giving a maximum value 0.3 of $|f'(z)|$) and therefore the optimum working temperature T_0 is given by

$$T_0 = 1.6\,H_0 \times 10^{-5}/1.47(m/m_0) \tag{A9.4}$$

Thus if $H_0 \sim 10^5$ G and $m/m_0 \sim 1.0$ (as for a typical large FS, such as that of a noble metal)

$$T_0 \sim 1\,\text{K}$$

Now we know (see table A7.1) that in these conditions $|\widetilde{M}_0| \sim 1$ G, so substituting $F = 5 \times 10^8$ G and $a = 10^3\,\text{erg cm}^{-3}\,\text{K}^{-2}$ (for Cu) into (A9.3) we find for a typical 'large' FS

$$|\widetilde{\Delta T}| \sim 10^{-3}\,\text{K} \tag{A9.5}$$

as the order of magnitude of the largest attainable amplitude. In practice some of our assumptions are probably rather too optimistic (especially if $\widetilde{M}$ is limited by magnetic interaction) and 10^{-4} K may be a more realistic estimate of what can be achieved in reasonably practical conditions.

Appendix 9

An exception to the assumption that the specific heat is predominantly electronic is provided by bismuth, in which it is the lattice term bT^3 which dominates down to about 0.1 K. If we again express $|\widetilde{\Delta T}|$ as the product of a function of z only and one of H only we find

$$|\widetilde{\Delta T}| = -\left(\frac{2\pi^2 k}{\beta}\right)^3 \frac{f'(z)}{2\pi Fbz^2} \frac{|\tilde{M}_0|}{H} \tag{A9.6}$$

Since the function $|f'(z)/z^2|$ increases without limit as z decreases, while $|M_0|/H$ is insensitive to H around its highest value (of order 10^{-5} for Bi), it is advantageous to reduce T and increase H as much as possible. However (A9.6) is valid only down to about 0.1 K, where the small electronic term aT in c begins to be important, and H cannot be increased to more than about 5×10^3 G because of the proximity of the quantum limit, so the lowest value of z is roughly 0.03 (assuming $m/m_0 = 10^{-2}$). The corresponding value of $|f'(z)/z^2|$ is about 10. If we now substitute $b = 9 \times 10^2 \, \mathrm{erg\,cm}^{-3}\,\mathrm{K}^{-4}$, $F = 1.4 \times 10^4$ G and $(2\pi^2 k/\beta) = 1.5 \times 10^3$, we find for $T = 0.1$ K and $H = 5 \times 10^3$ G

$$|\Delta T| \sim 4 \times 10^{-3} \, \mathrm{K} \tag{A9.7}$$

At the more convenient working temperature of 1 K, $|\widetilde{\Delta T}|$ would be ten times smaller.

We shall now estimate roughly how the magnetothermal oscillations compare with the dHvA effect, if regarded as a technique for detecting very weak oscillations in M. We suppose that the oscillations are weak because of a poor Dingle factor, but that H and T are chosen to give as large as possible an amplitude of T. The minimum detectable amplitudes are related by (A9.1), i.e.

$$|\tilde{M}|_{\mathrm{min}} = -\frac{2\pi Fc f(z)}{H^2 z f'(z)}|\widetilde{\Delta T}|_{\mathrm{min}} \tag{A9.8}$$

If (perhaps optimistically) we suppose $|\widetilde{\Delta T}|_{\mathrm{min}} \sim 10^{-8}$ K, then for the typical large FS metal we have $F = 5 \times 10^8$ G, $m/m_0 = 1$, $T = 1$ K, $H = 10^5$ G, $c \sim 10^3 \, \mathrm{erg\,cm}^{-3}\,\mathrm{K}^{-1}$ and $z = 1.6$, and we find

$$|\tilde{M}|_{\mathrm{min}} \sim 4 \times 10^{-6} \, \mathrm{G}$$

which compares rather poorly with the minimum detectable value of $|\tilde{M}|$ by dHvA methods (say $\sim 10^{-9}$ G). The comparison would be even less favourable at lower H and/or higher T.

For Bi, the magnetothermal oscillations show up a little better. Putting $F = 1.4 \times 10^4$ G, $m/m_0 = 10^{-2}$, $T = 0.1$ K, $H = 5 \times 10^3$ G and

516

$c \sim 2 \, \mathrm{erg \, cm^{-3} \, K^{-1}}$, we find

$$|\tilde{M}|_{\min} \sim 2 \times 10^{-7} \, \mathrm{G}$$

which however is still a good deal poorer limit of detection than can be achieved by the dHvA effect. With the more convenient temperature of $T = 1 \, \mathrm{K}$, $c \sim 10^{3} \, \mathrm{erg \, cm^{-3} \, K^{-1}}$ and we find $|\tilde{M}|_{\min} \sim 10^{-6} \, \mathrm{G}$.

Appendix 10
Oscillations of velocity of sound

As shown in standard text books (e.g. Pollard 1977), the velocity of sound, v, in a solid of density ρ, for an arbitrary direction of propagation having direction cosines (n_1, n_2, n_3) is given by the equation (cubic in v^2)

$$\begin{vmatrix} (\lambda_{11} - \rho v^2) & \lambda_{12} & \lambda_{13} \\ \lambda_{12} & (\lambda_{22} - \rho v^2) & \lambda_{23} \\ \lambda_{13} & \lambda_{23} & (\lambda_{33} - \rho v^2) \end{vmatrix} = 0 \qquad \text{(A10.1)}$$

where

$$\lambda_{im} = c_{iklm} n_k n_l \qquad \text{(A10.2)}$$

The ratios of the direction cosines $(\alpha_1, \alpha_2, \alpha_3)$ of the displacement vector are given (for each value of ρv^2) by the equations

$$\begin{aligned} \alpha_1(\lambda_{11} - \rho v^2) + \alpha_2 \lambda_{12} + \alpha_3 \lambda_{13} &= 0 \\ \alpha_1 \lambda_{12} + \alpha_2(\lambda_{22} - \rho v^2) + \alpha_3 \lambda_{33} &= 0 \\ \alpha_1 \lambda_{13} + \alpha_2 \lambda_{23} + \alpha_3(\lambda_{33} - \rho v^2) &= 0 \end{aligned} \qquad \text{(A10.3)}$$

and it can be shown that although the three possible displacement vectors are mutually perpendicular, they are not in general either parallel or perpendicular to the propagation direction (n_1, n_2, n_3). Thus for an arbitrary direction of propagation the three possible modes are neither longitudinal nor transverse. However, for special directions of propagation, such as directions of high symmetry, one of the three solutions coincides with (n_1, n_2, n_3) so that exactly longitudinal and transverse waves become possible (for a detailed discussion see Borgnis 1955). Since for the purpose of extracting strain derivatives it is usually sufficient to study propagation only along such special drections, we shall confine ourselves to the discussion of strictly longitudinal and transverse waves.

For this special case there is a rather convenient short cut to obtaining the sound velocities. This is to choose as axes 1, 2, 3 the direction of propagation of the sound and the two directions of transverse displacement, rather than the crystallographic axes which are conventionally used.

518

Oscillations of velocity of sound

The three modes of propagation will then have velocities given by

$$\rho v^2 = c'_{11} \quad \text{(longitudinal)}$$

$$\left.\begin{array}{l} \rho v^2 = c'_{55} \\ \rho v^2 = c'_{66} \end{array}\right\} \text{(transverse)} \tag{A10.4}$$

where c'_{ab} refers to our special axes. All that remains to be done now is to express c'_{ab} in terms of the conventional c_{ab}, referred to the crystallographic axes. This is easily done by standard matrix multiplication. If the direction cosines of the direction of propagation and the two transverse directions with respect to the crystallographic axes are (n_1^1, n_1^2, n_1^3), (n_2^1, n_2^2, n_2^3) and (n_3^1, n_3^2, n_3^3) respectively, we have that

$$c'_{ijkl} = n_i^p n_j^q n_k^r n_l^s c_{pqrs} \tag{A10.5}$$

(note that we must use the four suffix notation for the c tensor in order that the strict summation convention should be valid).

Since the special directions we wish to consider are usually ones of high symmetry, (A10.5) simplifies a great deal in practice. Thus with a cubic crystal, for propagation along [100] with transverse vibrations along [010] and [001], we obtain the obvious result

$$c'_{11} = c_{11}, \quad c'_{55} = c_{55} \quad \text{and} \quad c'_{66} = c_{66} \tag{A10.6}$$

For propagation along [110] and transverse vibration along [$\bar{1}$10] and [1$\bar{1}$0] (note that the direction cosines are $1/\sqrt{2}$ times these indices), we find

$$\left.\begin{array}{l} c'_{11} = \tfrac{1}{4}(c_{11} + c_{22} + 2c_{12} + 4c_{16} + 4c_{26} + 4c_{66}) \\ c'_{55} = \tfrac{1}{2}(c_{55} + c_{44} + 2c_{45}) \\ c'_{66} = \tfrac{1}{4}(c_{11} + c_{22} - 2c_{12}) \end{array}\right\} \tag{A10.7}$$

It will be noticed that, although we are considering a cubic crystal, moduli such as c_{16}, c_{26}, c_{45} appear which normally vanish, and distinction is made between the normally equal c_{11} and c_{22} and between the normally equal c_{44}, c_{55} and c_{66}. This is a rather subtle point: these features appear because the exact cubic symmetry has been destroyed by the presence of a magnetic field in an arbitrary direction*. The normally vanishing moduli and the

* At first sight it might appear that the lack of strict cubic symmetry invalidates our approach, since waves propagated along what were high symmetry directions in the absence of a field would no longer be strictly longitudinal or transverse. However since the distortions from cubic symmetry are very slight ($\sim 10^{-4}$) the displacement vectors differ from their assumed symmetry directions by angles sufficiently small ($\sim 10^{-4}$ radians) not to matter. The rigorous determinantal approach produces a complicated cubic equation, which however gives answers identical to those of our simple approach, if it is solved by treating the small oscillatory contributions to the moduli as perturbations of the solution in the absence of a field.

519

differences between normally equal moduli are the oscillatory components which are minutely small compared with the main moduli. Thus if we use our standard notation $\tilde{c}$ to denote only the oscillatory parts, (A10.6) and (A10.7) may be more meaningfully expressed as

$$c'_{11} = c_{11} + \tilde{c}_{11}, \quad c'_{55} = c_{44} + \tilde{c}_{55}, \quad c'_{66} = c_{44} + \tilde{c}_{66} \qquad \text{(A10.8)}$$

and

$$
\left.
\begin{aligned}
c'_{11} &= \tfrac{1}{2}(c_{11} + c_{12} + 2c_{44}) + \tfrac{1}{4}(\tilde{c}_{11} + \tilde{c}_{22} \\
&\quad + 2\tilde{c}_{12} + 4\tilde{c}_{16} + 4\tilde{c}_{26} + 4\tilde{c}_{66}) \\
c'_{55} &= c_{44} + \tfrac{1}{2}(\tilde{c}_{55} + \tilde{c}_{44} + 2\tilde{c}_{45}) \\
c'_{66} &= \tfrac{1}{2}(c_{11} - c_{12}) + \tfrac{1}{4}(\tilde{c}_{11} + \tilde{c}_{22} - 2\tilde{c}_{12})
\end{aligned}
\right\} \qquad \text{(A10.9)}
$$

For completeness we give also the solution for propagation along [111] with longitudinal displacement vectors along [111] and transverse ones along [01$\bar{1}$] and [$\bar{2}$11]. We find after some tedious enumeration of terms:

$$
\left.
\begin{aligned}
c'_{11} &= \tfrac{1}{3}(c_{11} + 2c_{12} + 4c_{44}) + \tfrac{1}{9}[(\tilde{c}_{11} + \tilde{c}_{22} + \tilde{c}_{33}) \\
&\quad + 2(\tilde{c}_{12} + \tilde{c}_{23} + \tilde{c}_{31}) + 4(\tilde{c}_{44} + \tilde{c}_{55} + \tilde{c}_{66}) \\
&\quad + 4(\tilde{c}_{14} + \tilde{c}_{15} + \tilde{c}_{16} + \tilde{c}_{24} + \tilde{c}_{25} + \tilde{c}_{26} + \tilde{c}_{34} \\
&\quad + \tilde{c}_{35} + \tilde{c}_{36}) + 8(\tilde{c}_{45} + \tilde{c}_{56} + \tilde{c}_{64})] \\
c'_{55} &= \tfrac{1}{3}(c_{11} - c_{12} + c_{44}) + \tfrac{1}{9}[(2\tilde{c}_{11} + \tfrac{1}{2}\tilde{c}_{22} + \tfrac{1}{2}\tilde{c}_{33}) \\
&\quad + (\tilde{c}_{23} - 2\tilde{c}_{12} - 2\tilde{c}_{13}) + (2\tilde{c}_{44} + \tfrac{1}{2}\tilde{c}_{55} + \tfrac{1}{2}\tilde{c}_{66}) \\
&\quad + (2\tilde{c}_{15} + 2\tilde{c}_{16} - 4\tilde{c}_{14} + 2\tilde{c}_{24} - \tilde{c}_{25} - \tilde{c}_{26} \\
&\quad + 2\tilde{c}_{34} - \tilde{c}_{35} - \tilde{c}_{36}) + (\tilde{c}_{56} - 2\tilde{c}_{45} - 2\tilde{c}_{46})] \\
c'_{66} &= \tfrac{1}{3}(c_{11} - c_{12} + c_{44}) + \tfrac{1}{6}[(\tilde{c}_{22} + \tilde{c}_{33} - 2\tilde{c}_{23} \\
&\quad + \tilde{c}_{55} + \tilde{c}_{56}) + 2(-\tilde{c}_{25} + \tilde{c}_{26} + \tilde{c}_{35} - \tilde{c}_{36} - \tilde{c}_{56})]
\end{aligned}
\right\} \qquad \text{(A10.10)}
$$

The leading terms in (A10.8)–(A10.10) give the correct expressions for the longitudinal and transverse sound velocities for $H = 0$, as of course they should. It should be noticed that the steady terms in c'_{55} and c'_{66} for propagation along [111] are identical, corresponding to the fact that any displacement vector is possible in the plane perpendicular to [111] and the transverse sound velocity is *independent* of the direction normal to [111]. The oscillatory components are, however, in general dependent on the transverse direction of displacement normal to [111]; thus they differ for instance as between displacement along [01$\bar{1}$] and [$\bar{2}$11]. It is instructive to check that for the *special* case of field along [111], which effectively makes the 1, 2, 3 cubic axes equivalent, the oscillatory components too become identical for the two transverse displacement directions considered.

Appendix 11
Damping of giant quantum oscillations by electron scattering

As was discussed in §2.3.7.2, if electrons are scattered with a relaxation time τ, the Landau levels are broadened, with the probability of the energy level lying between ε and $\varepsilon + d\varepsilon$ being

$$P(\varepsilon)\,d\varepsilon = \frac{(\hbar/2\tau)}{\pi}\,\frac{d\varepsilon}{(\varepsilon - \varepsilon_r)^2 + (\hbar/2\tau)^2} \tag{A11.1}$$

Thus the field for which the rth Landau tube crosses the Fermi surface at $\kappa = \kappa_0$ (where κ_0 is as given by (4.51) and is ordinarily negligibly small) is smeared out just as in the de Haas–van Alphen effect (where of course κ_0 vanishes exactly). The same arguments then apply as in §2.3.7.2 and lead to a phase smearing described by $D_2(z)$ as defined by (4.55) and (4.56).

However because of the broadening of the Landau levels the conservation of energy relation (4.46) is modified to

$$\varepsilon(a, \kappa) + \hbar\omega = \varepsilon(a, \kappa + q) + \Delta \tag{A11.2}$$

where Δ is an extra energy lying within the broadening of the level. With this modification we find

$$\kappa = \kappa_0 + 2\pi m\Delta/(\hbar^2 q A'') \tag{A11.3}$$

The ratio of the extra term to κ_0 is $\Delta/\hbar q s$ and this is usually considerable. Thus if we put $\Delta \sim \hbar/\tau$ as characteristic of the broadening, the ratio is of order $1/\omega\tau$ and even for rather pure metals with $\tau \sim 10^{-10}\,\text{s}$ we find for a typical ω, say $2\pi \times 10^8\,\text{s}^{-1}$, that the raio is > 10. For simplicity* we shall simply ignore the small κ_0 term in what follows. We may note also that the extra term can be an appreciable fraction of the Fermi surface size. Thus for a spherical FS we have

$$\kappa/k_F \simeq m\Delta/\hbar^2 q k_F \sim m\tau/\hbar q k_F = 1/ql \tag{A11.4}$$

where l is the electron mean free path, which might for a rather pure metal

---

* As pointed out by Svirskii (1963) this neglect is not always justified. In particular for an exceptionally pure metal, for which $\omega\tau \gtrsim 1$, κ_0 becomes relatively more important.

be 10^{-2} cm. Thus with $q \sim 2 \times 10^3$ cm^{-1}, $\kappa/k_F \sim 1/20$. For a less pure metal the ratio might well approach unity and some of the results that follow, which are based on the assumption that $ql \gg 1$, may cease to be quantitatively valid.

Due to the spread of possible Δs, of order $\hbar/\tau$, there is a corresponding spread of the phase at which absorption of a phonon occurs. The phase change ϕ due to absorption occurring at κ rather than at $\kappa = 0$ (the extremal section) is given by

$$\phi = \frac{c\hbar}{2eH} A'' \kappa^2 \tag{A11.5}$$

and substituting (A11.3), with the omission of κ_0, this becomes

$$\phi = \Delta^2 \left(\frac{2\pi}{A''}\right) \frac{\pi m}{\beta H \hbar^2 q^2} \tag{A11.6}$$

The distribution function of Δ is similar to, but not identical with, that of ε because it involves the convolution of two distributions over ε (both the start and the finish of Δ are governed by the distribution (A11.1)). Thus it is easily seen that the probability of Δ lying between Δ and $\Delta + d\Delta$ is given by

$$P'(\Delta)\,d\Delta = d\Delta \int_{-\infty}^{\infty} P(\varepsilon) P(\varepsilon + \Delta)\,d\varepsilon \tag{A11.7}$$

where $P(\varepsilon)$ is given by (A11.1). After some transformations this can be shown to reduce to

$$P'(\Delta)\,d\Delta = (\hbar/\tau)\,d\Delta/\pi[\Delta^2 + (\hbar/\tau)^2] \tag{A11.8}$$

i.e. again a Lorentzian distribution, but with just twice the characteristic spread of the original Lorentzian. If we translate the distribution function of Δ into one of the phase ϕ, we find

$$D(\phi/\lambda)\,d(\phi/\lambda) = \frac{1}{\pi} \frac{d(\phi/\lambda)}{(\phi/\lambda)^{1/2}(1 + \phi/\lambda)} \tag{A11.9}$$

with

$$\lambda = \frac{2\pi}{A''} \frac{\pi m}{\beta H q^2 \tau^2} \tag{A11.10}$$

Since ϕ/λ must be positive, ϕ must have the same sign as A'', i.e. only negative ϕs occur for a convex Fermi surface with A'' negative, but only positive ϕs occur if A'' is positive. The distribution function is in fact quite asymmetrical, as shown in fig. 4.7c. The D is the D_4 of (4.58) and the λ the λ_4 of (4.59). It should be noticed that $D(\phi/\lambda)$ has been normalized so that

$$\int_0^\infty D(\phi/\lambda)\,\mathrm{d}(\phi/\lambda) = 1 \qquad (A11.11)$$

since ϕ can only vary monotonically from zero (either to $-\infty$ or to ∞).

To calculate the amplitude of the pth harmonic in a Fourier analysis of the giant quantum oscillations, we need the Fourier transform $f(p|\lambda|)$ of $D(z)$ and the appropriate amplitude reduction factor R_Δ is then $|f(p|\lambda|)/f(0)|$; there is also a phase change equal to the phase of the complex quantity $f(p|\lambda|)/f(0)$. We have, putting $y^2 = \phi/p\lambda$, and for convenience of writing, omitting the mod signs round λ in what follows,

$$f(p\lambda)/f(0) = \frac{2}{\pi}\int_0^\infty e^{-ip\lambda y^2}\,\mathrm{d}y/(1 + y^2) \qquad (A11.12)$$

and this gives*

$$\begin{aligned} f(p\lambda)/f(0) = {}& [(1 - C - S)\cos p\lambda + (C - S)\sin p\lambda] \\ & - i[(C - S)\cos p\lambda - (1 - C - S)\sin p\lambda] \end{aligned} \qquad (A11.13)$$

where C and S are the Fresnel integrals with argument

$$v = (2p\lambda/\pi)^{1/2}$$

i.e. (as in (A5.3))

$$C(v) = \int_0^v \cos\tfrac{1}{2}\pi u^2\,\mathrm{d}u, \qquad S(v) = \int_0^v \sin\tfrac{1}{2}\pi u^2\,\mathrm{d}u \qquad (A11.14)$$

If $\pi\lambda \ll 1$, we find

$$R_\Delta = |f(p\lambda)/f(0)| = 1 - (2p\lambda/\pi)^{1/2} + p\lambda/\pi \dots \qquad (A11.15)$$

In the opposite limit $\lambda \gg 2\pi$, C and S are given approximately by

$$\begin{aligned} C(v) &= \frac{1}{2} + \frac{1}{\pi v}\sin\tfrac{1}{2}\pi v^2 \\[2mm] S(v) &= \frac{1}{2} - \frac{1}{\pi v}\cos\tfrac{1}{2}\pi v^2 \end{aligned} \qquad (A11.16)$$

---

* I am grateful to Professor T. Tsuzuki of Kyushu University for showing me how to calculate the integral. The minus sign in the exponent is needed for a convex Fermi surface since the phases ϕ are all negative. If A'' is positive the sign becomes plus and so reverses the sign of the $\pi/4$ phase change.

and (A11.12) reduces to

$$f(p\lambda)/f(0) = \frac{1}{\pi v}(1 - i) = \frac{1}{(\pi p\lambda)^{1/2}}\left(\frac{1}{\sqrt{2}} - \frac{i}{\sqrt{2}}\right)$$

Thus

$$R_\Delta = |f(p\lambda)/f(0)| = 1/(\pi p\lambda)^{1/2} \qquad\qquad \text{(A11.17)}$$

and the phase $p\psi$ of the harmonic is changed to $(p\psi - \pi/4)$.

These results can be obtained in a rather more direct way by a straight-forward integration of $e^{i(\psi + \phi)}$ over κ, with the elements $d\kappa$ weighted according to the normalized probability distribution. The significance of the approximation $\lambda \gg 1$ is then that the Lorentzian is effectively a constant over the whole range of integration for which the integrand remains appreciable. The approximation (A11.17) also follows directly from (A11.12), since if $\lambda \gg 2\pi$ the y^2 in the denominator can be ignored over the whole significant range and (A11.17) then follows immediately.

Appendix 12

Potential energy of array of strips of alternating charge

I am indebted to Professor Sir Brian Pippard for the following elegant derivation of (6.54). The problem is to calculate the potential energy per unit area of a plane array of infinitely long strips each of width $\frac{1}{2}X$ and carrying pole densities which are alternatively M_0 and $-M_0$. If x measures the distance across the strips, the pole density distribution $\sigma(x)$ can be Fourier analyzed as

$$\sigma(x) = \frac{4M_0}{\pi} \sum_{n=0}^{\infty} \frac{1}{2n+1} \sin \frac{2\pi(2n+1)x}{X} \tag{A12.1}$$

For a distribution

$$\sigma(x) = A \sin qx$$

the potential at a distance z out of the plane has the form

$$V(x, z) = v e^{\mp qz} \sin qx \tag{A12.2}$$

according as z is positive or negative, and the value of v is specified by the requirement that the discontinuity in the z component of field, i.e. of $-\partial V/\partial z$, in crossing the plane $z = 0$ should be $4\pi A \sin qx$. We find in this way

$$v = 2\pi A/q \tag{A12.3}$$

$$V(x, 0) = \frac{4M_0 X}{\pi} \sum_{n=0}^{\infty} \frac{1}{(2n+1)^2} \sin \frac{2\pi(2n+1)x}{X} \tag{A12.4}$$

The potential energy U per unit area is

$$U = \operatorname*{Lt}_{L \to \infty} \frac{1}{2L} \int_0^L V(x, 0)\sigma(x)\,\mathrm{d}x \tag{A12.5}$$

In multiplying the series (A12.1) and (A12.4) the products of terms with different n vanish on integration and we are left with

$$U = \operatorname*{Lt}_{L \to \infty} \frac{8M_0^2 X}{\pi^2 L} \sum_{n=0}^{\infty} \int_0^L \frac{\mathrm{d}x}{(2n+1)^3} \sin^2 \frac{2\pi(2n+1)x}{X}$$

$$= \frac{4M_0^2 X}{\pi^2} \sum_{n=0}^{\infty} \frac{1}{(2n+1)^3} \tag{A12.6}$$

Appendix 12

Now

$$\sum_0^\infty 1/(2n + 1)^3 = 1.0517$$

and the energy Δ_2 per unit volume, as defined in §6.4.2.1 is given by $\Delta_2 = 2U/Z$, since the plate is of thickness Z and there are surfaces of potential energy U on each face. Thus we find

$$\Delta_2 = 0.8525 \, M_0^2 \, X/Z \qquad\qquad \text{(A12.7)}$$

This agrees exactly with Kittel's (1949) result obtained by a slightly different approach. Putting $M_0 = 1/(4k)$ we obtain (6.54).

Appendix 13

Field modulation technique when HF and LF de Haas–van Alphen frequencies both present

We shall show that the modulation of an HF by an LF is considerably modified if in the field modulation technique large modulations are used. The value of field modulation amplitude for which the HF amplitude vanishes (Bessel zero) will be shown to provide a measure of the frequency modulation of the HF by the LF.

Our starting point will be equation (6.87) modified by the inclusion of the time dependent field $h_0 \cos \omega t$ i.e.

$$4\pi M = A\{\sin k(h + h_0 \cos \omega t) + \tfrac{1}{2}A'(k + k') \sin(k + k')(h + h_0 \cos \omega t)$$
$$- \tfrac{1}{2}A'(k - k') \sin(k - k')(h + h_0 \cos \omega t)\}$$

$$(A13.1)$$

If the sine functions are expanded in harmonics of ωt, we find that the nth harmonic amplitude v_n of the signal in a pick-up coil, which is proportional to dM_n/dt is given for odd n by

$$v_n = C\{J_n(z) \cos kh + \tfrac{1}{2}A'k(1 + \alpha)J_n[(1 + \alpha)z] \cos(k + k')h$$
$$- \tfrac{1}{2}A'k(1 - \alpha)J_n[(1 - \alpha)z] \cos(k - k')h\}$$

$$(A13.2)$$

where C is a coupling constant, which need not concern us and

$$\alpha = k'/k, \qquad z = kh_0 \tag{A13.3}$$

For even n the cosines in the bracket must be replaced by sines.

If $A'k \ll 1$, the maxima and minima of the amplitude of v_n with respect to h come very close to $kh = 2r\pi$ and $(2r + 1)\pi$ respectively, where r is an integer, and so for an arbitrary phase of the LF cycle (i.e. of $k'h$) we have

$$|v_n| = C\{J_n(z) + \tfrac{1}{2}A'k \cos k'h\{(1 + \alpha)J_n[(1 + \alpha)z]$$
$$- (1 - \alpha)J_n[(1 - \alpha)z]\}$$

$$(A13.4)$$

It is easily shown that this result is also valid for even n. Since $\alpha \ll 1$ we can assume $\alpha z \ll 1$, even though z itself may be large enough to make $J_n(z)$ small, and so we have

$$J_n[(1 \pm \alpha)z] = J_n(z) \pm \alpha z J_n'(z)$$

(A13.4) then reduces to

$$|v_n| = C\{J_n(z) + A'k'[J_n(z) + zJ'_n(z)]\cos k'h\} \qquad (A13.5)$$

If $z \ll 1$, $zJ'_1(z) = J_1(z) = \tfrac{1}{2}z$, and we find for $|v_1|$, which is proportional to dM/dH,

$$|v_1| = \tfrac{1}{2}Cz\{1 + 2A'k'\cos k'h\}$$

and we recover our earlier result (6.91) that the amplitude modulation m_a of dM/dH has only the feeble value $2A'k'$.

If however z approaches the value z_0 which makes $J_n(z)$ vanish ($z_0 = 3.83$ for $n = 1$), the term $zJ'_n(z)$ becomes much larger than $J_n(z)$ (notice too that it is of opposite sign) and the amplitude modulation can become considerable even though $A'k'$ is small. Indeed, very close to $z = z_0$, the r.h.s. of (A13.5) can change sign in the course of an LF cycle, so that the signal suffers a change of phase of π.

For given $k'h$, the value of z for which $|v_n|$ vanishes is somewhat modified from z_0 and it is convenient to speak of a small shift of Bessel zero. Since the shift is small we can put

$$J_n(z) = (z - z_0)J'_n(z)$$

for the z which makes $|v_n|$ vanish and this z is then determined to sufficient accuracy by

$$\frac{z - z_0}{z_0} = -A'k'\cos k'h = -m_f \cos k'h \qquad (A13.6)$$

Thus the relative shift of the Bessel zero measures the frequency modulation of the HF by the LF, as we intuitively assumed in the text. It is easy to show that this is still true to a fair approximation even with the more general formulation of (6.97) in which the side bands have arbitrary amplitudes. The factor of $A'k'\cos k'h$ in (A13.5) is then modified to

$$\frac{1}{2}\left[(p + q) + \frac{(p - q)}{\alpha}\right]J_n(z) + \tfrac{1}{2}[(p + q) + \alpha(p - q)]zJ'_n(z)$$

$$(A13.7)$$

and the shift of the Bessel zero is given to a good approximation by

$$\frac{z - z_0}{z_0} = \frac{-\tfrac{1}{2}[(p + q) + \alpha(p - q)]A'k'\cos k'h}{1 + \dfrac{1}{2}\left[(p + q) + \dfrac{(p - q)}{\alpha}\right]A'k'\cos k'h} \qquad (A13.8)$$

Field modulation technique

Provided the denominator is not too different from 1, this is a fair approximation to $-m_f \cos k'h$ as given by (6.99). However, various approximations have been made in the calculations leading to both (6.99) and (A13.8) and it is probably not really justified to treat the denominator of (A13.8) as significantly different from 1 within these approximations.

Appendix 14
Conductivity for two-dimensional network

The triangular orbits at O' etc. in fig. A14.1 (shown greatly enlarged in fig. A14.2) are treated as negligibly small switches and the calculation is made in three stages: (1) calculation of the switching probabilities of an electron arriving at a triangular switch, (2) calculation of the terminal point of an electron starting at an arbitrary point P, and hence of its average vector path L and (3) calculation of the conductivity tensor by use of equation (7.32).

A14.1. Switching probabilities
As with the one-dimensional network we can calculate either (a) the completely coherent situation or (b) the completely incoherent situation which is in fact a phase average of (a).

(a) coherent case

The various symbols in fig. A14.2a indicate the complex amplitudes of the electron wave which starts as 1 at U, after passing the junctions U, V, W. Since we shall need to calculate only $|a|^2$, $|b|^2$ and $|c|^2$, the phase change θ involved in changing circles need not be explicitly introduced, provided the phase change on each arc is taken as $\frac{1}{3}\psi$, where ψ is the de Haas–van Alphen phase $2\pi F/H$ of the triangular orbit, rather than the real phase change which is $\frac{1}{3}\psi + \theta$ (cf. similar situation in formu-

Fig. A14.1. Part of the hexagonal network in real space with the triangular intersections at O', O_1, O_2 etc. shrunk down to point 'switches'.

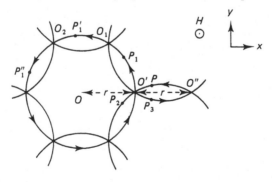

530

lating (7.38)). If we now do the same kind of bookkeeping as for (7.38) we obtain

$$
\left.\begin{aligned}
a &= q + ipwe^{i\psi/3} \\
u &= qwe^{i\psi/3} + ip \\
b &= ipue^{i\psi/3} \\
v &= que^{i\psi/3} \\
c &= ipve^{i\psi/3} \\
w &= qve^{i\psi/3}
\end{aligned}\right\} \tag{A14.1}
$$

From these we easily obtain (with $p^2 = P$, $q^2 = Q$, $P + Q = 1$)

$$
\left.\begin{aligned}
A = |a|^2 &= Q(1 - 2Q^{1/2}\cos\psi + Q)/(1 - 2Q^{3/2}\cos\psi + Q^3) \\
&= 1 - B(1 + Q) \\
B = |b|^2 &= P^2/(1 - 2Q^{3/2}\cos\psi + Q^3) \\
C = |c|^2 &= P^2 Q/(1 - 2Q^{3/2}\cos\psi + Q^3) = QB
\end{aligned}\right\} \tag{A14.2}
$$

(note that $A + B + C = 1$, as it must).

(b) *incoherent case*

The symbols in fig. A14.2b now represent particle fluxes and conservation of particles at U, V, W gives

$$
\begin{aligned}
A &= Q + P\gamma & \alpha &= Q\gamma + P \\
B &= P\alpha & \beta &= Q\alpha \\
C &= P\beta & \gamma &= Q\beta
\end{aligned} \tag{A14.3}
$$

Fig. A14.2. Blown-up schematic diagrams of the switches of fig. A14.1: (a) for full coherence over the internal paths in the switch; the symbols a, u etc. represent the complex amplitudes of the waves immediately adjacent to each of the junctions U, V, W where an incident wave can either break through or be Bragg reflected; (b) for complete incoherence; the symbols 1, α, β, γ, A, B, C now represent particle fluxes.

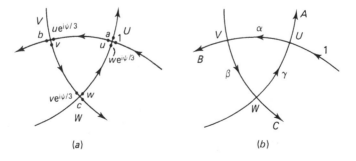

(a)   (b)

From these, we find

$$A = Q(1 + 2Q)/(1 + Q + Q^2) = 1 - B(1 + Q)$$
$$B = P/(1 + Q + Q^2) \tag{A14.4}$$
$$C = PQ/(1 + Q + Q^2) = QB$$

These results can also be obtained by averaging (A14.2) over ψ and using the identity

$$\frac{1}{2\pi} \int_0^{2\pi} dx/(1 + a\cos x) = (1 - a^2)^{-1/2} \tag{A14.5}$$

We may note also that as with the Fabry–Perot interferometer, both (A14.2) and (A14.4) can be obtained by following the fortunes of electrons repeatedly going round the triangular orbit and summing the appropriate infinite series. For the incoherent case, for instance, this gives $\alpha = P + PQ^3 + PQ^6 + \cdots = P/(1 - Q^3)$ and hence B, and similarly for β, C and γ, A.

A14.2. Terminal point of electron path

As before, we use complex notation to denote coordinates of points in the network; thus the point (x, y) is denoted by $z = x + iy$. An electron at any point P on the arc shown in fig. A14.1 will enter the switch O' and then wander through the network and arrive at some average terminal point Z relative to origin O'', or $Z + r$ relative to O'. After leaving O' it will appear at P_1, P_2 or P_3, with the probabilities A, B and C calculated above. An electron on the arc containing P_1 will evidently have a terminal point Z with respect to axes twisted 60° clockwise and origin O', so this terminal point z with respect to the x, y axes and origin O' is simply $Ze^{-i\pi/3}$. Similarly an electron on the arc containing P_2 will have a terminal point $Ze^{i\pi/3}$, again with respect to the x, y axes and origin O', while one on the arc P_3 will have terminal point $-Z$. Thus bearing in mind the relative probabilities, we must have

$$Z + r = AZe^{-i\pi/3} + BZe^{i\pi/3} - CZ$$

or

$$Z = -r/(1 - Ae^{-i\pi/3} - Be^{i\pi/3} + C) \tag{A14.6}$$

A14.3. Calculation of conductivity

If we now write down the terminal points for electrons starting on the arcs P_1, P_1' and P_1'' with respect to origin O, the effective path L to be put into

Conductivity for 2-D network

(7.32) is obtained by subtracting $re^{i\phi}$, where ϕ defines the starting point. Thus

$$\left. \begin{aligned} L(P_1) &= Ze^{-i\pi/3} + r - re^{i\phi} \\ L(P_1') &= Z + re^{i\pi/3} - re^{i\phi} \\ L(P_1'') &= Ze^{i\pi/3} + re^{2i\pi/3} - re^{i\phi} \end{aligned} \right\}$$ (A14.7)

The terminal points for the arcs below O are just the negative of those on the opposite arcs above and it is easily seen that the contribution to the integral of (7.32) from the lower semicircle is equal to that from the upper semicircle. In evaluating (7.32) we need not worry about E_y since for H along the hexad axis the form of the conductivity tensor must satisfy hexagonal symmetry. This implies that $J_x + iJ_y$ has the form $(\alpha + i\beta)E_x + (i\alpha - \beta)E_y$ or $(\alpha + i\beta)(E_x + iE_y)$, and thus if the complex coefficient of E_x is determined, it will also be that of $E = E_x + iE_y$. Thus, we can replace E_x by E in (7.32) and find

$$J = -\frac{2\,nec}{\pi\,rH}E\left\{ \int_0^{\pi/3} (Ze^{-i\pi/3} + r)\sin\phi\,d\phi \right.$$

$$+ \int_{\pi/3}^{2\pi/3} (Z + re^{i\pi/3})\sin\phi\,d\phi + \int_{2\pi/3}^{\pi} (Ze^{i\pi/3} + re^{2i\pi/3})\sin\phi\,d\phi$$

$$\left. - \int_0^{\pi} r(\cos\phi + i\sin\phi)\sin\phi\,d\phi \right\}.$$

which reduces to

$$J = \frac{3\,nec}{\pi\,H}E\left\{ -\left(\frac{Z}{r} + e^{i\pi/3}\right) + \frac{i\pi}{3} \right\}$$ (A14.8)

If we now use (A14.6) and the relations $C = QB$, $A = 1 - B(1 + Q)$ which are valid whatever the coherence condition, (A14.8) becomes

$$J = \frac{3nec}{\pi H}E\left\{ \left[-e^{i\pi/3} + \frac{e^{-i\pi/3}}{1 - i\sqrt{3}B(Q + e^{-i\pi/3})} \right] + \frac{i\pi}{3} \right\}$$ (A14.9)

This is essentially equivalent to the result of Pippard (1965a) or of Falicov, Pippard and Sievert (1966) (see their (3.4)) after allowance for the opposite handedness of Pippard's axes (so that the electrons rotate in the opposite sense to the one in Falicov et al. and here).

In order to compare with the properties of the real situation in Zn and Mg, (A14.9) needs some modifications. First, account should be taken of the electrons orbiting round the inside waist of the monster. For low fields

($H \ll H_0$) these electrons, together with those round the needle or cigar, exactly compensate the holes round the hexagon orbit. The transverse conductivity σ_{11} should then vary as a/H^2 (provided H is big enough to make $\omega_c\tau \gg 1$) and the Hall conductivity σ_{21} should go to zero. These limiting requirements can be achieved by adding a term a/H^2 (a real) to the coefficient of E in (A14.9) and replacing the term $i\pi/3$ by $i\sqrt{3}$. The other main adjustment that needs to be made is to allow for the fact the Fermi surface is not really cylindrical, as has been implicitly assumed up to now, because, of course, the idealized two-dimensional picture maintains its form for only a short range of k_H, indeed for a range considerably shorter than the full length of the needle or cigar. To deal with this properly an integration over k_H is required in which the probability B is itself a function of k_H, but a reasonable approximation is simply to replace the factor $3nec/\pi$ by an empirical factor b chosen to fit experiment. Finally the effect of finite τ should be taken into account; this is discussed qualitatively in the main text.

With these modifications (A14.9) becomes

$$J = E\left\{\frac{a}{H^2} + \frac{b}{H}\left[\frac{1}{1 - i\sqrt{3}B(Q + e^{-i\pi/3})} - 1\right]\right\} \qquad \text{(A14.10)}$$

If we use the 'incoherent' formula (A14.4) for B, we find, after some algebra, the conductivity tensor:

$$\sigma_{11} = \sigma_{22} = \frac{b}{H}\left[\frac{a}{bH} + \frac{3Q(1 - Q)}{1 - 2Q + 4Q^2}\right]$$

$$\sigma_{21} = -\sigma_{12} = \frac{\sqrt{3}b}{H}\frac{(1 - Q)^2}{1 - 2Q + 4Q^2} \qquad \text{(A14.11)}$$

For $Q \to 0$ (i.e. $H/H_0 \gg 1$) and for $P \to 0$ (i.e. $H/H_0 \ll 1$), this assumes the limiting forms (retaining only linear terms in Q and P respectively, except where the leading term is quadratic)

$$Q \to 0: \quad \sigma_{11} = \sigma_{22} = \frac{a}{H^2} + \frac{3Qb}{H},$$

$$\sigma_{21} = -\sigma_{12} = \frac{\sqrt{3}b}{H}$$

$$P \to 0: \quad \sigma_{11} = \sigma_{22} = \frac{a}{H^2} + \frac{Pb}{H}, \qquad \text{(A14.12)}$$

$$\sigma_{21} = -\sigma_{12} = \frac{\sqrt{3}}{3}\frac{P^2b}{H}$$

Use of the 'coherent' formula for B does not lead to any very transparent formula and the simplest procedure is to substitute values of Q and ψ directly into (A14.10) and then calculate numerically. The situation is complicated by the fact that the oscillations as ψ varies are very non-sinusoidal so it cannot be assumed that $\psi = 0$ and $\psi = \pi$ represent the extremes of the oscillation. If however the damping of the oscillations (by integration over k_H, by temperature and by other kinds of phase smearing) is taken into account their amplitude is much reduced and the higher harmonics are greatly reduced. Some discussion of the oscillations is given in the main text.

If the coherent formula for σ is averaged over ψ (this is most easily done directly from (A14.10) by casting the bracket into the form $\alpha + \beta/(1 + a\cos\psi)$ and using the identity (A14.5), we obtain:

$$J = E\left\{\frac{a}{H^2} + \frac{ib}{H}\right.$$

$$\times \left.\frac{\sqrt{3}e^{-i\pi/3}P^2(Q + e^{-i\pi/3})}{[(1 - Q^3)^2 - 3P^4(Q + e^{-i\pi/3})^2 - 2i\sqrt{3}P^2(1 + Q^3)(Q + e^{-i\pi/3})]^{1/2}}\right\}$$

(A14.13)

from which values of an average conductivity $\bar{\sigma}$ can be computed numerically as a function of Q (i.e. effectively of H). It is not difficult to calculate limiting values for $Q \to 0$ and $P \to 0$ and we find that for $Q \to 0$ we obtain exactly the result of (A14.12), but for $P \to 0$ we get a result which, though qualitatively similar to (A14.12), is appreciably different in detail:

$$P \to 0: \quad \bar{\sigma}_{11} = \bar{\sigma}_{22} = \frac{a}{H^2} + 0.729\frac{Pb}{H}$$

$$\bar{\sigma}_{21} = -\bar{\sigma}_{12} = 0.548\frac{Pb}{H} \qquad \text{(A14.14)}$$

A similar discrepancy (though in opposite sense) was already noted for the one-dimensional network. It is probably more correct in principle to average over phase in the final answer rather than in the switching probabilities, but it is by no means obvious that it is the *conductivity* rather than the *resistivity* which should be averaged and the answers will in general not be the same (though in the special case of the one-dimensional network they are). Since, however, this difficult point of principle does not affect the qualitative conclusions we shall simply ignore it and base the discussion (see main text) on a comparison of the experimentally observed resistivity tensor with the inverse of the theoretical average of the conductivity tensor.

To appreciate the relative importance of the various terms in the formulae we need to have estimates of a, b and H_0 for Zn and Mg. These are available from the comparisons made by Falicov *et al.* between theory and experiment and are collected in table A14.1. Falicov *et al.* point out that these values are in reasonable accord with estimates based on first principles.

Table A14.1

	Zn	Mg
$10^{-8}a(\text{kG})^2/\Omega\,\text{cm}$	0.17	0.52
$10^{-8}b\,\text{kG}/\Omega\,\text{cm}$	0.67	1.75
$a/b\,\text{kG}$	0.25	0.30
$H_0\,\text{kG}$	2.7	5.85

Appendix 15

Improvement of Dingle plot procedure
(Paul & Springford 1977)

Suppose we have a number of experimental points (x_r, y_r) on a Dingle plot where x denotes $1/H$ and y the appropriate logarithmic amplitude and r goes from 1 to n. The usual procedure is then to determine the slope a and intercept b of the best fit of the line

$$y = ax + b \qquad (A15.1)$$

to the points. The Dingle temperature is, of course, directly proportional to a. The least squares solution is

$$\left. \begin{array}{l} a = (n\sum x_r y_r - \sum x_r \sum y_r)/[n\sum x_r^2 - (\sum x_r)^2] \\ b = (\sum y_r \sum x_r^2 - \sum x_r \sum x_r y_r)/[n\sum x_r^2 - (\sum x_r)^2] \end{array} \right\} \qquad (A15.2)$$

and it can be shown that if notionally we imagine the procedure repeated many times, the mean square deviations from the true values a_0 and b_0 will be given by

$$\left. \begin{array}{l} \overline{(a - a_0)^2} = \sum v_r^2/[n\sum x_r^2 - (\sum x_r)^2] \\ \overline{(b - b_0)^2} = \sum v_r^2 \sum x_r^2/n[n\sum x_r^2 - (\sum x_r)^2] \end{array} \right\} \qquad (A15.3)$$

where the residuals v_r are defined as

$$v_r = y_r - ax_r - b \qquad (A15.4)$$

Strictly speaking factors of $n/(n-2)$ should be included on the r.h.s. of (A15.3) to allow for the slight difference between the *residuals* v_r from the fitted line and the *errors* e_r from the 'true' line. If, as is usually the case, $n \gg 1$ the distinction is unimportant.

If, however, we know in addition that the plot *must* go through a point $(0, y_0)$ on the y axis, we can make an estimate a' of the slope, which has less random error than given by (A15.3). Thus the least squares fit of the points to a line

$$y = a'x + y_0 \qquad (A15.5)$$

gives

$$a' = \sum x_r(y_r - y_0)/\sum x_r^2 \qquad (A15.6)$$

Appendix 15

and it can be shown that the uncertainty of a' is given by

$$\overline{(a' - a_0)^2} = \sum v_r'^2 / n \sum x_r^2 \tag{A15.7}$$

where once again we ignore a factor, this time $n/(n-1)$, on the right hand side, and

$$v_r' = y_r - a'x_r - y_0 \tag{A15.8}$$

We shall show in a moment that the random error (A15.7) can be considerably smaller than that of (A15.3), but it is important to realize that we have achieved this only by introducing a *systematic* error corresponding to whatever error Δy_0 we make in the choice of the fixed intercept y_0. This systematic error, as follows immediately from (A15.6) is

$$(a' - a_0)_s = (\Delta y_0) \sum x_r / \sum x_r^2 \tag{A15.9}$$

Indeed, if for $(\Delta y_0)^2$ we put the value of $(b - b_0)^2$ from (A15.3) and add $(a' - a_0)_s^2$ to the random error of (A15.7) we get back exactly our original error $\overline{(a - a_0)^2}$ as given by (A15.3).

The real value of reducing random error at the expense of introducing systematic error is when we wish to study *differences* of slope brought about by some variable which we *know* does not change y_0 appreciably. As can be seen immediately from (A15.6), provided the x_rs are much the same* in *different* Dingle plots, the *changes* of a' are independent of the assumed value of y_0.

We shall now assess the reduction in random error of slope achieved by force-fitting our line through the point $(0, y_0)$. The factor γ by which the error is reduced is

$$\gamma \equiv \left[\frac{\overline{(a - a_0)^2}}{\overline{(a' - a_0)^2}} \right]^{1/2} = \left[\frac{\sum v_r^2}{\sum v_r'^2} \frac{n \sum x_r^2}{[n \sum x_r^2 - (\sum x_r)^2]} \right]^{1/2} \tag{A15.10}$$

It is not difficult to show that

$$\sum v_r'^2 = \sum v_r^2 + (y_0 - b)^2 [n \sum x_r^2 - (\sum x_r)^2] / \sum x_r^2 \tag{A15.11}$$

where b is given (A15.2) and we see that if $(y_0 - b)^2$ is equated to the random error $\overline{(b - b_0)^2}$ of the least squares intercept as given by (A15.3) (and it could of course be made even smaller by taking the mean of several Dingle plot intercepts), the second term on the r.h.s. of (A15.11) is just $\sum v_r^2 / n$ and so can be safely ignored. Thus the random error improvement factor is

---

* The precise condition is that $\sum x_r / \sum x_r^2$ should not change; this condition is automatically satisfied if the x_rs have the same values in the different plots.

538

simply

$$\gamma = [n\sum x_r^2/(n\sum x_r^2 - (\sum x_r)^2)]^{1/2} \tag{A15.12}$$

This evidently becomes large if the points are 'bunched' over a range which is narrow compared with the smallest x_r, and this indeed often happens with Dingle plots because of the limited range of field over which the oscillations are appreciable. Some idea of the magnitude of γ can be obtained by considering a special example in which the x_rs are evenly spaced at intervals d, i.e.

$$x_r = (p + r)d \quad (r \text{ from } 0 \text{ to } n - 1) \tag{A15.13}$$

It is easily shown that for this spacing of points

$$\gamma = \left[1 + 12\left(\frac{p}{n} + \frac{1}{2}\right)^2\right]^{1/2} \tag{A15.14}$$

where we have assumed that $n \gg 1$. Typically p/n is at least 1, so γ would usually be greater than 5; for the example quoted in Paul and Springford, p/n was 3, so γ would have been 12.

It must now be pointed out that the procedure outlined here is in fact a little different from that described by Paul and Springford. Instead of making the Dingle plot go *exactly* through $(0, y_0)$ they merely add this as an $(n + 1)$th point to the others and then fit a linear regression to the $(n + 1)$ points using the same formulae (A15.2) and (A15.3) for a and $\overline{(a - a_0)^2}$. Their improvement factor γ' is not as large as that given by (A15.12): it is

$$\gamma' = \left[\frac{(n + 1)\sum x_r^2 - (\sum x_r)^2}{n\sum x_r^2 - (\sum x_r)^2}\right]^{1/2} \tag{A15.15}$$

which for the special example (A15.13) reduces for $n \gg 1$ to

$$\gamma' = \left\{1 + \frac{12}{n}\left[\left(\frac{p}{n}\right)^2 + \left(\frac{p}{n}\right) + \frac{1}{3}\right]\right\}^{1/2} \tag{A15.16}$$

For $p/n = 3$ and $n = 6$ (as in their experiment) this gives $\gamma' = 5$ (for the values of x_r actually used, it was 4.5) and their formula has the rather strange implication that the improvement would be reduced if each plot had a higher number of readings.

Finally we may note from (A15.6) that the criterion of constancy of y_0 with change of the relevant variable is that any change of y_0 must be small compared with the change of $\overline{y_r}$ brought about by changing the variable, where $\overline{y_r}$ is $\sum x_r y_r / \sum x_r$. In other words the percentage change of the constant C_p in (8.1) with the variable in question must be small compared with the observed percentage change of the weighted average ordinate.

Appendix 16*

Strain field of dislocations and estimate of $d \ln F / ds$

We shall consider only dislocations lying parallel to the field and ignore crystal anisotropy. The assumption of isotropy greatly simplifies the analysis and probably does not seriously change the results. The strain field of an edge dislocation in a plane normal to its direction and referred to axes (R, θ, z) which we label $(1, 2, 3)$ is given by

$$\varepsilon_{11} = \varepsilon_{22} = -\frac{b(1 - 2v)}{4\pi(1 - v)} \frac{\sin \theta}{R} \equiv \varepsilon_0 \qquad \text{(A16.1)}$$

$$\varepsilon_{12} = \frac{b}{4\pi(1 - v)} \frac{\cos \theta}{R} \qquad \text{(A16.2)}$$

where b is the magnitude of the Burgers vector, which is assumed to be perpendicular to the z direction and to lie along $\theta = 0$, and v is Poisson's ratio. For a screw dislocation the only strain is a shear strain ε_{23} in the θ, z plane, given by

$$\varepsilon_{23} = \frac{b}{4\pi R} \qquad \text{(A16.3)}$$

(we ignore pure rotations, which may, however, contribute to mosaic structure effects).

Since, however, the only experimental evidence relates to edge dislocations, we shall not discuss screw dislocations beyond pointing out that the analysis and orders of magnitude of the results would be similar to that for edge dislocations.

If the field (and the dislocation) is along the $\langle 111 \rangle$ direction[†] of a cubic crystal (the only case we shall consider explicitly), we would expect the ε_{12} strain of an edge dislocation to have little effect on the Fermi surface area in its own plane and we shall ignore it. The remaining strains ε_{11} and ε_{22} are

---

* I am indebted to Dr B. R. Watts for the result (A16.7) and for much helpful discussion.
† In reality dislocations cannot lie along $\langle 111 \rangle$ but only at an angle of about 20° to $\langle 111 \rangle$. Hopefully this angle is small enough not to upset our largely qualitative conclusions based on ignoring it.

540

equivalent to a pure dilation, together with a pure shear in each of the R, z and θ, z planes. We have in fact

$$
\begin{pmatrix} \varepsilon_0 & 0 & 0 \\ 0 & \varepsilon_0 & 0 \\ 0 & 0 & 0 \end{pmatrix} = \begin{pmatrix} \frac{2}{3}\varepsilon_0 & 0 & 0 \\ 0 & \frac{2}{3}\varepsilon_0 & 0 \\ 0 & 0 & \frac{2}{3}\varepsilon_0 \end{pmatrix} + \begin{pmatrix} 0 & 0 & 0 \\ 0 & \frac{1}{3}\varepsilon_0 & 0 \\ 0 & 0 & -\frac{1}{3}\varepsilon_0 \end{pmatrix}
$$
$$
+ \begin{pmatrix} \frac{1}{3}\varepsilon_0 & 0 & 0 \\ 0 & 0 & 0 \\ 0 & 0 & -\frac{1}{3}\varepsilon_0 \end{pmatrix} \tag{A16.4}
$$

in which the first matrix on the right describes a pure dilatation and each of the other two, a pure shear. We could then go on to estimate their effects on the dHvA frequency F of a noble metal belly or neck using the analysis[*] of Shoenberg and Watts (1967), but it is rather simpler to break up the strain system in a different way:

$$
\begin{pmatrix} \varepsilon_0 & 0 & 0 \\ 0 & \varepsilon_0 & 0 \\ 0 & 0 & 0 \end{pmatrix} = \begin{pmatrix} \varepsilon_0/(1+v) & 0 & 0 \\ 0 & \varepsilon_0/(1+v) & 0 \\ 0 & 0 & \varepsilon_0/(1+v) \end{pmatrix}
$$
$$
+ \begin{pmatrix} v\varepsilon_0/(1+v) & 0 & 0 \\ 0 & v\varepsilon_0/(1+v) & 0 \\ 0 & 0 & -\varepsilon_0/(1+v) \end{pmatrix} \tag{A16.5}
$$

which corresponds to superposing the effects of a hydrostatic pressure P and a uniaxial stress S along the z axis given by

$$
P = -3K\varepsilon_0/(1+v) \quad \text{and} \quad S = -Y\varepsilon_0(1+v) \tag{A16.6}
$$

where K is the bulk modulus and Y the Young's modulus.

The effect on F (or A, the FS area) then reduces to

$$
d\ln F = d\ln A = \left[\frac{\partial \ln F}{\partial P} + (1-2v)\frac{\partial \ln F}{\partial S} \right] \frac{Nb \sin \theta}{2\pi(1-v)R} \tag{A16.7}
$$

---

[*] In that analysis a shear ε appears, which is defined in an unconventional way and the numerical coefficients of equations (7) and (8) require modification if the shear is defined conventionally. For a uniaxial tension the strain system is equivalent to a pure dilatation and a pure shear in *each* of two crystallographically equivalent planes perpendicular to each other and containing the direction of tension. If *each* of these shears is denoted by ε, the coefficients $\frac{3}{4}$ and $\frac{3}{2}$ in (7) and (8) should be replaced by $\frac{1}{2}$ and 1 respectively. If the material is assumed isotropic rather than crystalline the coefficient of β in (7) and (8) becomes $(1+v)/(1-2v)$; for Cu, $v \simeq \frac{1}{3}$, so this coefficient is about 4 as compared with 2.7 for $\langle 111 \rangle$ (modified (7)) or with 8.3 for $\langle 100 \rangle$ (modified (8)).

where N is the rigidity modulus, related to Y and K by

$$N = \frac{Y}{2(1 + v)} = \frac{3K(1 - 2v)}{2(1 + v)}$$

and in this form the effect is expressed in terms of directly measured stress dependences. It is convenient to split (A16.7) into two factors, one of which is a strain s defined by

$$s = b \sin \theta / 2\pi R \qquad (A16.8)$$

and the other of which can be thought of as d ln F/ds, i.e.

$$\frac{d \ln F}{ds} = \frac{N}{(1 - v)} \left[\frac{\partial \ln F}{\partial P} + (1 - 2v) \frac{\partial \ln F}{\partial S} \right] \qquad \textbf{(A16.9)}$$

Equation (A16.7) then assumes the form

$$d \ln F = s \, d \ln F / ds, \qquad (A16.10)$$

on which the discussion in the text is based. The way the division is made is of course arbitrary, in the sense that it is only the product that matters, but our strain s is in fact of the same order as the actual strains in the sample.

For the purpose of estimating a 'characteristic' strain σ we calculate the r.m.s. of s over a circle centred on one dislocation, assuming that (A16.8) cuts off when R is comparable with the core radius of the dislocation. The answer is very insensitive to the value of this cut-off and it is reasonable to take the cut-off radius as equal to b. The contribution of the region within the core to the r.m.s. can be shown to be negligible. We also apply a cut-off at R_0, the upper end of the range of R, at such an R_0 that we can crudely suppose that for $R < R_0$ the strain is dominated by the central dislocation, while for $R > R_0$ the strain due to the central dislocation is on average cancelled out by the + and − fields of all the other dislocations. It is perhaps not unreasonable to take R_0 as something like half the average distance between dislocations and we assume $R_0 = \frac{1}{2}D^{-1/2}$, where D is the dislocation density (for a regular square array this means that we average over a circle of diameter equal to the side of the square unit cell). With these rather rough and ready assumptions, we find

$$\sigma = (\overline{s^2})^{1/2} = \left(\frac{\overline{\sin^2 \theta}}{\pi R_0^2} \int_b^{R_0} \frac{b^2 2\pi R \, dR}{4\pi^2 R^2} \right)^{1/2}$$

If we now put $R_0 = \frac{1}{2}D^{-1/2}$ and $\overline{\sin^2 \theta} = \frac{1}{2}$, this reduces to

$$\sigma = bD^{1/2} \left(\frac{1}{\pi^2} \ln \frac{1}{2bD^{1/2}} \right)^{1/2} \qquad \textbf{(A16.11)}$$

The factor in brackets varies only slightly over a wide range of D. With $b = 2.5 \times 10^{-8}$ cm (the value for Cu) it goes from 0.73 for $D = 10^{10}$ cm^{-2} to 1.00 for $D = 10^6$ cm^{-2}. We shall not make much error by treating it as a constant, and we shall adopt the value 1.0 for simplicity, so that

$$\sigma = bD^{1/2} \qquad\qquad (A16.12)$$

We shall now estimate the appropriate value of $d \ln F/ds$ as given by (A16.9) for the Cu belly and neck. The relevant data are those of Templeton (1966) for $d \ln F/dP$ and of Shoenberg and Watts (1967) for $d \ln F/dS$ and are shown in table A16.1, which also gives the resulting values of $d \ln F/ds$ based on (A16.9). We see that the first term of (A16.9) is dominant for the belly, but for the neck the two terms are comparable.

The value of $d \ln F/ds$ assumed in table 8.3 for Bi is hardly more than an order of magnitude. It is based on the oscillatory magnetostriction data of Aron and Chandrasekhar (1969). Since the type and orientation of the dislocations in the Bi samples was unknown, we took something like the geometric mean of the highest and lowest values of $d \ln F/ds$ for various geometries (these differ by a factor 3) and arrived at the value 150. We also assumed that (A16.12) is appropriate for estimating the characteristic strain which enters into H_0.

Table A16.1. *Stress and strain dependence of F for Cu*

	$10^{12} d \ln F/dP$	$10^{12} d \ln F/dS$	$d \ln F/ds$
Belly	0.43	-0.09	0.29
Neck	1.8	6.9	2.9

Notes: Here P is hydrostatic pressure and S uniaxial tension, both in dyne cm^{-2}, and $d \ln F/ds$ has been calculated from (A16.9) using the values $N = 4.8 \times 10^{11}$ dyne cm^{-2} and $\nu = 0.34$, and the entries in the first two columns (taken from Templeton (1966) and Shoenberg and Watts (1967)).

Appendix 17*

Statistical analysis of some phase addition problems.

A17.1. Addition of phases of equal magnitude, but of random sign

We suppose as in the argument based on (8.41) that the phase error Φ of an orbit is made up of N contributions of equal magnitude Δ but of random sign and we wish to calculate the phase smearing caused by the spread of Φ about zero. This is essentially equivalent to calculating the modulus of the expectation value $\langle e^{i\Phi} \rangle$ of $e^{i\Phi}$. Each of the contributions θ_k to Φ can be thought of as having a distribution function $D(\theta)$ independent of k, given by

$$D(\theta) = \tfrac{1}{2}[\delta(\theta - \Delta) + \delta(\theta + \Delta)]$$

Thus, since the contributions are all independent,

$$
\begin{aligned}
\langle e^{i\Phi} \rangle &= \int d\theta_1 \cdots d\theta_N \, D(\theta_1) \cdots D(\theta_N) e^{i\Sigma\theta_i} \\
&= \left\{ \int d\theta \, D(\theta) e^{i\theta} \right\}^N \\
&= [\tfrac{1}{2}(e^{i\Delta} + e^{-i\Delta})]^N \\
&= (\cos \Delta)^N
\end{aligned}
$$

Since this is entirely real its modulus has the same magnitude and so we find

$$R_D = |(\cos \Delta)^N| \tag{A17.1}$$

as the general form of the reduction factor according to this model. Alternatively this result can be derived by considering the probability distribution of Φ and applying the Fourier transform method. In fact the probability of Φ being the sum of r contributions Δ and $N - r$ contributions $-\Delta$, i.e. of $\Phi = (2r - N)\Delta$, is the binomial distribution

$$D(r) = N! \, 2^{-N}/(N - r)! \, r! \tag{A17.2}$$

---

* I am indebted to Dr G. G. Lonzarich for much of the analysis in this appendix.

Statistical analysis of phase addition problems

Thus the mean value of $e^{i\Phi}$ is

$$\langle e^{i\Phi} \rangle = \sum_r e^{i(2r-N)\Delta} D(r) = e^{-iN\Delta} \left[\sum e^{2ir\Delta} D(r) \right]$$

The quantity in brackets is the Fourier transform of $D(r)$ w.r.t. 2Δ and by a standard theorem we find, exactly as before,

$$\langle e^{i\Phi} \rangle = e^{-iN\Delta} \left(\frac{e^{2i\Delta} + 1}{2} \right)^N = (\cos \Delta)^N$$

If Δ (which is the same as $\mu/2\pi$) is small, we have approximately

$$R_D = \left(1 - \frac{\Delta^2}{2} \right)^N$$

and if N is large this becomes

$$R_D = \exp(-\tfrac{1}{2}N\Delta^2) \quad \text{or} \quad \ln R_D = -\frac{\mu p}{4\pi} \frac{H_0}{H} \tag{A17.3}$$

if N is given by (8.39) and Δ by (8.40), which is exactly the result (8.43). It is however only in this approximation (i.e. for Δ or $\mu/2\pi < 1$) that this result, which implies a Gaussian distribution of Φ, is valid. In general we can express (A17.1) as

$$\ln R_D = \frac{\ln|\cos \mu/2\pi|}{(\mu/2\pi)} \left(\frac{pH_0}{H} \right) \tag{A17.4}$$

which departs considerably from (A17.2) if $\mu/2\pi \gtrsim 1$, and gives absurd answers as $\mu/2\pi$ passes through values such as $\pi/2$ or π. These answers (which though physically absurd, are mathematically correct consequences of the model) come about because the model is too crude a representation of a continuously varying strain field if the changes of phase to be added are too large. Probably, however, (A17.3) still gives a reasonable estimate of $\ln R_D$ even for μ somewhat larger than 1.

A17.2. Superposition of N oscillations with statistically distributed phases

Suppose for simplicity that the grains of a polycrystal are all of the same size,* but have phases differing from the phase ψ of a perfect sample by ϕ_i for the ith grain. The oscillation of the whole sample can then be described

---

* If the grains are unequal, weights must be assigned to the ϕ_i, but provided the weights are randomly distributed and have no correlation with the ϕ_i, none of the results obtained by ignoring the weights are affected.

545

by

$$a \sin(\psi + \phi) = a_1 \sin \psi + a_2 \cos \psi$$

where

$$a_1 = \frac{1}{N} \sum_i \cos \phi_i, \qquad a_2 = \frac{1}{N} \sum_i \sin \phi_i$$

(A17.5)

With this formulation

$$a^2 = a_1^2 + a_2^2 \quad \text{and} \quad \tan \phi = a_2/a_1 \tag{A17.6}$$

The amplitude a has been normalized to be 1 for a 'perfect' sample, i.e. if each of the ϕ_i vanishes. We then have,

$$a^2 = \frac{1}{N} + \frac{1}{N^2} \sum_{i \neq j} (\cos \phi_i \cos \phi_j + \sin \phi_i \sin \phi_j) \tag{A17.7}$$

and if we use $\langle \rangle$ to denote expectation values, and assume that the ϕ_i are independent random variables drawn from an infinite 'reservoir' which has a continuous probability distribution function, we find

$$\langle a^2 \rangle = \frac{1}{N} + \frac{N-1}{N}(\langle \cos \phi \rangle^2 + \langle \sin \phi \rangle^2) \tag{A17.8}$$

since there are $N(N-1)$ terms in the summation of (A17.7).

If the probability distribution of ϕ is symmetrical (i.e. even in ϕ) $\langle \sin \phi \rangle$ vanishes, and in particular, for a Gaussian distribution proportional to $e^{-\phi^2/\lambda^2}$ or for a Lorentzian proportional to $1/[1 + (\phi^2/\lambda^2)]$ we find (see table 2.2) respectively

$$R = \langle \cos \phi \rangle = e^{-\lambda^2/4} \quad \text{or} \quad e^{-\lambda}$$

i.e. just the reduction factor R calculated for an infinite, rather than a finite population. Thus we have finally

$$R'^2 = \frac{1}{N} + \frac{N-1}{N} R^2 = R^2 \left(1 + \frac{1}{NR^2}\right) \tag{A17.9}$$

where we have written the expectation value of a^2 as the square of a reduction factor R' characteristic of the sample of N grains. We see then that as NR^2 gets smaller (by reduction of the field) the reduction factor R' will level off at a value $1/\sqrt{N}$ if $NR^2 \ll 1$. For $NR^2 \gg 1$, however, $R' = R$ is an excellent approximation.

A17.3. Variance of the reduction factor

The variance of the *square* of the amplitude a is more easily calculated than that of a itself. This is defined as

$$V(a^2) = \langle (a^2 - \langle a^2 \rangle)^2 \rangle = \langle a^4 \rangle - \langle a^2 \rangle^2 \qquad (A17.10)$$

where a^2 is given by (A17.7). We have then

$$\frac{V(a^2)}{\langle a^2 \rangle^2} = \frac{\langle [N + \sum c_i c_j + \sum s_i s_j]^2 \rangle - N^2 [1 + (N-1)(\langle c \rangle^2 + \langle s \rangle^2)]^2}{N^2 [1 + (N-1)(\langle c \rangle^2 + \langle s \rangle^2)]^2}$$

$$(A17.11)$$

where we have abbreviated $\cos \phi_i$ and $\sin \phi_i$ to c_i and s_i.

For simplicity we shall assume that the distribution function of ϕ_i is even, so that terms like $\langle s_i \rangle$ and $\langle s_i s_j \rangle$ vanish. After careful enumeration of terms and again assuming that all the ϕ_is are statistically independent, we find for $N \gg 1$

$$\frac{V(a^2)}{\langle a^2 \rangle^2} = \frac{2[2N(\langle c^2 \rangle \langle c \rangle^2 - \langle c \rangle^4) + \langle c^2 \rangle^2 + \langle s^2 \rangle^2]}{(1 + N \langle c \rangle^2)^2} \qquad (A17.12)$$

To calculate $\langle c^2 \rangle$ and $\langle s^2 \rangle$ we notice that they are $\frac{1}{2}(1 \pm \langle \cos 2\phi_i \rangle)$ and that $\langle \cos 2\phi_i \rangle$ is just the reduction factor for *twice* the relevant parameter. Thus for the Gaussian $\langle \cos 2\phi_i \rangle = e^{-\lambda^2} = R^4$, and for the Lorentzian $\langle \cos 2\phi_i \rangle = e^{-2\lambda} = R^2$. We also have for both distributions that $\langle c \rangle = R$, so (A17.12) reduces to

$$\frac{V(a^2)}{\langle a^2 \rangle^2} = \frac{2NR^2(1 + R^y) - 4NR^4 + 1 + R^{2y}}{1 + 2NR^2 + N^2 R^4} \qquad (A17.13)$$

where $y = 2$ for the Lorentzian distribution and $y = 4$ for the Gaussian. If, as is the case for R not very small, $NR^2 \gg 1$, we find

$$\frac{V(a^2)}{\langle a^2 \rangle^2} = \frac{2(1 - R^2)}{NR^2} \text{ for the Lorentzian and}$$

$$= \frac{2(1 - R^2)^2}{NR^2} \text{ for the Gaussian} \qquad (A17.14)$$

If, however, the factor R is small enough to make $NR^2 < 1$ (though N is still $\gg 1$), we find for either distribution

$$\frac{V(a^2)}{\langle a^2 \rangle^2} = \frac{1 + 2NR^2}{(1 + NR^2)^2} \qquad (A17.15)$$

Thus as NR^2 gets smaller (either by reduction of field or because N is not very large) the ratio approaches unity, i.e. the square root of the variance of a^2 becomes equal to a^2. This means that although the expectation value of the reduction factor levels out at $1/\sqrt{N}$ when NR^2 is small, the actual reduction factor will vary as between samples of the same N by an amount comparable to the reduction factor itself.

Appendix 18

Long beat in amplitude

Consider two oscillations of slightly different frequencies and amplitudes. Their resultant is

$$y = \tfrac{1}{2}(1 + \varepsilon)\sin(1 + \alpha)\theta + \tfrac{1}{2}(1 - \varepsilon)\sin(1 - \alpha)\theta \qquad (\text{A}18.1)$$

(here θ represents the mean phase $2\pi F/H$ and $2\alpha = \Delta F/F$). If α were to vanish the amplitude of y would be just unity. If α is small we can think of y as a single oscillation of approximately the mean frequency and of slowly varying amplitude and phase

$$y = (\cos^2 \alpha\theta + \varepsilon^2 \sin^2 \alpha\theta)^{1/2} \sin(\theta + \phi) \qquad (\text{A}18.2)$$

where the phase ϕ is such that $\tan \phi = \varepsilon \tan \alpha\theta$.

The minimum amplitude is for $\alpha\theta = \pi/2$ and is just ε. The value of ε can be estimated adequately from the depth of the minimum of fig. 8.7 below the straight line, but away from the minimum the exact value of ε need not be accurately known. Thus if $\varepsilon \tan \alpha\theta \ll 1$, (A18.2) is approximately

$$y = \cos \alpha\theta(1 + \tfrac{1}{2}\varepsilon^2 \tan^2\alpha\theta)\sin(\theta + \phi) \qquad (\text{A}18.3)$$

and for $\varepsilon = 0.047$ (the value for fig. 8.7), the correction term in the brackets is less than 1% for $\alpha\theta < 0.4\pi$ and for $\alpha\theta > 0.6\pi$.

The position of the minimum determines α. If the minimum is taken at the 81st oscillation (allowing for the damping of the Dingle factor, which makes the minimum of (A18.2) appear slightly to the left of the minimum observed amplitude), $\alpha\theta = \pi/2$ for $\theta = 81 \times 2\pi$. Therefore $\alpha = 1/(4 \times 81)$ and $\Delta F/F = 1/162$; a misalignment of $0.3°$ in the binary-bisectrix plane would be sufficient to account for this frequency splitting. To correct for the beat we need to divide each observed amplitude by the factor

$$\cos\left(\frac{n\pi}{162}\right)\left(1 + \varepsilon^2 \tan^2 \frac{n\pi}{162}\right)^{1/2}$$

where n is the serial number of the oscillation.

This analysis is a little oversimplified in assuming that ε is a constant independent of field (i.e. of θ). In fact the relative amplitudes of the two frequencies do vary slightly with field, but a more elaborate analysis shows

549

that this variation is unimportant: by the time ε has departed appreciably from its value at the minimum the contribution of the term in ε^2 has become negligible.

The origin of ε is, of course, that the two slightly different frequencies come from slightly non-equivalent extremal cross-sections because of slight misalignment of the binary axis. These then have slightly different cyclotron masses and slightly different Dingle temperatures resulting in slightly different amplitudes. A detailed analysis shows that the observed value of $\varepsilon = 0.047$ at the minimum of fig. 8.7 is reasonably consistent with the misalignment responsible for the frequency splitting.

Appendix 19

Ambiguities in the HR method

We discuss first the derivation of S from measurement of the ratio R defined as

$$R = |\cos 2\pi S/\cos \pi S| \qquad (A19.1)$$

For given observed R it is easily shown that

$$\left.\begin{array}{l}
|\cos \pi S| = \left(\dfrac{1}{2} + \dfrac{R^2}{16}\right)^{1/2} \pm \dfrac{R}{4} \\[4mm]
\cos 2\pi S = \dfrac{R^2}{4} \pm R\left(\dfrac{1}{2} + \dfrac{R^2}{16}\right)^{1/2}
\end{array}\right\} \qquad (A19.2)$$

If $R > 1$, the $+$ sign is impossible, since a cosine cannot exceed 1; the value of $|\cos \pi S|$ is then uniquely specified and only the ambiguity of determining S from $|\cos \pi S|$ remains. As discussed in the text, this ambiguity can be halved if the sign of $\cos \pi S$ is determined by a measurement of absolute phase.

If $R < 1$ the choice between $+$ and $-$ can be settled by determining the sign of $\cos 2\pi S$ (or of $\cos 2l\pi S$ for any integral l). This sign in turn can be determined by measuring the relative phase $\phi_2 - 2\phi_1$ (or more generally $\phi_{2l} - 2l\phi_1$). To demonstrate this we calculate the value of $\phi_p - p\phi_1$. Thus using (9.8), we have according to the signs of $\cos \pi S$ and $\cos \pi S$

$$\frac{\phi_p}{2\pi} - \frac{p\phi_1}{2\pi} = \left\{\begin{array}{ccc}
 & \cos \pi S & \cos \pi S \\[2mm]
\frac{1}{2}(1 - p) \pm \frac{1}{8}(1 - p) & + & + \\[4mm]
\frac{1}{2} \pm \frac{1}{8}(1 - p) & - & + \\[4mm]
-\frac{1}{2}p \pm \frac{1}{8}(1 - p) & + & - \\[4mm]
0 \pm \frac{1}{8}(1 - p) & - & -
\end{array}\right. \qquad (A19.3)$$

where the $\pm$ refers as usual to minimum $(+)$ and maximum $(-)$ extremal areas, and of course the usual convention holds that an appropriate integer should be added or subtracted from the r.h.s. to bring it into the range 0 to 1.

551

It can be seen that

$$\frac{\phi_p}{2\pi} - \frac{p\phi_1}{2\pi} = \begin{cases} \begin{rcases} \tfrac{1}{2} \pm \tfrac{1}{8}(1-p) & \text{if} \quad \cos p\pi S \text{ is } + \\ 0 \pm \tfrac{1}{8}(1-p) & \text{if} \quad \cos p\pi S \text{ is } - \end{rcases} \text{for } p \text{ even} \\[2em] \begin{rcases} \tfrac{1}{2} \pm \tfrac{1}{8}(1-p) & \text{if} \quad \cos p\pi S/\cos \pi S \text{ is } - \\ 0 \pm \tfrac{1}{8}(1-p) & \text{if} \quad \cos p\pi S/\cos \pi S \text{ is } + \end{rcases} \text{for } p \text{ odd} \end{cases}$$

$$(A19.4)$$

To give a concrete example, consider a noble metal neck, for which the LK harmonics can be made relatively large without MI being important. The neck has a minimum area, so that the $+$ sign should be chosen above, and (A19.4) gives

$$\frac{\phi_2}{2\pi} - \frac{2\phi_1}{2\pi} = \begin{cases} \tfrac{3}{8} & \text{if} \quad \cos 2\pi S \text{ is } + \\ \tfrac{7}{8} & \text{if} \quad \cos 2\pi S \text{ is } - \end{cases}$$

Thus measurement of $\phi_2 - 2\phi_1$ will immediately settle the sign of $\cos 2\pi S$ and so resolve the ambiguity of (A19.2) if $R < 1$.

If the $p = 3$ harmonic ratio, say $|\rho|$ is measured, where ρ is defined as

$$\rho = \cos 3\pi S/\cos \pi S \, (= 4\cos^2 \pi S - 3)$$

the sign of ρ is at once determined by the relative phase, since from (A19.4) we have

$$\frac{\phi_3}{2\pi} - \frac{3\phi_1}{2\pi} = \begin{cases} \tfrac{1}{4} & \text{if} \quad \rho \text{ is } - \\ \tfrac{3}{4} & \text{if} \quad \rho \text{ is } + \end{cases}$$

and so once again $|\cos \pi S|$ is unambiguously determined. Similar arguments may be developed for higher harmonic ratios and relative phases and their measurement may sometimes provide useful confirmation on the reliability of the measurements and their interpretation.

Bibliography and author index

The following abbreviations are used below

EFS   Springford, M. (ed.) (1980) *Electrons at the Fermi Surface*, Cambridge University Press.

FS   Harrison, W. A. & Webb, M. B. (eds.) (1960) *The Fermi Surface*, Proceedings of an international conference held at Cooperstown, N.Y., Wiley, New York.

LT9   (1965) *Proc. 9th Int. Conf. L.T. Phys.*, Columbus, Ohio 1964, Plenum, New York.

LT10 (1967) *Proc. 10th Int. Conf. L.T. Phys.*, Vol. 3, Moscow 1966, VINITI.

LT14 (1975) *Proc. 14th Int. Conf. L.T. Phys.*, Vol. 3, Otaniemi, Finland 1975, North-Holland, Amsterdam.

LT16 (1981) *Proc. 16th Int. Conf. L.T. Phys.*, Los Angeles 1981, *Physica* **108**.

PLT   Gorter, C. J. (ed.) *Progress in Low Temperature Physics*, North-Holland, Amsterdam.

PM   Ziman, J. M. (ed.) (1969) *The Physics of Metals* 1. *Electrons*, Cambridge University Press.

SF   Cochran, J. F. and Haering, R. R. (ed.) (1968) *Solid State Physics*, Vol. 1, *Electrons in metals*, The Simon Fraser University lectures.

SSP   Seitz, F. & Turnbull, D. (eds.) *Solid State Physics*, Academic Press, New York.

The journals *Zh. eksp. teor. fiz.*, *Dokl. Akad. Nauk SSSR*, *Pribory i tekhnika eksperimenta*, *Fiz. niz. temp.* and *Fiz. tver. tela* are in Russian; where possible reference is given to the English translation (which sometimes has appeared only in the following year). Italic numbers in square brackets indicate pages in the text.

Abrikosov, A. A. & Falkovsky, L. A. (1962), *Zh. eksp. teor. fiz.* **43**, 1089 (*Sov. Phys. JETP* **16**, 769, 1963) [*234*]

Adams E. N. & Holstein, T. D. (1959), *J. Phys. Chem. Sol.* **10**, 254 [*153–4*]

Akhieser, A. (1939), *C. R. Acad. Sci. URSS* **23**, 874 [*10, 66*]

Alekseevski, N. E. & Gaidukov, Yu. P. (1959), *Zh eksp. teor. fiz.* **37**, 672 (*Sov. Phys. JETP* **10**, 481, 1960) [*16*]

Alers, G. A. & Swim, R. T. (1963), *Phys. Rev. Lett.* **11**, 72 [*145*]

Alers, P. B. (1956), *Phys. Rev.* **101**, 41 [*156*]

Alers, P. B. (1957), *Phys. Rev.* **107**, 959 [*156*]

Allen, S. J., Rupp, L. W. & Schmidt, P. H. (1973), *Phys. Rev. B* **7**, 5121 [*193*]

Alles, H. G. & Higgins, R. J. (1973), *Rev. Sci. Inst.* **44**, 1646 [*119*]

Bibliography

Alles, H. G. & Higgins, R. J. (1974), *Phys. Rev. B* **9**, 158 [*465, 467*]
Alles, H. G. & Higgins, R. J. (1975), *Rev. Sci. Inst.* **46**, 963 [*119*]
Alles, H. G., Higgins, R. J. & Lowndes, D. H. (1973), *Phys. Rev. Lett.* **30**, 705 [*465*]
Alles, H. G. & Lowndes, D. H. (1973), *J. Phys. E* **6**, 895 [*301*]
Alpher, R. A. & Rubin, R. J. (1954), *J. Low. Temp. Phys.* **9**, 67 [*150*]
Altounian, Z. & Datars, W. R. (1980), *Canad. J. Phys.* **58**, 370 [*237–8*]
Andersen, O. K. & Mackintosh, A. R. (1968), *Sol. State Comm.* **6**, 285 [*222*]
Anderson, J. R. & Gold A. V. (1963), *Phys. Rev. Lett.* **10**, 277 [*225–6*]
Anderson, J. R. & Gold, A. V. (1965), *Phys. Rev.* **139**, A1459 [*13, 215, 240*]
Anderson, J. R. & Lane, S. S. (1970), *Phys. Rev. B* **2**, 298 [*214, 300*]
Anderson, J. R., O'Sullivan, W. J. and Schirber, J. E. (1972), *Phys. Rev. B* **5**, 4683 [*215, 239*]
Anderson, M. S., Gutman, E. J., Packard, J. R. & Swensen, C. A. (1969), *J. Phys. Chem. Solids* **30**, 1587 [*238*]
Ando, T., Fowler, A. B. & Stern, F. (1982), *Rev. Mod. Phys.*, **54**, 437 [*48*]
Animalu, A. O. E. & Heine, V. (1965), *Phil. Mag.* **11**, 379 [*214*]
Aoki, H. & Ogawa, K. (1978), *J. Low Temp. Phys.* **32**, 131 [*297–9*]
Aoki, H. & Ogawa, K. (1979), *J. Low Temp. Phys.* **35**, 329 [*297*]
Arko, A. J., Lowndes, D. H., Muller, F. A., Roeland, L. W., Wolfrat, J., van Kassel, A. T., Myron, H. W., Mueller, F. M. & Webb, G. W. (1978), *Phys. Rev. Lett.* **40**, 1590 [*102, 235*]
Aron, P. R. (1972), *J. Low Temp. Phys.* **9**, 67 [*243*]
Aron, P. R. & Chandrasekhar (1969), *Phys. Lett.* **30A**, 86 [*543*]
Ashcroft, N. W. (1963), *Phil. Mag.* **8**, 2055 [*211–4*]
Ashcroft, N. W. (1965), *Phys. Rev.* **140**, A935 [*194*]
Ashcroft, N. W. & Guild, L. J. (1965), *Phys. Lett.* **14**, 23 [*214*]
Azbel, M. Ya & Kaner, E. A. (1956), *Zh. eksp. teor. fiz.* **30**, 811 (*Sov. Phys. JETP* **3**, 772) [*16*]
Baraff, G. A. & Tsui, D. C. (1981), *Phys. Rev. B* **24**, 2274 [*157*]
Barklie, R. C. & Shoenberg, D. (1975), *Phys. Cond. Matter* **19**, 175 [*231, 398–9, 405–6*]
Basinski, Z. S., Howie, A., Lonzarich, G. G. & Sigfusson, T. (1981), Private communication [*402*]
Basinski, Z. S. & Saimoto, S. (1967), *Canad. J. Phys.* **45**, 1161 [*389*]
Bate, R. T. & Einspruch N. G. (1967), *Phys. Rev.* **153**, 796 [*249*]
Beardsley, G. M. & Schirber, J. E. (1972), *J. Low Temp. Phys.* **8**, 421 [*238*]
Beck, A., Jan, J.-P., Pearson, W. B. & Templeton, I. M. (1963), *Phil. Mag.* **8**, 351 [*234*]
Bellessa, G. (1973), *Phys. Rev. B* **7**, 2400 [*161, 170–1*]
Benedek, R. & Baratoff, A. (1973), *Sol. State Comm.* **13**, 385 [*386*]
Berko, S. (1978), *J. de Physique* **39**, Suppl. C6, 1568 [*17, 209*]
Berko, S. (1979) in *Electrons in Disordered Metals at Metallic Surfaces*, ed. P. Phariseau, B. L. Györffy & L. Scheire, Plenum Press, London and New York, p. 239 [*209*]
Berlincourt, T. G. & Steele, M. C. (1954), *Phys. Rev.* **95**, 1421 [*241, 245, 436, 443*]
Bhargava, R. N. (1966), *Phys Rev.* **156**, 785 [*228, 231, 249–50*]
Bibby, W. M. (1976), Ph.D. Thesis, University of Cambridge (as quoted by Shoenberg D. (1976), see fig. 3 and table II). [*228, 294*]

Bibliography

Bibby, W. M., Coleridge, P. T., Cooper, N. S., Nex, C. M. M. & Shoenberg, D. (1979), *J. Low Temp. Phys.* **34**, 681 [*204, 451*]
Bibby, W. M. & Shoenberg, D. (1979), *J. Low Temp. Phys.* **34**, 659 [*279, 372, 400, 459–61*]
Blackman, M. (1938), *Proc. Roy. Soc.* A**166**, 1 [*8*]
Bliek, L. M. & Landwehr, G. (1969), *Z. Naturf.* **23A**, 1861 [*156*]
Bloch, F. (1928), *Z. Phys.* **52**, 555 [*2*]
Blount, E. I. (1962), *Phys. Rev.* **126**, 1636 [*18, 331, 334–5*]
Bohm, H. V. & Easterling, V. J. (1962), *Phys. Rev.* **128**, 1021 [*208*]
Bömmel, H. E. (1955), *Phys. Rev.* **100**, 758 [*16*]
Borgnis, F. E. (1955), *Phys. Rev.* **98**, 1000 [*518*]
Bosacchi, B. & Franzosi, P. (1976), *J. Phys.* F **6**, L99 [*202*]
Brailsford, A. D. (1966), *Phys. Rev.* **149**, 456 [*61, 77, 375*]
Brandt, N. B. (1960), *Zh. eksp. teor. fiz.* **38**, 1355 (*Sov. Phys. JETP* **11**, 975) [*228*]
Brandt, N. B. & Lyubitina, L.G. (1964), *Zh. eksp. teor. fiz.* **47**, 1711 (*Sov. Phys. JETP* **20**, 1150, 1965) [*153*]
Brandt, N. B. & Razumenko, M. V. (1960), *Zh. eksp. teor. fiz.* **39**, 276 (*Sov. Phys. JETP* **12**, 198) [*249*]
Brandt, N. B. & Ryabenko, G. A. (1959), *Zh. eksp. teor. fiz.* **37**, 389 (*Sov. Phys. JETP* **10**, 278, 1960) [*142*]
Brignall, N. L. (1974), *J. Phys. C* **7**, 4266 [*95, 156*]
Brignall, N. L. & Shoenberg, D. (1974), *J. Phys. C* **7**, 1499 [*95*]
Broshar, W., McCombe, B. & Seidel, G. (1966), *Phys. Rev. Lett.* **16**, 325 [*277, 281–2*]
Brown, H. R. & Myers A. (1972), *J. Phys.* F **2**, 683 [*377, 383*]
Brown, R. N., Hartman, R. L. & Koenig, S. H. (1968), *Phys. Rev.* **172**, 598 [*229–30*]
Brown, R. N., Mavroides, J. G. & Lax, B. (1963), *Phys. Rev.* **129**, 2055 [*173, 232, 475*]
Caplin, A. D. & Shoenberg, D. (1965), *Phys. Lett.* **18**, 238 [*151*]
Casimir, H. B. G. (1968), *Helv. Phys. Acta.* **41**, 741 [*xiv*]
Chambers, R. G. (1956), *Canad. J. Phys.* **34**, 1395 [*22–3, 33*]
Chambers, R. G. (1966), *Proc. Phys. Soc.* **88**, 701 [*335–6, 338, 343, 345, 477*]
Chambers, R. G. (1969), *PM*, p. 175 [*174*]
Chambers, W. G. (1965), *Phys. Rev.* **140**, A135 [*347*]
Chandrasekhar, B. S. & Fawcett, E. (1971), *Adv. in Phys.* **20**, 775 [*144*]
Chang, Y. K. & Higgins, R. J. (1975), *Phys. Rev.* B **12**, 4261 [*393–6, 398–9, 401–5, 415–6*]
Chung, Y. & Lowndes, D. H. (1977), *Sol. State Comm.* **21**, 647 [*467*]
Chung, Y., Lowndes, D. H. & Hendel, C. L. (1978), *J. Low Temp. Phys.* **32**, 599 [*465*]
Chung, Y., Lowndes, D. H., Hendel, C. L. & Rode, J. P. (1976), *Sol. State Comm.* **20**, 101 [*467*]
Cochran, J. F. & Haering, R. R. (ed.) (1968), *SF* [*21*]
Cohen, M. H. & Blount, E. I. (1960), *Phil. Mag.* **5**, 115 [*11, 436*]
Cohen, M. H. & Falicov, L. M. (1961), *Phys. Rev. Lett.* **7**, 231 [*18, 331, 336*]
Coleridge, P. T. (1966), *Proc. Roy. Soc.* A **295**, 476 [*222*]
Coleridge, P. T. (1972), *J. Phys.* F **2**, 1016 [*377, 382*]
Coleridge, P. T. (1979), *J. Phys.* F **9**, 473 [*382–3*]
Coleridge, P. T. (1980) in *EFS*, p. 321 [*63, 248, 253, 375, 381, 386*]
Coleridge, P. T. & Holtham, P.M. (1977), *J. Phys.* F **7**, 1891 [*214, 437*]

Coleridge, P. T., Holzwarth, N. A. W. & Lee, M. J. G. (1974), *Phys. Rev. B* **10**, 1213 [*381–3*]

Coleridge, P. T., Scott, G. B. & Templeton, I. M. (1972), *Canad. J. Phys.* **50**, 1999 [*462–5*]

Coleridge, P. T. & Templeton, I. M. (1968), *Phys. Lett. A* **27**, 344 [*462*]

Coleridge, P. T. & Templeton, I. M. (1971), *Phys. Rev. Lett.* **27**, 507 [*437*]

Coleridge, P. T. & Templeton, I. M. (1972), *J. Phys. F* **2**, 643 [*202, 204, 431–3, 436–7*]

Coleridge, P. T. & Templeton, I. M. (1982) *Phys. Rev. B* **25**, 7818 [*202*]

Coleridge, P. T., Templeton, I. M. & Toyoda, T. (1981), *J. Phys. F* **11**, 2345 [*377, 383*]

Coleridge, P. T., Templeton, I. M. & Vašek, P. (1983), *J. Phys. F* to be published. [*377, 383*]

Coleridge, P. T. & Watts, B. R. (1971*a*), *Canad. J. Phys.* **49**, 2379 [*205*]

Coleridge, P. T. & Watts, B. R. (1971*b*), *Phil. Mag.* **24**, 1163 [*398, 401–3, 416*]

Condon, J. H. (1966), *Phys. Rev.* **145**, 526 [*255, 264, 267, 273–4, 277, 281–2, 302, 313, 316, 318*]

Condon, J. H. & Marcus, J. A. (1964), *Phys. Rev.* **134**, A446 [*91–2*]

Condon, J. H. & Walstedt, R. E. (1968), *Phys. Rev. Lett.* **21**, 612 [*275–6*]

Cooper, N. S. (1979), Ph.D. thesis, University of Cambridge [*247*]

Coulter, P. G. & Datars, W. R. (1980), *Phys. Rev. Lett.* **45**, 1021 [*196*]

Crabtree, G. W. (1977), *Phys. Rev. B* **16**, 1117 [*263*]

Crabtree, G. W., Dye. D. H., Karim, D. P. & Ketterson, J. B. (1979), *J. Magnetism Mag. Mat.* **11**, 236 [*222*]

Cracknell, A. P. & Wong, K. C. (1973), *The Fermi Surface*, Clarendon Press, Oxford [*21, 179*]

Croft, G. T., Donahoe, F. J. & Love, W. F. (1955), *Rev. Sci Inst.* **26**, 360 [*91*]

Davis, H. L. (1970), *Colloques Int. du Centre Nationale de la Recherche Scientifique*, No. 188, Grenoble, 1969 [*243*]

Davis, H. L., Faulkner, J. S. & Joy, H. W. (1968), *Phys. Rev.* **167**, 601 [*242–3*]

Dean, C. N. (1976), D. Phil. thesis, University of Sussex [*156*]

de Haas, W. J. (1914), *Proc. Netherlands Roy. Acad. Sci.* **16**, 1110 [*4*]

de Haas, W. J. & van Alphen, P. M. (1930*a*), *Proc. Netherlands Roy. Acad. Sci.* **33**, 680 [*3, 87*]

de Haas, W. J. & van Alphen, P. M. (1930*b*), *Proc. Netherlands Roy. Acad. Sci.* **33**, 1106 [*3–5, 87*]

de Haas, W. J. & van Alphen, P. M. (1932), *Proc. Netherlands Roy. Acad. Sci.* **35**, 454 [*4, 87*]

Deimel, P. & Doezema, R. E. (1974), *Phys. Rev. B* **10**, 4897 [*206–7*]

de Wilde, J. (1978), Doctoral thesis, Free University of Amsterdam [*150*]

de Wilde, J. & Meredith, D. J. (1976), *J. Phys. E* **9**, 62 [*96*]

Dhillon, J. S. & Shoenberg, D. (1955), *Phil. Trans. Roy. Soc. A* **248**, 1 [*230, 245, 350–1, 431, 435–6, 441*]

Dimmock, J. O. (1971), *SSP* **26**, 103 [*204*]

Dingle, R. B. (1952*a*), *Proc. Roy. Soc. A* **211**, 500 [*10, 65, 76*]

Dingle, R. B. (1952*b*), *Proc. Roy. Soc. A* **211**, 517 [*10, 61, 369*]

Doezema, R. E. & Koch, J. F. (1972), *Phys. Rev. B* **5**, 3866 [*206–7*]

Doezema, R. E. & Koch, J. F. (1975), *Phys. Cond. Matter* **19**, 17 [*176*]

Bibliography

Donaghy, J. J. & Stewart, A. T. (1967), *Phys. Rev.* **164**, 391 [*194*]

Dresselhaus, M. S. & Mavroides, J. G. (1964), *Sol. State Comm.* **2**, 297 [*173*]

Drude, P. (1900), *Ann. Phys. Lpz.* (4) **1**, 566 [*1*]

Dunsworth, A. E. & Datars, W. R. (1973), *Phys. Rev. B* **7**, 3435 [*250*]

Dupree, R. & Holland, B. W. (1967), *Phys. Stat. Sol.* **24**, 275 [*455*]

Dye, D. H., Campbell, S. A., Crabtree, G. W., Ketterson, J. B., Sandesara, N. B. & Vuillemin, J. J. (1981), *Phys. Rev. B* **23**, 462 [*224*]

Ebert, G., von Klitzing, K., Probst, C. and Ploog, K. (1982), *Sol. State Comm.* **44**, 95 [*157, 159*]

Eddy, J. W, Jr. & Stark, R. W. (1982), *Phys. Rev. Lett.* **48**, 275 [*351, 353*]

Edelman, V. S. (1973), *Zh. eksp. teor. fiz.* **64**, 1734 (*Sov. Phys. JETP* **37**, 875) [*228*]

Edelman, V. S. (1976), *Adv. in Phys.* **25**, 555 [*179, 228, 233, 440*]

Edelman, V. S., Volsky, E. P. & Khaikin, M. S. (1966), *Pribory i tekhnika eksperimenta* No. 3, 179 [*108*]

Edwards, G. J., Springford, M. & Saito, Y. (1969), *J. Phys. Chem. Solids* **30**, 2527 [*235*]

Elliott, M. (1979), D. Phil. thesis, University of Sussex [*75*]

Elliott, M. & Datars, W. R. (1982) *J. Phys. F* **12**, 465 [*238*]

Elliott, M., Ellis, T. & Springford, M. (1978), *Phys. Rev. Lett.* **41**, 709 [*424*]

Elliott, M., Ellis, T. & Springford, M. (1980), *J. Phys. F.* **10**, 2681 [*76, 81, 371, 423–4, 501, 505*]

Engelsberg, S. & Simpson, G. (1970), *Phys. Rev. B* **2**, 1657 [*79, 81, 422*]

Everett, P. M. & Grenier, C. G. (1977), *Phys. Rev. B* **15**, 3826 [*299*]

Falicov, L. M. (1960), *FS*, p. 39 [*17*]

Falicov, L. M. (1962), *Phil. Trans. Roy. Soc. A* **255**, 55 [*221*]

Falicov, L. M., Pippard, A. B. & Sievert, P. R. (1966), *Phys. Rev.* **151**, 498 [*359–62, 533, 536*]

Falicov, L. M. & Sievert, P. R. (1964), *Phys. Rev. Lett.* **12**, 558 [*357*]

Falicov, L. M. & Stachowiak, H. (1966), *Phys. Rev.* **147**, 505 [*62, 341, 348–9*]

Fawcett, E. (1956), *Phys. Rev.* **103**, 1582 [*16*]

Fawcett, E., Griessen, R., Joss, W., Lee, M. J. G. & Perz, J. M. (1980), *EFS*, p. 278 [*140, 144, 237, 242–4*]

Fletcher, R. (1981), *J. Low Temp. Phys.* **43**, 363 [*174*]

Foldy, L. L. (1968), *Phys. Rev.* **170**, 670 [*185*]

Foner, S. (1959), *Rev. Sci. Inst.* **30**, 548 [*95*]

Foner, S. (1975), *Rev. Sci. Inst.* **46**, 1425 [*95*]

Fowler, A. B., Fang, F. F., Howard, W. E. & Stiles, P. J. (1966), *Phys. Rev. Lett.* **16**, 901 [*50, 157*]

Fowler, M. & Prange, R. (1965), *Physics* **1**, 315 [*79, 81*]

Friedberg, C. B. (1974), *J. Low Temp. Phys.* **14**, 147 [*368*]

Fujimori, Y. (1969), *Mem. Fac. Sci. Kyushu Univ. B* **4**, 47 [*160*]

Gaertner, A. A. & Templeton, I. M. (1977), *J. Low Temp. Phys.* **29**, 205 [*185, 188, 191, 193–4, 437*]

Gallagher, B. L. & Greig, D. (1981), *LT*16, p. 899 [*381*]

Gamble, D. & Watts, B. R. (1972), *Phys. Lett.* **40A**, 22 [*242–3*]

Gamble, D. & Watts, B. R. (1973), *J. Phys. F* **3**, 98 [*131, 240*]

Gantmakher, V. F. (1962), *Zh. eksp. teor. fiz.* **43**, 345 (*Sov. Phys. JETP* **16**, 247, 1963) [*17*]

Bibliography

Gantmakher, V. F. (1972), *Zh. eksp. teor. fiz. pis. red.* **16**, 256 (*Sov. Phys. JETP Letters*, **16**, 180) [*422*]

Gantmakher, V. F., Lebech, J. & Bak, C. K. (1979), *Phys. Rev.* B **20**, 5111 [*175*]

Gasparov, V. A. & Harutunian, M. H. (1976), *Phys. Stat. Sol.* (b) **74**, K107 [*208*]

Gerstein, I. I. & Elbaum, C. (1973), *Phys. Stat. Sol.* (b) **57**, 157 [*131*]

Ginzburg, Vl. L. & Maksimov, A. O. (1977), *Fiz. Niz. Temp.* **3**, 1285 [*393*]

Girvan, R. F., Gold, A. V. & Phillips, R. A. (1968), *J. Phys. Chem. Solids* **29**, 1485 [*222–4*]

Giuliani, G. F. & Overhauser, A. W. (1980), *Phys. Rev.* B **22**, 3639 [*195*]

Glinski, R. & Templeton, I. M. (1969), *J. Low Temp. Phys.* **1**, 223 [*238*]

Gold, A. V. (1958), *Phil. Trans. Roy. Soc.* A **251**, 85 [*13, 211*]

Gold, A. V. (1968), *SF*, p. 39 [*21, 23, 477*]

Gold, A. V. (1974), *J. Low Temp. Phys.* **16**, 3 [*179, 228*]

Gold, A. V. & Schmor, P. W. (1976), *Canad. J. Phys.* **54**, 2445 [*96, 445*]

Gold, A. V. & Van Schyndel, A. J. (1981), *J. Low Temp. Phys.* **44**, 73 [*101, 327, 445, 447, 451, 454*]

Goldstein, A., Williamson, S. J. & Foner, S. (1965), *Rev. Sci. Inst.* **36**, 1356 [*103*]

Goodrich, R. G., Khan, S. A. & Reynolds, J. M. (1971), *Phys. Rev.* B **3**, 2379 [*174*]

Goy, P. & Castaing, B. (1973), *Phys. Rev.* B **7**, 4409 [*80*]

Gradshteyn, E. S. & Ryzhik, I. W. (1965), *Tables of Integrals, Series and Products*, 4th edn, Academic Press, New York & London [*497*]

Graebner, J. E. (1977), *Sol. State Comm.* **21**, 353 [*235*]

Graebner, J. E. & Robbins, M. (1976), *Phys. Rev. Lett.* **36**, 422 [*235*]

Grassie, A. D. C. (1963), Ph.D. thesis, University of Cambridge [*156*]

Gray, A., Matthews, G. B. & MacRobert, T. M. (1952), *A treatise on Bessel Functions and their applications to mathematical physics*, Macmillan & Co., London [*259*]

Green, B. A. & Chandrasekhar, B. S. (1963), *Phys. Rev. Lett.* **11**, 331 [*140*]

Grenier, C. G., Reynolds, J. M. & Zebouni, N. H. (1963), *Phys. Rev.* **129**, 1088 [*174*]

Griessen, R. (1973), *Cryogenics* **13**, 375 [*90*]

Griessen, R., Lee, M. J. G. & Stanley, D. J. (1977), *Phys. Rev.* B **16**, 4385 [*146*]

Griessen, R. & Sorbello, R. S. (1972), *Phys. Rev.* B **6**, 2198 [*146, 239*]

Griessen, R. & Sorbello, R. S. (1974), *J. Low Temp. Phys.* **16**, 237 [*240*]

Grimes, C. C. (1978), *Surf. Sci.* **73**, 379 [*48*]

Grimes, C. C., Adams, G. & Schmidt, P. H. (1967), *Bull. Am Phys. Soc.* **12**, 414 [*193*]

Grimes, C. C. & Kip, A. F. (1963), *Phys. Rev.* **132**, 1991 [*192*]

Grimvall, G. (1981) *The electron–phonon interaction in metals*, North-Holland, Amsterdam [*73*]

Grodski, J. J. & Dixon, A. E. (1976), private communication [*418*]

Gunnersen, E. M. (1957), *Phil. Trans. Roy. Soc.* A **249**, 299 [*12, 214*]

Gurevich, V. L. & Firsov, Yu.A. (1961), *Zh. eksp. teor. fiz.* **40**, 198 (*Sov. Phys. JETP* **13**, 137) [*176*]

Gurevich, V. L., Skobov, V. G. & Firsov, Yu.A. (1961), *Zh. eksp. teor. fiz.* **40**, 786 (*Sov. Phys. JETP*, **13**, 552) [*160*]

Gutman, E. J. & Trivisonno, J. (1967), *J. Phys. Chem. Solids* **28**, 805 [*238*]

Halloran, M. H. & Hsu, F. S. L. (1965), *Bull. Am. Phys. Soc.* **10**, 350 [*316*]

Halloran, M. H. & Kunzler, J. E. (1968), *Rev. Sci. Inst.* **39**, 1501 [*137*]

Halse, M. R. (1969), *Phil. Trans. Roy. Soc.* A **265**, 507 [*199–204, 209–10, 310, 448*]

Hamman, D. & Overhauser, A. W. (1966), *Phys. Rev.* **143**, 183 [*461*]

Bibliography

Harper, P. G., Hodby, J. W. & Stradling, R. A. (1973), *Rep. Prog. Phys.* **36**, 1 [*176*]
Harrison, W. A. (1960), *Phys. Rev.* **118**, 1190 [*211*]
Harrison, W. A. & Webb, M. B. (eds) (1960), *FS* [*17*]
Harte, G. A., Priestley, M. G. & Vuillemin, J. J. (1978), *J. Low Temp. Phys.* **31**, 879 [*249–51*]
Hebborn, J. E. & Sondheimer, E. H. (1960), *J. Phys. Chem. Solids* **13**, 105 [*490*]
Heine, V. (1957), *Proc. Roy. Soc. A* **240**, 340, 354 and 361 [*13*]
Heine, V. & Abarenkov, I. (1964), *Phil. Mag.* **9**, 451 [*195*]
Heine, V., Cohen, Marvin L. & Weaire, D. (1970), *SSP* **24**, 1, 38 and 250 [*13, 211*]
Higgins, R. J. & Lowndes, D. H. (1980), *EFS*, p. 393 [*463, 467*]
Hill, P. H. and Vanderkooy, J. (1978), *Phys. Rev. B* **17**, 1563 [*444*]
Hoekstra, J. A. & Stanford, J. L. (1973), *Phys. Rev. B* **8**, 1416 [*222*]
Holstein, T. D., Norton, R. E. & Pincus, P. (1973), *Phys. Rev. B* **8**, 2649 [*255*]
Holtham, P. M. & Parsons, D. (1976), *J. Phys. F* **6**, 1481 [*252*]
Holzwarth, N. A. W. (1975), *Phys. Rev. B* **11**, 3718 [*381*]
Hörnfeldt, S. P., Ketterson, J. B. & Windmiller, L. R. (1973), *J. Phys. E* **6**, 265 [*301*]
Hsiang, T. Y., Reister, J. W., Weinstock, H., Crabtree, G. W. & Vuillemin, J. J. (1981) *Phys. Rev. Lett.* **47**, 523 [*224*]
Huberman, M. & Overhauser, A. W. (1981), *Phys. Rev. Lett.* **47**, 682 [*196*]
Hughes, A. J. & Shepherd, J. P. G. (1969), *J. Phys. C* **2**, 661 [*214*]
Hulbert, J. K. (1976), *J. Phys. E* **9**, 283 [*121*]
Jan, J.-P., MacDonald, A. H. & Skriver, H. L. (1980), *Phys. Rev. B* **21**, 5584 [*238–9*]
Jan, J.-P., Pearson, W. B. & Saito, Y. (1967), *Proc. Roy. Soc. A* **297**, 275 [*235*]
Jan, J.-P. & Skriver, H. L. (1977), *J. Phys. F* **7**, 957 [*236*]
Joseph, A. S. & Thorsen, A. C. (1964a), *Phys. Rev.* **134**, A979 [*449*]
Joseph, A. S. & Thorsen, A. C. (1964b), *Phys. Rev. Lett.* **13**, 9 [*202, 450*]
Joseph, A. S. & Thorsen, A. C. (1965), *Phys. Rev.* **138**, A1159 [*93, 302, 304*]
Joseph, A. S., Thorsen, A. C., Gertner, E. & Valby, L. E. (1966), *Phys. Rev.* **148**, 569 [*450*]
Joss, W. (1981), *Phys. Rev. B* **23**, 4913 [*239*]
Kaganov, M. I., Lifshitz, I. M. & Sinelnikov, K. D. (1957), *Zh. eksp. teor. fiz.* **32**, 605 (*Sov. Phys. JETP* **5**, 500) [*68, 151*]
Kahn, A. H. & Frederikse, H. P. R. (1959), *SSP* **9**, 257 [*156*]
Kane, E. O. (1960), *J. Phys. Chem. Solids* **12**, 181 [*334*]
Kaplan, J. A. & Glasser, M. L. (1969), *Phys. Rev.* **186**, 958 [*456*]
Kawaji, S. & Wakabayashi, J. (1976), *Surf. Sci.* **58**, 238 [*157*]
Ketterson, J. B. & Stark, R. W. (1967), *Phys. Rev.* **156**, 748 [*217–8*]
Khaikin, M. S. (1969), *Adv. in Phys.* **18**, 1 [*176*]
Kimball, J. C., Stark, R. W. & Mueller, F. M. (1967), *Phys. Rev.* **162**, 600 [*221*]
King-Smith, P. E. (1965), *Phil. Mag.* **12**, 1123 [*252, 375, 381*]
Kip, A. F. (1960), *FS* p. 147 [*16*]
Kittel, C. (1949), *Rev. Mod. Phys.* **21**, 541 [*526*]
Knecht, B. (1975), *J. Low Temp. Phys.* **21**, 619 [*112, 115, 444–50, 455, 458–9*]
Knecht, B., Lonzarich, G. G., Perz, J. M. & Shoenberg, D. (1977), *J. Low Temp. Phys.* **29**, 499 [*109, 156, 304*]
Koch, J. F., Stradling, R. A. & Kip, A. F. (1964), *Phys. Rev.* **133**, A240 [*204–5*]
Kollarits, F. J. & Trivisonno, J. (1968), *J. Phys. Chem. Solids* **29**, 2133 [*238*]

Bibliography

Korolyuk, A. P. & Prushchak, T. A. (1961), *Zh. eksp. teor. fiz.* **41**, 1689 (*Sov. Phys.* **14**, 1201, 1962) [*170*]

Krishnan, K. S. & Banerjee, S. (1935), *Phil. Trans. Roy. Soc. Lond.* **234**, 265 [*6*]

Kulik, I. O. & Gogadze, G. A. (1963), *Zh eksp. teor. fiz.* **44**, 530 (*Sov. Phys. JETP* **17**, 361) [*151*]

Kumar, S. & Lee, H. J. (1970), unpublished report [*330*]

Kunzler, J. E., Hsu, F, S. L. & Boyle, W. S. (1962), *Phys. Rev.* **128**, 1084 [*132*]

Kunzler, J. E. & Klauder, J. R. (1961), *Phil. Mag.* **6**, 1045 [*208*]

Landau, L. (1930), *Z. Phys.* **64**, 629 [*1, 38*]

Landau, L. D. (1939), see Appendix to Shoenberg, 1939 [*38, 57, 228*]

Landau, L. D. & Lifshitz, E. M. (1980), *Statistical Physics* Pt. I, 3rd edn, Pergamon Press, Oxford [*37*]

Larson, C. O. & Gordon, W. L. (1967), *Phys. Rev.* **156**, 703 [*213–4*]

Lee, H. J. (1970), *Phys. Rev. B* **2**, 4618 [*328, 330*]

Lee, H. J., Canuto, V., Chiu, H. Y. & Chiuderi, C. (1969), *Phys. Rev. Lett.* **23**, 390 [*328*]

Lee, M. J. G. (1966), *Proc. Roy. Soc. A* **295**, 440 [*123, 191, 194, 421*]

Lee, M. J. G. (1969a), *Phys. Rev.* **178**, 953 [*194–5*]

Lee, M. J. G. (1969b), *Phys. Rev.* **187**, 901 [*199, 204*]

Lee, M. J. G. (1970), *Phys. Rev. B* **2**, 250 [*206, 208*]

Lee, M. J. G. & Falicov, L. M. (1968), *Proc. Roy. Soc. A* **304**, 319 [*185, 188, 195*]

Lee. M. J. G., Holzwarth, N. A. W. & Coleridge, P. T. (1976), *Phys. Rev. B* **13**, 3249 [*381, 383*]

Legkostupov, M. S. (1971), *Zh. eksp. teor. fiz.* **61**, 262 (*Sov. Phys. JETP* **34**, 136, 1972) [*128*]

Lengeler, B. (1977), *Phys. Rev. B* **15**, 5504 [*377, 386*]

Lengeler, B. (1978), *Electronic structure of noble metals*, Springer Tracts in Modern Physics, Berlin [*21*]

Lengeler, B., Wampler, W. R., Bourassa, R. R., Mika, K., Wingerath, K. & Uelhoff, W. (1977), *Phys. Rev. B* **15**, 5493 [*204–8*]

Lifshitz, I. M., Azbel, M. Ya & Kaganov, M. I. (1956), *Zh. eksp. teor. fiz.* **31**, 63 (*Sov. Phys. JETP* **4**, 41, 1957) [*16*]

Lifshitz, I. M. & Kosevich, A. M. (1954) *Dokl. Akad. Nauk SSR* **96**, 963 [*11, 33, 67*]

Lifshitz, I. M. & Kosevich, A. M. (1955), *Zh. eksp. teor. fiz.* **29**, 730 (*Sov. Phys. JETP* **2**, 636 (1956)) [*11, 24, 35, 67, 484*]

Lifshitz, I. M. & Pogorelov, A. V. (1954), *Dokl. Akad. Nauk SSR* **96**, 1143 [*12, 179, 183*]

Lipson, S. G. (1966), *Proc. Roy. Soc. A* **293**, 275 [*208*]

Llewellyn, B., Paul, D. McK., Randles, D. L. & Springford, M. (1977), *J. Phys. F* **7**, 2531 and 2545 [*375, 377, 379–80*]

Lomer, W. M. (1962), *Proc. Phys. Soc.* **80**, 489 [*221*]

Lonzarich, G. G. (1980), *EFS* p. 225 [*179, 183, 227–8, 244–5, 247*]

Lonzarich, G. G. & Gold, A. V. (1974), *Canad. J. Phys.* **52**, 694 [*246–7*]

Lonzarich, G. G. & Holtham, P. M. (1975), *Phys. Canada* **31**, 17 [*111, 156, 352*]

Loucks, T. L. & Cutler, P. H. (1964), *Phys. Rev.* **133**, A819 [*219, 221*]

Lowndes, D. H., Miller, K. M., Poulsen, R. G. & Springford, M. (1973), *Proc. Roy. Soc. A* **331**, 497 [*374, 377–8, 383*]

Luttinger, J. M. (1960a), *FS* pp. 2 and 67 [*17*]

Bibliography

Luttinger, J. M. (1960*b*), *Phys. Rev.* **119**, 1153 [*78*]
McClure, J. W. (1976) *J. Low Temp. Phys.* **25**, 523 [*234*]
McClure, J. W. & Choi, K. H. (1977), *Solid State Comm.* **21**, 1015 [*234*]
McClure, J. W. & Shoenberg, D. (1976), *J. Low Temp. Phys.* **22**, 233 [*492*]
MacDonald, A. H. (1980), *J. Phys. F* **10**, 1737 [*194*]
MacDonald, A. H. & Vosko, S. H. (1976), *J. Low Temp. Phys.* **25**, 27 [*458*]
Mackinnon, L. (1949), *Proc. Phys. Soc. B* **62**, 170 [*9*]
Mackintosh, A. R. & Andersen, O. K. (1980), *EFS* p. 149 [*221*]
Mahan, G. D. (1981) *Many-particle physics*, Plenum Press, New York [*73*]
Maltz, M. & Dresselhaus, M. S. (1970), *Phys. Rev. B* **2**, 2877 [*174*]
Marcus, J. A. (1947), *Phys. Rev.* **71**, 559 [*9, 87*]
Marquardt, W. R. & Trivisonno, J. (1965), *J. Phys. Chem. Solids* **26**, 273 [*238*]
Martin, D. L. (1970), *Canad. J. Phys.* **48**, 1327 [*193*]
Martin, D. L. (1973), *Phys. Rev. B* **8**, 5357 [*206*]
Mase, S. (1958) *J. Phys. Soc. Japan* **13**, 434 [*234*]
Mase, S., Fujimori, Y. & Mori, H. (1966), *J. Phys. Soc. Japan* **21**, 1744 [*170*]
Mavroides, J. G., Lax, B., Button, K. J. & Shapira, Y. (1962), *Phys. Rev. Lett.* **9**, 451 [*140, 145*]
Mayers, J. & Watts, B. R. (1978), *J. Phys. F* **8**, 799 [*131*]
Melz, P. J. (1966), *Phys. Rev.* **152**, 540 [*130, 239*]
Mercouroff, W. (1967) *La surface de Fermi des metaux*, Masson et cie, Paris [*21*]
Miller, K. M., Poulsen, R. G. & Springford, M. (1972), *J. Low Temp. Phys.* **6**, 411 [*417*]
Molenaar, J., Coleridge, P. T. & Lodder, A. (1981), *J. Phys. F* **11**, 1437 [*386*]
Morgan, G. J. (1966), *Proc. Phys. Soc.* **89**, 365 [*381*]
Morton, V. M. (1960), Ph.D. thesis, University of Cambridge [*210*]
Mueller, F. M. (1966), *Phys. Rev.* **148**, 636 [*185*]
Mueller, F. M. & Myron, H. W. (1976), *Comm. Phys.* **1**, 99 [*424*]
Mueller, F. M. & Priestley, M. G. (1966), *Phys. Rev.* **148**, 638 [*185*]
Mueller, F. M., Windmiller, L. R. & Ketterson, J. B. (1970), *J. App. Phys.* **41**, 2312 [*182, 188*]
Myers, A. & Bosnell, J. R. (1966), *Phil. Mag.* **13**, 1273 [*441*]
Nakhimovich, N. M. (1941), *J. Phys. USSR* **6**, 11 [*9*]
Narita, S., Takeyama, S., Luo W. B., Hiyamizu, S., Nanbu, K. & Hashimoto, H. (1981), *Jap. J. App. Phys.* **20**, L443 [*48*]
Nowak, D. & Lee, M. J. G. (1972), *Phys. Rev. B* **5**, 2851 [*204*]
O'Connell, R. F. & Roussel, K. M. (1971), *Nature, Physical Science* **231**, 32 [*328, 330*]
Oder, R. R. & Maxwell, E. (1965), *Phys. Lett.* **19**, 108 [*102, 124*]
Onsager, L. (1952), *Phil. Mag.* **43**, 1006 [*11, 33*]
O'Shea, M. J. & Springford, M. (1981), *Phys. Rev. Lett.* **46**, 1303 [*196*]
O'Sullivan, W. J. & Schirber, J. E. (1966), *Phys. Rev.* **151**, 484 [*218, 241, 245*]
O'Sullivan, W. J. & Schirber, J. E. (1967), *Phys. Rev.* **162**, 519 [*436, 441–3*]
Overhauser, A. W. (1964), *Phys. Rev. Lett.* **13**, 190 [*195*]
Overhauser, A. W. (1968), *Phys. Rev.* **167**, 691 [*195*]
Overhauser, A. W. (1978), *Adv. in Phys.* **27**, 343 [*195*]
Paalanen, M. A., Tsui, D. C. & Gossard, A. C. (1982), *Phys. Rev. B* **25**, 5566 [*157*]
Paciga, J. J. & Williams, D. L. (1971), *Canad. J. Phys.* **49**, 3227 [*194*]

Bibliography

Palin, C. J. (1970), Ph.D. thesis, University of Cambridge [422]
Palin, C. J. (1972), Proc. Roy. Soc. A 329, 17 [80, 245, 371, 422]
Panousis, P. T. & Gold, A. V. (1969), Rev. Sci. Inst. 40, 123 [101]
Paul, D. McK. & Springford, M. (1977), J. Low Temp. Phys. 27, 561 [373, 537, 539]
Pauli, W. (1927), Z. Phys. 41, 81 [1]
Peierls, R. (1933a), Z. Phys. 80, 763 [486, 489]
Peierls, R. (1933b), Z. Phys. 81, 186 [4, 5, 9]
Pelikh, L. N. (1966), Fiz. tver. tela 8, 1954 (Sov. Phys. Solid State 8, 1550) [151–2]
Pepper, M. (1977), Contemp. Phys. 18, 423 [48]
Pepper, M. (1978), Phil. Mag. 37, 83 [157]
Perz, J. M. & Hum, R. H. (1971), Canad. J. Phys. 49, 1 [131]
Perz, J. M. & Shoenberg, D. (1976), J. Low Temp. Phys. 25, 275 [312, 400, 421, 438, 447, 459]
Peter, M., Randles, D. L. & Shoenberg, D. (1970), Phys. Lett. 33A, 357 [152]
Phillips, R. A. & Gold, A. V. (1969), Phys. Rev. 178, 932 [100, 290, 446, 451, 467]
Piccini, A. & Whitten, W. B. (1966), Phys. Lett. An. Acad. Brasil Cienc. 38, 253 [151]
Pidduck, F. B. (1925), A Treatise on Electricity (2nd edn) Cambridge University Press [116]
Pines, D. & Nozières, P. (1966) The Theory of Quantum Liquids, Benjamin, New York [73, 456, 462]
Pippard, A. B. (1954), Proc. Roy. Soc. A 224, 273 [15]
Pippard, A. B. (1957a), Phil. Trans. Roy. Soc. A 250, 325 [15, 196, 209–10]
Pippard, A. B. (1957b), Phil. Mag. 2, 1147 [16]
Pippard, A. B. (1962), Proc. Roy. Soc. A 270, 1 [333, 335, 337, 342]
Pippard, A. B. (1963), Proc. Roy. Soc. A 272, 192 [256, 260]
Pippard, A. B. (1964), Phil. Trans. Roy. Soc. A 256, 317 [333, 338, 346, 350, 354, 363]
Pippard, A. B. (1965a), Proc. Roy. Soc. A 287, 165 [333, 337–8, 354–5, 363, 533]
Pippard, A. B. (1965b), The Dynamics of Conduction Electrons, Blackie & Son, Glasgow [153, 154]
Pippard, A. B. (1968), SF, p. 1 [25, 43, 453]
Pippard, A. B. (1969), PM, p. 113 [333, 453]
Pippard, A. B. (1980), EFS, p. 124 [255, 260, 269, 273, 322]
Platzmann, P. M. & Wolf, P. A. (1973), SSP Suppl. 13, 1 [456]
Plummer, R. D. & Gordon, W. L. (1964), Phys. Rev. Lett. 13, 432 [277, 281, 302–3]
Plummer, R. D. & Gordon, W. L. (1966), Phys. Lett. 20, 612 [302, 304]
Pollard, H. F. (1977), Sound Waves in Solids, Pion Ltd, London [518]
Poulsen, R. G., Randles, D. L. & Springford, M. (1974), J. Phys. F 4, 981 [379–80]
Priestley, M. G. (1960), Phil. Mag. 5, 111 [208]
Priestley, M. G. (1962), Phil. Mag. 7, 1205 [214, 411, 413, 416]
Priestley, M. G. (1963), Proc. Roy. Soc. A 276, 258 [219]
Priestley, M. G., Falicov, L. M. & Weiss, G. (1963), Phys. Rev. 131, 617 [336]
Pudalov, V. M. & Khaikin, M. S. (1974), Zh. eksp. teor. fiz. 67, 2260 (Sov. Phys. JETP 40, 1121) [319]
Randles, D. L. (1972), Proc. Roy. Soc. A 331, 85 [444, 450–1, 455]
Randles, D. L. & Springford, M. (1973), J. Phys. F 3, L185 [421]
Randles, D. L. & Springford, M. (1976), J. Phys. F 6, 1827 [189, 191–4, 421–2]
Reed, W. A. & Condon, J. H. (1970), Phys. Rev. B 1, 3504 [324]
Reifenberger, R. (1977), J. Low Temp. Phys. 26, 827 [368]

Bibliography

Reifenberger, R. & Stark, R. W. (1975), *Phys. Cond. Matt.* **19**, 361 [*368*]
Roaf, D. J. (1962), *Phil. Trans. Roy. Soc. A* **255**, 135 [*199, 200, 209*]
Rode, J. P. & Lowndes, D. H. (1977), *Phys. Rev. B* **16**, 2792 [*411*]
Roth, L. M. (1966), *Phys. Rev.* **145**, 434 [*33, 40, 436, 486, 489–90*]
Ruesink, D. W. & Perz, J. M. (1983), *Canad. J. Phys.* **61**, 177 [*242–3*]
Sandesara, N. B. & Stark, R. W. (1983) to be published
Schirber, J. E. (1965), *Phys. Rev.* **140**, A2065 [*241*]
Schirber, J. E. (1970) *Cryogenics* **10**, 418 [*13, 241*]
Schubnikow, L. W. & de Haas, W. J. (1930), *Proc. Netherlands Roy. Acad. Sci.* **33**, 130 and 163 [*4, 153*]
Schultz, S. & Dunifer, G. (1967), *Phys. Rev. Lett.* **18**, 283 [*458*]
Sellmyer, D. J. (1978), *SSP* **33**, 83 [*19, 179, 235*]
Shapira, Y. (1964), *Phys. Rev. Lett.* **13**, 162 [*439*]
Shapira, Y. (1968), *Phys. Acoust.* **5**, 1 [*161*]
Shapira, Y. & Lax, B. (1965), *Phys. Rev.* **138**, A1191 [*439*]
Shapira, Y. & Neuringer, L. J. (1967), *Phys. Rev. Lett.* **18**, 1133 [*170, 172*]
Shoenberg, D. (1938), *Zh. eksp. teor. fiz.* **8**, 1178 [*9*]
Shoenberg, D. (1939), *Proc. Roy. Soc. A* **170**, 341 [*7, 9, 88, 228, 369, 431, 435*]
Shoenberg, D. (1949), *Nature* **164**, 225 [*10*]
Shoenberg, D. (1952a), *Phil. Trans. Roy. Soc. Lond.* **245**, 1 [*10, 89, 214, 370, 374–5, 431, 436*]
Shoenberg, D. (1952b), *Nature* **170**, 569 [*13, 96*]
Shoenberg, D. (1953), *Physica* **19**, 791 [*13, 96*]
Shoenberg, D. (1955), *Conference de Physique des Basses Temperatures*, Paris (suppl. au *Bulletin de l'Inst. du Froid*) p. 212 [*101*]
Shoenberg, D. (1957), *PLT* **2**, 226 [*13, 21*]
Shoenberg, D. (1959), *Nature* **183**, 171 [*15*]
Shoenberg, D. (1960a), *Phil. Mag.* **5**, 105 [*17*]
Shoenberg, D. (1960b), *FS*, p. 74 [*17, 331*]
Shoenberg, D. (1962), *Phil. Trans. Roy. Soc. A* **255**, 85 [*17, 18, 96, 99, 100, 102, 199, 254, 277, 411*]
Shoenberg, D. (1965), *LT* 9, p. 665 [*21*]
Shoenberg, D. (1968), *Canad. J. Phys.* **46**, 1915 [*290–1, 304, 310*]
Shoenberg, D. (1969a), *Phys. Kond. Mat.* **9**, 1 [*57, 369*]
Shoenberg, D. (1969b), *PM*, p. 62 [*15, 17, 180, 194*]
Shoenberg, D. (1976), *J. Low Temp. Phys.* **25**, 755 [*283, 286, 288, 294, 296, 300, 400*]
Shoenberg, D. & Stiles, P. J. (1964), *Proc. Roy. Soc. A* **281**, 62 [*19, 103, 112, 185–8, 417–8*]
Shoenberg, D. & Templeton, I. M. (1968), *Canad. J. Phys.* **46**, 1925 [*289, 307*]
Shoenberg, D. & Templeton, I. M. (1973), *Physica* **69**, 293 [*304, 496, 498*]
Shoenberg, D. & Uddin, M. Z. (1936), *Proc. Roy. Soc. A* **156**, 701 [*6, 87, 248, 369*]
Shoenberg, D. & Vanderkooy, J. (1970), *J. Low Temp. Phys.* **2**, 483 [*94, 450*]
Shoenberg, D. & Vuillemin, J. J. (1966), *LT* 10 p. 67 [*142, 277–81, 293, 444*]
Shoenberg, D. & Watts, B. R. (1967), *Phil. Mag.* **15**, 1275 [*131, 142, 242–3, 541, 543*]
Simpson, G. & Paton, B. E. (1971), *Phys. Lett. A* **34**, 180 [*465*]
Skobov, V. G. (1961), *Zh. eksp. teor. fiz.* **40**, 1446 (*Sov. Phys. JETP* **13**, 1014) [*169*]
Slavin, A. J. (1973), *Phil. Mag.* **27**, 65 [*146, 242–3*]
Smith, G. E., Baraff, G. A. & Rowell, J. M. (1964), *Phys. Rev.* **135**, A1118 [*440*]

Bibliography

Smith, G. E., Galt, J. K. & Merritt, F. R. (1960), *Phys. Rev. Lett.* 4, 276 [*455*]
Smith, J. L. (1974), Ph.D. thesis, Brown University, Providence R. I. [*280*]
Sommerfeld, A. (1928), *Z. Phys.* 47, 1 [*1*]
Sondheimer, E. H. (1954), *Proc. Roy. Soc. A* 224, 160 [*15*]
Sondheimer, E. H. & Wilson, A. H. (1951), *Proc. Roy. Soc. A* 210, 173 [*10, 65*]
Sorbello, R. S. & Griessen, R. (1973), *Sol. State Comm.* 12, 689 [*239–40*]
Sparlin, D. M. & Marcus, J. A. (1966), *Phys. Rev.* 144, 484 [*222*]
Springford, M. (1971), *Adv. in Phys.* 20, 493 [*19*]
Springford, M. (ed) (1980), *EFS* [*21*]
Springford, M., Stockton J. R. & Templeton, I. M. (1969), *Phys. Kond. Mat.* 9, 15 [*377*]
Stark, R. W. (1964), *Phys. Rev.* 135, A1698 [*436, 441*]
Stark, R. W. (1967), *Phys. Rev.* 162, 589 [*218, 349–51*]
Stark, R. W. & Falicov, L. M. (1967a), *PLT* 5, 235 [*18, 332–3, 335, 344–5, 350–1, 353, 360*]
Stark, R. W. & Falicov, L. M. (1967b), *Phys. Rev. Lett.* 19, 795 [*221*]
Stark, R. W. & Friedberg, C. B. (1971), *Phys. Rev. Lett.* 26, 556 [*331*]
Stark, R. W. & Friedberg, C. B. (1974), *J. Low Temp. Phys.* 14, 111 [*364–6*]
Stark, R. W. & Reifenberger, R. (1977), *J. Low Temp. Phys.* 26, 763 and 819 [*364, 368*]
Stark, R. W. & Windmiller, L. R. (1968), *Cryogenics* 8, 272 [*103*]
Středa, P. (1974), *Czech. J. Phys. B* 24, 794 [*63*]
Stern, E. A. (1960), *FS*, p. 50 [*17*]
Sullivan, P. & Seidel, G. (1967), *Phys. Lett.* 25A, 229 [*139*]
Svirskii, M. S. (1963), *Zh. eksp. teor. fiz.* 44, 628 (*Sov. Phys. JETP* 17, 426) [*521*]
Sydoriak, S. G. & Robinson, J. E. (1949), *Phys. Rev.* 75, 118 [*9*]
Templeton, I. M. (1966), *Proc. Roy. Soc. A* 292, 413 [*130–1, 240, 543*]
Templeton, I. M. (1972), *Phys. Rev. B* 5, 3819 [*437*]
Templeton, I. M. (1974), *Canad. J. Phys.* 52, 1628 [*130, 240, 243*]
Templeton, I. M. (1975), *Rev. Sci. Inst.* 46, 302 [*119*]
Templeton, I. M. (1981), *J. Low Temp. Phys.* 43, 293 [*130, 185–8, 191, 238*]
Templeton, I. M. & Coleridge, P. T. (1975a), *J. Phys. F* 5, 1307 [*253*]
Templeton, I. M. & Coleridge, P. T. (1975b), *LT* 14 p. 143 [*383*]
Terwilliger, D. W. & Higgins, R. J. (1973), *Phys. Rev. B* 7, 667 [*398, 401–3*]
Testardi, L. R. & Condon, J. H. (1970), *Phys. Rev. B* 1, 3928 [*150, 319–20, 327*]
Thorsen, A. C. & Berlincourt, T. G. (1961), *Nature* 192, 959 [*234*]
Tobin, P. J., Sellmyer, D. J. & Averbach, B. L. (1969), *Phys. Lett.* 28A, 723 [*156*]
Tripp, J. H., Everett, P. M., Gordon, W. L. & Stark, R. W. (1969), *Phys. Rev.* 180, 669 [*221*]
Tsui, D. C. (1967), *Phys. Rev.* 164, 669 [*221*]
Tsui, D. C. & Stark, R. W. (1966), *Phys. Rev. Lett.* 16, 19 [*219*]
van Alphen, P. M. (1933), Doctoral Thesis, Leiden University
Vanderkooy, J. (1969), *J. Phys. E* 2, 718 [*91*]
Vanderkooy, J. & Datars, W. R. (1968), *Canad. J. Phys.* 46, 1215 [*89, 306–7*]
Van Dyke, J. P., McClure, J. W. & Doar, J. F. (1970), *Phys. Rev. B* 1, 2511 [*455*]
Van Vleck, J. H. (1932), *The Theory of Electric and Magnetic Susceptibilities*, Clarendon Press, Oxford, p. 102 [*255*]
van Weeren, J. H. P. & Anderson, J. R. (1973), *J. Phys. F* 3, 2109 [*214, 252, 300*]

Bibliography

Vecchi, M. P. & Dresselhaus, M. S. (1974), *Phys. Rev.* B **9**, 3257 [*174, 232*]

Verkin, B. I. & Dmitrenko, I. M. (1958), *Zh. eksp. teor. fiz.* **35**, 291 (*Sov. Phys. JETP* **35**, 200, 1959) [*131*]

Verkin, B. I., Lazarev, B. G. & Rudenko, N. S. (1949), *Dokl. Akad. Nauk SSSR* **69**, 773 [*9*]

Verkin, B. I., Pelikh, L. N. & Eremenko, V. Y. (1964), *Dokl. Akad. Nauk SSSR* **159**, 771 (*Sov. Phys. Doklady* **9**, 1076) [*151–2*]

Volski, E. P. (1962), *Zh. eksp. teor. fiz.* **43**, 1120 (*Sov. Phys. JETP* **16**, 791, 1963) [*123*]

Volski, E. P. (1964a), *Zh. eksp. teor. fiz.* **46**, 123 (*Sov. Phys. JETP* **19**, 89) [*123, 212, 214*]

Volski, E. P. (1964b), *Zh. eksp. teor. fiz.* **46**, 2035 (*Sov. Phys. JETP* **19**, 1371) [*153*]

Volski, E. P. & Petrashov, V. T. (1970), *Zh. eksp. teor. fiz.* **59**, 96 (*Sov. Phys. JETP* **32**, 55 (1971) [*123*]

Volski, E. P. & Teplinsky, V. M. (1970), *Pribory i tekhnika eksperimenta* No. 3, 195 [*108*]

von Klitzing, K., Dorda, G. and Pepper M. (1980) *Phys. Rev. Lett.* **45**, 449 [*159*]

Vosko, S. H., Perdew, J. P. & MacDonald, A. H. (1975), *Phys. Rev. Lett.* **35**, 1725 [*458–9*]

Wakabayashi, J., Myron, H. W. & Pepper, M. (1982), *16th Int. Conf. on Physics of Semiconductors*, Montpellier [*157–8*]

Wampler, W. R. & Lengeler, B. (1977), *Phys. Rev.* B **15**, 4614 [*377*]

Wampler, W. R. & Springford, M. (1972), *J. Phys.* C **5**, 2345 [*95, 156*]

Wang, C. S. & Callaway, J. (1977), *Phys. Rev.* B **15**, 298 [*227*]

Watts, B. R. (1963), *Phys. Lett.* **3**, 284 [*220*]

Watts, B. R. (1964), *Proc. Roy. Soc.* A **282**, 521 [*218, 220*]

Watts, B. R. (1972), *J. Phys.* C **5**, 1930 [*393*]

Watts, B. R. (1973), *J. Phys.* F **3**, 1345 [*381*]

Watts, B. R. (1974), *J. Phys.* F **4**, 1371 [*392–3*]

Watts, B. R. (1977), *J. Phys.* F **7**, 939 [*393–6*]

Watts, B. R. & Mayers, J. (1979), *J. Phys.* F **9**, 831 [*240*]

Watts, B. R. & Sundström, L. J. (1979), *J. Phys.* F **9**, 849 [*240*]

Whitten, W. B. & Piccini, A. (1966), *Phys. Lett.* **20**, 248 [*151*]

Wilkins, J. W. (1980) *EFS*, p. 72 [*74*]

Wilson, A. H. (1953a), *The Theory of Metals*, Cambridge University Press [*1*]

Wilson, A. H. (1953b), *Proc. Camb. Phil. Soc.* **49**, 292 [*489*]

Windmiller, L. R. (1966), *Phys. Rev.* **149**, 472 [*250, 450*]

Windmiller, L. R., Ketterson, J. B., Hörnfeldt, S. (1970), *J. App. Phys.* **41**, 1232 [*449*]

Yafet, Y. (1963), *SSP* **14**, 1 [*453*]

Yahia, J. & Marcus, J. A. (1959), *Phys. Rev.* **113**, 137 [*156*]

Young, R. C. (1977), *Rep. Prog. Phys.* **40**, 1123 [*179, 221*]

Ziman, J. M. (1964), *Principles of the Theory of Solids*, Cambridge University Press [*334*]

Ziman, J. M. (ed) (1969), *PM* [*21*]

565

Notes added in proof June 1983

pp. 50, 123   The experiment on 2-D dHvA oscillations in a Si inversion
layer has proved successful (F. Fang and P. J. Stiles, *Bull. Am.
Phys. Soc.* 1983, **28**, 858 and submitted to *Phys. Rev.*).
Oscillations have also been observed in a 2-D GaAs–
(AlGa)As heterostructure by H. L. Störmer, T. Haavasoja,
V. Narayanamurti, A. C. Gossard and W. Wiegmann (*Bull.
Am. Phys. Soc.* 1983, **28**, 408 and to be published in *J. Vac. Sci.
Tech.*) using a SQUID magnetometer; sufficient sensitivity was
achieved by superimposing 4000 layers, equivalent to an area
of 240 cm^2.

pp. 191, 194   Recent positron annihalation experiments on Li (S. L.
Basinski, R. J. Douglas and A. T. Stewart, 1979, in *Positron
Annihalation*, p. 665, R. R. Hasiguti and K. Fujiwara (eds.)
Japan Institute of Metals, Sendai and A. A. Manuel,
L. Oberli, T. Jarlborg, R. Sachot, P. Descouts and M. Peter,
1982, in *Positron Annihalation*, p. 281, P. G. Coleman,
S. C. Sharma and L. M. Diane (eds.) North-Holland,
Amsterdam) agree within about 20% and suggest values of
$10^2 \times \Delta k/k_0$ of -0.5 for $\langle 100 \rangle$, 1.3 for $\langle 110 \rangle$ and -0.5 for
$\langle 111 \rangle$. The shape of the distortion of the FS is much as in
fig. 5.5*a*, but the magnitude of the distortion is rather less
than indicated by the data of RS.

p. 196   Further experiments of M. J. O'Shea and M. Springford (*J.
Phys. F.* 1983, **13**, 357) support the idea that dHvA amplitude
anomalies in K are due to sample inhomogeneity.

p. 214   P. T. Coleridge (*J. Phys. F.* 1982, **12**, 2563) has successfully
fitted the FS of Al to a KKR band structure calculation with
three phase shift parameters.

p. 238   I. M. Templeton (*J. Low Temp. Phys.* 1983, **52**, 361) has
measured the anisotropy of the pressure dependence of the FS
of Rb and Cs. This proves to be much greater than the zero
pressure anisotropy of the FS (about 4.7 times for Rb and 5.6
times for Cs). The pressure dependences averaged over the FS
are in excellent agreement with the ultrasonic compressibility
values of table 5.5.

p. 247   G. G. Lonzarich and N. S. Cooper (*J. Phys. F.* to be published)
have prepared a detailed account of the experiments on the
T-dependence of the neck area in Au together with a
phenomenological interpretation in terms of the electron–
phonon interaction.

p. 381   P. T. Coleridge and I. M. Templeton (*Phys. Rev. Lett.* 1982, **49**,
940) show that for dilute alloys of Cu with Z above 1, the rigid

band model can be made to work fairly well if an effective ΔZ is assumed which is less than the actual ΔZ (by a factor of about 0.9 for $\Delta Z = 2$, 0.75 for $\Delta Z = 3$ and 0.45 for $\Delta Z = 4$).

p. 385   Some of the figures in table 8.2(c) for Cu(Zn) dilute alloys are slightly modified in the final version of Coleridge *et al.* (1983).

pp. 393–6   (a) Dr Watts has discovered an error in his 1977 theory which tends to exaggerate the slope of Dingle plots such as fig. 8.3 by a factor which increases to $\pi/2$ in the limit of high H_0/H. The qualitative predictions of his theory are however still valid.

(b) Two other theoretical calculations (by R. A. Brown, *J. Phys. F.* 1978, **8**, 1159 and by E. Mann, *J. Phys. F.*, 1979, **9**, L135) should have been discussed, but were inadvertently overlooked.

p. 411   Implicit in our analysis of mosaic structure is the assumption that the specification of the mosaic spread around any particular crystal direction is independent of which direction this is. It can be shown that this assumption is valid provided that (a) there is no 'texture' to the mosaic spread, (b) the spread can be described by the probability distribution of a single angle ψ which specifies the departure of a crystal direction in some grain from its mean orientation (e.g. for a Gaussian $dP = 2\alpha\psi \, d\psi \, \exp(-\alpha\psi^2)$ and (c) the angles of spread can be treated as small. A more general specification of textured structures is given by R.-J. Roe, *J. App. Phys.* 1965, **36**, 2024.

p. 459   M. Springford, I. M. Templeton and P. T. Coleridge (to appear in *J. Low Temp. Phys.* 1983, **53**) have determined the g-factor in Cs as $g = 2.96 \pm 0.03$, with no appreciable dependence on orientation.

Subject index

References to individual metals are intended to be fairly comprehensive so that the various studies of any particular metal can be easily discovered. For other subjects, however, particularly those which occur frequently, the coverage is less thorough, though enough references are usually given to indicate where there is significant discussion of the subject.